I0056471

Wireless Communications: Designs, Circuits and Optics

Wireless Communications: Designs, Circuits and Optics

Edited by **Rapheal Dagget**

Willford Press
New York

Published by Willford Press,
118-35 Queens Blvd., Suite 400,
Forest Hills, NY 11375, USA
www.willfordpress.com

Wireless Communications: Designs, Circuits and Optics
Edited by Rapheal Dagget

© 2016 Willford Press

International Standard Book Number: 978-1-68285-065-7 (Hardback)

This book contains information obtained from authentic and highly regarded sources. Copyright for all individual chapters remain with the respective authors as indicated. All chapters are published with permission under the Creative Commons Attribution License or equivalent. A wide variety of references are listed. Permission and sources are indicated; for detailed attributions, please refer to the permissions page and list of contributors. Reasonable efforts have been made to publish reliable data and information, but the authors, editors and publisher cannot assume any responsibility for the validity of all materials or the consequences of their use.

The publisher's policy is to use permanent paper from mills that operate a sustainable forestry policy. Furthermore, the publisher ensures that the text paper and cover boards used have met acceptable environmental accreditation standards.

Trademark Notice: Registered trademark of products or corporate names are used only for explanation and identification without intent to infringe.

Printed in the United States of America.

Contents

Preface

I am honored to present to you this unique book which encompasses the most up-to-date data in the field. I was extremely pleased to get this opportunity of editing the work of experts from across the globe. I have also written papers in this field and researched the various aspects revolving around the progress of the discipline. I have tried to unify my knowledge along with that of stalwarts from every corner of the world, to produce a text which not only benefits the readers but also facilitates the growth of the field.

Wireless communications are an integral part of modern telecommunication systems and engineering. They rely on radio frequencies, electromagnetic waves and signals for transmission of information. The aim of this book is to provide an understanding of the multiple aspects of wireless communications like designing and modeling of circuits, wireless sensor networks, electromagnetic wave transmission, etc. Those with an interest in wireless communications would find this book insightful.

Finally, I would like to thank all the contributing authors for their valuable time and contributions. This book would not have been possible without their efforts. I would also like to thank my friends and family for their constant support.

Editor

Bounds on Minimum Energy per Bit for Optical Wireless Relay Channels

A. D. RAZA, S. Sheikh MUHAMMAD

Department of Electrical Engineering, National University of Computer and Emerging Sciences (FAST-NU), Lahore, Pakistan

ad.raza@nu.edu.pk, sm.sajid@nu.edu.pk

Abstract. *An optical wireless relay channel (OWRC) is a classical three node network consisting of source, relay and destination nodes with optical wireless connectivity. The channel law is assumed Gaussian. This paper studies the bounds on minimum energy per bit required for reliable communication over an OWRC. It is shown that capacity of an OWRC is concave and energy per bit is monotonically increasing in square of the peak optical signal power, and consequently the minimum energy per bit is inversely proportional to the square root of asymptotic capacity at low signal to noise ratio. This has been used to develop upper and lower bound on energy per bit as a function of peak signal power, mean to peak power ratio, and variance of channel noise. The upper and lower bounds on minimum energy per bit derived in this paper correspond respectively to the decode and forward lower bound and the min-max cut upper bound on OWRC capacity.*

Keywords

Optical wireless (OW), network information theory (NIT), optical wireless relay channel (OWRC), channel capacity, energy per bit E_b.

1. Introduction

Information theory provides the scientific and theoretical foundation for the development of today's most beloved computers, smart phones and the Internet. Channel capacity is the central concept within information theory and draws the boundary between the physically possible and impossible in terms of reliable data rates. "A mathematical theory of communication" [1] laid the foundations of information theory that focused on determining and achieving the capacity of single input single output (SISO) channel.

"Two-way communication channel" [2] initiated another new field of study, the network information theory (NIT). NIT is a field that has been evolving to answer the questions that are not directly answerable by the link based classical information theory. NIT shifted the focus to studying the capacity of networks comprising multiple transmitters and receivers competing and cooperating for the capacity of underlying SISO channels to communicate to one another simultaneously. The problem though simple to formulate has defied a general solution till date. However, a lot of work has been done to find out capacity regions for fundamental network structures like broadcast channel, multiple access channel, relay channel, multiple input multiple output (MIMO) channel amongst many others [3].

Due to higher achievable bit rate and absence of regulatory controls and cost optical wireless is attracting attention for use in access network. This is despite of handicap of short coverage distance and constraint on peak signal power due to concern for safety of human eye. Capacity of optical wireless SISO is studied by a number of researchers [4–6]. Research in OW systems and in particular terrestrial OW links has for a long time attempted at increasing the availability and the reliability of the links, but recently it has been realized that probably, the better way to design systems is to attempt throughput maximization [7]. To overcome the degradation of OW channel due to scintillation Chatzidiamantis et al. [8] proposed using relay channels. Presently the study of OW network structures like relay and MIMO channels have been attracting attention [9–11].

Whereas capacity has been the dominant measure for a channel or network performance, minimum energy per bit needed for reliable communication has evolved into an alternative metric [12–14]. This metric becomes specially relevant in case of sensor relay networks where battery life is a critical design factor.

This paper studies minimum energy per bit requirement for reliable communication over a Gaussian optical wireless relay channel (OWRC). OWRC is a network comprising three nodes, source, relay and destination, connected through optical wireless links. The channel law for OWRC is assumed to be Gaussian. This study finds its relevance in view of the increasing use of wireless relay networks that could possibly be optical. The energy per bit bounds developed in this paper correspond to the bounds on the capacity of an OWRC in [15]. These bounds on the OWRC capacity have been briefly discussed in Sections 2.1.1 and 2.1.2.

The paper is organized as follows. Section 2 defines the optical wireless relay channel and its channel capacity and other preliminaries related to the study of energy per bit. Sections 3 and 4 provide original results on energy per bit for the optical wireless relay channel (OWRC). Upper and lower bounds on the energy per bit are derived.

2. Optical Wireless Relay Channel

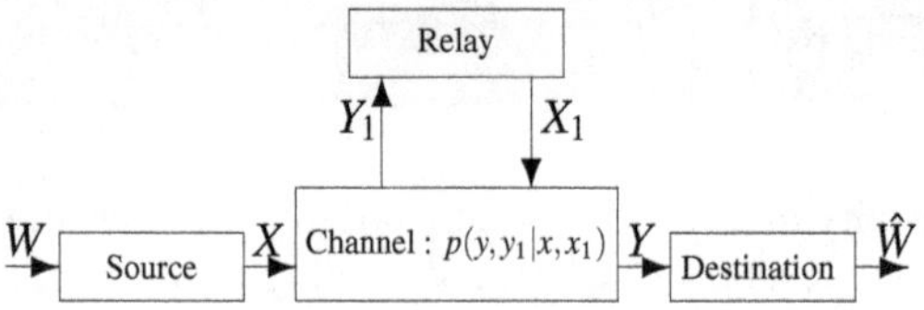

Fig. 1. The relay channel.

Van Der Meulen [16] introduced the relay channel that is a three node network: a source, a relay and a destination node; as shown in Fig. 1. Source node is transmit only while destination node is receive only node. The relay node receives the signal from the source node and transmits it to the destination node. A discrete memory-less relay channel is characterized by the triplet $(X \times X_1, p(Y,Y_1 \mid X,X_1), Y \times Y_1)$. There are four sets of alphabet, sets of input alphabet X and X_1 and output alphabet $Y$and Y_1, and a collection of probability distribution functions $p(\cdot,\cdot \mid x,x_1)$ on $Y \times Y_1$ space one for each $(x,x_1) \in X \times X_1$. The channel law $p(\cdot,\cdot \mid x,x_1)$ is assumed Gaussian.

$W = \{1,2,\cdots,2^{nR}\}$ is the set of messages $\{$indices$\}$ to be sent to the destination by the source node, where R is the feasible rate and n is the number of bit in the code.

$X = (x^n(w))$ belongs to the code book $\{x^n(1), x^n(2), \cdots, X^n(2^{nR})\}$ at the source node containing n bit code words for $\forall w \in W$. At time k the source transmits $X_k = (x_k^n(w))$.

$Y_{1,k}$ is the output of the source-relay link at time k.

Relay function f_1^n such that $X_{1.k} = f_1^n(Y_{1,1}, Y_{1,2}, \cdots, Y_{1,k-1})$ is the relay output at time k. However, generally block Markov coding is employed which implies $X_{1.k} = f_1^n(Y_{1,k-1})$. The relay transmits in time slot k depending only on what it received in time slot $k-1$.

Decoding rule d: $d(Y_k) = \hat{w} \in W$, where Y_k is the signal received by the destination node at time k.

Error occurs when $\hat{w} \neq w$. Average probability of error $P_e^{(n)}$ is defined as

$$P_e^{(n)} = \frac{\sum_{w=1}^{2^{n\times R}} P[\hat{w} \neq w]}{2^{n\times R}} \tag{1}$$

for $\forall w \in W$. R is the feasible rate, and capacity C is the supremum of the set of achievable rates.

The minimum energy per bit E_b is the infimum of the set of achievable energy per bit $E^{(n)}$ that is defined as

$$E^{(n)} = \frac{1}{nR_n}\big(\max_k E_s^{(n)}(k) + E_r^{(n)}\big) \tag{2}$$

where energy $E^{(n)}(k)$ for codeword k expended by the source node is

$$E_s^{(n)}(k) = \sum_{i=1}^{n} x_i(k) \tag{3}$$

and the energy spent by the relay $E_r^{(n)}(k)$ for the code word k is:

$$E_r^{(n)}(k) = \max_{y_1^n}\big(\sum_{i=1}^{n} x_{1i}\big). \tag{4}$$

The energy per bit $E^{(n)}$ is achievable if there exist a sequence of $(2^{nR}, n)$ codes such that probability of error $P_e^{(n)} \to 0$ as $n \to \infty$. The minimum energy per bit E_b is greater than $\limsup E^{(n)}$.

The OWRC input signals X and X_1 are inherently power signals and non-negative. They are further subject to both mean and peak power constraints dictated by the concerns of source power conservation and safety of human eye. These limitations translate to the following conditions on the optical intensity signal X and X_1.

$$X, X_1 \geq 0, \tag{5}$$

$$\mathrm{E}[X], \mathrm{E}[X_1] \leq \mathcal{E}, \tag{6}$$

$$\mathrm{Prob}[X > A], \mathrm{Prob}[X_1 > A] = 0. \tag{7}$$

A Gaussian OWRC is defined by the following equations:

$$Y_1 = g_1 x + Z_1, \tag{8}$$

$$Y = g_0 x + g_2 x_1 + Z \tag{9}$$

where g_0, g_1, g_2 are link gain as shown in Fig. 2. Z and Z_1 are zero mean Gaussian random variables with variance σ^2 depicting noise.

2.1 Bounds on the OWRC Capacity

The capacity theorems for general relay channels have been established in [17]. Upper bound is based on the max-min cut capacity [17, Theorem 4]:

$$C = \max_{p(x,x_1)} \min(I(X,X_1;Y), I(X;Y,Y_1 \mid X_1)) \tag{10}$$

where $I(\cdot;\cdot)$ is the mutual information. $I(X,X_1;Y)$ is the mutual information of the cut at the destination node (multi-access cut); whereas $I(X;Y,Y_1 \mid X_1)$ is the mutual information of the cut at source node (broadcast cut). The inner bound given below is based on the concept of a degraded relay channel modifying mutual information of broadcast cut as $I(X:Y_1 \mid X_1)$ [17, Theorem 1];

$$C \leq \max_{p(x,x_1)} \min(I(X,X_1;Y), I(X;Y_1 \mid X_1)). \tag{11}$$

Upper and lower bounds on the capacity of a Gaussian OWRC have been derived [15]. The min-max cut upper bound is derived through evaluation of (10) using the concept of duality [6] considering gaussian measure on the input X and X_1 with mean $\mathcal{E}$ and variance $(1-\alpha)A^2$ [15] . The lower bounds are obtained by applying entropy power inequality [3, Chap. 16.7] to 11. The lower bounds are optimised by the choice of maximum entropy approaching probability measure on the input alphabet X and X_1 and decode and forward relay function. This relaying strategy is known to yield the maximal lower bound [13]. Separate bounds for $\alpha \in (0, \frac{1}{2})$ and $\alpha \in (\frac{1}{2}, 1]$ have been derived in view of different maxentropic measures applicable in these two ranges of α.

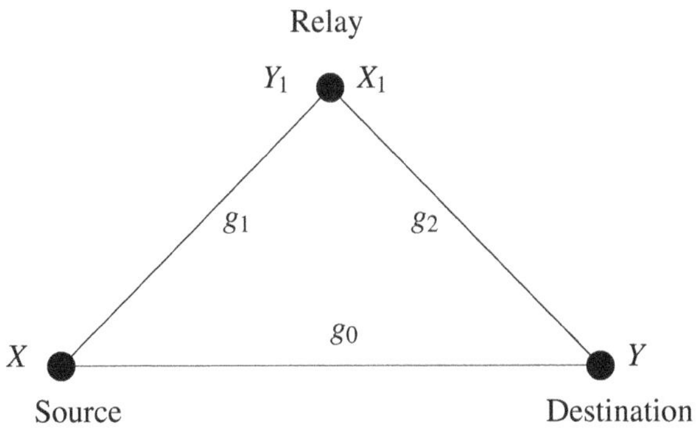

Fig. 2. Link gain coefficients for the relay channel.

2.1.1 Capacity Bounds for $0 < \alpha < \frac{1}{2}$

For $0 < \alpha < \frac{1}{2}$ the capacity of an OWRC, operating under peak power A and mean to peak power ratio α is upper bounded by [15]

$$C(A,\alpha A) \leq \begin{cases} \mathfrak{C}\left(\frac{\left(g_1 g_2+g_0\sqrt{g_0^2+g_1^2-g_2^2}\right)^2(1-\alpha)A^2}{(g_0^2+g_1^2)\sigma^2}\right) & \text{if } \frac{g_0^2+g_1^2}{g_2^2} > 1, \\ \mathfrak{C}\left(\frac{(g_0^2+g_1^2)(1-\alpha)A^2}{\sigma^2}\right) & \text{otherwise} \end{cases} \quad (12)$$

where $\mathfrak{C}(x) = \frac{1}{2}\log(1+x)$ and $C(A,\alpha A)$ is the capacity when peak signal is A and mean signal power is αA and it is lower bounded by

$$C(A,\alpha A) \leq \begin{cases} \mathfrak{C}\left(\frac{\left(g_0\sqrt{g_1^2-g_2^2}+g_2\sqrt{g_1^2-g_0^2}\right)^2 e^{2\mu(1-\alpha)}A^2}{2\pi e\sigma^2 g_1^2(1-\mu\alpha)^2}\right); \\ \text{if } g_0,\ g_2 > g_1 \\ \mathfrak{C}\left(\frac{g_1^2 e^{2\mu(1-\alpha)}A^2}{2\pi e\sigma^2(1-\mu\alpha)^2}\right); \text{ otherwise} \end{cases} \quad (13)$$

where μ is the unique solution of the following equation,

$$\alpha = \frac{1}{\mu*} - \frac{e^{-\mu*}}{1-e^{-\mu*}}.$$

2.1.2 Capacity Bounds for $\frac{1}{2} \leq \alpha \leq 1$

When $\frac{1}{2} \leq \alpha \leq 1$ the capacity of an OWRC with peak signal A is upper bounded by [15],

$$C(A,\alpha A) \leq \begin{cases} \mathfrak{C}\left(\frac{\left(g_1 g_2+g_0\sqrt{g_0^2+g_1^2-g_2^2}\right)^2 A^2}{4(g_0^2+g_1^2)\sigma^2}\right) & \text{if } \frac{g_1^2}{g_0^2+g_2^2} > 1, \\ \mathfrak{C}\left(\frac{(g_0^2+g_1^2)A^2}{4\sigma^2}\right) & \text{otherwise,} \end{cases} \quad (14)$$

is lower bounded by

$$C(A,\alpha A) \leq \begin{cases} \mathfrak{C}\left(\frac{g_1^2 A^2}{2\pi e\sigma^2}\right); & \text{if } g_0,\ g_2 \leq g_1, \\ \mathfrak{C}\left(\frac{\left(g_0\sqrt{g_1^2-g_2^2}+g_2\sqrt{g_1^2-g_0^2}\right)^2 A^2}{2\pi e\sigma^2 g_1^2}\right); & \text{otherwise.} \end{cases}$$

From the above expressions for the capacity of an OWRC in equations (12), (13), (14), (15) it is obvious that in general the capacity can be expressed as

$$C(A,\alpha A) = \mathfrak{C}\left(\frac{\beta A^2}{\sigma^2}\right) \geq \mathfrak{C}\left(\frac{\alpha^2 A^2}{\sigma^2}\right) \quad (15)$$

where $\beta > 0$ is the the coefficient of A^2. β is a function of link gains g_0, g_1, g_2 and mean to peak power ratio α. We will use these capacity bounds to work out energy per bit requirements in the next section.

3. Energy per Bit for OWRC

For bounding the energy per bit E_b we need to establish it is a non decreasing function of A^2. To do this it is required to be shown that the capacity of an ORWC is concave in A^2. Before we proceed to prove concavity of capacity of an OWRC let us define it in general terms as.

Definition 1 (Capacity of OWRC)**.** With peak power A and mean $\mathcal{E}$ to peak power ratio $\alpha = \frac{\mathcal{E}}{A}$ at the source and relay node, capacity of the OWRC is

$$C_k(A,\ \alpha A) = \frac{1}{k} \sup_{\substack{E(X)\leq\mathcal{E} \\ E(X_1)\leq\mathcal{E} \\ P(X>A=0) \\ P(X_1>A=0)}} I(X^k;Y^k), \quad (16)$$

$$C(A,\ \alpha A) = \sup_k C_k(A,\ \alpha A) \quad (17)$$

$$= \lim_{k\to\infty} C_k(A,\ \alpha A). \quad (18)$$

Lemma 2 (Concavity of Capacity of OWRC)**.** The capacity of an OWRC under average and peak power constraints (5-7) satisfies the following:

1. $C(A,\ \alpha A) \geq 0$ if $A > 0$ and tends to ∞ as $A \to \infty$.
2. $C(A,\ \alpha A) \to 0$ as $A \to 0$.
3. $C(A,\ \alpha A)$ is concave and strictly increasing in A^2.
4. $C^2(A,\ \alpha A)$ is concave and strictly increasing in A^2.
5. $\frac{A^2}{C^2(A,\ \alpha A)}$ is non decreasing in A, $\forall A > 0$.

Proof. 1. Since $C(A,\ \alpha A)$ is greater than or equal to $\mathfrak{C}(\frac{\alpha^2 A^2}{\sigma^2})$ which is strictly larger than zero for $\forall A > 0$ and approaches ∞ as $A \to \infty$.

2. Since the upper and lower bounds [15] on $C(A, \alpha A)$ go to zero as $A \to 0$.

3. $C_k(A,\alpha A)$ is concave in A^2. Therefore, $C(A,\alpha A) = \sup_k C_k(A,\alpha A)$ is concave being a supremum of a concave function [18, Theorem D, p. 16]. Because of concavity and prepositions (1) and (2) of this lemma, that is $C(A,\alpha A) = 0$ at $A = 0$ and $C(A,\alpha A) \to \infty$ when $A \to \infty$, $C(A,\alpha A)$ is monotonically non decreasing function in A.

4. As $C(A,\alpha A)$ is non-negative, increasing and concave function in A^2 so $C^2(A,\alpha A)$ is also concave in A^2 [18, Theorem C, p. 16].

5. It follows from the concavity of $C^2(A,\alpha A)$ that for any $0 < A_1 < A_2$

$$\frac{A_1^2}{A_2^2} C^2(A_2,\alpha A_2) + \frac{A_2^2 - A_1^2}{A_2^2} C^2(0,0) \leq C^2(A_1,\alpha A_1).$$

Because $C(0,0) = 0$, the above relation translates to

$$\begin{aligned} \frac{A_1^2}{A_2^2} C^2(A_2,\alpha A_2) &\leq C^2(A_12,\alpha A_12), \\ \frac{A_1^2}{C^2(A_12,\alpha A_1)} &\leq \frac{A_2^2}{C^2(A_2,\alpha A_2)}. \end{aligned} \tag{19}$$

Equation (19) shows that $\frac{A^2}{C^2(A,\alpha A)}$ is a non decreasing function in A^2.

□

Lemma 3 (Minimum Energy per Bit for OWRC)**.** When the source and relay nodes have same peak power A and mean power $\mathcal{E}$ constraints and $A \geq 0$ and $0 < \alpha \leq 1$, the minimum energy per bit E_b for OWRC is given by

$$E_b^2 = \lim_{A \to 0} \frac{2\alpha^2 A^2}{C^2(A,\alpha A)}. \tag{20}$$

Proof. The achievability and weak converse can be established by showing that

$$E_b^2 = \inf_{A>0} \frac{2\alpha^2 A^2}{C^2(A,\alpha A)}. \tag{21}$$

The proposition (5) of Lemma 2 allows replacement of inf by lim.

Achievability: There exists $E' > 0$ and $\varepsilon > 0$ such that

$$\begin{aligned} E &> \sqrt{\frac{2\alpha^2 A'^2}{C^2(A',\alpha A')}} \\ &= \inf_{A>0} \frac{2\alpha A}{C(A,\alpha A)} + \varepsilon. \end{aligned} \tag{22}$$

Thus there exists $R < C(A',\alpha A')$ that can be achieved using random coding with average and peak power constraints. This proves achievability of E.

Weak Converse: We need to prove that for any sequence $(2^{nR_n}, n)$ of codes with $P_e^{(n)} \to 0$

$$\begin{aligned} \liminf E^{(n)^2} &\geq E_b^2 \\ &= \inf_{A>0} \frac{2\alpha^2 A^2}{C^2(A,\alpha A)}. \end{aligned}$$

Fano's inequality yields

$$R_n \leq C(A_n,\alpha A_n) + \frac{1}{n}(P_e^{(n)}) + R_n P_e^{(n)}. \tag{23}$$

Therefore,

$$R_n \geq \frac{C(A_n,\alpha A_n) + \frac{1}{n}(P_e^{(n)})}{(1 - P_e^{(n)})}.$$

Now by applying definition of energy per bit (2)

$$\begin{aligned} E^{(n)^2} &\geq \frac{2\alpha^2 A^2}{R_n^2} \\ &\geq \frac{2\alpha^2 A^2 \left(1 - P_e^{(n)}\right)^2}{\left(C(A_n,\alpha A_n) + \frac{1}{n} H(P_e^{(n)})\right)^2} \\ &= \frac{2\alpha^2 A^2}{C^2(A_n,\alpha A_n)} \times \frac{\left(1 - P^{(n)}\right)^2}{\left(1 + \frac{\frac{1}{n} H(P_e^{(n)})}{C(A_n,\alpha A_n)}\right)^2} \\ &\geq E_b \frac{\left(1 - P^{(n)}\right)^2}{\left(1 + \frac{\frac{1}{n} H(P_e^{(n)})}{C(A_n,\alpha A_n)}\right)^2}, \end{aligned} \tag{24}$$

$P_e^{(n)} \to 0$, $C(A_n,\alpha A_n) > 0$ and $H(P_e^{(n)}) > 0$ yields $\liminf E^{(n)} \geq E_b$.

□

4. Bounds on Energy per Bit

4.1 Energy Per Bit

From (20), energy per nat is

$$E_b = \sqrt{\lim_{A^2 \to 0} \frac{2\alpha^2 A^2}{C^2(A,\alpha A)}} \tag{25}$$

and $C(A,\alpha A)$ can be expressed in the generic form as

$$C(A,\alpha A) = \frac{1}{2}\log(1 + \beta A^2). \tag{26}$$

This yields the energy per nat E_{nat} as

$$E_{nat} = \frac{2\alpha}{\sqrt{\beta}} \tag{27}$$

and energy per bit E_b is

$$E_b = \frac{2\alpha}{\sqrt{\beta}}\log 2. \tag{28}$$

4.2 Lower Bound on Energy per Bit

Energy per bit for the upper bound on capacity of OWRC (12), (14) will yield the lower bound on energy per bit. The bounds are as follows:

Proposition 4. For $0 < \alpha < \frac{1}{2}$ the lower bound on E_b is

$$E_b \geq \frac{2\,\alpha\,\sigma\log 2}{1-\alpha}\,\frac{\sqrt{g_0^2+g_1^2}}{g_1g_2+g_0\sqrt{g_0^2+g_1^2}}. \tag{29}$$

Proof. Applying (28) to upper bound on OWRC capacity (12) we have

$$E_b \geq \frac{\alpha\,\sigma\log 2}{(1-\alpha)}\min\Bigg(\underbrace{\min_{g_o^2+g_1^2>g_2^2}\frac{\sqrt{g_0^2+g_1^2}}{g_1\,g_2+g_0\sqrt{g_0^2+g_1^2-g_2^2}}}_{k_1},\ \underbrace{\min_{g_o^2+g_1^2\leq g_2^2}\frac{1}{\sqrt{g_0^2+g_1^2}}}_{k_2}\Bigg). \tag{30}$$

Assuming $g_o^2+g_1^2 \leq g_2^2$,

$$k_1 = \frac{\sqrt{g_0^2+g_1^2}}{g_1\,g_2+g_0\sqrt{g_0^2+g_1^2-g_2^2}}$$

$$\leq^f \frac{\sqrt{g_0^2+g_1^2}}{g_1\sqrt{g_0^2+g_1^2}+g_0\sqrt{g_0^2+g_1^2}} \tag{31}$$

$$= \frac{1}{\sqrt{g_0^2+g_1^2}} = k_2 \tag{32}$$

where (f) stems from i) replacement of g_2 by $\sqrt{g_o^2+g_1^2}$, and ii) dropping of g_2^2 from $\sqrt{g_0^2+g_1^2-g_2^2}$ in the denominator. □

Proposition 5. For $\frac{1}{2} < \alpha \leq 1$ the lower bound on E_b is

$$E_b \geq \frac{2\,\sigma\log 2\,\sqrt{g_0^2+g_1^2}}{g_1g_2+g_0\sqrt{g_0^2+g_1^2}}. \tag{33}$$

Proof. This is obtained using (28) and (14) in a fashion similar to that of proposition 4. □

4.3 Upper Bound on Energy per Bit

This bound corresponds to decode and forward lower bound on capacity of OWRC (13) and (15).

Proposition 6. For $0 < \alpha < \frac{1}{2}$ the upper bound on E_b is

$$E_b \leq 2\alpha\,\sigma\,\lambda(\alpha)\log 2\,\min\left(\frac{1}{g_0+g_2}, \frac{1}{g_1}\right) \tag{34}$$

where $\lambda(\alpha) = \frac{\sqrt{2\pi e}(1-\mu\alpha)}{e^{\mu(1-\mu\alpha)}}$.

Proof. Using the decode forward lower bound on capacity (13) and (28) we get

$$E_b \leq 2\alpha\,\sigma\,\lambda(\alpha)\log 2\,\min\Bigg(\min_{g_0,g_2\geq g_1}\frac{1}{g_1},\ \underbrace{\min_{g_0,g_2<g_1}\frac{g_1}{g_0\sqrt{g_1^2-g_2^2}+g_2\sqrt{g_1^2-g_0^2}}}_{k3}\Bigg). \tag{35}$$

Now if k_3 is closely observed in the light of above condition and assume lowest possible values of g_0 and g_2 that is negligibly small compared to g_1 we get

$$\begin{aligned} k_3 &= \frac{g_1}{g_0\sqrt{g_1^2-g_2^2}+g_2\sqrt{g_1^2-g_0^2}} \\ &< \frac{g_1}{g_0g_1+g_2g_1} \\ &= \frac{1}{g_0+g_2}. \end{aligned} \tag{36}$$

That results in the upper bound on E_b given in this proposition. □

Proposition 7. For $0 < \alpha < \frac{1}{2}$ the upper bound on E_b is

$$E_b \leq \sqrt{2\pi e}\,\sigma\,\log 2\,\min\left(\frac{1}{g_0+g_2}, \frac{1}{g_1}\right). \tag{37}$$

Proof. This can be proved on the same lines as the proof of proposition 6. □

Fig. 3. Setup of relay channel for simulation.

Energy per bit as a function of source-relay distance for the relay channel set up in Fig. 3 is shown in Fig. 4. The source-destination distance is set equal to 1. The relay node is positioned anywhere between the source and destination. Let the source-relay distance be w, $0 \leq w \leq 1$. The link gain g_i, $(i = 0, 1, 2)$ is inversely proportional to the square of link distance as per free space path loss principle. Without loss of generality assuming $g_0 = 1$, the normalised source-relay and relay-destination link gains are $g_1 = w^{-2}$ and $g_2 = (1-w)^{-2}$. With these assumptions bounds on minimum energy per bit (29), (33), (34), (37) as function of relay location w are plotted. Minimum energy per bit E_B has been normalised to the standard deviation of noise σ in the plot. The upper and lower bounds on E_b diverge if relay node is in the vicinity of source node. The bounds are convergent when the relay node is near the destination. If the relay is placed midway between the source and destination nodes normalised minimum energy per bit $\frac{E_b}{\sigma}$ for reliable communication under decode and forward strategy is -2.9 dB and 0.7 dB for $\alpha = 0.3$ and $0.5 \leq \alpha \leq 1$ respectively.

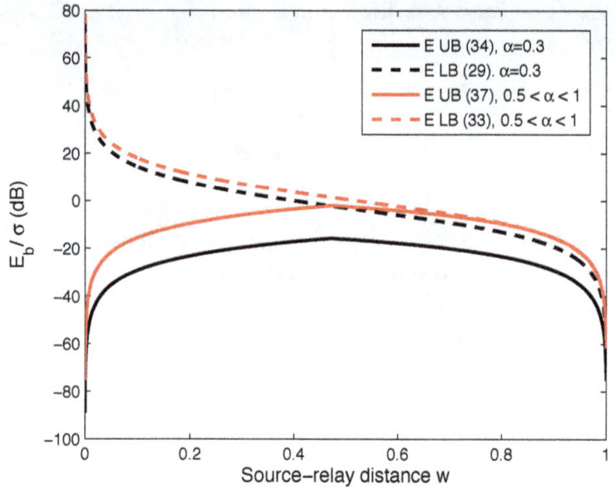

Fig. 4. Lower and upper bounds on energy per bit for an OWRC for $\alpha = 0.3$ (29, 34) and $0.5 \leq \alpha < 1$ (33, 37).

5. Conclusion

It has been proven that the capacity of a Gaussian OWRC is a monotonically increasing concave function in the square of peak signal A^2. It is also shown that energy per bit is a non decreasing function in A^2. Using these results upper and lower bounds on minimum energy per bit (29), (33), (34), (37) required for reliable communication over an OWRC are derived. Minimum energy per bit being an important performance metric will help develop better theoretical understanding of optical wireless relay channels and networks.

References

[1] SHANNON, C. E. A methamatical theory of communication. *Bell System Technical Journal* , 1948, vol. 27, p. 379 - 423, 623 - 656.

[2] SHANNON, C. E. Two-way communication channels. In *Proceedings of the Fourth Berkeley Symposium on Mathematical Statistics and Probability*. 1961, p. 611 - 644.

[3] COVER, T., THOMAS, J. *Elements of Information Theory*. New York: John Willey & Sons, 1991.

[4] HRANILOVIC, S., KSCHISCHANG, F. R. Capacity bounds for power and band-limited optical intensity channels corrupted by Gaussain noise. *IEEE Transactions on Information Theory*, 2004, vol. 50, no. 5, p. 784 - 795.

[5] LAPIDOTH, A., MOSER, S. M., WIGGER, M. A. On the capacity of free-space optical intensity channels. *IEEE Transactions on Information Theory*, 2009, vol. 55, no. 10, p. 4449 - 4461.

[6] LAPIDOTH, A., MOSER, S. M. Capacity bounds via duality with applications to multiple-antenna systems on flat fading channels. *IEEE Transactions on Information Theory*, 2003, vol. 49, no. 10, p. 2426 - 2467.

[7] CZAPUTA, M., HRANILOVIC, S., MUHAMMAD, S. S., LEITGEB, E. Free-space optical links for latency-tolerant traffic. *IET Communications*, 2012, vol. 6, no. 5, p. 507 - 513.

[8] CHATZIDIAMANTIS, N. D., MICHALOPOULOS, D. S., KRIEZIS, E. E., KARAGIANNIDIS, G. K., SCHOBER, R. Relay selection protocols for relay-assisted free-space optical systems. *Journal of Optical Communications and Networking*, 2013, vol. 5, no. 1, p. 92 - 103.

[9] DEMERS, F., YANIKOMEROGLU, H., ST-HILAIRE, M. A survey of opportunities for free space optics in next generation cellular networks. In *Ninth IEEE Annual Communication Networks and Services Research Conference (CNSR)*. 2011, p. 210 - 216.

[10] LETZEPIS, N., HOLLAND, I., COWLEY, W. The Gaussian free space optical MIMO channel with q-ary pulse position modulation. *IEEE Transactions on Wireless Communications*, 2008, vol. 7, no. 5, p. 1744 - 1753.

[11] LI, X., VUCIC, J., JUNGNICKEL, V., ARMSTRONG, J. On the capacity of intensity-modulated direct-detection systems and the information rate of ACO-OFDM for indoor optical wireless applications. *IEEE Transactions on Communications*, 2012, vol. 60, no. 3, p. 799 - 809.

[12] GAMAL, A. E., MOHSENI, M., ZAHEDI, S. Bounds on capacity and minimum energy-per-bit for AWGN relay channels. *IEEE Transactions on Information Theory*, 2006, vol. 52, no. 4, p. 1545 - 1561.

[13] GASTPAR, M., VETTERLI, M. On the capacity of large Gaussian relay networks. *IEEE Transactions on Information Theory*, 2005, vol. 51, no. 3, p. 765 - 779.

[14] VERDU, S. On channel capacity per unit cost. *IEEE Transactions on Information Theory*, 1990, vol. 36, no. 5, p. 1019 - 1030.

[15] RAZA, A. D., MUHAMMAD, S. S. Capacity bounds for a Gaussian optical wireless relay channel. *Transactions on Emerging Telecommunications Technologies*, 2014 (to appear).

[16] VAN DER MEULEN, E. C. Three-terminal communication channel. *Advances in Applied Probability*, 1971, vol. 3, no. 1, p. 120 - 154.

[17] COVER, T. M., GAMAL, A. E. Capacity theorems for the relay channels. *IEEE Transactions on Information Theory* , 1979, vol. 23, no. 5, p. 572 - 584.

[18] ROBERTS, A. W., VARBERG, D. E. Convex functions. *Pure and Applied Mathematics*, 1973, vol. 57.

About Authors ...

A. D. RAZA completed his Bachelor's degree in Electrical Engineering (Hons.) in 1973 and Master's degree in Electrical Engineering in 1984 at the University of Engineering and Technology, Lahore, Pakistan. He remained with Pakistan's state owned Telecom operator from 1975 to 2008 and rose to the level of Executive Vice President. He joined the National University of Computer and Emerging Sciences in

2008 and is pursuing his doctoral studies besides his teaching responsibilities as an Assistant Professor in the Department of Electrical Engineering. His current research interest revolves around application of network information theory to optical wireless networks.

S. Sheikh MUHAMMAD completed his Bachelor's degree in Electrical Engineering (Hons.) in 2001 and Master's degree in Electrical Engineering in 2003 at the University of Engineering and Technology, Lahore, Pakistan whereby he also remained in the faculty for around 3 years. He completed his Ph.D. in Electrical Engineering in 2007 with Excellence at the Graz University of Technology conducting research on coded modulation techniques for free space optical systems. He has guest edited 2 special issues in Optical Wireless Communications in 2009 and 2012 and has organized regular IEEE Colloquiums on Optical Wireless in 2008 (Austria), 2010 (UK) and 2012 (Poland). He has published more than 50 peer reviewed papers in journals and conferences of repute and has chaired a number of international conference sessions. He is currently working as an Associate Professor of Electrical Engineering at the National University of Computer and Emerging Sciences (FAST-NU) in Lahore and his current research revolves around application of network information theory to optical wireless and optical wave propagation through random media.

Flexible Polymer Planar Optical Waveguides

Václav PRAJZLER [1], Pavla NEKVINDOVÁ [2], Petr HYPŠ [1], Oleksiy LYUTAKOV [2], Vítězslav JEŘÁBEK [1]

[1] Dept. of Microelectronics, Czech Technical University, Technická 2, 168 27 Prague, Czech Republic
[2] Institute of Chemical Technology, Technická 5, 166 27 Prague, Czech Republic

xprajzlv@feld.cvut.cz

Abstract. *We report about design, fabrication and properties of flexible polymer optical planar waveguides made of epoxy novolak resin as planar waveguides deposited on various foil substrates. The design of the presented planar waveguides was realized on the bases of modified dispersion equation and was schemed for 633 nm, 850 nm, 1310 nm and 1550 nm wavelength. Propagation optical loss measurements were done by the fibre probe technique at wavelegnth 633 nm (He-Ne laser) and samples have optical losses lower than 2 dB·cm^{-1}. Unlike the up-to-now presented structures our constructin is fully flexible what makes it possible to be used in innovative photonics structures.*

Keywords

Optical planar flexible waveguide, polymer, epoxy novolak resin.

1. Introduction

In recent years, there has been a continuing growth of the demand for data communications link capacity. Existing interconnection technologies for shorter distance used mainly metal copper wiring connection, but due to the rising data-rates and their sensitivity to electromagnetic interference, they soon will be unable to keep up [1], [2]. Therefore it seems that light as a transmission medium for the future interconnections (rack-to-rack, board-to-board, multi-chip modules, on-board) is a right choice. Optical interconnects have many advantages over wire tracks: higher bandwidth, immunity from crosstalk and electromagnetic interference, light weight, low skew, jitter, etc. [3], [4].

Conventional optical link consists of glass optical fiber and traditional photonics planar structures and devices have been made of semiconductors, inorganic crystals and glasses. Though these materials are good candidates for common photonics structures they are not enough flexible and it is difficult to use them for new photonics devices which are continuously miniaturized and integrated [5].

Polymer materials for the fabrication of flexible planar optical waveguides appeared to be a good choice for their excellent optical properties such as their high transparency from visible to infra-red wavelengths, well-controlled refractive indices, reasonable temporal and temperature stability, low optical losses, easy fabrication process and low costs, and, last but not least, their mechanical properties [6-13].

There are a number of different polymers that can be considered for use in new photonics structures and devices. Most of the early work was focused to the Polymethylmethacrylate as the waveguide material [14], [15]. Recently quite a lot of researches groups examined a new type of polymers for photonics applications and as many companies are very active in this field such polymers are nowadays commercially available. It concerns, e.g., Acrylate (AlliedSignal), Acrylate Polyguide™ (DuPont), Acrylate Benzocyclobutene (Dow Chemical), Chloro-fluorinated polyimides (Samsung), Deuterated polysiloxane (NTT), Epoxy novolak resin (Micro Resist Technology), Fluorinated polyimide and Ultradel 9000 series polyimide (Amoco Chemicals), Halogenated acrylate, Polyetherimide (General Electric), Polycarbonate with CLD-1 chromophore (PacificWave), Polycarbonate (JDS Uniphase), Polyurethane (Lumera), ZPU resin (ChemOptics Exguide™), etc. [16-19].

Integration of optical waveguides and opto-electronic components inside a flexible foil introduces a complete new concept of flexibility into the on-board optical communications [5]. For our research, we chose two types of epoxy novolak resin (ENR) Su8-5 and Su8-50 supported by Micro resist technology GmbH as a core waveguide material. This polymer was chosen for its excellent properties (optical losses 2 dB·cm^{-1} at 980 nm, 0.77 dB·cm^{-1} at 1310 nm, and 1.71 dB·cm^{-1} at 1500 nm) [20-22] and feasible fabrication process. For a substrate, we used Polymethylmethacrylate (PMMA) and commercially available CL400 and PET foil supported by Omniplast because of their suitable properties, mainly low value of the refractive indices.

2. Design of the Planar Waveguides

The optical planar waveguide is a fundamental element for realization of optical ridge or channel waveguides that can be used for interconnection of various devices of optical integrated circuits and photonics structures.

In our case planar optical waveguide is a step-index structure and consists of a high-index dielectric layer surrounded on upper and lower sides with lower index materials (Fig. 1). If the cover and substrate materials have the same refractive index, the waveguide is called symmetric; otherwise the waveguide is called asymmetric.

Here we are going to design optical planar waveguides with polymer foil substrate, Su8 polymer waveguides; the upper side will be left open so that the air will act as a "cover" (n_c).

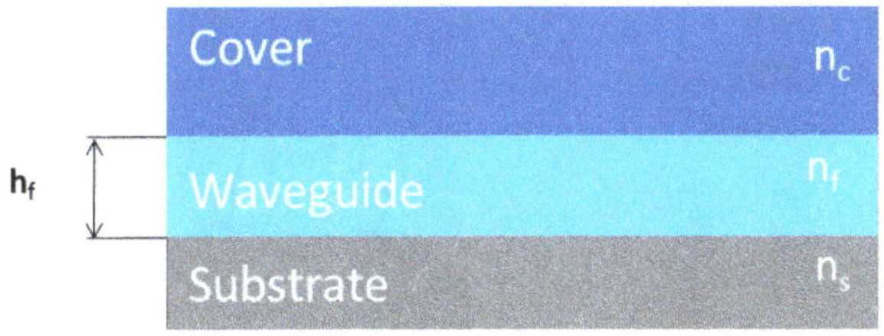

Fig. 1. Schema of an optical planar waveguide.

The index of refraction of the guiding slab n_f must be higher than that of the substrate materials n_s, or cover materials n_c in order to ensure total internal reflection occurring at the interfaces [23].

$$n_f > n_s, n_f > n_c. \quad (1)$$

Thickness h_f of the core of the optical waveguide film was calculated by using modification of dispersion equation (2), number of guided modes m is determined from equation (3) [24]:

$$h_f = \frac{\lambda_0}{2\pi\sqrt{n_f^2 - n_s^2}}\left\{n\pi + \text{arctg}\left[p\sqrt{\frac{n_s^2 - n_c^2}{n_f^2 - n_s^2}}\right]\right\}, \quad (2)$$

$$m = INT\left\{\frac{2}{\lambda_0}h_f\sqrt{n_f^2 - n_c^2} - \frac{1}{\pi}\text{arctg}\left[p\sqrt{\frac{n_s^2 - n_c^2}{n_f^2 - n_s^2}}\right]\right\} \quad (3)$$

where λ_0 is operating wavelength, n is an integer number $n = 0, 1, 2 \ldots$, and p is for the TE mode

$$p = 1 \quad (4)$$

and for the TM mode

$$p = \left(\frac{n_f}{n_s}\right)^2. \quad (5)$$

Before the actual proposal the optical waveguide layer (Su8-50 and Su8-5) were deposited on a glass substrate and then refractive indices of Su8 polymers and substrate foils that are needed for the calculation were measured by prism coupling method (Fig. 2). The figure shows that the values of the refractive indices decreased with the increasing wavelengths and also that the foils used for the substrate had lower refractive indices than Su8 waveguide materials.

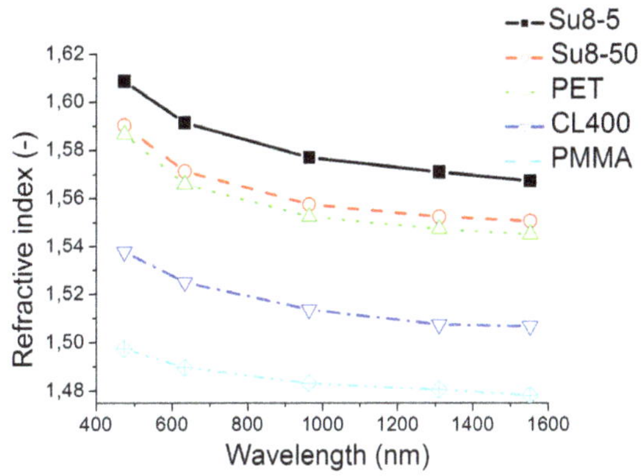

Fig. 2. Refractive indices of Su8 waveguide layer and PET, CL400, PMMA substrate.

Refractive indices for the substrate PMMA, CL400, PET and Su8-50, Su8-5 waveguides layers used for the design of the planar waveguides are listed in Tab. 1.

Wavelength (nm)	Refractive indices (-)				
	Substrates foil n_s			Waveguides layer n_f	
	PMMA	CL400	PET	Su8-50	Su8-5
633	1.4898	1.5251	1.5660	1.5713	1.5914
850	1.4855	1.5175	1.5573	1.5622	1.5816
1310	1.4807	1.5076	1.5474	1.5525	1.5709
1550	1.4783	1.5068	1.5453	1.5508	1.5673

Tab. 1. Refractive indices used for the design of the PMMA substrate, Su8-50 and Su8-5 waveguide layer.

The minimal calculated thickness for four wavelengths (633, 850, 1310, 1550 nm) of the designed single mode Su8-50 and Su8-5 planar optical waveguides are listed in Tab. 2. The results of mode calculations performed for 633 nm for TE for the first 20 modes concerning the waveguide structure described above are shown in Fig. 3.

Wavelength (nm)	mode	PMMA Su8-50	PMMA Su8-5	CL400 Su8-5	PET Su8-5
		h_f (μm)			
633	TE_0	0.21	0.20	0.33	0.48
	TE_1	0.86	0.76	1.17	1.59
850	TE_0	0.32	0.28	0.46	0.66
	TE_1	1.20	1.06	1.60	2.20
1310	TE_0	0.52	0.45	0.62	1.04
	TE_1	1.92	1.69	2.22	3.46
1550	TE_0	0.61	0.53	0.83	1.27
	TE_1	2.27	2.02	2.90	4.24

Tab. 2. Calculated minimum thicknesses for planar waveguides for PMMA. CL400. PET substrates and Su8-50. Su8-5 waveguide and air cover layer.

For example for PMMA/Su8-50 single mode waveguide structure we achieved the thickness of the waveguide layer h_f 0.21 μm for 633 nm and the thickness h_f 0.32 μm for 850 nm. For bigger thickness h_f than 0.52 μm for 1310 nm and 0.61 μm for 1550 nm waveguides became multimode (for more details see Tab. 2).

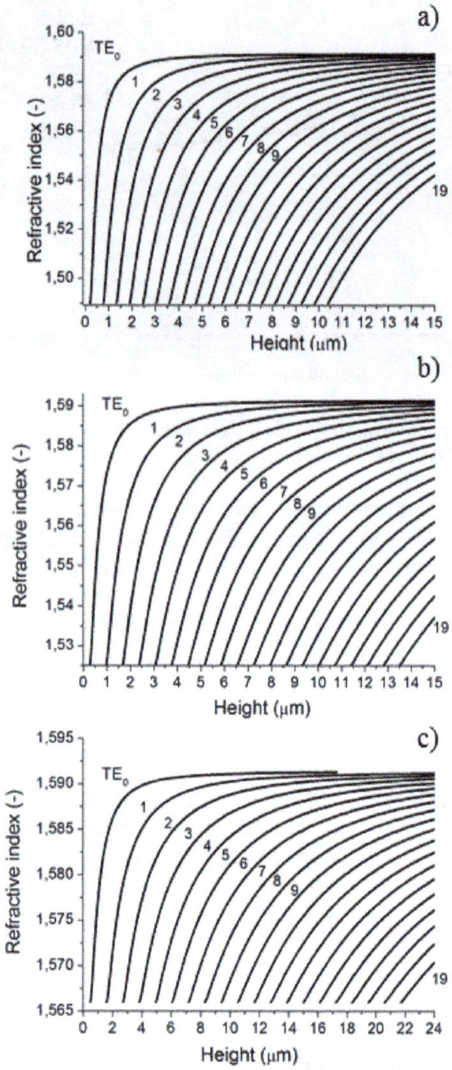

Fig. 3. TE mode calculation of the polymer planar waveguides for operation wavelength 633 nm for structures: a) PMMA/Su8-5, b) CL400/Su8-5 and c) PET/Su8-5.

3. Fabrication of the Waveguides

The experiments were performed on three types of substrates PMMA, CL400 and PET foils and two types of waveguide layers Su8-50 and Su8-5 (epoxy novolak resin). Fabrication process of the planar polymer flexible waveguides is illustrated in Fig. 4 step by step. PMMA foils for the substrates were made by dissolving pieces of PMMA in dichloroethane; this process needed some four to five days. The obtained solutions were let to dry for few days in petri dishes having different diameters. The dried substrates were removed from the petri dishes and cut for desired dimensions. Then polymer ENR waveguide layers were deposited on PMMA substrate by using spin coating (Fig. 4a); after that step soft bake process was applied at 50°C for 30 min in order to evaporate the remaining solvent (Fig. 4b). Then we applied UV curing process (Fig. 4c) and finally post exposure bake was done (Fig. 4d). Similar processes (except the step a) were applied for used CL400 and PET substrates.

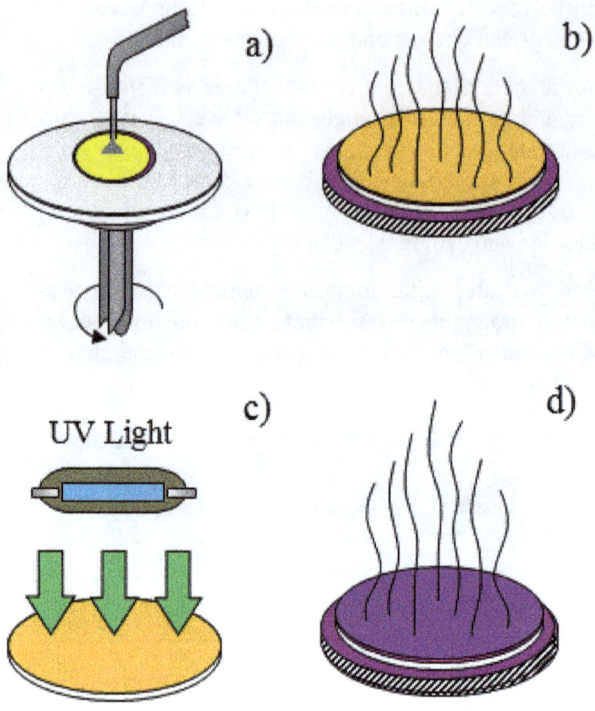

Fig. 4. Fabrication process for flexible planar optical waveguides: a) deposition of Su8 core waveguide layer, b) soft bake process, c) UV curing process, d) post exposure bake.

4. Results

The thicknesses of the fabricated PMMA flexible substrate were measured by dial thickness gauge LIMIT12.5/0.001 mm, while the thicknesses of the waveguides core layers were measured by profile-meters Talystep Hommel Tester 1000. The experimentally found thicknesses of the structure were as follows: flexible polymer PMMA substrates 30 μm to 500 μm depending on the amount of the polymer casted into the mould; CL400 foil 500 μm and PET substrate 1000 μm. The thicknesses of the polymer waveguide layers were from units to 60 μm, depending on the rate of spinning of the coater during the deposition.

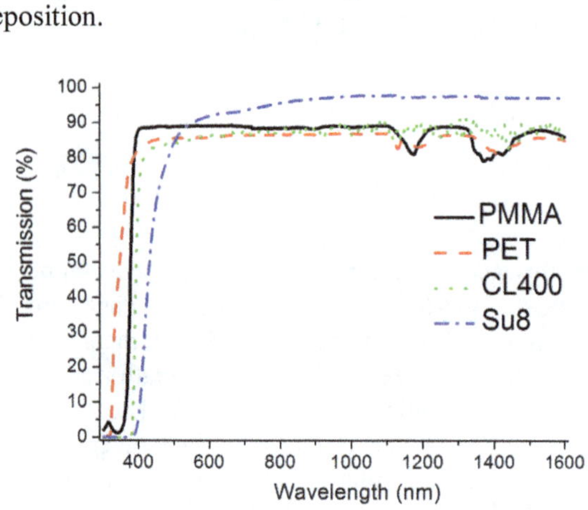

Fig. 5. Transmission spectra of PMMA, PET, CL400 substrates and Su8 waveguide layers.

Transmission spectra of the used substrates and Su8 waveguide layers were collected by UV-VIS-NIR Spectrometer (UV-3600 Shimadzu) in the spectral range from 300 to 1600 nm and are given in Fig. 5. Obviously the waveguide layer is transparent within the whole range of the measured wavelengths. Polymer substrates revealed two absorption peaks in the near infrared wavelength region, which can be attributed to the vibrational overtones of C–H bonds.

Waveguiding properties of the flexible ENR planar waveguides were examined by dark mode spectroscopy using Metricon 2010 prism-coupler system [25-27] (Fig. 6).

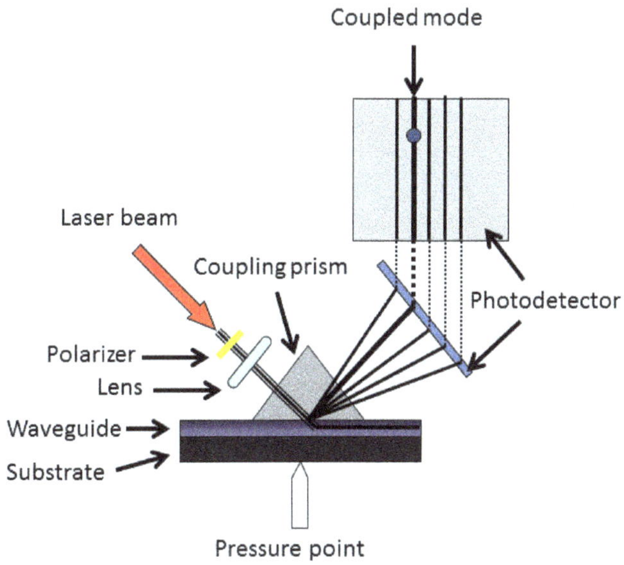

Fig. 6. Schematic view of the dark mode spectroscopy measurement.

The measured sample is brought into contact with the base of a couple prisms by means of a pneumatically-operated coupling head leaving narrow air gap between the waveguide film and the prism. Laser beam strikes the base of the prism and is totally reflected at the prism base onto a photodetector at certain discrete values of the incident angle Θ called mode angles. Photons can tunnel across the air gap into the waveguide film and enter into a guided optical propagation mode causing a sharp drop of the intensity of light reaching the detector [25].

The waveguiding properties were measured at five wavelengths (473, 633, 964, 1311 and 1552 nm). Fig. 7a gives an example of measured mode spectra of the multimode CL400/Su8-50 waveguide and the particular modes are signified by the arrows (26 modes in whole), designative for calculation of pertinent refractive index depth profile. All the refractive index depth profiles are then illustrated in Fig. 7b giving confirmation of a step-like character of all the profiles of 18 μm thick waveguiding layer. They also showed decrease of the refractive index values with the increasing wavelengths (see also Fig. 2).

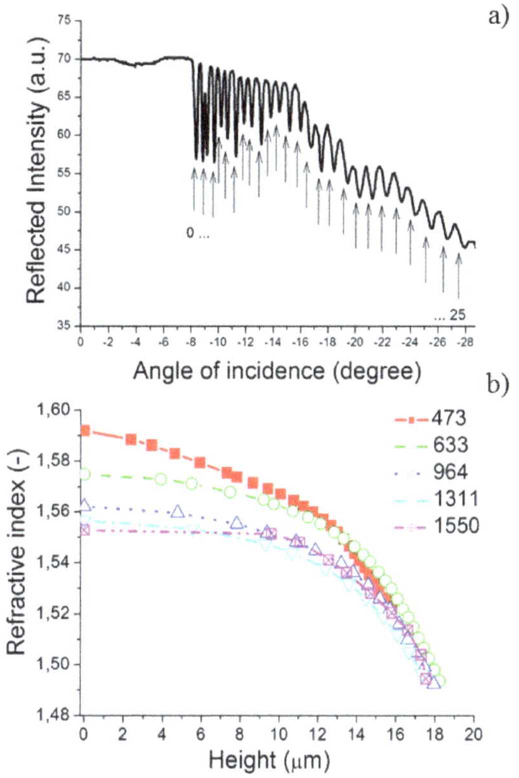

Fig. 7. Evaluation of the refractive indices depth profile of Su8-50 waveguide for various wavelengths for TE modes.

Mode pattern for Su8-50 optical planar waveguide deposited on PMMA foil substrate for three wavelengths (473, 633 and 1552 nm) is shown in Fig. 8. The arrows ① denote the first mode of the Su8 waveguide layer respective to the actual wavelengths, the arrows ② close to the edges of strong peaks show where the PMMA substrate begins. Incident angles ① for particular wavelengths are as follows: -7°15′ for 473 nm corresponds to refractive index of Su8 polymer 1.6116; -7°12′ for 633 nm corresponds to refractive index of Su8 polymer 1.5925 and finally angle -6°19′ for 1552 nm corresponds to refractive index of Su8 polymer 1.5702.

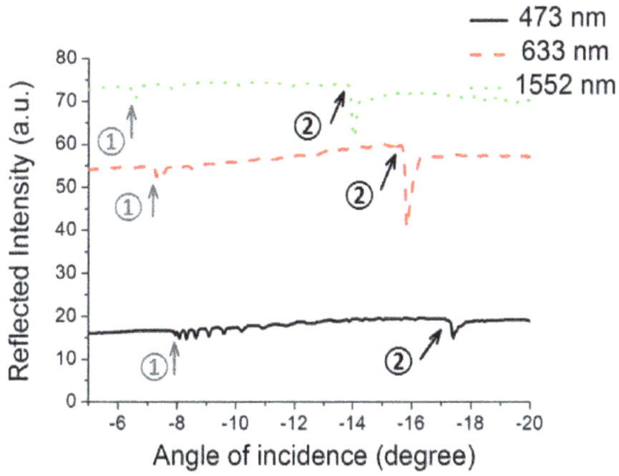

Fig. 8. Mode pattern of Su8-50/PMMA foil planar waveguides (TE modes). For easy orientation the curves for only three wavelengths are presented.

In the case of determination of the refractive indices of the PMMA substrate foil (concerning the incident angle ②) we used a similar procedure: -17°55′ at 473 nm corresponds to refractive index 1.5009; -15°40′ at 633 nm corresponds to refractive index 1.4923 and -13°40′ at 1552 nm corresponds to refractive index 1.4819.

Optical losses of the planar waveguides were measured by a technique involving measurement of transmitted and scattered light intensity as a function of propagation distance along the waveguide [28]. Actually it follows losses of optical waveguides by scanning with a fiber optic probe and a photodetector down the length of a propagating streak to measure intensity of the light scattered from the surface of the guide. The optical fiber method is similar to a concept of a CCD camera used to measure decay of the propagating streak with the advantage that our approach does not need the camera that should have very uniform response sensitivity over the full array. With the scanning fiber method, only a small single-element silicon detector is used and spatial uniformity is not an issue [25], [29]. We measured optical losses by using He-Ne laser at 633 nm and the principle of the method is shown in Fig. 9.

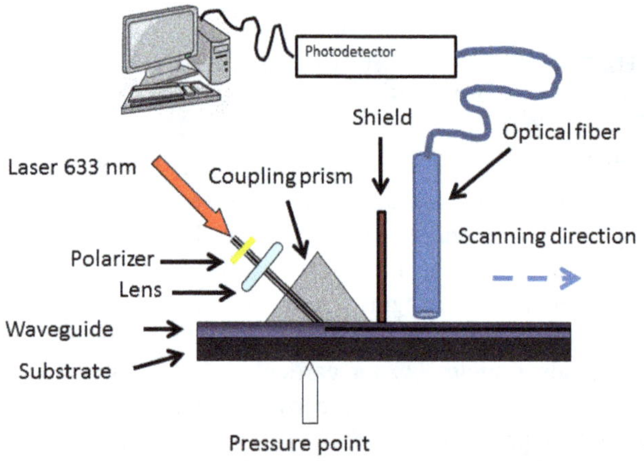

Fig. 9. Schematic view of the optical planar loss measurement.

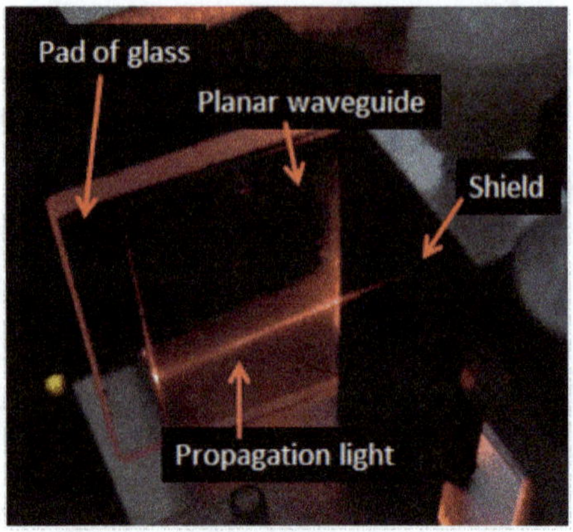

Fig. 10. Coupling of the optical signal (633 nm) into flexible polymer waveguides for optical loss measurements.

Fig. 10 shows an image of flexible waveguides supporting optical light at 633 nm. The flexible polymer sample must be fixed to a glass pad; otherwise the planar waveguides might bend, which would make achieving of quality optical contact difficult and thus worsen the coupling of the laser beam into the waveguide.

Optical loss measurements are given in Fig. 11 showing results for planar waveguide PMMA/Su8-5 in Fig. 11a and the results for PET/Su8-5 waveguide in Fig. 11b.

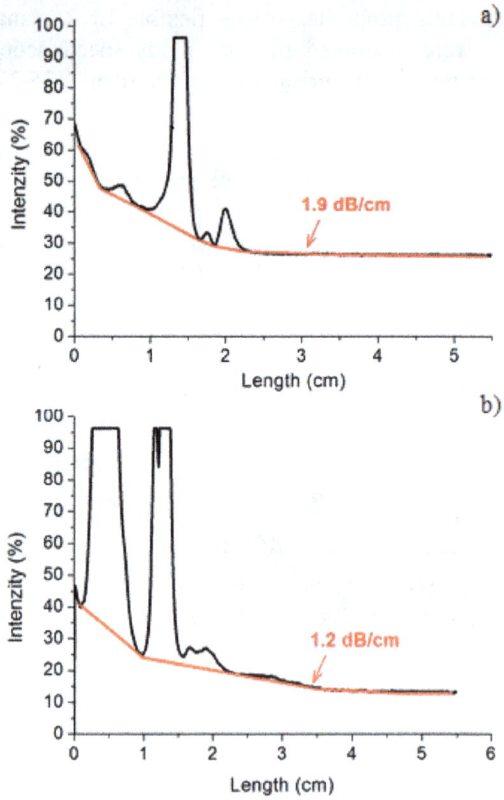

Fig. 11. Optical losses of the flexible waveguides for wavelength 633 nm a) PMMA/Su8-5, b) PET/Su8-5.

Our optical planar waveguides had optical losses lower than 2 $dB \cdot cm^{-1}$ with the best sample having optical losses as low as 1.2 $dB \cdot cm^{-1}$. This value is similar to the values reported in [20], [22], [30] regarding to the facts that in [20] a single-mode waveguide Si/SiO_2/Su8/PMMA (1.36 for TE and 2.01 for TM modes) with the losses taken at longer (980 nm) wavelength is mentioned; the results in [22] (1.5 $dB \cdot cm^{-1}$ at 650 nm) are for a ridge waveguide and [30] (0.19 $dB \cdot cm^{-1}$ at 633 nm) concern much more demanding technology (proton beam writing).

5. Conclusion

We report about design, fabrication and properties of flexible polymer planar waveguides made of epoxy novolak resin (Su8) polymer as a core waveguide layer deposited on PMMA, CL400, or PET foil substrates.

Optical waveguiding properties of our planar waveguides samples were characterized by Metricon 2010 prism-coupler system for five wavelength (473, 633, 964, 1311 and 1552 nm) and optical losses were measured by collecting the scattered light using fiber scanning along the waveguide read by the Si photodetector at 633 nm. Our best sample had optical losses around 1.2 dB·cm^{-1}.

The main advantage of our samples is that they are deposited on flexible substrates which makes them suitable for advanced sophisticated interconnection devices. Next we are going to design and construct multimode flexible ridge waveguides based on the same principle.

Acknowledgements

Our research is supported by grant P 108/12/6108 GAČR.

References

[1] BAMIEDAKIS, N., BEALS, J., PENTY, R.V., WHITE, I.H., DE GROOT, J.V., CLAPP, T.V. Cost-effective multimode polymer waveguides for high-speed on-board optical interconnects. *IEEE Journal of Quantum Electronics*, 2009, vol. 45, no. 4, p. 415–424.

[2] YOSHITAKE, N., TERAKAWA, Y., HOSOKAWA, H. Polymer optical waveguide devices for FTTH. In *Proceedings of the Opto-Electronics and Communications Conference. 2008 and the 2008 Australian Conference on Optical Fibre Technology*. Sydney (Australia), 2008, p. 1–2.

[3] CHOI, C., LIN, L., LIU, Y., CHOI, J., WANG, L., HAAS, D., MAGERA, J., CHEN, R.T. Flexible optical waveguide film fabrications and optoelectronic devices integration for fully embedded board-level optical interconnects. *Journal of Lightwave Technology*, 2004, vol. 22, no. 9, p. 2168–2176.

[4] ISHIDA, Y., HOSOKAWA, H. Optical link utilizing polymer optical waveguides application in multimedia device. *Proceedings SPIE. Photonics in Multimedia II,* 2008, vol. 7001, p. 70010J-1 to 70010J-9.

[5] BOSMAN, E., VAN STEENBERGE, G., VAN HOE, B., VAN MISSINNE, J., VAN FETEREN, J., VAN DAELE, P. Highly reliable flexible active optical links. *IEEE Photonics Technology Letters*, 2010, vol. 22, no. 5, p. 287–289.

[6] BOOTH, B.L. Low-loss channel waveguides in polymers. *Journal of Lightwave Technology*, 1989, vol. 7, no. 10, p. 1445–1453.

[7] WONG, W.H., LIU, K.K., CHAN, K.S., PUN, E.Y.B. Polymer devices for photonics applications. *Journal of Crystal Growth*, 2006, vol. 288, no. 1, p. 100–104.

[8] TUNG, K.K., WONG, W.H., PUN, E.Y.B. Polymeric optical waveguides using direct ultraviolet photolithography process. *Applied Physics A-Materials Science & Processing*, 2005, vol. 80, p. 621–626.

[9] LYUTAKOV, O., TUMA, J., PRAJZLER, V., HÜTTEL, I., HNATOWICZ, V., SVORCIK, V. Preparation of rib channel waveguides on polymer in electric field. *Thin Solid Films*, 2010, vol. 519, no. 4, p. 1452–1457.

[10] PRAJZLER, V., KLAPUCH, J., LYUTAKOV, O., HÜTTEL, I., ŠPIRKOVÁ, J., NEKVINDOVÁ, P., JEŘÁBEK, V. Design, fabrication and properties of rib poly (methylmethacrylimide) optical waveguides. *Radioengineering*, 2011, vol. 20, p. 479–485.

[11] PRAJZLER, V., NERUDA, M., ŠPIRKOVÁ, J. Planar large core polymer optical 1x2 and 1x4 splitters connectable to plastic optical fiber. *Radioengineering*, 2013, vol. 22, p. 751–757.

[12] HIKITA, M., YOSHIMURA, R., USUI, M., TOMARU, S., IMAMURA, S. Polymeric optical waveguides for optical interconnections. *Thin Solid Films*, 1998, vol. 331, no. 1-2, p. 303–308.

[13] LYUTAKOV, O., HÜTTEL, I., PRAJZLER, V., JEŘÁBEK, V., JANCAREK, A., HNATPOWICZ, V., SVORCIK, V. Pattern formation in PMMA film induced by electric field. *Journal of Polymer Science B Polymer Physics*, 2009, vol. 47, no. 12, p. 1131–1135.

[14] FISCHBEC, G., MOOSBURGER, R., TOPPER, M., PETERMANN, K. Design concept for singlemode polymer waveguides. *Electronics Letters,* 1996, vol. 32, no. 3, p. 212–213.

[15] IMANURA, S., YOSHIMURA, R., IZAWA, T. Polymer channel waveguides with low loss at 1.3 µm. *Electronics Letters*, 1991, vol. 27, p. 1342–1343.

[16] MA, H., JEN, A.K.Y., DALTON, L.R. Polymer based optical waveguides: Materials, processing and devices. *Advanced Materials*, 2002, vol. 14, no. 19, p. 1339–1365.

[17] ELDADA, L. Optical communication components. *Review of Scientific Instruments*, 2004, vol. 75, no. 3, p. 575–593.

[18] ELDADA, L., SHACKLETTE, L.W. Advances in polymer integrated optics. *IEEE Journal of Selected Topics in Quantum Electronics*, 2000, vol. 6, p. 54–68.

[19] YENIAY, A., GAO, R.Y., TAKAYAMA, K., GAO, R.F., GARITO, A.F. Ultra-low-loss polymer waveguides. *Journal of Lightwave Technology*, 2004, vol. 22, no. 1, p. 154–158.

[20] BECHE, B., PELLETIER, N., GAVIOT, E., ZYSS, J. Single-mode TE_{00}-TM_{00} optical waveguides on SU-8 polymer. *Optics Communications*, 2004, vol. 230, no. 1-3, p. 91–94.

[21] YANG, B., YANG, L., HU, R., SHENG, Z., DAI, D., LIU, Q., HE, S. Fabrication and characterization of small optical ridge waveguides based on SU-8 polymer. *Journal of Lightwave Technology*, 2009, vol. 27, no. 18, p. 4091–4096.

[22] PRAJZLER, V., LYUTAKOV, O., HÜTTEL, I., BARNA, J., ŠPIRKOVÁ, J., NEKVINDOVÁ, P., JEŘÁBEK, V. Simple way of fabrication of Epoxy Novolak Resin optical waveguides on silicon substrate. *Physica Status Solidi C-Current Topics in Solid State Physics*, 2011, vol. 8, no. 9, p. 2942–2945.

[23] POLLOCK, C., LIPSON, M. *Integrated Photonics.* Kluwer Academic Publishers, 2003.

[24] ADAMS, M.J. *An Introduction to Optical Waveguides.* Toronto: JohnWiley&Sons Ltd., 1981.

[25] METRICON CORPORATION. www.metricon.com

[26] ULRICH, R., TORGE, R. Measurement of thin film parameters with a prism coupler. *Applied Optics,* 1973, vol. 12, no. 12, p. 2901–2908.

[27] KERSTEN, R.T. A new method for measuring refractive index and thickness of liquid and deposited solid thin films. *Optics Communications*, 1975, vol. 13, p. 327–329.

[28] NOURSHARGH, N., STARR, E.M., FOX, N.I., JONES, S.G. Simple technique for measuring attenuation of integrated optical waveguides. *Electronics Letters,* 1985, vol. 21, no. 18, p. 818–820.

[29] OKAMURA, Y. YOSHINAKA, S. YAMAMOTO, S. Measuring mode propagation losses of integrated optical waveguides: a simple method. *Applied Optics*, 1983, vol. 22, no. 23, p. 3892–3894.

[30] SUM, T.C., BETTIOL, A.A., VAN KAN, J.A., WATT, F., PUN, E.Y.B., TUNG, K.K. Proton beam writing of low-loss polymer optical waveguides. *Applied Physics Letters*, 2003, vol. 83, no. 9, p. 1707–1709.

About Authors ...

Václav PRAJZLER was born in 1976 in Prague, Czech Republic. In 2001 he graduated from the Faculty of Electrical Engineering, Czech Technical University in Prague, Department of Microelectronics. Since 2005 he has been working at the same department as a research fellow. In 2007 he obtained the PhD degree from the same university. His current research is focused on design, fabrication and investigation of properties of photonics structures.

Pavla NEKVINDOVÁ was born in 1972 in Kolín, Czech Republic. She graduated from the Institute of Chemical Technology, Prague (ICTP) in 1999. Now she is the Assistant Professor at the ICTP giving lectures on general and inorganic chemistry. She has worked there continuously in materials chemistry research. She has a long-term experience with fabrication and characterization of optical waveguiding structures in single-crystalline and glass materials.

Petr HYPŠ was born in 1989 in Pelhřimov Czech Republic. He has studied at the Faculty of Electrical Engineering. Czech Technical University in Prague. In June 2013 he obtained a bachelor's degree in Communication, Multimedia and Electronics. Now his field of study is polymer photonics structures.

Oleksiy LYUTAKOV was born in 1982 in Kramatorsk, Ukraine. He studied at the Chemistry Faculty of Donetsk National University. He obtained the PhD degree at the Institute of Chemical Technology in 2009. At present he is working at the same institute at the Department of Solid State Engineering as a research fellow.

Vítězslav JEŘÁBEK was born in Prague in 1951. He received his M.Sc. and Ph.D. degrees in Microelectronics from the Czech Technical University in Prague in 1975 and 1987. He is currently an Associate Professor of Electronics and head of optoelectronic group at the Microelectronic Department, Czech Technical University in Prague.

3

Modeling of Luneburg Lenses with the Use of Integral Equation Macromodels

Andrzej A. KUCHARSKI

Telecommunications and Teleinformatics Department, Wrocław University of Technology, Wybrzeze Wyspianskiego 27, 50-370 Wrocław, Poland

andrzej.kucharski@pwr.wroc.pl

Abstract. *The so-called integral equation macromodel allowing to efficiently include Luneburg lens into the body-of-revolution method-of-moments (BoR-MoM) computational scheme is described. In the process of the macromodel construction, we make use of the equivalence-principle domain-decomposition-method (EP-DDM) and the asymptotic waveform evaluation (AWE) method. By the use of the macromodel, the number of unknowns in the final system of equations is reduced to those describing sources on the equivalent surface surrounding the lens. Moreover, thanks to the macromodel being valid in a certain frequency interval, the domain decomposition procedure does not have to be repeated for every frequency of interest, but it should only be done in some specified frequency points. However, the range of validity of the macromodel should be carefully investigated on the basis of full radiation pattern rather than on the basis of a single direction of observation.*

Keywords

Method of moments, macromodel, Luneburg lens.

1. Introduction

The Luneburg lens [1] is a well-know type of spherical dielectric lens with dielectric permittivity profile depending on the distance from the center, according to the formula:

$$\varepsilon_r(r) = 2 - \left(\frac{r}{a}\right)^2, \quad 0 \leq r \leq a, \tag{1}$$

where r is the distance from the sphere center, a denotes the radius of the sphere.

In the optical regime, it possesses an interesting property of collimating rays originating from the point source placed at the lens surface, into parallel rays leaving the lens on its other side. Although this property does not transform fully into radio-frequency bands [2], the lens plays an important role in the design of antennas [3]. Although the interest in this technique has been abandoned for many years [4] mainly due to problems with production of inhomogeneous dielectrics, it is recently re-awakened because of possible applications in future High Altitude Platform (HAP) systems. Those applications include obtaining the whole cellular pattern from the single platform [5], as well as the construction of ground-based antennas with mechanical scanning possibilities [6]. Modeling of antenna configurations incorporating Luneburg lenses provides some challenge for computational methods, as their inhomogeneous nature enforces either applying differential-equation/grid methods or using volume sources within integral-equation/method-of-moments (MoM) solutions. In the first case one has to either use proper absorbing conditions to model the radiation [2], or to use hybrid solution [7]. The second choice (MoM) leads to a large number of unknowns in the linear equation system, described by a dense matrix. In both cases, the situation is complicated by the fact that the lens is usually electrically large.

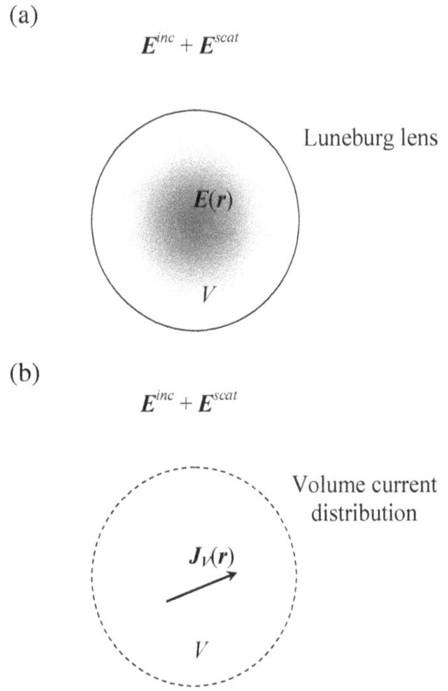

Fig. 1. Luneburg lens immersed in the incident field (a) and the equivalent situation (b).

Alternatively, one may apply solutions of Maxwell's equations for a concentric multishell dielectric sphere [3], [6], which is however outside the scope of this paper. The need for large computational resources may be relaxed, if we note that the lens possesses rotational symmetry. Thus, we may apply the well-known body-of-revolution (BoR) technique, in which solution scheme is decomposed into a number of azimuthal modes, analyzed separately (cf. [7–9]). For some excitations, like plane wave traveling along the BoR symmetry axis, it is even possible to limit the analysis to one mode, thus reducing the problem from three-dimensional to two-dimensional. Even if this is the case, the computation may require a relatively large number of unknowns [9], which makes it quite time-consuming to perform for instance wideband feed optimizations. Here, we give a simple solution to this problem, applying previously introduced technique of integral equation macromodels [10]. The idea used in this paper was initially given in [11] for wideband analysis of partially inhomogeneous bodies of revolution. However, examples given in [11] were confined to simple two-layered spheres. Here, we show its usefulness for more sophisticated case of Luneburg lenses. The main purpose of the paper is twofold: first, we show that it is really possible to describe the Luneburg lens as a wideband "black box" with the number of unknowns in the final equation set equal to that of a homogeneous object, second, we investigate carefully the "bandwidth" of the macromodel based on the full radiation pattern analysis.

Throughout the paper we assume $e^{j\omega t}$ time convention.

2. Integral Equation Macromodel Based on Volume-Surface Integral Equation

2.1 Classical Volume Integral Equation (VIE) Formulation

The original situation to be analyzed is depicted in Fig. 1. The usual procedure relies on replacing the dielectric by the volume distribution of polarization current, radiating in free-space [12], [13]:

$$\mathbf{J}_V(\mathbf{r}) = j\omega\left[\varepsilon(\mathbf{r}) - \varepsilon_0\right]\mathbf{E}(\mathbf{r}). \tag{2}$$

Then we formulate the volume integral equation, which states that the total electric field in the body volume (as at any other point in space) is a sum of incident and scattered fields (cf. [13]):

$$\mathbf{E}^{inc}(\mathbf{r}) + \mathbf{E}^{scat}(\mathbf{r}) = \frac{\mathbf{J}_V(\mathbf{r})}{j\omega\left[\varepsilon(\mathbf{r}) - \varepsilon_0\right]}, \quad \mathbf{r} \in V. \tag{3}$$

Above, vector $\mathbf{r}$ indicates the observation point. Electric field due to polarization current may be obtained from standard mixed-potential formula:

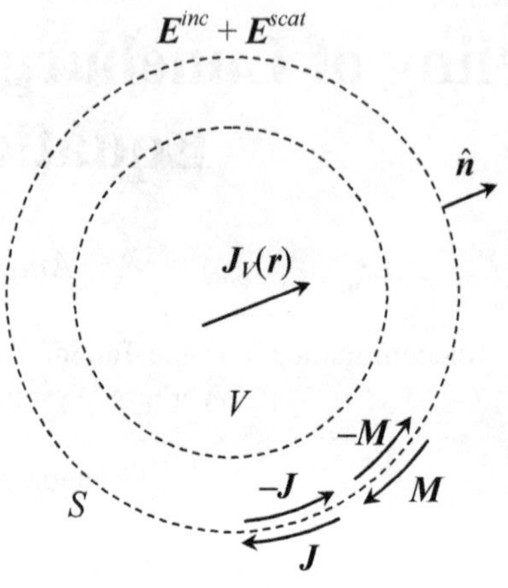

Fig. 2. Additional surface S with equivalent currents surrounding the lens (external and internal equivalent situations combined into the single picture).

$$\mathbf{E}^{scat}(\mathbf{r}) = \mathbf{E}^{scat}(\mathbf{J}_V) = -\frac{j\omega\mu_0}{4\pi}\int_V \mathbf{J}_V(\mathbf{r}')\frac{e^{-jk|\mathbf{r}-\mathbf{r}'|}}{|\mathbf{r}-\mathbf{r}'|}dV' + \frac{1}{4\pi\varepsilon_0}\nabla\int_V \frac{\nabla\cdot\mathbf{J}_V(\mathbf{r}')}{j\omega}\frac{e^{-jk|\mathbf{r}-\mathbf{r}'|}}{|\mathbf{r}-\mathbf{r}'|}dV' \tag{4}$$

where $\mathbf{r}'$ indicates source point, and $k = \omega\sqrt{\varepsilon_0\mu_0}$ is a wavenumber.

It is to be noted that, usually, it is the electric flux density used as the unknown quantity [13], which enables application of roof-top-like basis and testing functions in the method-of-moments (MoM) solution.

For bodies with rotational symmetry, the scheme proposed in [9] may be used to efficiently solve equation (3). The method utilizes conventional azimuthal mode decoupling together with specially constructed divergence-less basis functions. Sample results obtained with this basic approach will be shown in the next section as the comparison data.

2.2 Volume-Surface (VIE-SIE) Formulation and Domain Decomposition Method (DDM)

Another solution is to apply domain decomposition scheme based on the equivalence principle [14]. Thus, we surround the inhomogeneous part by the artificial surface S, which may or may not coincide with the outer surface of the analyzed body (Fig. 2). Then, introducing artificial electric and magnetic currents, and applying usual internal and external equivalent situations, we arrive at the system of equations [11], [15]:

$$\hat{\mathbf{n}} \times \left(\mathbf{E}^{inc} + \mathbf{E}_e^+(\mathbf{J},\mathbf{M})\right) = \hat{\mathbf{n}} \times \mathbf{E}_i^-(-\mathbf{J},-\mathbf{M},\mathbf{J}_V) \quad \text{on } S, \tag{5}$$

$$\hat{\mathbf{n}} \times \left(\mathbf{H}^{inc} + \mathbf{H}_e^+(\mathbf{J},\mathbf{M})\right) = \hat{\mathbf{n}} \times \mathbf{H}_i^-(-\mathbf{J},-\mathbf{M},\mathbf{J}_V) \quad \text{on } S, \tag{6}$$

$$\mathbf{E}_i = \mathbf{E}_i^-(-\mathbf{J},-\mathbf{M},\mathbf{J}_V) \quad \text{in } V. \tag{7}$$

Above, sub-scripts e and i denote environments external and internal to S, respectively, while super-scripts "+" and "–" show that corresponding fields are calculated, as the observation point approaches S from outside and inside, respectively. Note that although in sub-sequent calculations, external and internal environments are assumed to be the same (free space), in general it may not be the case. For example, different Green's functions, accounting for other configurations, may be used for the outer environment. Note also that the above equations are the same as in the case of partially inhomogeneous dielectric bodies [16]. Obviously, fields in (5) – (7) produced by several currents are understood as superpositions of fields produced by individual currents.

Formulas to obtain fields due to surface electric current for the internal interactions are standard ones:

$$\mathbf{E}_i(\mathbf{r}) = \mathbf{E}_i(\mathbf{J}) = -\frac{j\omega\mu_0}{4\pi}\int_S \mathbf{J}(\mathbf{r}')\frac{e^{-jk|\mathbf{r}-\mathbf{r}'|}}{|\mathbf{r}-\mathbf{r}'|}dS' + \frac{1}{4\pi\varepsilon_0}\nabla\int_S \frac{\nabla\cdot\mathbf{J}(\mathbf{r}')}{j\omega}\frac{e^{-jk|\mathbf{r}-\mathbf{r}'|}}{|\mathbf{r}-\mathbf{r}'|}dS', \quad (8)$$

$$\mathbf{H}_i(\mathbf{r}) = \mathbf{H}_i(\mathbf{J}) = -\frac{1}{4\pi}\nabla\times\int_S \mathbf{J}(\mathbf{r}')\frac{e^{-jk|\mathbf{r}-\mathbf{r}'|}}{|\mathbf{r}-\mathbf{r}'|}dS', \quad (9)$$

while the formulas for fields due to magnetic currents may be easily obtained by duality, and fields due to the volume electric current are the counterparts of (8), (9), with integration over surface S replaced with integration over volume V. For external interactions, the situation depends on the environment, in which the lens (together with surrounding surface S, is immersed. If the medium is again the free space, obviously (8) and (9) are applied.

After application of MoM, the system (5) – (7) is transformed into the matrix equation:

$$\begin{bmatrix} \mathbf{Z}_{EJ}^{SSe}+\mathbf{Z}_{EJ}^{SSi} & \mathbf{Z}_{EM}^{SSe}+\mathbf{Z}_{EM}^{SSi} & \mathbf{Z}_{EJ}^{SVi} \\ \mathbf{Z}_{HJ}^{SSe}+\mathbf{Z}_{HJ}^{SSi} & \mathbf{Z}_{HM}^{SSe}+\mathbf{Z}_{HM}^{SSi} & \mathbf{Z}_{HJ}^{SVi} \\ \mathbf{Z}_{EJ}^{VSi} & \mathbf{Z}_{EM}^{VSi} & \mathbf{Z}_{EJ}^{VVi} \end{bmatrix} \begin{bmatrix} \mathbf{J} \\ \mathbf{M} \\ \mathbf{J}_V \end{bmatrix} = \begin{bmatrix} \mathbf{E}^{inc} \\ \mathbf{H}^{inc} \\ \mathbf{0} \end{bmatrix} \quad (10)$$

where $\mathbf{J}$, $\mathbf{M}$, $\mathbf{J}_V$ now denote vectors of coefficients describing approximations of respective currents. Similarly, $\mathbf{E}^{inc}$, $\mathbf{H}^{inc}$ are vectors resulting from testing of incident fields at S. $\mathbf{Z}_{PQ}^{KLr}$ denotes sub-matrix with elements corresponding to field P (electric E or magnetic H) from currents Q (electric J or magnetic M) flowing in the domain L (surface S or volume V) tested over domain K (also S or V), while the radiation takes place in the environment r (external e or internal i).

In the system, we obviously have more unknowns, than in the matrix counterpart of (3), because in addition to unknowns resulting from the volume current distribution, we have coefficients describing surface electric in magnetic currents. However, we can replace the set by [11]:

$$\left[\mathbf{Z}^{SSe} + \left\{\mathbf{Z}^{SSi} - \mathbf{Z}^{SVi}\left(\mathbf{Z}_{EJ}^{VVi}\right)^{-1}\mathbf{Z}^{VSi}\right\}\right]\begin{bmatrix} \mathbf{J} \\ \mathbf{M} \end{bmatrix} = \begin{bmatrix} \mathbf{E}^{inc} \\ \mathbf{H}^{inc} \end{bmatrix} \quad (11)$$

with

$$\mathbf{Z}^{SSe} = \begin{bmatrix} \mathbf{Z}_{EJ}^{SSe} & \mathbf{Z}_{EM}^{SSe} \\ \mathbf{Z}_{HJ}^{SSe} & \mathbf{Z}_{HM}^{SSe} \end{bmatrix}, \quad (12)$$

$$\mathbf{Z}^{SSi} = \begin{bmatrix} \mathbf{Z}_{EJ}^{SSi} & \mathbf{Z}_{EM}^{SSi} \\ \mathbf{Z}_{HJ}^{SSi} & \mathbf{Z}_{HM}^{SSi} \end{bmatrix}, \quad (13)$$

$$\mathbf{Z}^{SVi} = \begin{bmatrix} \mathbf{Z}_{EJ}^{SVi} \\ \mathbf{Z}_{HJ}^{SVi} \end{bmatrix}, \quad (14)$$

$$\mathbf{Z}^{VSi} = \begin{bmatrix} \mathbf{Z}_{EJ}^{VSi} & \mathbf{Z}_{EM}^{VSi} \end{bmatrix}. \quad (15)$$

Note, that the matrix term in curly braces:

$$\mathbf{Z}_M = \left\{\mathbf{Z}^{SSi} - \mathbf{Z}^{SVi}\left(\mathbf{Z}_{EJ}^{VVi}\right)^{-1}\mathbf{Z}^{VSi}\right\} \quad (16)$$

may be calculated independently on the outer environment and/or excitation. Thus, once $\mathbf{Z}_M$ is obtained, the final system to be solved is that incorporating only surface sources with associated unknown coefficients. Taking into account that the number of unknowns associated with surface sources is usually many times smaller, than the number of unknowns corresponding to volume sources, solving (11) is much quicker that solving the original equation (3). Also, pre-calculated and stored $\mathbf{Z}_M$ may be re-used several times for different situations, and even for single problem incorporating cloning of the structure (like in analysis of antenna arrays, or other periodic geometries). The usefulness of above procedure was proven in [15], where pre-calculated $\mathbf{Z}_M$ was applied to analysis of Luneburg lenses for various types of excitations.

2.3 Integral-Equation Macromodel (IEM) of Luneburg Lens

The drawback of the domain decomposition technique outlined above is that the whole procedure must be repeated for every frequency of interest. One solution is to perform the computation at certain sample frequencies and then apply matrix interpolation [17] to $\mathbf{Z}_M$ for other frequencies. However, this matrix represents the partial problem solution, and therefore the elements of $\mathbf{Z}_M$ may exhibit resonant behavior (cf. [17]) leading to large numbers of frequency points to be analyzed directly, contrary to standard MoM techniques, where matrix elements represent simple source-field interactions. Therefore, in this work we make use of integral equation macromodels [10]. This technique, introduced previously by the author, allows to get the wideband approximation for $\mathbf{Z}_M$ making use of asymptotic waveform evaluation (AWE) technique (cf. [18]). The procedure consists of the following steps (cf. [10], [11]):

1. We find Taylor approximation to $\left(\mathbf{Z}_{EJ}^{VVi}\right)^{-1}\mathbf{Z}^{VSi}$ term, present in (16). The procedure relies on applying AWE method to the matrix equation:
$$\mathbf{Z}_{EJ}^{VVi}\cdot\mathbf{X}=\mathbf{U}, \qquad (17)$$
vector $\mathbf{U}$ standing for successive columns of $\mathbf{Z}^{VSi}$. Note that we don't need to find the approximation to $\left(\mathbf{Z}_{EJ}^{VVi}\right)^{-1}$. The procedure has to be repeated only the number of times equal to the number of unknowns on the surface interface S, corresponding to the number of columns in $\mathbf{Z}^{VSi}$. This may be understood as exciting the structure by a limited number of "ports" at the interface.

2. Next, we obtain the Taylor coefficients of $\mathbf{Z}_M$ equating corresponding powers of left and right hand side expansions of (16).

3. Finally, we get Padé approximation for each element of $\mathbf{Z}_M$ utilising a usual procedure [18]. This final approximation, which is valid over some frequency interval, we call here the Integral Equation Macromodel (IEM) [10].

As seen above, the AWE method, and what follows integral equation macromodel technique, is based on Taylor expansions, which require computing of derivatives of the integral equation kernels, with respect to frequency. Luckily, here all kernels in (5) – (7) are the free-space ones, so the computation does not present difficulties. Proper formulas needed to compute derivatives of the kernels associated with surface sources may be found in [18]. The situation is a little bit more complicated for the case of kernels associated with volume sources, as the polarization current definition (2) itself includes the dependence on the frequency. Thus, we have to take it into account while computing the derivatives of proper matrix elements. The formulas for derivatives found in expressions for electric field due to polarization current may be found in author's paper [19], while the remaining formula for n-th derivative of the magnetic field kernel, multiplied by k, is:

$$\left(k(1+jkR)\frac{e^{-jkR}}{R^3}\right)^{(n)} =$$
$$\left\{kR(2n-1)+j\left[n(n-2)-k^2R^2\right]\right\}(-jR)^{n-2}\frac{e^{-jkR}}{R^2}. \qquad (18)$$

Having the approximation to $\mathbf{Z}_M$, we compute both the sub-matrix of external interactions $\mathbf{Z}^{SSe}$ and the excitation vectors $\mathbf{E}^{inc}$, $\mathbf{H}^{inc}$, for a frequency of interest, and then solve (11), which is much quicker than solving original VIE. We can also introduce other objects external to S, which changes (11) into a larger linear set incorporating additional matrix blocks corresponding to interactions between the object and the surface S. This however does not change $\mathbf{Z}_M$ allowing it to be computed once and reused for different outer environments, excitation mechanisms and possible objects. As mentioned in the Introduction, the first check of the above procedure was described in [11], where simple case of two-layer dielectric sphere was analyzed. Here, we present application of the method to more demanding case of Luneburg lenses.

3. Sample Results

First, we have applied the algorithm to model Luneburg lenses with moderate sizes ka ranging fro 5 to 20, where a is the lens radius. We have applied two BoR models: VIE model with the resolution 30×60 (see [9]) and VIE-SIE macromodel obtained with the same resolution for VIE part, and outer surface with radius $b = 1.5\ a$, with BoR generating arc divided into 150 segments. The geometrical details of basis functions used in the BoR model are those presented in author's earlier paper [16]. The quantity of interest was bistatic radar cross section (RCS) of the Luneburg lens.

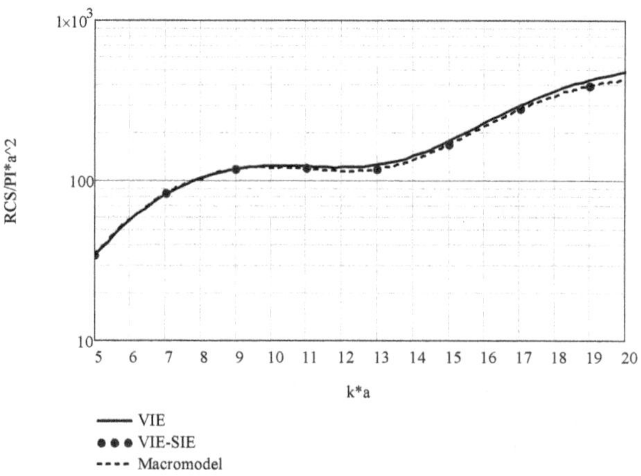

Fig. 3. Normalized forward RCS of the Luneburg lens – comparison of results.

The expansion points of the macromodel were values of ka = 5, 7, 9, 11, 13, 15, 17 and 19. The validity intervals associated with each expansion point were chosen to start and end in the midpoints between the expansion points, for instance for the interval from ka = 6 to 8, the computations were done using approximation obtained for the expansion point ka = 7. In the computations, orders of numerator and denominator of Padé approximations were $L = 5$ and $M = 5$, respectively.

In Fig. 3, it is shown the dependence of forward RCS with respect to electrical size of the lens. It is to be noted that the macromodel solution gives exact VIE-SIE values at the expansion frequencies of the macromodel, so those values, shown with black circles, allow to judge the exactness of the encapsulation of the lens with the additional surface, without involving wideband AWE approximations. From the figure, it is seen that the macromodel solution by definition ideally agrees with VIE-SIE solution at the expansion points. The agreement with the original VIE model for higher frequencies is less favorable.

Let's confine our interest to the ka interval from 5 to 10, where the agreement between VIE and VIE-SIE (and macromodel) solution seems almost ideal. Questions that arise are: whether it is possible to use a single expansion point to construct macromodels valid in the whole range, and what should be the order of Padé approximations in the macromodel construction. To answer those questions we performed calculations for the specified ka interval, choosing as the expansion point the mid value $ka = 7.5$, and applying different values of L and M, assuming however (see [18]) that $L = M$. The results are shown in Fig. 4.

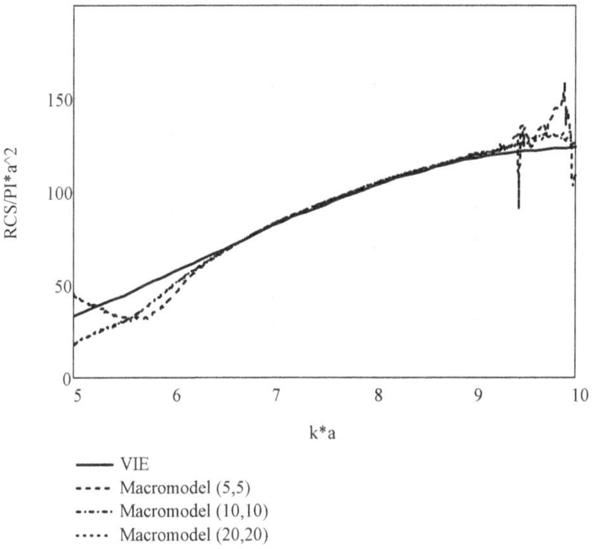

Fig. 4. Normalized forward RCS of the Luneburg lens – results for different macromodel orders.

One can see that in general increasing the order of approximation does not dramatically improve the frequency range of the macromodel. For the expansion point at $ka = 7.5$ we can judge the validity interval as ranging from $ka = 6.5$ to 9.0. This is however based on Fig. 4, which shows only forward RCS, not complete pattern. Therefore it is useful to compare also full bistatic RCS patterns at chosen frequencies. We performed comparisons for ka = 6.5; 7.0; 7.5; 8.0; 8.5 and 9.0. All plots denoted as "macromodel" were obtained using macromodels constructed around $ka = 7.5$. Results are shown in Fig. 5. From the figure it can be seen that real range of validity of the macromodel is from $ka = 7.0$ to 8.0. Increasing of the approximation order also in this case does not improve the quality of the macromodel. Thus, we may draw the conclusion that it is better to use more expansion points and lower orders of approximations.

The main purpose of the "black-boxing" the Luneburg lens was to decrease the number of unknowns in the final solution. In the above examples, this corresponds to the reduction of the set of 3630 linear equations (original VIE) to 598 linear equations (number of unknowns in (11)). Thus, in the example, we reduced the number of unknowns about six times. In direct solvers, using for example Gaussian elimination or lower-upper-triangular decomposition, the number of operations is proportional to $O(N^3)$ [18], where N is the number of unknowns. So, the six-times reduction of N is equivalent to more than 200 times reduction of the number of operations, at the stage of final solution of the linear equation set.

Of course, we do not count here the time needed for the macromodel construction, but the original idea of integral equation macromodels lies in the possibility of reusing once constructed macromodel in several different configurations/environments [10].

4. Conclusions

In this paper we have applied integral equation macromodels to speed-up calculations involving modeling of Luneburg lenses. It has been shown that although resulting macromodels are not very wideband, it still makes sense to apply them for more efficient integral-equation/method-of –moments analyzes. Once a macromodel is constructed for a given frequency interval, the computation time needed for solving the final system of equations may be considerably reduced. The constructed macromodel remains within the BoR scheme, this is however not the limitation, in view of the existence of convenient scheme enabling the analysis of mixed BoR-3D geometries with the use of so-called characteristic basis functions (CBFs) [20].

References

[1] LUNEBURG, R. K., HERZBERGER, M. *The Mathematical Theory of Optics*. Providence (RI, USA): Brown University Press, 1944.

[2] GREENWOOD, A. D., JIN, J.-M. A field picture of wave propagation in inhomogeneous dielectric lenses. *IEEE Antennas and Propagation Magazine*, 1999, vol. 41, no. 5, p. 9 - 18.

[3] MOSALLAEI, H., RAHMAT-SAMII, Y. Nonuniform Luneburg and two-shell lens antennas: Radiation characteristics and design optimization. *IEEE Transactions on Antennas and Propagation*, 2001, vol. 49, no. 1, p. 60 - 69.

[4] LIANG, C. S., STREATER, D. A., JIN, J.-M., DUNN, E., ROZENDAL, T. A quantitative study of Luneburg-lens reflectors. *IEEE Antennas and Propagation Magazine*, 2005, vol. 47, no. 7, p. 30 - 41.

[5] THORNTON, J. Properties of spherical lens antennas for High Altitude Platform communications. In *6th European Workshop on Mobile/Personal Satcoms & 2nd Advanced Satellite Mobile Systems (EMPS & ASMS)*. 2004.

[6] THORNTON, J. Wide-scanning multi-layer hemisphere lens antenna for Ka band. *IEE Proceedings – Microwaves, Antennas and Propagation*, 2006, vol. 153, no. 6, p. 573 - 578.

[7] MORGAN, M. A., MEI, K. K., Finite-element computation of scattering by inhomogeneous penetrable bodies of revolution. *IEEE Transactions on Antennas and Propagation*, 1979, vol. 27, no. 2, p. 202 - 214.

[8] GREENWOOD, A. D., JIN, J.-M. A novel, efficient algorithm for scattering from a complex BOR using mixed finite elements and

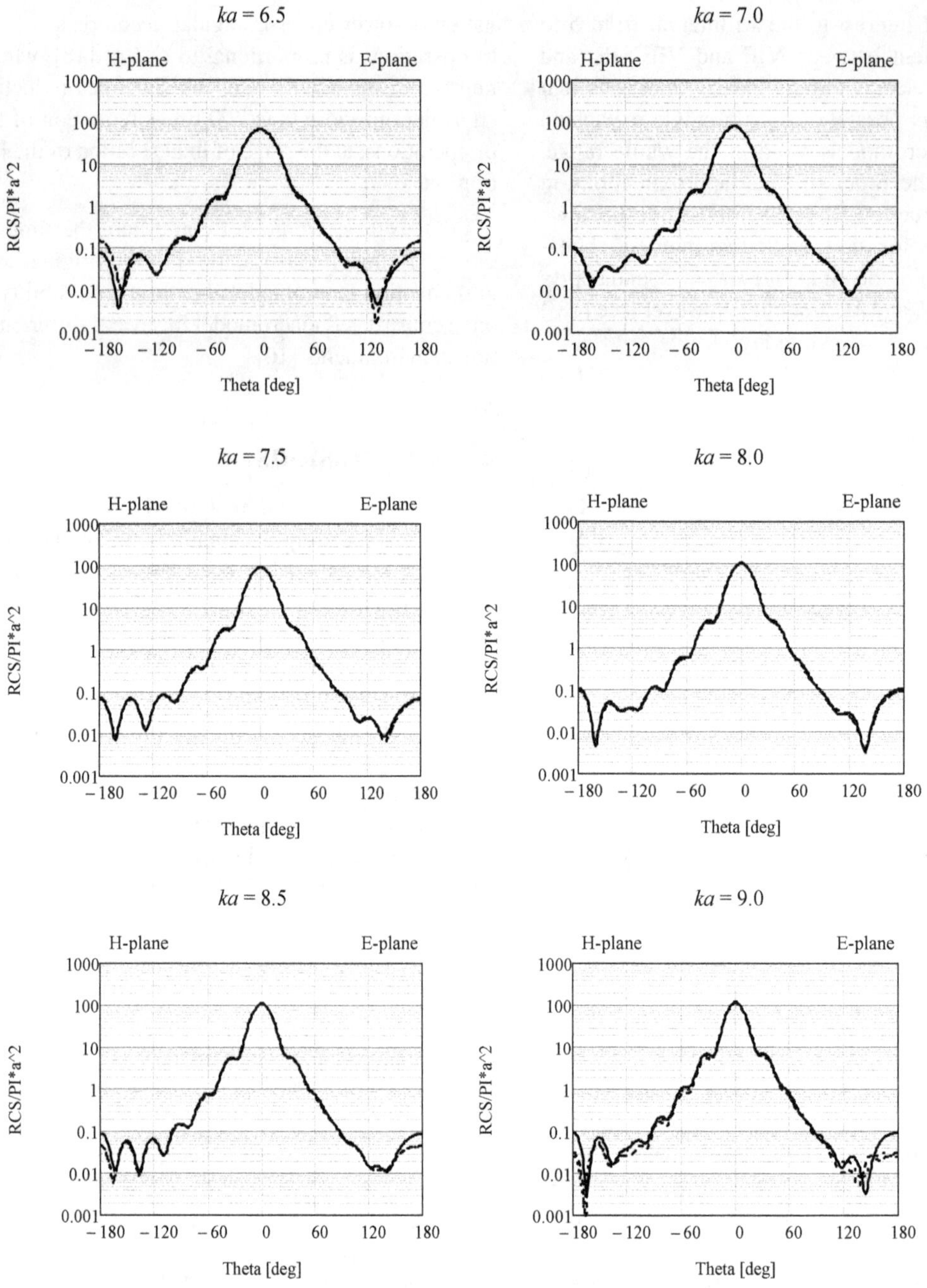

Fig. 5. Bistatic RCS of the Luneburg lens computed using VIE (solid lines), macromodel with $L = 5$, $M = 5$ (dashed lines), and macromodel with $L = 10$, $M = 10$ (dotted lines).

cylindrical PML. *IEEE Transactions on Antennas and Propagation*, 1999, vol. 47, no. 4, p. 620 - 629.

[9] KUCHARSKI, A. A. A method of moments solution for electromagnetic scattering by inhomogeneous dielectric bodies of revolution. *IEEE Transactions on Antennas and Propagation*, 2000, vol. 48, no. 8, p. 1202 - 1210.

[10] KUCHARSKI, A. A. Efficient solution of integral-equation based electromagnetic problems with the use of macromodels. *IEEE Transactions on Antennas and Propagation*, 2008, vol. 56, no. 5, p. 1482 - 1487.

[11] KUCHARSKI, A. A. Wideband analysis of electromagnetic scattering from partially inhomogeneous dielectric bodies-of-revolution with the use of macromodels. In *3rd European Conference on Antennas and Propagation (EuCAP)* Berlin (Germany), 2009, p. 93 - 96.

[12] HARRINGTON, R. F. *Time-Harmonic Electromagnetic Fields*. McGraw-Hill Book Company, 1961.

[13] SCHAUBERT, D. H., WILTON, D. R., GLISSON, A. W. A tetrahedral modeling method for electromagnetic scattering by arbitrarily shaped inhomogeneous dielectric bodies. *IEEE Transactions on Antennas and Propagation*, 1984, vol. 32, no. 1, p. 77 - 85.

[14] LI, M.-K., CHEW, W. C., JIANG, L. J. A domain decomposition scheme based on equivalence theorem. *Microwave and Optical Technology Letters*, 2006, vol. 48, no. 9, p. 1853 - 1857.

[15] KUCHARSKI, A. A. Analysis of Luneburg lenses using domain decomposition method. In *17th International Conference on Microwaves, Radar and Wireless Communications (MIKON)*. 2008.

[16] KUCHARSKI, A. A. Electromagnetic scattering by partially inhomogeneous dielectric bodies of revolution. *Microwave and Optical Technology Letters*, 2005, vol. 44, no. 3, p. 275 - 281.

[17] NEWMAN, E. H. Generation of wide-band data from the method of moments by interpolating the impedance matrix. *IEEE Transactions on Antennas and Propagation*, 1988, vol. 36, no. 12, p. 1820 - 1824.

[18] CHEW, W. C., MICHIELSSEN, E., SONG, J. M., JIN, J. M. *Fast and Efficient Algorithms in Computational Electromagnetics*. Boston, London: Artech House, 2001.

[19] KUCHARSKI, A. A. Asymptotic waveform evaluation for scattering by inhomogeneous dielectric bodies of revolution, *Microwave and Optical Technology Letters*, 2007, vol. 49, no. 5, p. 1028 - 1031.

[20] KUCHARSKI, A. A. Application of the CBF method to the scattering by combinations of bodies of revolution and arbitrarily shaped structures. *Radioengineering*, 2013, vol. 22, no. 3, p. 665 - 671.

About Author ...

Andrzej A. KUCHARSKI was born in 1964. He received his M.Sc., Ph.D. and D.Sc. from Wroclaw University of Technology in 1988, 1994, and 2001, respectively. He is the Head of the Antenna Theory and Computational Electromagnetics Group in the Telecommunications and Teleinformatics Department, Faculty of Electronics, Wroclaw University of Technology, Poland. His research interests include computational electromagnetics in radiation and scattering problems.

4

Performance of Cross-layer Design with Multiple Outdated Estimates in Multiuser MIMO System

Xiangbin YU[1,2], *Yan LIU*[1], *Yang LI*[1], *Qiuming ZHU*[1], *Xin YIN*[1], *Kecang QIAN*[1]

[1] Key Laboratory of Radar Imaging and Microwave Photonics, Ministry of Education,
College of Electronic and Information Engineering, Nanjing University of Aeronautics and Astronautics, Nanjing, China
[2] National Mobile Communications Research Laboratory, Southeast University, Nanjing, China

yxbxwy@gmail.com

Abstract. *By combining adaptive modulation (AM) and automatic repeat request (ARQ) protocol as well as user scheduling, the cross-layer design scheme of multiuser MIMO system with imperfect feedback is presented, and multiple outdated estimates method is proposed to improve the system performance. Based on this method and imperfect feedback information, the closed-form expressions of spectral efficiency (SE) and packet error rate (PER) of the system subject to the target PER constraint are respectively derived. With these expressions, the system performance can be effectively evaluated. To mitigate the effect of delayed feedback, the variable thresholds (VTs) are also derived by means of the maximum a posteriori method, and these VTs include the conventional fixed thresholds (FTs) as special cases. Simulation results show that the theoretical SE and PER are in good agreement with the corresponding simulation. The proposed CLD scheme with multiple estimates can obtain higher SE than the existing CLD scheme with single estimate, especially for large delay. Moreover, the CLD scheme with VTs outperforms that with conventional FTs.*

Keywords

Multiuser MIMO, cross-layer design, imperfect feedback, multiple estimates, variable thresholds.

1. Introduction

The ultimate objectives of wireless communication systems are to satisfy the quality of service (QoS) and rate requirements. Therefore, in order to ensure high data rates, low latency and increased link throughput, the advanced technique schemes are often employed, such as cross-layer design (CLD) and multiple-input multiple-output (MIMO) are adopted to obtain high spectral efficiency (SE) and capacity [1]. Moreover, CLD combining adaptive modulation (AM) and automatic repeat request (ARQ) is widely accepted as an efficient means to improve the overall performance of transmission in fading channels [2-6]. In [2], a cross-layer design combined adaptive modulation and coding at the physical layer (PHY) and ARQ protocol at the data link layer (DLL) over Nakagami fading channel is developed. By employing different space-time coding schemes, the performance of MIMO system with CLD is studied in [3]. The performance of CLD with STBC is analyzed in optical MIMO system in [4]. Considering that prefect feedback is hard to achieve, [5] and [6] study the performance of CLD with STBC under delayed feedback and imperfect estimation information, respectively. However, the above CLD schemes are designed for single-user MIMO environment, and thus they will be not suitable for multiuser scenario in practice. For this, the performance of AM scheme for multiuser system with imperfect feedback is studied in [7], but the analysis is only suitable for single antenna system. Under the feedback constraint, a CLD scheme is presented for multiuser MIMO systems, and the corresponding performance is investigated in Rayleigh fading channel [8]. Unfortunately, these CLD or AM schemes for imperfect feedback basically consider single outdated estimate only, and thus the performance improvement is limited. Although [9] presents an AM scheme based on multiple channel estimations to improve SE, the scheme is only designed for continuous-rate modulation. Thus, the practicability is not strong because the modulation is often based on discrete rate in practice.

According to the analysis above, the system performances of single user MIMO with CLD scheme are well studied. Moreover, most of CLD schemes are based on perfect CSI, but the perfect CSI is difficult to obtain due to channel estimation error or feedback delay. Although some schemes are designed for imperfect feedback, they are limited in single outdated estimate (SOE) information and fixed switching thresholds, and thus the system performance is hard to be improved effectively. Motivated by the reason above, we will develop a CLD scheme for multiuser MIMO system with imperfect feedback information by combining the AM at the PHY and ARQ as well as user scheduling at the DLL, where multiple outdated estimates (MOE) method is presented to improve the system performance. Based on this MOE method and performance analysis under imperfect channel state information (CSI), the probability density function (PDF) of effective signal-to-noise-ratio (SNR) and the fading gain switching thresholds for AM are respectively derived. According to these results, the closed-form expressions of average packet error rate (PER) and SE of the system are obtained. These expressions include the ones under conventional SOE or under perfect CSI as special cases. With these expressions, the system performance under imperfect CSI can be effectively assessed. Besides, subject to target packet loss rate (PLR) constraint, the variable switching thresholds are derived by using the maximum a posteriori (MAP) method. Simulation results show that the proposed CLD scheme with MOE can obtain much higher SE than the conventional CLD scheme with SOE because of more available outdated information. Moreover, the derived variable thresholds (VTs) can further reduce the effect of imperfect CSI on the system performance. Namely, the system performance with VTs is superior to that with conventional fixed thresholds (FTs).

The notations throughout this paper are as follows. Bold upper case and lower case letters denote matrices and column vectors, respectively. The superscripts $(\cdot)^H$, $(\cdot)^T$ and $(\cdot)^*$ denote the Hermitian transposition, transposition and complex conjugation, respectively.

2. System Model

We address downlink transmission in a multiuser MIMO system employing antenna selection (AS) shared by K users, and the system operates in a flat Rayleigh fading channel and homogenous case. There are n_T transmit antennas at the base station (BS) and n_R receive antennas at each user side. For a homogeneous case, the statistics of users are the same and their SNRs are assumed to be independent, identically distributed (*i.i.d*) random variables [10], [7]. The scheduler of the BS will select the user with the maximum absolute SNR, that is, the absolute SNR-based scheduling scheme (greedy scheduling) is used, which means maximizing the throughput of a multiuser system [7].

The channel between the transmitter and the kth user is characterized by a $n_R \times n_T$ matrix such that $\mathbf{H}^k = [h_{j,i}^k]_{n_R \times n_T} = [\mathbf{h}_1^k, ..., \mathbf{h}_i^k ..., \mathbf{h}_{n_T}^k]$, whose elements are *i.i.d* complex Gaussian random variables (*r.v.s*) with zero mean and unit variance. $\mathbf{h}_i^k = [h_{1,i}^k ..., h_{j,i}^k ..., h_{n_R,i}^k]^T$ is the channel vector of the kth user. $h_{j,i}^k$ denotes the channel gain from the ith transmit antenna to the jth receive antenna for the kth user. The channel is assumed to be perfectly known at the receiver, and is fed back to the transmitter with delay τ. Each receiver tracks its own instantaneous channel SNR and feeds back the CSI to the BS over the outdated channel. For the current channel $h_{j,i}^k(t)$, we use its single delayed version $h_{j,i}^k(t-\tau)$ as its estimate $\hat{h}_{j,i}^k(t)$. The power correlation coefficient between the channel $h_{j,i}^k(t)$ and its estimation $\hat{h}_{j,i}^k(t)$ is given by $\rho = J_0^2(2\pi f_d \tau)$ [11], where $J_0(\cdot)$ is the zeroth-order Bessel function of the first kind, and f_d is the maximum Doppler frequency. According to the CSI feedback, the BS scheduler selects the user that has the best link quality among all the K active users based on a greedy scheduling scheme, then the BS adapts the transmission rate to the signal of the scheduled user, and after that, the BS selects the best transmit antenna which has the maximum feedback SNR to perform the data transmission. If the transmitter selects the ith antenna for data transmission, the $n_R \times 1$ received signal vector $\mathbf{x}^k$ of user k can be written as [10]

$$\mathbf{x}^k = \mathbf{h}_i^k s^k + \mathbf{w}^k \tag{1}$$

where s^k is transmitted data symbol of user k with average energy E_s, and $\mathbf{w}^k$ is a $n_R \times 1$ noise vector, whose elements are *i.i.d* complex Gaussian random variables with mean zero and variance σ^2. The average SNR per data symbol is $\bar{\gamma} = E_s / \sigma^2$ [10]. When maximal-ratio combining (MRC) is used at the receivers, the instantaneous SNR with ith selected transmit antenna for the kth user can be expressed as $\gamma_i^k = \bar{\gamma}\sum_{j=1}^{n_R}|h_{j,i}^k|^2$, where $\sum_{j=1}^{n_R}|h_{j,i}^k|^2$ is chi-square distributed with $2n_R$ degrees of freedom. Hence, the probability density function (PDF) of γ_i^k can be given by

$$f_i^k(\gamma) = \frac{1}{\bar{\gamma}\Gamma(n_R)}(\gamma/\bar{\gamma})^{n_R-1}\exp(-\gamma/\bar{\gamma}) \tag{2}$$

where $\Gamma(\cdot)$ is the gamma function. This is also the PDF of the feedback SNR for user k, $\hat{\gamma}_i^k = \bar{\gamma}\sum_{j=1}^{n_R}|\hat{h}_{j,i}^k|^2$. Considering that $\hat{h}_{j,i}^k$ and $h_{j,i}^k$ are from the same random process, they have the same probability distribution. The transmitter allocates the radio resource to a user who can achieve the largest feedback SNR when selecting the "best" transmit antenna, and thus the effective SNR, $\hat{\gamma}$ is expressed as

$$\hat{\gamma} = \max_{i=1,2,...,n_T, k=1,2,...,K}\{\hat{\gamma}_i^k\}. \tag{3}$$

According to the order statistics [12] and (3), and using transformation of variables, the CDF and PDF of $\hat{\gamma}$ can be obtained as follows:

$$\begin{aligned} F(\hat{\gamma}) &= [\int_0^{\hat{\gamma}} f_i^k(t)dt]^L \\ &= [1-\exp(-\hat{\gamma}/\bar{\gamma})\sum_{n=0}^{n_R-1}\frac{1}{n!}(\hat{\gamma}/\bar{\gamma})]^L \end{aligned}, \quad \hat{\gamma} \geq 0 \tag{4}$$

and

$$\begin{aligned} f(\hat{\gamma}) = &\frac{L(\hat{\gamma})^{n_R-1}}{(n_R-1)!}.(\frac{1}{\bar{\gamma}})^{n_R}\exp(-\frac{\hat{\gamma}}{\bar{\gamma}})\sum_{m=0}^{L-1}\binom{L-1}{m} \\ &\times(-1)^m\exp(-m\hat{\gamma}/\bar{\gamma})\sum_{c=0}^{m(n_R-1)}\omega_{c,m}(\hat{\gamma}/\bar{\gamma})^c \end{aligned} \tag{5}$$

where $L = n_T K$, and $\omega_{c,m}$ is the coefficient of $\left(\hat{\gamma}/\bar{\gamma}\right)^c$ in the expansion of $(\sum_{n=0}^{n_R-1}(\hat{\gamma}/\bar{\gamma})^n/n!)^m$.

Based on the analysis above, only a single delay channel is used for predicting the real channel. Although this single estimation method has a pervasive application, the estimated channel is not accurate enough since other outdated channel information is not utilized. For this reason, multiple outdated estimates will be employed for accurate estimation in the following CLD.

3. Cross-layer Design with Multiple Outdated Estimates

Considering that existing CLD schemes with imperfect feedback are basically based on single outdated estimation method in Section 2, the performance improvement is limited. Based on this, we will present the CLD scheme for multiuser MIMO based on multiple outdated estimates method so that the SE and PER performance will be effectively improved.

3.1 Multiple Outdated Estimates

In this subsection, we will give the MOE method. As analyzed in Section 2, $\hat{\gamma}$ is derived from the estimate channel. However, conventional estimation only considers a single delay channel $h_{j,i}^k(t-\tau)$. For this, we use multiple previous channel estimates, which are produced prior to $h_{j,i}^k(t-\tau)$, in combination to reduce the uncertainty in $\hat{h}_{j,i}^k(t)$. Assuming Z estimates, $h_{j,i}^k(t-\tau)$, $h_{j,i}^k(t-2\tau)$, ..., $h_{j,i}^k(t-Z\tau)$ are available, then the correlation coefficient is given by

$$E\left\{h_{j,i}^k(t-u\tau)h_{j,i}^{k*}(t-v\tau)\right\} = J_0(2\pi f_d(v-u)\tau) \quad (6)$$

Let the estimated channel vector be $\hat{\mathbf{h}}_{j,i}^k = [h_{j,i}^k(t-\tau), h_{j,i}^k(t-2\tau), \ldots, h_{j,i}^k(t-Z\tau)]^T$, then $\hat{\mathbf{h}}_{j,i}^k$ is Gaussian distributed with zero mean, and the covariance matrix is given by

$$E\left\{\begin{pmatrix} h_{j,i}^k(t) \\ \hat{\mathbf{h}}_{j,i}^k \end{pmatrix}\begin{pmatrix} h_{j,i}^{k*}(t) & (\hat{\mathbf{h}}_{j,i}^k)^H \end{pmatrix}\right\} = \begin{bmatrix} 1 & \mathbf{a}^H \\ \mathbf{a} & \mathbf{B} \end{bmatrix} \quad (7)$$

where $\mathbf{a} = E\left\{\hat{\mathbf{h}}_{j,i}^k h_{j,i}^{k*}(t)\right\}$, $\mathbf{B} = E\left\{\hat{\mathbf{h}}_{j,i}^k(\hat{\mathbf{h}}_{j,i}^k)^H\right\}$. $h_{j,i}^k(t)$ given $\hat{\mathbf{h}}_{j,i}^k$ is a Gaussian distributed with mean $\hat{h}_{j,i}^k(t) = \mathbf{a}^H\mathbf{B}^{-1}\hat{\mathbf{h}}_{j,i}^k$ and variance $\delta^2 = 1 - \mathbf{a}^H\mathbf{B}^{-1}\mathbf{a}$. $h_{j,i}^k(t)$ and $\hat{h}_{j,i}^k(t)$ are jointly complex Gaussian with power correlation coefficient $\rho = 1 - \delta^2$ [9]. Since the multiple previous channel estimates are utilized, the correlation coefficient ρ will be large. As a result, $\hat{h}_{j,i}^k$ will be accurate enough to estimate the real $h_{j,i}^k$. Correspondingly, the effective SNR $\hat{\gamma}$ is very close to the real SNR γ, and the feedback information will be reliable. This multiple estimates method includes single outdated estimate (i.e., $Z = 1$) as a special case. Based on the correlation coefficient of multiple outdated estimates, the conditional PDF of γ given $\hat{\gamma}$ is expressed as [11]

$$p_{\gamma|\hat{\gamma}}\left(\gamma|\hat{\gamma}\right) = \frac{1}{(1-\rho)\bar{\gamma}} I_{n_R-1}\left(\frac{2\sqrt{\rho\hat{\gamma}\gamma}}{(1-\rho)\bar{\gamma}}\right) \times \left(\frac{\gamma}{\rho\hat{\gamma}}\right)^{\frac{n_R-1}{2}} \exp\left(-\frac{\gamma+\rho\hat{\gamma}}{(1-\rho)\bar{\gamma}}\right), \gamma \geq 0 \quad (8)$$

where $I_n(\cdot)$ is the n-order modified Bessel function of the first kind [13]. With (5) and (8), the joint pdf of $\hat{\gamma}$ and γ can be obtained as

$$p_{\gamma,\hat{\gamma}}\left(\gamma,\hat{\gamma}\right) = \frac{L\bar{\gamma}^{-(n_R+1)}}{(1-\rho)(n_R-1)!}\left(\frac{\gamma\hat{\gamma}}{\rho}\right)^{\frac{n_R-1}{2}} I_{n_R-1}\left(\frac{2\sqrt{\rho\hat{\gamma}\gamma}}{(1-\rho)\bar{\gamma}}\right) \times \exp\left(-\frac{\gamma+\hat{\gamma}}{(1-\rho)\bar{\gamma}}\right)\sum_{m=0}^{L-1}\binom{L-1}{m}(-1)^m \times \exp\left(-\frac{m\hat{\gamma}}{\bar{\gamma}}\right)\sum_{c=0}^{m(n_R-1)}\omega_{c,m}\left(\frac{\hat{\gamma}}{\bar{\gamma}}\right)^c \quad (9)$$

3.2 Cross-layer Design

By combining AM at the PHY and ARQ protocol as well as user scheduling at the DLL, we will give a cross-layer design scheme based on the MOE above for multiuser MIMO system with antenna selection, and MQAM is considered for modulation in the system.

At the transmitter, according to the delayed CSI feedback from the receiver, the BS scheduler selects the user that has the best link quality among all the active users, and then the BS selects the best transmit antenna which has the maximum feedback SNR. After that, the modulator performs adaptive modulation in terms of the feedback CSI, and subsequently, the modulated symbols are transmitted by the selected transmit antenna. At the receiver, the channel states are estimated and measured for controlling the modulation mode by using MOE method, and the estimated CSI is fed back to the transmitter via a feedback path. With the obtained CSI, the adaptive demodulation is performed and the resultant decoded bits are obtained. Then, these bit streams are mapped to packets, which are pushed upwards to the data link layer. According to [2], at the data link layer, the selective repeat ARQ protocol is implemented. If an error is detected in a packet, a retransmission request is generated by the ARQ generator, and is communicated to the ARQ controller at the transmitter. Otherwise, no retransmission request is sent.

For discrete-rate MQAM, the constellation size M_n is defined as $\{M_0 = 0, M_n = 2^n, n = 1,\ldots,N\}$, where M_0 means no data transmission. The MQAM of constellation size M_n is used for modulation when the SNR $\hat{\gamma}$ falls in the nth region $[\gamma_n, \gamma_{n+1})$. According to [2], the PER of MQAM with two dimensional Gray code over additive white Gaussian noise (AWGN) channel for the received SNR $\hat{\gamma}$ and constellation size M_n is approximately given by

$$Per_n(\hat{\gamma}) \cong \begin{cases} 1, & if\ \hat{\gamma} < \gamma_{pn} \\ a_n \exp(-g_n \hat{\gamma}), & if\ \hat{\gamma} \geq \gamma_{pn} \end{cases} \quad (10)$$

where $\{a_n, g_n, \gamma_{pn}\}$ are constellation and packet-size dependent constants, and they can be obtained by fitting (10) to the exact PER.

We first define the target packet loss rate for the data link layer as P_{loss}. Since truncated ARQ is used at the data link layer, the packets in error may be retransmitted up to $N_r^{\max}$ (i.e., maximum number of retransmissions). Hence, the target PER is $P_o = P_{loss}^{1/(N_r^{\max}+1)}$ at the physical layer, which is generally limited as $P_o < 1$. The switching thresholds $\{\gamma_n\}$ can be set to be the required SNR to achieve the target PER, P_o, over an AWGN channel. By inverting the P_o in (10), we can obtain the switching threshold values as

$$\gamma_n = -\ln(P_o / a_n)/g_n, \quad n = 1,\ldots,N,$$
$$\gamma_0 = 0, \quad \gamma_{N+1} = +\infty. \quad (11)$$

The above thresholds do not consider the impact of delayed feedback information, so they are referred as fixed thresholds (FTs). Because our CLD scheme is based on MOE, the resulting system performance will perform better than the conventional CLD scheme based on SOE only, which will be confirmed in the following numerical results.

4. Performance Analysis of CLD with Multiple Estimates

In this section, we will give the performance analysis of CLD with antenna selection (referred to as CLD-AS) for multiuser MIMO with imperfect feedback. Based on the switching thresholds described in (11), we can calculate the probability that the effective SNR $\hat{\gamma}$ falls in the n-th region $[\gamma_n, \gamma_{n+1})$, denoted by Pr_n, as

$$Pr_n = \frac{L}{(n_R-1)!}\sum_{m=0}^{L-1}\binom{L-1}{m}(-1)^m \sum_{c=0}^{m(n_R-1)} \frac{\omega_{c,m}}{(m+1)^{n_R+c}} \times[\Gamma(n_R+c,\frac{m+1}{\bar{\gamma}}\gamma_n)\text{-}\Gamma(n_R+c,\frac{m+1}{\bar{\gamma}}\gamma_{n+1})] \quad (12)$$

where $\Gamma(\cdot,\cdot)$ is incomplete Gamma function [13]. For discrete-rate adaptive scheme, the SE at the PHY is defined as the ensemble average of valid transmission rate. So the average SE of the system at the PHY can be given by

$$\overline{Se}_{phy} = \sum_{n=1}^{N} R_n Pr_n$$
$$= \sum_{n=1}^{N}\frac{R_n L}{(n_R-1)!}\sum_{m=0}^{L-1}\binom{L-1}{m}(-1)^m \sum_{c=0}^{m(n_R-1)} \frac{\omega_{c,m}}{(m+1)^{n_R+c}} \quad (13)$$
$$\times[\Gamma(n_R+c,\frac{m+1}{\bar{\gamma}}\gamma_n)\text{-}\Gamma(n_R+c,\frac{m+1}{\bar{\gamma}}\gamma_{n+1})]$$

We defined ensemble average PER at the PHY for multiuser MIMO system with CLD-AS and feedback delay as

$$\overline{Per} = (\sum_{n=1}^{N} R_n \overline{Per}_n)/(\sum_{n=1}^{N} R_n Pr_n) \quad (14)$$

where $\overline{Per}_n$ denote the average PER for constellation size M_n, and it can be obtained as:

$$\overline{Per}_n = \int_{\gamma_n}^{\gamma_{n+1}}\left(\int_0^{\infty} Per_n(\gamma)\, p_{\gamma|\hat{\gamma}}(\gamma|\hat{\gamma})\,d\gamma\right) f(\hat{\gamma})\,d\hat{\gamma} \quad (15)$$

where

$$Per_n(\gamma) \cong \begin{cases} 1, & if\ \gamma < \gamma_{pn} \\ a_n \exp(-g_n \gamma), & if\ \gamma \geq \gamma_{pn} \end{cases} \quad (16)$$

Substituting (16), (5) and (8) into (15) gives

$$\overline{Per}_n = I1_n + I2_n - I3_n \quad (17)$$

where $I1_n$, $I2_n$, $I3_n$ are written as (18)-(20) at the bottom of the paper due to their long expressions. Equation (17) is a closed-form expression of the average PER for multiuser MIMO with CLD-AS and MOE under imperfect CSI. With (17), the average PLR at the data link layer with the maximum number of retransmissions $N_r^{\max}$ is

$$\overline{Plr} = \overline{Per}^{N_r^{\max}+1}. \quad (21)$$

Thus, the average number of transmissions per packet can be calculated as:

$$\overline{N} = 1 + \overline{Per} + \ldots + \overline{Per}^{N_r^{\max}} = (1-\overline{Per}^{N_r^{\max}+1})/(1-\overline{Per}). \quad (22)$$

Using (22) and (14) as well as (13), the overall average SE of multiuser MIMO system with CLD-AS and multiple estimates under imperfect CSI can be obtained as:

$$\overline{Se} = \overline{Se}_{phy}/\overline{N}. \quad (23)$$

When $N_r^{\max}$ is set to be zero, then $\overline{N} = 1$, and (23) is reduced to h_{phy}, which corresponds to the average SE at the physical layer only.

5. Variable Thresholds for CLD with Multiple Estimates

Considering that the switching thresholds in (11) can not adapt to the change of delayed feedback information, the performance loss will happen when feedback has delay. For this, we will present a variable threshold method for imperfect feedback by employing the Bayes' theorem [14] and MAP criterion [14], [5]. With the variable thresholds, the system performance will be effectively improved, and outperforms that with fixed thresholds (11).

In what follows, we employ the estimated instantaneous PER, $\hat{Per}_n$, to calculate VTs. $\hat{Per}_n$ can be obtained by maximizing the conditional PDF of $p_{Per_n|\hat{\gamma}}(Per_n\,|\,\hat{\gamma}) = p_{Per_n,\hat{\gamma}}(Per_n,\hat{\gamma}) / p_{\hat{\gamma}}(\hat{\gamma})$. From the functional relationship between Per_n and γ in (16), using the analysis method in [14], the joint PDF of Per_n and $\hat{\gamma}$ is derived as,

$$
\begin{aligned}
p_{Per_n,\hat{\gamma}}\left(Per_n,\hat{\gamma}\right) &= \left|\frac{\partial\gamma}{\partial Per_n}\right| f_{\gamma,\hat{\gamma}}\left(Per_n^{-1},\hat{\gamma}\right) \\
&= \sum_{m=0}^{L-1}\binom{L-1}{m}(-1)^m \sum_{c=0}^{m(n_R-1)} \frac{L\omega_{c,m} e^{\varepsilon^2}\hat{\gamma}^c}{g_n a_n \bar{\gamma}^{n_R+c+1}(1-\rho)\Gamma(n_R)} \\
&\times\left(\frac{\hat{\gamma}}{g_n\rho}\right)^{\frac{n_R-1}{2}} \exp\left(-\frac{(m-m\rho+1)\hat{\gamma}}{(1-\rho)\bar{\gamma}} - \frac{\varepsilon^2}{g_n(1-\rho)\bar{\gamma}}\right) \\
&\times\varepsilon^{n_R-1} I_{n_R-1}\left(\frac{2\varepsilon\sqrt{g_n\rho\hat{\gamma}}}{g_n(1-\rho)\bar{\gamma}}\right) \\
&\approx \sum_{m=0}^{L-1}\binom{L-1}{m}(-1)^m \sum_{c=0}^{m(n_R-1)} \frac{L\omega_{c,m} D\varepsilon^{n_R-1.5} e^{\beta(\varepsilon)}}{2a_n\sqrt{\pi b_n\sqrt{g_n\rho\hat{\gamma}}}}
\end{aligned}
\tag{24}
$$

where

$b_n = [g_n(1-\rho)\bar{\gamma}]^{-1}$, $\varepsilon = \sqrt{\ln(a_n / Per_n)}$,

$\beta(\varepsilon) = (1-b_n)\varepsilon^2 + 2\varepsilon b_n\sqrt{\rho\hat{\gamma}g_n}$, and

$$D = \frac{(1/\bar{\gamma})^{n_R+c+1}\hat{\gamma}^c}{g_n(1-\rho)\Gamma(n_R)}\left(\frac{\hat{\gamma}}{g_n\rho}\right)^{(n_R-1)/2} \exp\left(-\frac{(m-m\rho+1)\hat{\gamma}}{(1-\rho)\bar{\gamma}}\right).$$

According to (24), by omitting the terms not related to ε, the MAP function can be defined as

$$I_{MAP} = \varepsilon^{n_R-1.5} e^{\beta(\varepsilon)} \ . \tag{25}$$

By solving the equation $\frac{\partial I_{MAP}}{\partial Per_n} = \frac{\partial I_{MAP}}{\partial\varepsilon}\cdot\frac{\partial\varepsilon}{\partial Per_n} = 0$, we can obtain the approximated expression of the estimated PER as

$$\hat{Per}_n \approx a_n \exp\left(-n_R^2\rho g_n\hat{\gamma} / \left(n_R - g_n(1-\rho)\right)^2\right) \tag{26}$$

Let $\hat{Per}_n = P_o$, then with (26) we can achieve the VTs dependent on the feedback delay as

$$\gamma_{An} = \ln\left(\frac{a_n}{P_o}\right)\cdot\left(1 - \frac{g_n(1-\rho)}{n_R}\right)^2 \Big/ (\rho g_n) . \tag{27}$$

By submitting the obtained VTs into (14) and (23) in Section 4, we can calculate the average PER and SE, and obtain the corresponding closed-form expressions of multiuser MIMO system with CLD-AS and MOE under imperfect CSI. Unlike the FTs in (11), the VTs in (27) consider the effect of delayed feedback. Thus, with (27), the impact of feedback delay may be reduced, and reliable PER performance will be realized. When the feedback is perfect (i.e. $\rho = 1$), (27) is reduced to (11), and thus the derived VTs include the FTs as special cases.

6. Numerical Results and Analysis

In this section, we will use the derived formulae to assess the performance of average PER and SE of multiuser MIMO system with CLD and multiple outdated estimates over Rayleigh fading channel. The channel is assumed to be flat fading. The Gray code is employed to map the data bits to MQAM constellations, and 10^6 Monte-Carlo simulations are employed for the performance evaluation. The set of MQAM constellations is $\{M_n\}_{n=0,1,\ldots,7}=\{0, 2, 4, 8, 16, 32, 64, 128\}$. The target packet loss rate at the data link layer, $P_{loss} = 10^{-3}$, and the maximum number of ARQ retransmissions $N_r^{\max} = 2$. Thus, the target PER, P_o, is equal to 0.1. In the simulation results below, *x*T*y*R*z*U denotes the multiuser MIMO system with *x* transmit antennas and *y* receive antennas and *z* users. Unless other mentioned, when the impact of feedback delay on the performance is evaluated, the average SNR, $\bar{\gamma}$, is set to 20 dB.

Fig. 1 shows the average SE versus the normalized time delay ($f_d\tau$) for 2T2R3U system with CLD under imperfect CSI, where MOE and SOE are used for comparison.

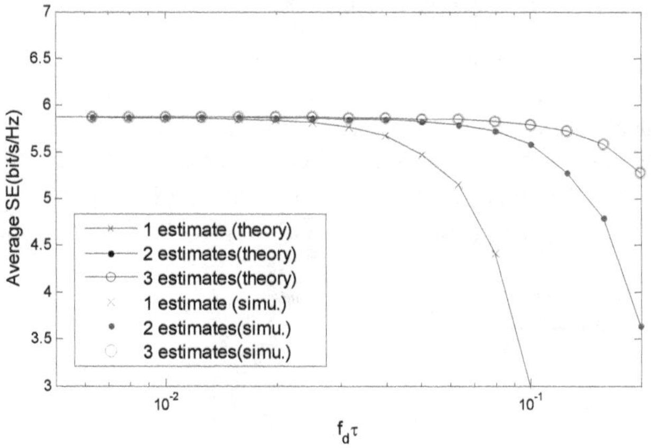

Fig. 1. Average SE for 2T2R3U system with multiple outdated estimates.

The theoretical average SE is calculated by using (23) with the switching thresholds defined in (11). It is observed that the theoretical SE is in good agreement with the simulation result, and thus the derived theoretical expression (23) is valid. It is shown that multiple outdated estimates method can obviously reduce the effect of the time delay. This is because the former makes fully use of multiple outdated channel information to decrease the uncertainty of feedback information. As a result, the system with MOE can obtain much higher SE than that with the conventional SOE, especially for large time delay.

In Fig. 2, we give the average PER versus $f_d\tau$ for 2T2R3U system with CLD under imperfect CSI, where MOE and SOE are used for comparison. The theoretical average PER is calculated by using (17) with the switching thresholds defined in (11). From this figure, we can see that multiple outdated estimates method enables the system to tolerate larger delay. It is shown that the CLD with MOE method can tolerate the normalized time delay up to about 0.01 with a slight degradation in the average PER. But when $f_d\tau$ increases beyond 0.01, the PER performance will degrade increasingly. The CLD with single estimate fails to meet the target PER ($P_0 = 0.1$) at $f_d\tau = 0.062$, the CLD with two estimates fails at $f_d\tau = 0.12$, while the CLD with three estimates fails at $f_d\tau = 0.21$, which means that three-estimate method can tolerate larger delay than the former two due to relatively more feedback information. Besides, the theoretical average PER agrees with the simulation for different time delays, and thus the derived average PER expressions of CLD with imperfect feedback are also valid.

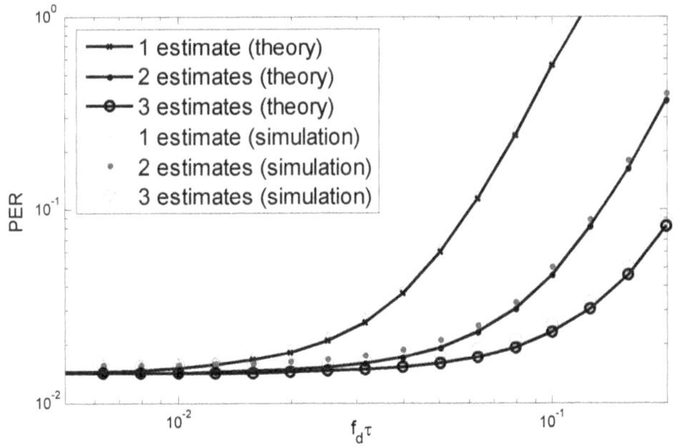

Fig. 2. Average PER for 2T2R3U system with multiple outdated estimates.

In Fig. 3, we plot the average SE versus average SNR $\bar{\gamma}$ for multiuser MIMO system with CLD and MOE under imperfect CSI, where 2T1R1U and 2T1R3U systems are considered for comparison, and $f_d\tau = 0.1$. The average SE is calculated by using (23) with the switching thresholds defined in (11). It is shown that CLD is able to increase SE with SNR, and multiuser 2T1R3U system can obtain higher SE than the corresponding single user 2T1R1U system due to the available multiuser diversity. From this figure, we can see that the systems with MOE have higher SE than those with the conventional SOE due to the reason analyzed in Fig. 1. The system SE with 3 estimates is higher than that with 2 estimates, and the system SE with 2 estimates is higher than that with 1 estimate. Moreover, with the outdated channel information increase, the obtained SE increment will decrease. This is because the added outdated channel information is too old to provide more reliable channel information for the system.

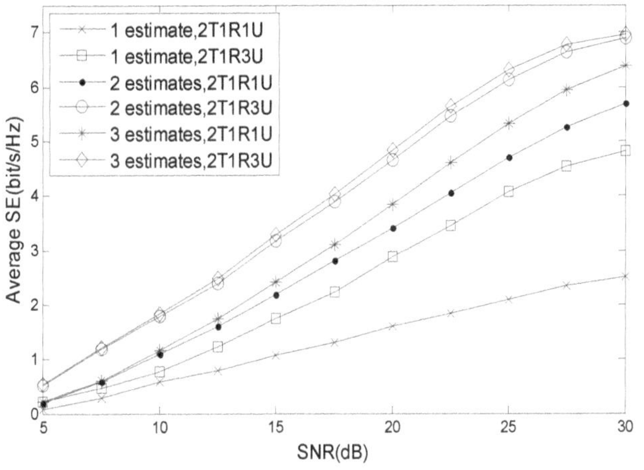

Fig. 3. Average SE for multiuser MIMO system with multiple outdated estimates.

In Fig. 4, we give the average PER versus average SNR $\bar{\gamma}$ for multiuser MIMO system with CLD and MOE under imperfect CSI, the system configurations are the same as Fig. 3, and $f_d\tau = 0.1$. The average PER is calculated by using (17) with the switching thresholds defined in (11). Some results similar to Fig. 3 can be found. Namely, multiuser 2T1R3U system has lower PER than the single user 2T1R1U system, and the systems with MOE have lower PER than those with the conventional SOE. Moreover, from Figs. 1-4, it is found that two outdated estimates can obtain obvious performance superiority over single outdated estimate, and has less complexity than other multiple outdated estimates (more than two). Thus, it provides a tradeoff between the performance and complexity. Based on this, in the following figures, we use it for the performance evaluation and comparison.

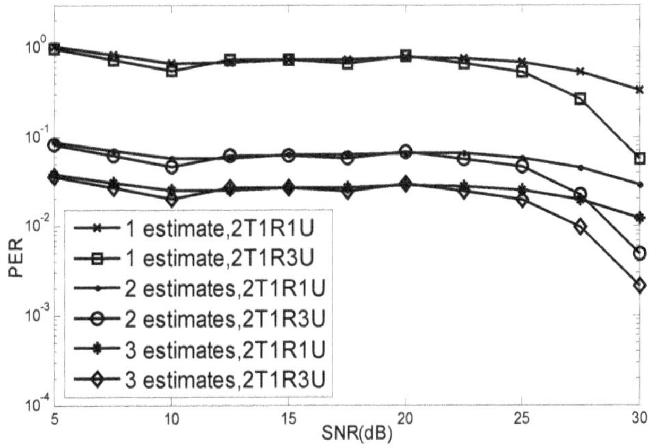

Fig. 4. Average PER for multiuser MIMO system with multiple outdated estimates.

Considering the influence of imperfect CSI on the switching thresholds, we use the VTs instead of conventional FTs to improve the system performance further. Fig. 5 and Fig. 6 show the average SE and PER of 3T1R2U with CLD under imperfect feedback, where FTs and VTs are used for the performance evaluation. In Fig. 5, we plot the average SE vs. the normalized time delay ($f_d\tau$), the average SE is calculated by using (23) with VTs defined in (27). It is found that the SE with VTs is higher than that with FTs, especially for large time delay, the SE increment is obvious. This is because the application of VTs can lower the system PER (as shown Fig. 6), which will bring about the decrease of the average number of transmissions $\bar{N}$ according to (22). As a result, the increase of overall average SE is obtained. Moreover, the derived theoretical SE and the corresponding simulation are very close for both VTs and FTs. Besides, the SE with two estimates is obviously higher than that with conventional single estimate as expected.

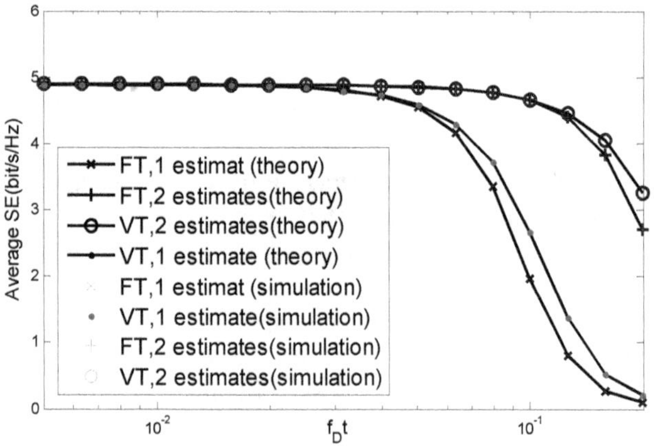

Fig. 5. Average SE for multiuser MIMO system with FTs and VTs for different estimates.

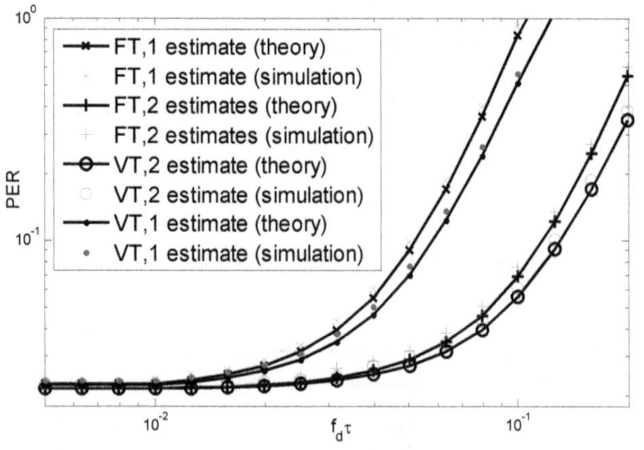

Fig. 6. Average PER for multiuser MIMO system with FTs and VTs for different estimates.

In Fig. 6, we plot the average PER vs. the normalized time delay ($f_d\tau$) for 3T1R2U with CLD, where VTs are from (27). It is found that the PER with VTs is lower than that with FTs, especially for large time delay, the performance superiority becomes obvious. Moreover, the derived theoretical PER can match the corresponding simulation well for both VTs and FTs, which testifies that the derived theoretical expression is valid for PER performance evaluation. Besides, the average PER with two estimates is obviously lower than that with single estimate, which indicates the two outdated estimates is a practical method in view of performance and complexity.

7. Conclusions

Based on multiple outdated channel information, we have studied multiuser MIMO system with CLD-AS in Rayleigh fading channel. By the performance analysis, the average PER and SE of the system subject to target PLR have been derived. As a result, closed-form expressions of average PER and SE are achieved. They include the existing PER and SE expressions employing one outdated estimate as special cases, and can match the corresponding simulations very well. Thus, with these expressions, CLD performance in multiuser MIMO system can be effectively assessed, and the impact of delayed CSI on the system performance can be analyzed well. Moreover, multiple estimates method can bring about the obvious performance improvement of the system when compared to conventional single estimate method. By means of the maximum a posteriori method and approximate PER expression, we have derived the variable switching thresholds as well. These VTs include the conventional FTs as special cases, can adapt to the delayed feedback information, and reduce the effect of feedback delay. Simulation results show the presented CLD with MOE can obtain better SE and PER performance than the existing CLD with SOE because multiple outdated channel information is fully utilized. It is shown that two outdated estimates can implement effective tradeoff between the performance and complexity. Furthermore, the proposed variable thresholds method can also improve the system performance effectively, and it performs better than conventional fixed thresholds method, especially for large time delay.

In this paper, we mainly address the performance study of multiuser CLD with AS under homogeneous case. In practice, however, the statistics of the SNRs of users may not be identical since individual users may locate at different distances from the BS. So the users may have different path loss and average SNRs, which is referred to as heterogeneous case. For this reason, in the future work, we will further study the performance of the multiuser CLD system under heterogeneous case where both path loss and small-scale fading (Rayleigh fading) are considered. It is expected that some practical results can be achieved.

$$I1_n = \frac{L}{\Gamma(n_R)} \sum_{m=0}^{L-1} \binom{L-1}{m} (-1)^m \sum_{c=0}^{m(n_R-1)} \omega_{c,m} \sum_{\xi=0}^{\infty} \rho^{\xi} (1-\rho)^{n_R+c} \left[1 - \frac{\Gamma\left(\xi+n_R, \gamma_{pn}/\left((1-\rho)\bar{\gamma}\right)\right)}{\Gamma(i+n_R)} \right] \times \frac{\left[\Gamma\left(n_R+c+\xi, \gamma_n \left(\frac{m+1}{\bar{\gamma}} + \frac{\rho}{(1-\rho)\bar{\gamma}} \right) \right) - \Gamma\left(n_R+c+\xi, \gamma_{n+1} \left(\frac{m+1}{\bar{\gamma}} + \frac{\rho}{(1-\rho)\bar{\gamma}} \right) \right) \right]}{\Gamma(\xi+1)\left[(1-\rho)(m+1)+\rho\right]^{n_R+c+\xi}} \tag{18}$$

$$I2_n = \frac{L}{\Gamma(n_R)\bar{\gamma}^{n_R}} \frac{a_n}{\left[g_n(1-\rho)\bar{\gamma}+1\right]^{n_R}} \sum_{m=0}^{L-1} \binom{L-1}{m} (-1)^m \sum_{c=0}^{m(n_R-1)} \frac{\omega_{c,m}}{\bar{\gamma}^c} \left(m+1+\frac{\rho g_n \bar{\gamma}}{g_n(1-\rho)\bar{\gamma}+1} \right)^{-(n_R+c)} \times \left[\Gamma\left(n_R+c, \gamma_n \left(\frac{m+1}{\bar{\gamma}} + \frac{\rho g_n}{g_n(1-\rho)\bar{\gamma}+1} \right) \right) - \Gamma\left(n_R+c, \gamma_{n+1} \left(\frac{m+1}{\bar{\gamma}} + \frac{\rho g_n}{g_n(1-\rho)\bar{\gamma}+1} \right) \right) \right] \tag{19}$$

$$I3_n = \frac{La_n}{\Gamma(n_R)} \sum_{m=0}^{L-1} \binom{L-1}{m} (-1)^m \sum_{c=0}^{m(n_R-1)} \omega_{c,m} \sum_{\xi=0}^{\infty} \left[g_n(1-\rho)\bar{\gamma}+1\right]^{\xi+n_R} \left[1 - \frac{\Gamma\left(\xi+n_R, \gamma_{pn}/\left((1-\rho)\bar{\gamma}\right)\right)}{\Gamma(\xi+n_R)} \right] \times \frac{\rho^{\xi}(1-\rho)^{n_R+c} \left[\Gamma\left(n_R+c+\xi, \gamma_n \left(\frac{m+1}{\bar{\gamma}} + \frac{\rho}{(1-\rho)\bar{\gamma}} \right) \right) - \Gamma\left(n_R+c+\xi, \gamma_{n+1} \left(\frac{m+1}{\bar{\gamma}} + \frac{\rho}{(1-\rho)\bar{\gamma}} \right) \right) \right]}{\Gamma(\xi+1)\left[(1-\rho)(m+1)+\rho\right]^{n_R+c+\xi}} \tag{20}$$

Acknowledgements

The authors would like to thank the anonymous reviewers for their valuable comments and suggestions which improve the quality of this paper greatly. This work is supported in part by Open Research Fund of National Mobile Communications Research Laboratory of Southeast University (2012D17), Aeronautical Science Foundation of China (20120152001), National Natural Science Foundation of China (61172077), and China Postdoctoral Science Foundation (2013M541661).

References

[1] RAMIS, J., FEMENIAS, G. Cross-layer design of adaptive multi-rate wireless networks using truncated HARQ. *IEEE Transactions on Vehicular Technology*, 2011, vol. 60, no. 3, p. 944–954.

[2] LIU, Q.W., ZHOU, S., GIANNAKIS, G. B. Cross-layer combining of adaptive modulation and coding with truncated ARQ over wireless links. *IEEE Transactions on Wireless Communication*, 2004, vol. 3, p. 1746–1755.

[3] LU, X.F., ZHU, G.X., LIU, G., LI, L. A cross-layer design over MIMO Rayleigh fading channels. In *Proceedings of IEEE International Conf. on Wireless Communications, Networking and Mobile Computing* (*WiCOM*' 2005). Wuhan (China), 2005, p. 30–33.

[4] ZAIDI, S.A.R., HAFEEZ, M. Cross layer design for orthogonal space time block coded optical MIMO systems. In *Proceedings of IEEE Fifth International Conference on Wireless and Optical Communications Networks* (*WOCN* '08). Surabaya (Indonesia), 2008, p. 1–5.

[5] ZHOU, T., YU, X., LI, Y., JIAO, Y. Cross-layer design with feedback delay over MIMO Nakagami-m fading channels. In *Proceedings of IEEE International Conference on Communication Techniques* (*ICCT*'2010). Nanjing (China), 2010, p. 1370–1373.

[6] YU, X., TAN, W., YIN, X. Cross-layer design for space-time block coded MIMO systems with imperfect channel state information. In *Proceedings of IEEE International Conference on Computer Science and Automation Engineering* (*CSAE*'2012). Zhangjiajie (China), 2012, p. 535– 539.

[7] MA, Q., TEPEDELENLIOGLU, C. Practical multiuser diversity with outdated channel feedback. *IEEE Transactions on Vehicular Technology*, 2005, vol. 54, no. 4, p. 1334–1345.

[8] QI, J., AISSA, S. Cross-layer design for multiuser MIMO MRC systems with feedback constraints. *IEEE Transactions on Vehicular Technology*, 2009, vol. 58, p. 3347–3360.

[9] YE, S., BLUM, R. S., CIMINI, L. J. Adaptive OFDM systems with imperfect channel state information. *IEEE Transactions on Wireless Communication*, 2006, vol. 5, no. 11, p. 3255–3265.

[10] ZHANG, X., LV, Z., WANG, W. Performance analysis of multiuser diversity in MIMO systems with antenna selection. *IEEE Trans. on Wireless Communication*, 2008, vol. 7, no. 1, p. 15–21.

[11] ALOUINI, M. S., GOLDSMITH, A. J. Adaptive modulation over Nakagami fading channels. *Wireless Personal Communication*, 2000, vol. 13, p. 119–143.

[12] DAVID, H. A. *Order Statistics*. 2nd ed. New York: Wiley, 1981.

[13] GRADSHTEYN, I. S., RYZHIK, I. M. *Table of Integrals, Series, and Products*. 7th ed. San Diego, CA: Academic, 2007.

[14] PAPOULIS, A., PILLAI, S. U. *Probability, Random Variables and Stochastic Processes*. 4th ed. New York: McGraw Hill, 2002.

[15] ONG, L. T., SHIKH-BAHAEI, M., CHAMBERS, J. A. Variable rate and variable power MQAM system based on Bayesian bit error rate and channel estimation techniques. *IEEE Transactions on Wireless Communication*, 2008, vol. 56, p. 177–182.

About Authors ...

XIANGBIN YU was born in Jiangsu, China. He received his Ph.D. in Communication and Information Systems in 2004 from National Mobile Communications Research Laboratory at Southeast University, China. He is a full Professor of Information and Communication Engineering at Nanjing University of Aeronautics and Astronautics. Currently, he also works as a Visiting Scholar at University of Delaware, USA. He has served as a technical program committee of Globecom'2006, International Conference on Communications Systems (ICCS'2008, ICCS'10), International Conference on Communications and Networking in China (Chinacom'2010, Chinacom'2014) and International Conference on Wireless Communications and Signal Processing 2011. He has been a member of IEEE ComSoc Radio Communications Committee (RCC) since May 2007. His research interests include multi-carrier CDMA, multi-antenna technique, distribute antenna systems, adaptive modulation and cross-layer design.

YAN LIU was born in Shandong, China. He is currently working towards the M.Sc. degree at Nanjing University of Aeronautics and Astronautics.

YANG LI was born in Jiangsu, China. He is currently working towards the M.Sc. degree at Nanjing University of Aeronautics and Astronautics.

QIUMING ZHU was born in Jiangsu, China. He received his Ph.D. in Communication and Information Systems in 2012 from Nanjing University of Aeronautics and Astronautics, China. He is an Associate Professor of Information and Communication Engineering at Nanjing University of Aeronautics and Astronautics.

XIN YIN was born in Jiangsu, China. She is currently working towards the M.Sc. degree at Nanjing University of Aeronautics and Astronautics.

KECANG QIAN was born in Zhejiang, China. He is currently working towards the B.S. degree at Nanjing University of Aeronautics and Astronautics.

5

Maximum Bandwidth Enhancement of Current Mirror using Series-Resistor and Dynamic Body Bias Technique

Vandana NIRANJAN, Ashwani KUMAR, Shail Bala JAIN

Dept. of Electronics & Communication Engineering, Indira Gandhi Delhi Technical University for Women, Kashmere Gate, Delhi-110006, India

vandana7379@gmail.com, drashwnikumar@yahoo.com, shail_jain@rediffmail.com

Abstract. *This paper introduces a new approach for enhancing the bandwidth of a low voltage CMOS current mirror. The proposed approach is based on utilizing body effect in a MOS transistor by connecting its gate and bulk terminals together for signal input. This results in boosting the effective transconductance of MOS transistor along with reduction of the threshold voltage. The proposed approach does not affect the DC gain of the current mirror. We demonstrate that the proposed approach features compatibility with widely used series-resistor technique for enhancing the current mirror bandwidth and both techniques have been employed simultaneously for maximum bandwidth enhancement. An important consequence of using both techniques simultaneously is the reduction of the series-resistor value for achieving the same bandwidth. This reduction in value is very attractive because a smaller resistor results in smaller chip area and less noise. PSpice simulation results using 180 nm CMOS technology from TSMC are included to prove the unique results. The proposed current mirror operates at 1 Volt consuming only 102 μW and maximum bandwidth extension ratio of 1.85 has been obtained using the proposed approach. Simulation results are in good agreement with analytical predictions.*

Keywords

Body effect, dynamic body bias, cascode current mirror, low voltage, high bandwidth, low power.

1. Introduction

The current mirrors (CM) are most commonly used in signal processing and conditioning analog circuits. Low voltage current mirrors having simple circuitry are very attractive for analog integrated circuit applications where large bandwidth and low power consumption are required. The current design trend of industry to reduce the supply voltage to sub-volt supplies and minimum gate size, places challenge to analog circuit designers. The shrinking sizes of semiconductor devices in CMOS technology entail the simultaneous reduction of supply voltage and threshold voltage of MOS transistor. Since the threshold voltage of a MOS transistor is not reduced at the same rate as the power supply, it becomes increasingly difficult to design wideband current mirror circuits because of the reduced voltage headroom. Analog signal processing circuits now operating from single 1.5 V supplies and dropping, are constantly demanding higher bandwidth and speed performances of current mirrors. The current mirror bandwidth performance degrades as power supply voltage is reduced. For a given bias point, the DC accuracy requirement fixes the power-speed ratio of the current mirror circuit and the tradeoff between the bandwidth, the accuracy and the power consumption is set by technology constants [1], [2].

Various techniques have been reported for bandwidth extension of low voltage current mirrors [3–8]. Throughout the bibliography many references [9–13] can be found considering series-resistor technique [3]. This technique improves the bandwidth of a current mirror by inserting a series-resistor between the gates of the primary pair transistors of current mirror. By appropriately choosing the resistor value, zero cancels a dominant pole thereby enhancing current mirror's bandwidth. The added resistor however increases the overall noise of the circuit owing to the additional thermal noise. Also, bandwidth enhancement is much smaller if the current gain is large. Hence this technique is not suitable for low noise high frequency applications. For full monolithic integration, resistor can be made of polysilicon or a diffusion resistor. The main drawback of integrated resistor is the large tolerance of its absolute values. Even with mature process, the passive and active components can have more than 10% variations. Although a MOS transistor can be used for active implementation of series-resistor but this results in penalty of extra power consumption and increase in chip area. Under very low supply voltages, the reduction of voltage headroom poses an important limitation on the power consumption in additional passive component/circuitry for improving bandwidth. Hence other techniques must be investigated.

In this work, we have proposed a new approach for bandwidth enhancement of a low voltage CMOS current mirror. The proposed approach is based on boosting the transconductance of a MOS transistor using dynamic body bias technique, while maintaining low power consumption. This approach also results in reduction of threshold voltage

(V_{TH}) of MOS transistor utilizing body effect. We have used the proposed approach along with series-resistor technique in a low voltage cascode current mirror circuit for maximum bandwidth enhancement. The parasitic resistance and capacitance due to dynamic body biasing introduces a zero in the transfer function of the proposed current mirror. This zero cancels one of the poles thereby enhancing the bandwidth of the proposed current mirror.

The dynamic body bias technique is implemented using triple well CMOS technology which is compatible with standard CMOS process. Triple well process can be achieved at the slightly higher cost process but without increase in any chip area. A triple well structure reduces the cross-talk in mixed systems-on-chip designs and is more robust to process and well junction capacitance variations [14]. The paper is organized as follows: Section 2 introduces conventional current mirror where series-resistor technique has been used for enhancing its bandwidth. Section 3 introduces dynamic body bias technique and circuit implementation of proposed current mirror. Bandwidth analysis is carried out using small signal analysis to predict its maximum theoretical bandwidth. Simulation results are presented in Section 4. Conclusions are summarized in Section 5.

2. Conventional Current Mirror

In modern sub-micron CMOS process the g_m/g_{ds} ratio is less than 100 and consequently a significant gain error results when more of the current amplifier stages are cascaded. This gain error is usually reduced by increasing the output impedance using different cascode topologies [15]. Fig. 1 shows a low voltage cascode current mirror. Here, the input voltage (V_{in}) depends solely on the biasing conditions of M1 and minimum V_{in} required to pump input current (I_{in}) into the input port of the mirror is given as [16].

$$V_{inmin} = V_{TH1} \,. \tag{1}$$

The other important factors that establish its capability to operate in a low-voltage environment are the minimum supply voltage and output voltage given by the following expressions

$$V_{DDmin} = V_{TH1} + V_{DSsat5} \,, \tag{2}$$

$$V_{outmin} = V_{DSsat2} + V_{DSsat3} \,. \tag{3}$$

Transistors M2 and M3 form output transistor cascode pair and due to the independent biasing of the transistor M3, provided by M4, the output voltage swing is not affected. Hence the output impedance of the structure can be increased to have high gain structures at low voltage levels. The input resistance of the mirror is decided by the transconductance (g_{m1}) of M1.The expression for input and output resistances is given as

$$R_{in} \cong \frac{1}{g_{m1}} \,, \tag{4}$$

$$R_{out} \cong g_{m3} \cdot r_{o2} \cdot r_{o3} \,. \tag{5}$$

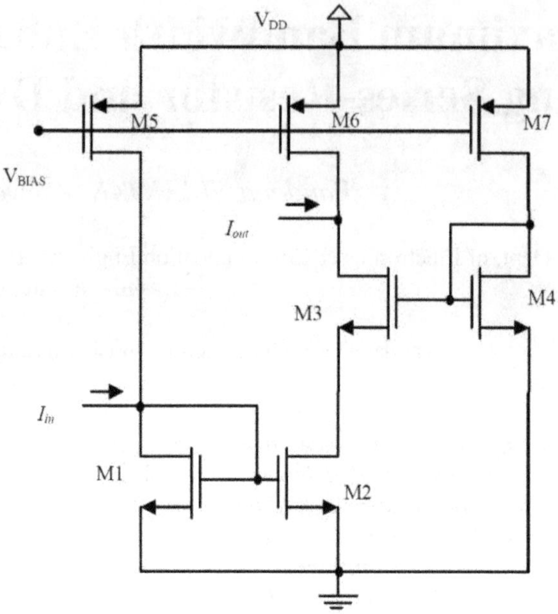

Fig. 1. Low voltage cascode current mirror.

Cascode current mirror suppress the effect of channel-length modulation and improves input-output isolation as there is no direct coupling from the output to input. The simple circuit structure of current mirror in Fig. 1 ensures a higher bandwidth. Its DC gain and -3dB frequency is given by the following expression [13]

$$Gain = \frac{g_{m2}}{g_{m1}} \,, \tag{6}$$

$$\omega_0 = \sqrt{\frac{g_{m1}(g_{m3})}{2\, C_{gs1} C_{gs3}}} \,. \tag{7}$$

From (7) we observe that the ratio between gate transconductance and gate-to-source capacitance of a MOS transistor dominates the frequency behavior. Thus g_m and C_{gs} are the main parameters that have major influence over the current mirror bandwidth.

The series-resistor technique to improve the bandwidth performance of low voltage cascode current mirror is shown in Fig. 2. Henceforth throughout this paper, we have considered current mirror in Fig. 2 as conventional current mirror under study. The introduction of R to the primary pair transistors of the current mirror results in a zero which cancels one of the poles when $R = 1/g_{m1}$ is selected, thereby enhancing its bandwidth. The expressions for DC gain and -3dB frequency is given as [13]

$$Gain_{CCM} = \frac{g_{m2}}{g_{m1}} \,, \tag{8}$$

$$\omega_{0,CCM} = \sqrt{\frac{g_{m1}(g_{m3})}{C_{gs1} C_{gs3}}} \,. \tag{9}$$

Comparing (7) and (9) we obtain

$$\omega_{0,CCM} = \sqrt{2}\, \omega_0 \,. \tag{10}$$

Thus from (6), (8) and (9), it is evident that series-resistor technique enhances the bandwidth by factor $\sqrt{2}$ without compromising DC gain.

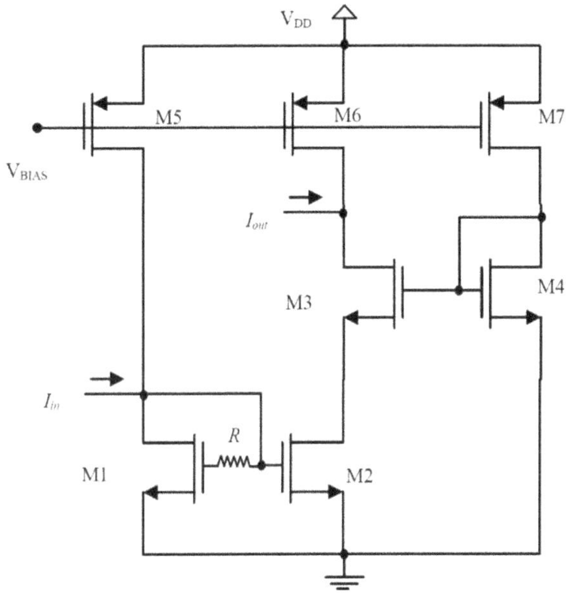

Fig. 2. Conventional current mirror.

We observe from Fig. 2 that in conventional CM, M3 suffers from body effect. In most of the current mirror analysis, generally body effect is neglected either for the sake of simplicity or due to its negligible effect, at higher power supply voltage. Some inaccuracy nevertheless arises due to the body effect present in a MOS transistor. Due to body effect, the ideal square-law behavior of MOS transistor in saturation region of operation approaches more closely to an ideal linear transfer function. As a result, high-order harmonics introduced to the drain current expression results in an error in its transconductance [17]. Due to this effect, there is non-negligible attenuation, which may not be neglected in the submicron circuits operating at low power supply voltage.

With the progressive reduction of MOS transistors minimum dimension and their associated supply voltages, use of body effect is becoming an attractive opportunity for improving the performance of low voltage analog integrated circuits [18]. In this paper we have utilized body effect in M3 using dynamic body bias technique for improving bandwidth and output resistance of conventional CM.

3. Proposed Current Mirror

In this section, we have briefly discussed dynamic body bias technique and a small signal model of MOS transistor biased with this technique is proposed. The circuit implementation of proposed current mirror using dynamic body biasing is suggested. Bandwidth analysis of proposed CM shows that the dynamic body bias technique has enhanced its bandwidth and output resistance.

3.1 Dynamic Body Bias Technique

For low voltage low power design, various techniques have been reported in the literature to overcome the difficulties introduced by the relatively high threshold voltage of a MOS transistor. One such reported technique in literature is bulk-driven technique, where input signal is applied at body/bulk terminal instead of gate terminal, after biasing the gate terminal to a sufficient voltage; thus the threshold voltage in this setup is removed from signal path. Unfortunately, the transconductance of a bulk-driven MOS transistor is substantially smaller and the equivalent input referred noise is higher than that of a conventional gate-driven MOS transistor, which may result in lower gain bandwidth and worse frequency response [19]. Another circuit technique which provides an important solution to the threshold voltage scaling limitation is body bias technique, which modulates the threshold voltage of a MOS transistor electronically using body effect, without any technology modification. Body effect enables a variety of effective body bias techniques [20].

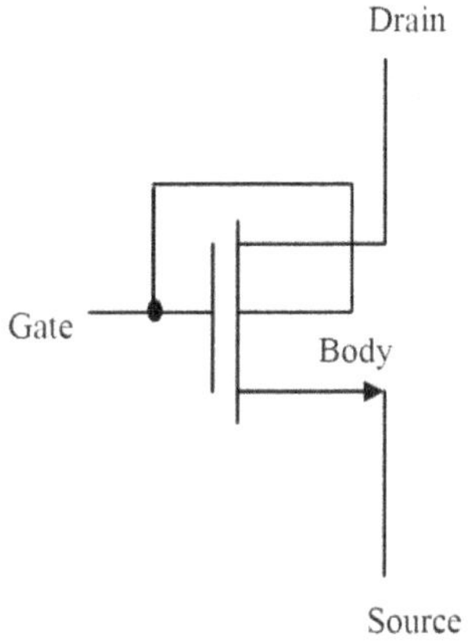

Fig. 3. MOS transistor using dynamic body bias technique.

In this work, our proposed approach is based on using both bulk/body and gate terminals of a MOS transistor as signal input. This concept, first presented in [21] is shown in Fig. 3. Due to gate and body terminals shorted together, threshold voltage (V_{TH}) of the transistor becomes function of input signal. With change in input, bias voltage at body terminal also changes dynamically as input changes hence known as dynamic body bias technique. The relation between input signal and V_{TH} is described using the following equation [22]

$$V_{TH} = V_{TH0} + \gamma\left(\sqrt{\psi_S + V_{SB}} - \sqrt{\psi_S}\right). \qquad (11)$$

where V_{TH} is threshold voltage due to body effect i.e. applied V_{SB} , V_{TH0} is the threshold voltage when V_{SB} is zero and mainly depends on the manufacturing process. γ is typically equal to 0.4 $V^{0.5}$ and depends on the gate oxide capacitance, silicon permittivity, doping level and other parameters. ψ_s is surface potential in strong inversion and typically is 0.6 V. ψ_s in (11) is assumed to $|\ 2\Phi_F\ |$, where Φ_F is Fermi potential.

In dynamic body bias technique, $V_{GS} = V_{BS}$ is maintained all the time and same bias voltage is applied at gate and body terminals. Here source-body junction gets slightly forward biased when input signal increases. Although source-body junction parasitic diode is slightly forward biased but any substantial conducting pn junction current is absent as V_{TH} decreases due to the body effect as predicted by (11). The potential in the channel region is strongly controlled by the gate and body terminals, leading to a high transconductance owing to faster current transport.

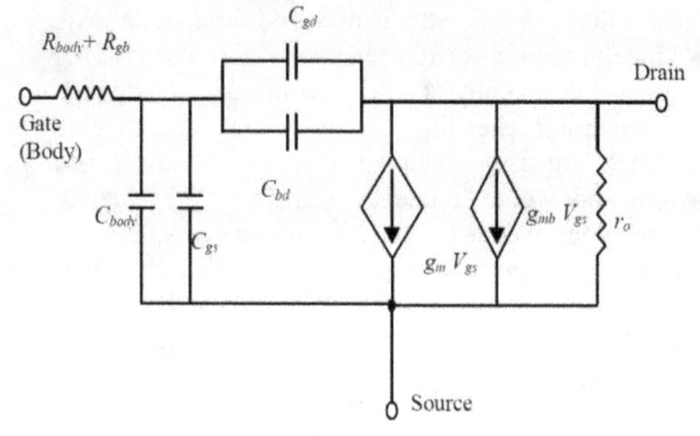

Fig. 4. Proposed small signal model of MOS transistor using dynamic body bias technique.

The proposed small signal equivalent circuit of MOS transistor using dynamic body biasing is shown in Fig. 4. It has two transconductances, the gate transconductance g_m and body transconductance g_{mb}. The relation between both transconductances is [22]

$$\frac{g_{mb}}{g_m} = \eta \approx (0.2 - 0.4) \quad (12)$$

where η is the specific parameter and its value depends on bias conditions and on the technology used. Dynamic body bias increases the effective transconductance from g_m to (g_m+ g_{mb}) as V_{SB} = V_{GS} is maintained all the time and both transconductance contributes to the conduction current. The effective transconductance is obtained as

$$g_{m,eff} = g_m(1 + \eta)\ . \quad (13)$$

Due to higher effective transconductance, the input referred noise power spectral density is also reduced, defined as

$$v^2_{noise}(f) = \frac{i^2_{ni}}{(g_m+g_{mb})^2} \quad (14)$$

where i^2_{ni} is the total drain current generated by noise sources.

The bulk of a MOS transistor has finite resistance and additional parasitic capacitance. The effective input capacitance from Fig. 4 is defined as

$$C_{eff} \approx C_{gs} + C_{body} \quad (15)$$

where C_{body} is body capacitance and C_{gs} is gate capacitance. The effective input resistance from Fig. 4 is defined as

$$R_{eff} \approx R_{gb} + R_{body} \quad (16)$$

where R_{gb} is gate-body contact resistance and R_{body} is bulk/body resistance.

Dynamic body bias technique is implemented using triple well CMOS technology hence latch-up is absent. This technique exhibits merits over other body bias techniques in terms of higher transconductance-to-drain current ratio and elimination of additional circuitry for bias voltage generation.

3.2 Circuit Implementation

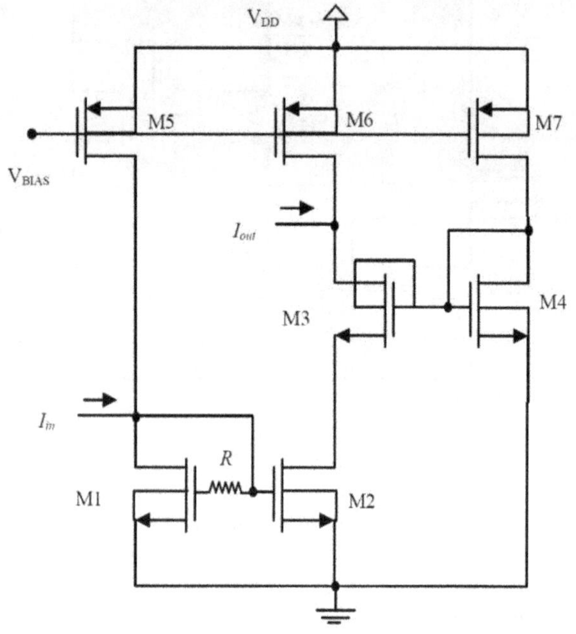

Fig. 5. Proposed current mirror.

The current mirrors using MOS transistors operating in saturation mode have higher transconductance than those based on MOS transistors operating in triode or sub-threshold mode. This leads to higher bandwidth and better input and output impedances. A convenient way to bring MOS transistor into saturation is by using body effect [23]. Fig. 5 shows the circuit implementation of proposed CM in which we have applied dynamic body bias technique in M3 and thus for any variation in load, M3 always remains in saturation, utilizing body effect. For remaining transistors, body and source terminals are connected together.

A current mirror may be characterized as having output impedance which affects the accuracy of the current replicated in the current mirror. High output impedance in current mirrors is required for accurate replication of currents. It can be seen from (4), (5) and (13) that due to dynamic body bias in M3, input resistance of proposed CM remains unaffected whereas output resistance is increased, given as

$$R_{out} \cong g_{eff}\,.r_{o2}.r_{o3}\ . \quad (17)$$

3.3 Bandwidth Analysis

The small signal model for bandwidth calculation of proposed CM is shown in Fig. 6. In the small signal analysis of proposed current mirror, short channel effects like drain-induced barrier lowering, hot carriers effects, velocity saturation are neglected for the sake of simplicity of hand calculations with the goal to give an idea on the order of magnitude of bandwidth improvement rather than finding an exact value. The notations used in the analysis are as follows: C_{gsi} is gate-source capacitance, V_{gsi} is gate-source voltage and g_{mi} is gate transconductance for Mi where $i = 1$ to 4. Here, g_{mb3} is body transconductance, C_{eff} is total input capacitance and R_{eff} is total input resistance of M3. R is series-resistor between gates of M1 and M2.

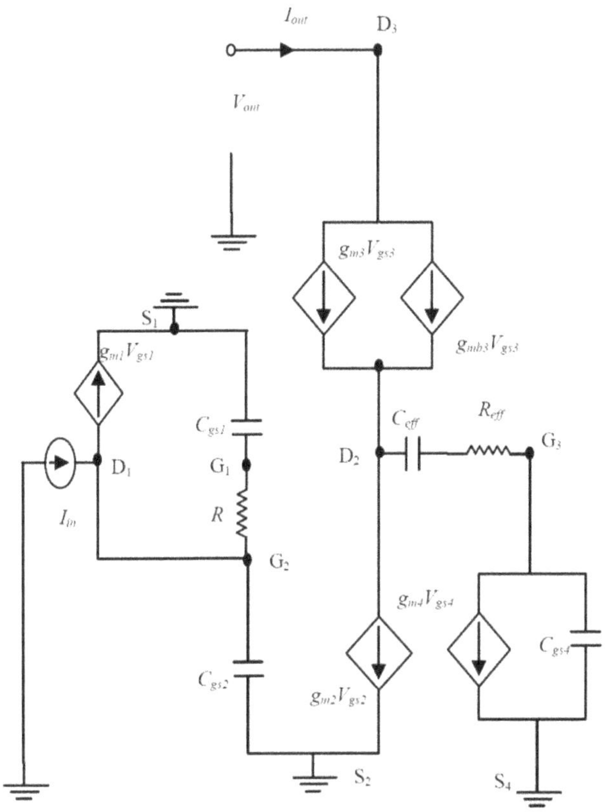

Fig. 6. Small signal model for calculating bandwidth of proposed CM.

Neglecting output conductance and capacitance in Fig. 6, equation for node D1 is given as

$$I_{in}(s) = g_{m1}V_{gs1} + s\,C_{gs2}V_{G2} + \frac{V_{G2}-V_{G1}}{R}. \tag{18}$$

Since both R and C_{gs1} are in series we get

$$\frac{V_{G2}-V_{G1}}{R} = V_{G1}s\,C_{gs1}. \tag{19}$$

Simplifying (19) we get

$$V_{G1} = \frac{V_{G2}}{(1+sRC_{gs1})}. \tag{20}$$

Substituting (19) in (18) we get

$$I_{in}(s) = V_{G1}\left(g_{m1} + s\,C_{gs1}\right) + s\,C_{gs2}V_{G2}. \tag{21}$$

Substituting (20) in (21) we get

$$V_{G2} = V_{gs2} = I_{in}(s)\left[\frac{(1+sRC_{gs1})}{g_{m1}+s\,C_{gs1}+s\,C_{gs2}(1+sRC_{gs1})}\right] \tag{22}$$

Writing equation at node S3

$$g_{m2}V_{gs2} = (g_{m3}+g_{mb3})V_{gs3} + \frac{V_{gs3}}{R_{eff}+\frac{1}{sC_{eff}}} \tag{23}$$

Because of mirror action $V_{gs1} = V_{gs2}$ and substituting (22) in (23) we get

$$I_{in}(s)\left[\frac{g_{m2}(1+sRC_{gs1})}{g_{m1}+s\,C_{gs1}+s\,C_{gs2}(1+sRC_{gs1})}\right] = \left(g_{m3}+g_{mb3}+\frac{1}{R_{eff}+\frac{1}{sC_{eff}}}\right)V_{gs3} \tag{24}$$

Writing equation for node D3 we get

$$I_{out}(s) = g_{m3}V_{gs3} + g_{mb3}V_{gs3}. \tag{25}$$

Simplifying (25) we get

$$V_{gs3} = \frac{I_{out}(s)}{g_{m3}+g_{mb3}}. \tag{26}$$

Substituting (26) in (24) we obtain transfer function as

$$\frac{I_{out}(s)}{I_{in}(s)} = \frac{g_{m2}(1+sRC_{gs1})(g_{m3}+g_{mb3})}{\left[g_{m1}+s\,C_{gs1}+s\,C_{gs2}(1+sRC_{gs1})\right]\left(g_{m3}+g_{mb3}+\frac{1}{R_{eff}+\frac{1}{sC_{eff}}}\right)} \tag{27}$$

Simplifying and rearranging (27) we get

$$\frac{I_{out}(s)}{I_{in}(s)} = \left[\frac{g_{m2}(g_{m3}+g_{mb3})\,RC_{gs1}\left(s+\frac{1}{RC_{gs1}}\right)}{RC_{gs1}C_{gs2}\left\{s^2+s\left(\frac{C_{gs1}+C_{gs2}}{RC_{gs1}C_{gs2}}\right)+\frac{g_{m1}}{RC_{gs1}C_{gs2}}\right\}}\right] \times \left[\frac{R_{eff}C_{eff}\left(s+\frac{1}{R_{eff}C_{eff}}\right)}{\{C_{eff}+R_{eff}C_{eff}(g_{m3}+g_{mb3})\}\left[s+\frac{g_{m3}+g_{mb3}}{C_{eff}+R_{eff}C_{eff}(g_{m3}+g_{mb3})}\right]}\right]. \tag{28}$$

Assuming M1 and M2 to be matched and substituting $C_{gs1} = C_{gs2}$ in (28), we get

$$\frac{I_{out}(s)}{I_{in}(s)} = \left[\frac{g_{m2}(g_{m3}+g_{mb3})\,RC_{gs1}\left(s+\frac{1}{RC_{gs1}}\right)}{RC_{gs1}{}^2\left\{s+\frac{1}{RC_{gs1}}\right\}^2}\right] \times \left[\frac{R_{eff}C_{eff}\left(s+\frac{1}{R_{eff}C_{eff}}\right)}{\{C_{eff}+R_{eff}C_{eff}(g_{m3}+g_{mb3})\}\left[s+\frac{g_{m3}+g_{mb3}}{C_{eff}+R_{eff}C_{eff}(g_{m3}+g_{mb3})}\right]}\right] \tag{29}$$

We observe from (29) that the introduction of R in the circuit has resulted in a zero at $Z_R = -1/RC_{gs1}$ and dynamic body bias technique has resulted in a zero at $Z = -1/R_{eff}C_{eff}$.

Substituting $R = 1/g_{m1}$ in (29) then zero cancels one of the poles and simplified transfer function is written as

$$\frac{I_{out}(s)}{I_{in}(s)} = \left[\frac{g_{m2}(g_{m3}+g_{mb3})}{C_{gs1}\left(s+\frac{g_{m1}}{C_{gs1}}\right)}\right] \times \left[\frac{R_{eff}\left(s+\frac{1}{R_{eff}C_{eff}}\right)}{\{1+R_{eff}(g_{m3}+g_{mb3})\}\left[s+\frac{g_{m3}+g_{mb3}}{C_{eff}\{1+R_{eff}(g_{m3}+g_{mb3})\}}\right]}\right]. \tag{30}$$

Assuming $R_{eff}\,(g_{m3}+g_{mb3}) >> 1$, (30) can be further simplified as

$$\frac{I_{out}(s)}{I_{in}(s)} = \left[\frac{g_{m2}(g_{m3}+g_{mb3})}{C_{gs1}\left(s+\frac{g_{m1}}{C_{gs1}}\right)}\right]\left[\frac{\left(s+\frac{1}{R_{eff}C_{eff}}\right)}{(g_{m3}+g_{mb3})\left(s+\frac{1}{R_{eff}C_{eff}}\right)}\right]. \quad (31)$$

We observe from (31) that zero cancels one of the poles and transfer function becomes

$$\frac{I_{out}(s)}{I_{in}(s)} = \left(\frac{g_{m2}}{g_{m1}}\right)\frac{\frac{g_{m1}}{C_{gs1}}}{\left(s+\frac{g_{m1}}{C_{gs1}}\right)}. \quad (32)$$

Comparing (32) with standard equation of first order transfer function given as

$$T(s) = \frac{K}{1+\frac{s}{\omega_0}} \quad (33)$$

where ω_0 is -3dB frequency. DC gain and -3dB frequency of proposed CM is obtained as

$$Gain_{PCM} = \frac{g_{m2}}{g_{m1}}, \quad (34)$$

$$\omega_{0,PCM} = \frac{g_{m1}}{C_{gs1}}. \quad (35)$$

On the basis of analytical predictions for bandwidth from (35) and comparing it with (9), it is observed that the proposed approach enhances the bandwidth of current mirror. It is evident from the analysis that the bandwidth enhancement mechanism of dynamic body bias technique and series-resistor techniques differs fundamentally. This suggests that both techniques can be employed simultaneously for maximum bandwidth enhancement. From (5) and (17) we see that output resistance is also improved using the proposed approach as effective transconductance of M3 has increased. The dynamic body bias technique does not affect the current mirror's DC gain as evident from (8) and (34).

4. Simulation Results

Conventional current mirror in Fig. 2 and proposed current mirror in Fig. 5 have been designed in 180 nm CMOS technology from TSMC. The current mirror (CM) circuits have been simulated at $V_{DD} = 1$ V and $V_{BIAS} = 0.35$ V. All the transistors have the same channel length $L = 180$ nm and width parameters are given in Tab. 1.

MOS transistor	Type	W(µm)
M1	NMOS	8.3
M2,M3	NMOS	9
M4	NMOS	0.36
M5	PMOS	0.36
M6,M7	PMOS	9

Tab. 1. Width parameters for various transistors.

Fig. 7 shows simulated frequency response of conventional and proposed current mirror for $R = 1$ kΩ. -3dB frequency of conventional CM and proposed CM is obtained as 344.224 MHz and 538.987 MHz respectively with 0 dB peaking. Thus dynamic body bias technique boosts the bandwidth by a factor of 1.56 ($R = 1$ kΩ). Fig. 8 shows transfer characteristics. The approx. range of proposed CM is 0 to 250 µA. The simulated static power consumption of proposed CM is 0.102e-4 W.

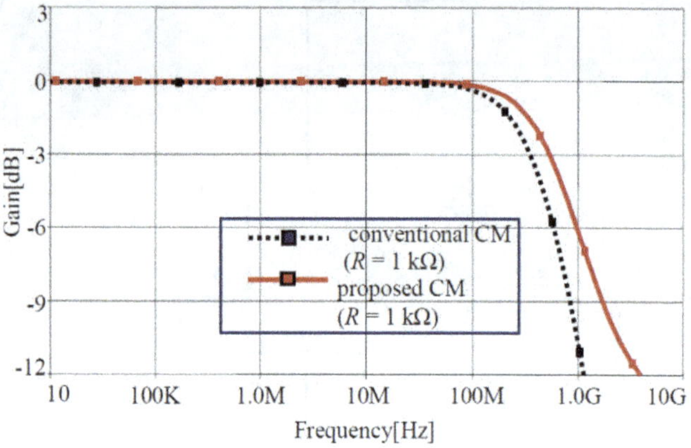

Fig. 7. Frequency response of conventional and proposed current mirror.

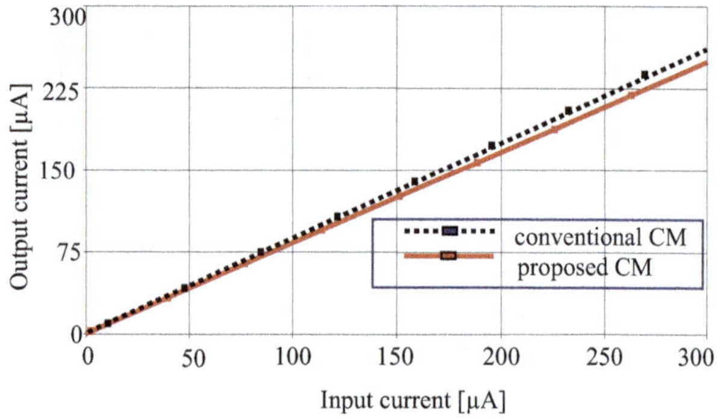

Fig. 8. Transfer characteristics of proposed current mirror.

The parasitic capacitances in CMOS circuits introduce zeros which results in peaks in frequency response. To maximize the bandwidth and avoid ringing in time domain response, the series-resistor is sized based on the criterion of the critical damping. Fig. 9 shows frequency behavior of conventional CM when R is varied from 1 kΩ to 60 kΩ, detailed in Tab. 2.

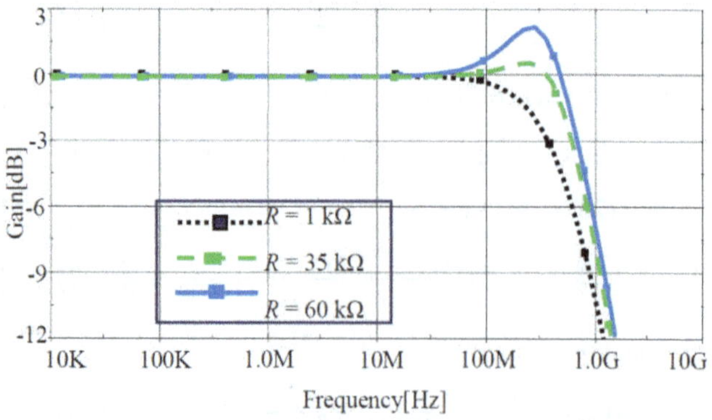

Fig. 9. Variation in frequency response of conventional current mirror with R.

R (kΩ)	Conventional CM (MHz)	Peaking (dB)	Proposed CM (MHz)	Peaking (dB)	BWER (bandwidth extension ratio)
1	344.224	0	538.987	0	1.565
5	370.186	0	608.429	0	1.643
10	417.881	0	765.741	0	1.832
15	471.719	0	875.195	0	1.855
20	526.081	0	941.204	0.245	1.789
25	565.759	0	964.296	0.766	1.704
30	586.706	0.221	976.053	1.263	1.663
35	601.100	0.580	976.053	1.727	1.623
40	608.429	0.961	976.053	2.148	1.604
45	615.848	1.331	976.053	2.528	1.585
50	615.848	1.658	976.053	2.835	1.585
55	615.848	1.947	976.053	>3	1.585
60	615.848	2.216	976.053	>3	1.585

Tab. 2. Comparison of -3dB frequency of conventional and proposed current mirror(CM) for different values of *R*.

It is observed from the table that the bandwidth increases as value of *R* increases and there is no peaking in frequency response up to $R = 25$ kΩ. The maximum bandwidth obtained without peaking and with peaking for conventional CM is 565.759 MHz and 615.848 MHz respectively.

Fig. 10 shows frequency behavior of proposed CM when *R* is varied from 1 kΩ to 50 kΩ, detailed in Tab. 2. It is observed from the table that the bandwidth increases as value of *R* increases and there is no peaking in frequency response up to $R = 15$ kΩ. The maximum bandwidth obtained without peaking and with peaking for proposed CM is 875.195 MHz and 976.053 MHz respectively.

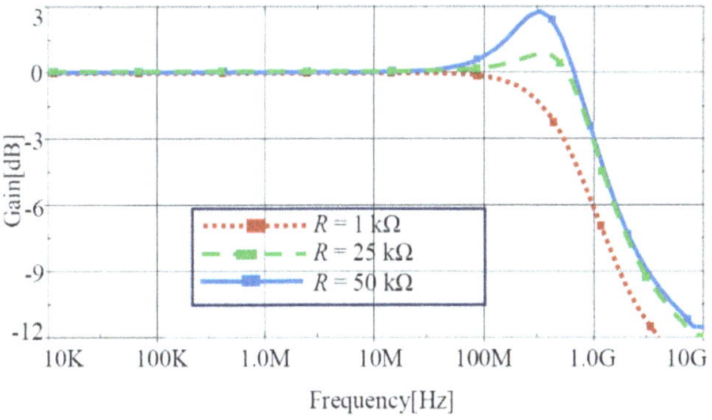

Fig. 10. Variation in frequency response of proposed current mirror with *R*.

It is seen from Tab. 2 that about 600 MHz bandwidth in conventional CM is obtained when $R = 45$ kΩ is used. In proposed CM the same bandwidth of about 600 MHz is achieved for $R = 5$ kΩ only. This reduction in value of *R* is very attractive because a smaller resistor results in smaller chip area and less noise. Further resistors with a smaller parasitic capacitance, such as non-silicide poly resistors, may be preferred over diffusion resistors for reducing peaking and achieving higher bandwidth.

It is observed from Tab. 2 that maximum bandwidth which can be achieved using dynamic body bias technique, without peaking, in the proposed CM is 875.195 MHz for $R = 15$ kΩ. For conventional CM, bandwidth of 471.719 MHz is obtained for $R = 15$ kΩ. Thus maximum BWER of 1.85 is obtained using dynamic body bias technique without peaking.

A current mirror having excellent static performance is more preferable for biasing applications. For signal processing applications, current mirrors lie along the signal path of a circuit and must have good dynamic performance. Fig. 11 shows simulated responses of the proposed and conventional current mirror to a step change in I_{in} from 25 μA to 175 μA. From this plot, it is evident that the step response of the proposed CM is much faster than that of the conventional CM.

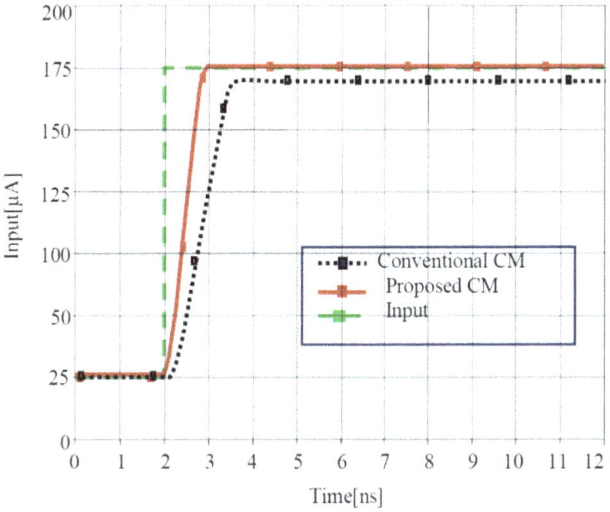

Fig. 11. Transient response of proposed current mirror.

Fig. 12 shows simulated frequency response of CM in Fig. 1, i.e. without any bandwidth extension technique and -3dB frequency of 340.341 MHz is obtained.

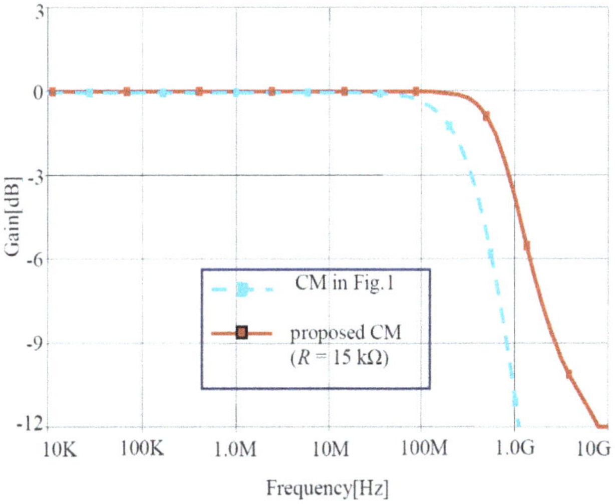

Fig. 12. Frequency response of proposed current mirror and current mirror (Fig. 1) without any bandwidth extension technique.

Tab. 4 compares the bandwidth extension ratio (BWER) achieved using series-resistor technique and combination of series resistor and dynamic body bias technique for $R = 15$ kΩ, without peaking (as observed in Tab. 2). It is evident from Tab. 4 that BWER of 1.38 is obtained using series-resistor technique and BWER of 2.57 is obtained when both dynamic body bias technique and series-resistor technique are used simultaneously in low voltage cascode current mirror shown in Fig. 1.

Parameters	Conventional current mirror (Fig. 2)	Proposed current mirror(Fig. 5)
Supply voltage (V)	1	1
CMOS technology (nm)	180	180
Simulated maximum bandwidth without peaking (MHz)	565.759 (for R = 25 kΩ)	875.195 (for R = 15 kΩ)
Range (μA)	262	250
Input resistance (kΩ)	1.381	1.381
Output resistance (kΩ)	561.382	817.331
Power dissipation (W)	0.104e-4	0.102e-4

Tab. 3. Comparison of simulated parameters for conventional and proposed current mirror.

Parameter	Current mirror in Fig. 1	Bandwidth extension using only series-resistor technique (Fig. 2)	Bandwidth extension using both dynamic body bias technique and series-resistor technique simultaneously (Fig. 5)
-3dB frequency (MHz)	340.341	471.719	875.195
BWER obtained for CM in Fig. 1 without peaking for R = 15 kΩ		1.38	2.57

Tab. 4. Comparison of BWER using series-resistor and proposed technique.

5. Conclusion

In this paper, the authors have revealed a new approach of enhancing the bandwidth of a low voltage CMOS current mirror using dynamic body bias technique. The unique feature of this biasing technique is that no additional circuitry is required for bias voltage generation. The proposed approach boosts the bandwidth of the current mirror by a factor of 1.85 and output resistance by factor 1.45 at low supply voltage of 1 volt. The proposed approach does not add any noise owing to higher transconductance of MOS transistor. In emerging CMOS technologies, analog integrated circuits containing passive components are less preferred. It is due to the difficulty faced in fabricating high quality passive devices with tightly-controlled values or a reasonable physical size. This proposed approach reduces the value of series-resistor for achieving same bandwidth performance. This reduction in value of R is very attractive because a smaller resistor results in smaller chip area and less noise. It is pertinent to mention that the dynamic body bias technique virtually has no effects on the overall power consumption of the proposed current mirror and therefore any improvement is not at the expense of increased power consumption. The proposed approach is particularly attractive for low voltage CMOS current mirrors for higher bandwidth performance without degrading overall noise.

References

[1] RAMÍREZ-ANGULO, J., CARVAJAL, R. G., TORRALBA, A. Low supply voltage high-performance CMOS current mirror with low input and output voltage requirements. *IEEE Transactions on Circuits and Systems—II: Express Briefs*, 2004, vol. 51, no. 3, p. 124-129.

[2] PETERSON, K. D., GEIGER, R. L. Area/bandwidth tradeoffs for CMOS current mirrors. *IEEE Transactions on Circuits and Systems*, 1986, vol. CAS-33, no. 1, p. 667-669.

[3] VOO, T., TOUMAZOU, C. High-speed current mirror resistive compensation technique. *IEE Electronics Letters*, 1995, vol. 31, no. 4, p. 248–250.

[4] VOO, T., TOUMAZOU, C. Precision temperature stabilized tunable CMOS current-mirror for filter applications. *IEE Electronics Letters*, 1996, vol. 32, no. 2, p. 105–106.

[5] BENDONG, S., YUAN, F. A new inductor series-peaking technique for bandwidth enhancement of CMOS current-mode circuits. *Analog Integrated Circuits and Signal Processing*, 2003, vol. 37, p. 259–264.

[6] YUAN, F. Low voltage CMOS current-mode circuits: topology and characteristics. *IEE Proc-Circuits Devices Systems*, 2006, vol. 153, no. 3, p. 219–230.

[7] ITAKURA, T., IIDA, T. A feedforward technique with frequency dependent current mirrors for low-voltage wideband amplifier. *IEEE Journal of Solid State Circuits*, 1996, vol. 31, no. 6, p. 847 to 850.

[8] RAJ, N., SINGH, A. K., GUPTA, A. K. Low-voltage bulk-driven self-biased cascode current mirror with bandwidth enhancement. *IEE Electronics Letters*, 2014, vol. 50, no. 1, p. 23–25.

[9] TIKYANI, M., PANDEY, R. A new low-voltage current mirror circuit with enhanced bandwidth. In *Proceedings of the International Conference on Computational Intelligence and Communication Networks*. 2011, p. 42–46.

[10] GUPTA, M., MALHOTRA, A., MALIK, A. Low-voltage current mirror with extended bandwidth. In *Proc. of the IEEE 5th India International Conference on Power Electronics*, 2012, p. 1–5.

[11] SHARMA, S., RAJPUT, S. S., MANGOTRA, L. K., JAMUAR, S. S. FGMOS current mirror: behaviour and bandwidth enhancement. *Analog Integrated Circuits and Signal Processing*, 2006, vol. 46, no. 3, p. 281–286.

[12] GUPTA, M., SINGH, U., SRIVASTAVA, R. Bandwidth extension of high compliance current mirror by using compensation methods. *Active and Passive Electronic Components*, vol. 2014, Article ID 274795, 8 pages.

[13] GUPTA, M., AGGARWAL, P., SINGH, P., JINDAL, N. K. Low voltage current mirrors with enhanced bandwidth. *Analog Integrated Circuits and Signal Processing*, 2009, vol. 59, p. 97–103.

[14] NIRANJAN, V., KUMAR, A., JAIN, S. B. Triple well subthreshold CMOS logic using body-bias technique. In *Proceedings of the IEEE International Conference on Signal Processing, Computing and Control*, 2013, p. 1–6.

[15] RAZAVI, B. *Design of Analog CMOS Integrated Circuits.* 2nd ed., Tata McGraw-Hill Publishing Company Limited, 2002.

[16] KOLIOPOULOS, C., PSYCHALINOS, C. A comparative study of the performance of the flipped voltage follower based low voltage current mirror. In *Proceedings of the IEEE International Symposium on Signals, Circuits and Systems*, 2007, p. 1–4.

[17] ZHU, X., SUN, Y. Low-distortion low-voltage operational transconductance amplifier. *IEE Electronics Letters*, 2008, vol. 44, no. 25, p. 1434–1436.

[18] MONSURRÒ, P., PENNISI, S., SCOTTI, G., TRIFILETTI, A. Exploiting the body of MOS devices for high performance analog design. *IEEE Circuits and Systems Magazine*, Fourth Quarter, 2011, p. 8–23.

[19] KHATEB, F., DABBOUS, S. B. A., VLASSIS, S. A survey of non-conventional techniques or low-voltage low-power analog circuit design. *Radioengineering*, 2013, vol. 22, no. 2, p. 415–427.

[20] NIRANJAN, V., GUPTA, M. Body biasing - a circuit level approach to reduce leakage in low power CMOS circuits. *Journal of Active and Passive Electronic Devices*, 2011, vol. 6, no. 1-2, p. 89–99.

[21] ASSADERAGHI, F., SINITSKY, D., PARKE, S. A., ET AL. Dynamic threshold voltage MOSFET (DTMOS) for ultra low voltage VLSI. *IEEE Transactions on Electron Devices*, 1997, vol. 44, no. 3, p. 414–422.

[22] TSIVIDIS, Y. P. *Operation and Modeling of the MOS Transistor*. New York: Mc-Graw Hill, 1987.

[23] LEDESMA, F., GARCIA, R., RAMIREZ ANGULO, J. Comparison of new and conventional low voltage current mirrors. In *Proceedings of the 45th Midwest Symposium on Circuits and Systems*, 2002, p. 49–52.

About Authors ...

Vandana NIRANJAN was born in 1979 and received her B.Tech. in Electronics & Communication Engineering from Government Engineering College, Bhopal, India in 2000 and her M.Tech degree in Semiconductor Devices & VLSI Technology from Indian Institute of Technology (IIT), Roorkee in March 2002. She joined the Department of Electronics and Communication Engineering, Indira Gandhi Delhi Technical University for Women, New Delhi, in July 2002 as Asst. Professor. At present she is pursuing research in the area of MOSFET body bias techniques. Her teaching interest includes VLSI circuit design and low voltage CMOS analog integrated circuits. She is a member of IEEE and Women in Engineering, USA.

Ashwani KUMAR was born in 1969 and received his B.E. in Electronics & Communication Engineering from Delhi College of Engineering, Delhi University in the year 1991 and his M.E. in Electronics & Communication Engineering from Delhi College of Engineering, Delhi University in the year 1993. He received his Ph.D. in the field of Signal Processing from Delhi College of Engineering, Delhi University in the year 2000. From October 1991 to January 2008, he has rich industry experience as Senior Program Manager in Centre for Development of Telematics, Delhi, Telecom Technology Centre of Government of India Centre, Ministry of Communication and Information Technology, Government of India. Since February 2008, he joined Indira Gandhi Delhi Technical University for Women, New Delhi as Professor. At present he is the Head of the Electronics and Communication Engineering Department. He has published a large number of papers in national and international journals. His teaching and research interests are signal processing and low voltage analog circuits. He is a member of IEEE, USA.

Shail Bala JAIN was born in 1950 and received her B.E in Electronics & Communication Engineering from Delhi College of Engineering, Delhi University and her M.Tech. in Electronics and Communication Engineering from Indian Institute of Technology (IIT), Delhi. She received her Ph.D in the field of Signal Processing from Delhi University. From 1971 till Oct. 2002 she has 31 years of teaching experience in the Department of Electronics and Communication Engineering, Delhi College of Engineering. At present she is a Professor, Electronics and Communication Engineering, Indira Gandhi Delhi Technical University for Women, New Delhi. A life fellow of IETE and a senior member of IEEE, USA, she is co-author of two books, (i) Linear Integrated Circuits, (ii) Electronics Devices and Circuits. She has published a large number of papers in national and international journals. Her field of interest is integrated circuits and signal processing.

Modeling the Flux-Charge Relation of Memristor with Neural Network of Smooth Hinge Functions

Xiaomu MU, Juntang YU, Shuning WANG

Dept. of Automation, TNList, Tsinghua University, Qinghuayuan 1, 100084 Beijing, People's Republic of China

mxm09@mails.tsinghua.edu.cn, yjt11@mails.tsinghua.edu.cn, swang@mail.tsinghua.edu.cn

Abstract. *The memristor was proposed to characterize the flux-charge relation. We propose the generalized flux-charge relation model of memristor with neural network of smooth hinge functions. There is effective identification algorithm for the neural network of smooth hinge functions. The representation capability of this model is theoretically guaranteed. Any functional flux-charge relation of a memristor can be approximated by the model. We also give application examples to show that the given model can approximate the flux-charge relation of existing piecewise linear memristor model, the window function memristor model, and a physical memristor device.*

Keywords

Memristor, mathematical model, smooth hinge function, neural network.

1. Introduction

Memristor was proposed to be the fourth basic circuit element, in order to characterize the relation between the flux φ and the charge q [1], [2]. The other three basic circuit elements are the resistor (current-voltage relation), the inductor (current-flux relation) and the capacitor (charge-voltage relation). According to [1], [3], the flux-charge relation of a charge-controlled memristor is expressed as

$$\varphi = \varphi(q) \tag{1}$$

and the current and voltage relation is described by (note that $\mathrm{d}q = i\mathrm{d}t$ and $\mathrm{d}\varphi = v\mathrm{d}t$)

$$v(t) = M(q(t))i(t) \tag{2}$$

where

$$M(q) = \mathrm{d}\varphi(q)/\mathrm{d}q. \tag{3}$$

Similarly, a flux-controlled memristor is expressed as:

$$q = q(\varphi) \tag{4}$$

and the current and voltage relation is described by

$$i(t) = W(\varphi(t))v(t) \tag{5}$$

where

$$W(\varphi(t)) = \mathrm{d}q(\varphi)/\mathrm{d}\varphi. \tag{6}$$

The charge-controlled memristor and the flux-controlled memristor are also referred as the ideal memristor [2], [3].

The concept of memristor was generalized to memristive systems [4], [5]. A current-controlled memristive system is described by

$$\dot{x} = f(x,i,t), \tag{7a}$$
$$v = R(x,i,t)i \tag{7b}$$

where x is the state variable of the memristive system. The ideal memristor is a special case of the memristive system (letting $x = q$, $f(x,i,t) = \dot{q} = i$ and $R(x,i,t) = M(q(t))$). Similarly, a voltage-controlled memristive system is described by

$$\dot{x} = f(x,v,t) \tag{8a}$$
$$i = G(x,v,t)v. \tag{8b}$$

The first physical memristor device was found by [6]. Then many other memristor devices with different physical mechanisms have been proposed by researchers, see [7], [8], [9] for examples. The mechanisms of these devices are very complex. The memristive system is widely used to model physical memristor devices [6], [10], [11], [12], since the ideal memristor (1) and (4) usually can not fully describe the behavior of the physical memristor devices. For example, voltage dependence is observed in many physical memristor devices [8], [10], [12]. The flux-controlled memristor (4) can not model the voltage dependence of the device. There is only one variable φ in (4), therefore the conductance $W(\varphi(t))$ in (5) can not represent the influence of the voltage $v(t)$ at a certain time t.

The memristive system is a generalization of the memristor in the perspective of current and voltage relation, as $v = M(q)i$ is generalized to $v = R(x,i,t)i$ and charge q is generalized to state variable x. To our best knowledge, the generalization of the memristor in the perspective of flux-charge relation is little studied. In this paper, we focus on the flux-charge relation of the memristor, considering that the memristor was originally proposed to characterize the missing relation between flux and charge [2]. We propose a generalized model to describe the flux-charge relation of the memristor based on the neural network of smooth hinge functions. Our model has good representation capability and thus is a suitable choice for modeling memristor devices. We will show that with theoretical analysis and three examples.

2. Generalized Flux-Charge Relation Model

Besides extending memristor to memristive system in the perspective of current and voltage relation, we believe that we can directly generalize the flux-charge relation of the memristor. Note that current $i(t)$ and voltage $v(t)$ are added as parametric variables into the memristive system to form a more general and complex system. We add the current $i(t)$ and voltage $v(t)$ into the flux-charge relation of the memristor as a generalization. The $\varphi = \varphi(q)$ relation of a charge-controlled memristor can be generalized to $\varphi = \varphi(q,i)$, and the $q = q(\varphi)$ relation of a flux-controlled memristor can be generalized to $q = q(\varphi, v)$. We use the neural network of smooth hinge functions to represent the generalized flux-charge relation of the memristor.

Specifically, a generalized charge-controlled memristor is given by

$$\begin{aligned} \varphi(q,i) &= a_0 q(t) + b_0 i(t) + c_0 + \\ &\quad \Sigma_{k=1}^{m} \eta_k \ln(1+\exp(a_k q(t) + b_k i(t) + c_k)). \end{aligned} \tag{9}$$

In (9), $\ln(1+\exp(a_k q(t) + b_k i(t) + c_k)$ is the base function of the neural network of smooth hinge functions [13], m is the number of base functions, a_k, b_k, c_k, η_k are parameters. Similar to $v(t) = M(q(t))i(t)$, $M(q) = \mathrm{d}\varphi(q)/\mathrm{d}q$ of (1), the current and voltage relation of (9) is given by $v(t) = \frac{\mathrm{d}\varphi(q,i)}{\mathrm{d}t} = \frac{\partial \varphi(q,i)}{\partial q} i(t) + \frac{\partial \varphi(q,i)}{\partial i} \frac{\mathrm{d}i(t)}{\mathrm{d}t}$. Similarly, a generalized flux-controlled memristor is given by

$$\begin{aligned} q(\varphi, v) &= a_0 \varphi(t) + b_0 v(t) + c_0 + \\ &\quad \Sigma_{k=1}^{m} \eta_k \ln(1+\exp(a_k \varphi(t) + b_k v(t) + c_k)). \end{aligned} \tag{10}$$

The representation capability of our model is theoretically guaranteed by using the neural network of smooth hinge functions. Any continuous function can be approximated by the neural network of smooth hinge functions to arbitrary precision with a sufficient number of base functions [13]. The memristor was proposed to characterize the relation between flux and charge. Therefore as long as there is a function relation between φ and q (in the form of $\varphi = \varphi(q,i)$ or $q = q(\varphi, v)$) of the memristor device, our model can approximate such flux-charge relation well. As far as we know, no existing memristor model can guarantee such representation capability. Due to the good approximation capability of the neural network of smooth hinge functions [13], the generalized flux-charge memristor model can properly fit the experimental data of the memristor device. The smooth hinge function $\ln(1+\exp(a_k q(t) + b_k i(t) + c_k)$ is differentiable in the domain. This advantage may make further analysis of the memristor easier, compared with the existing piecewise linear model, as will be shown in the following example.

Our generalized flux-charge memristor model also has good extensibility. Parametric Variables other than i or v, such as power p [8], can be easily added into the neural network of smooth hinge functions, in order to get a more precise description of the memristor device. At this situation, the base function becomes $\ln(1+\exp(a_k q(t) + b_k i(t) + c_k p(t) + d_k)$. Adding new variables will not change the aforementioned representation capability and smooth characteristics of the model [13]. For possible multi valued flux-charge relation, using the masked input technique given in [14], our model may still be able to describe such flux-charge relation.

3. Application Examples of Generalized Flux-Charge Relation Model

As analyzed in the last section, the representation capability of our generalized flux-charge relation model is theoretically guaranteed by the property of the neural network of smooth hinge functions. In this section we give three examples to show the representation capability of our model. We use the model to approximate two existing memristor models, a piecewise linear flux-charge relation model and a memristive system model. We also use our model to fit the voltage dependent flux-charge relation of physical memristor device based on experimental data.

In [1], [3], [15], the following piecewise linear flux-charge relation is used to characterize a memristor

$$\varphi(q) = bq + 0.5(a-b)(|q+1| - |q-1|). \tag{11}$$

a, b are the parameters of the model, and we can suppose $a \neq b$ for general cases [1]. We show that our smooth model can properly approximate the piecewise linear flux-charge relation given by (11). First, we equivalently represent (11) with hinge functions [16]

$$\varphi(q) = aq + (b-a)\max\{0, q-1\} - (b-a)\max\{0, -q-1\}. \tag{12}$$

Then according to [13], (12) can be approximated by

$$\begin{aligned} \varphi(q) &= aq + (b-a)\ln(1+\exp\alpha(q-1))/\alpha \\ &\quad -(b-a)\ln(1+\exp\alpha(-q-1))/\alpha, \end{aligned} \tag{13}$$

as shown in Fig. 1. In (13), increasing parameter α can reduce the differences between a hinge function $\max\{0, x\}$ and a smooth hinge function $\ln(1+\exp\alpha x)/\alpha$ around the hinge $x = 0$ [13]. The pinched hysteresis loops of both models are also shown in Fig. 1. It can be seen that our model can properly approximate the existing piecewise linear model.

There are limitations of the existing piecewise linear model. One is that (12) is not differentiable at $q = \pm 1$, which will cause a sudden memristance change at $q = \pm 1$ (i.e., $M(q) = \mathrm{d}\varphi(q)/\mathrm{d}q$ in (2),(3) will discontinuously change between a and b). Our model does not have such limitation since we use the neural network of smooth hinge functions. A smooth hinge function $\ln(1+\exp x)$ has a continuous derivative $1/1+\exp(-x)$, and therefore (13) has a continuous derivative (i.e., $M(q)$) over the domain including $q = \pm 1$. Another limitation of the existing piecewise linear flux-charge relation model is that it may not properly represent the flux-charge relation of physical memristor device (we will show that in the following example).

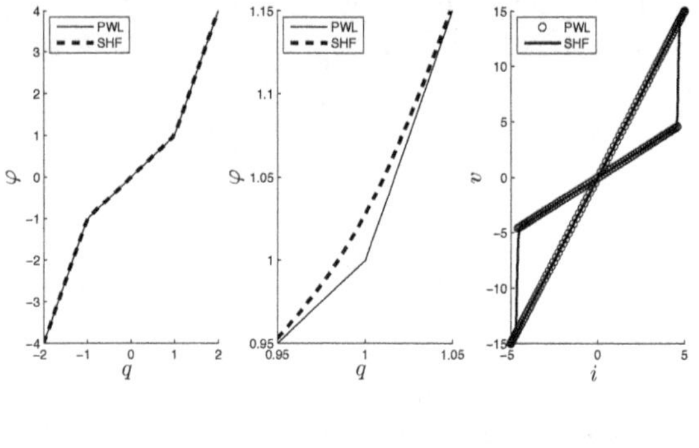

Fig. 1. Approximating existing piecewise linear (PWL) model (12) with smooth hinge function (SHF) model (13). $a = 1$, $b = 3$, $\alpha = 50$. The left figure shows the flux-charge relations of both models. The middle figure shows the local details of both flux-charge relations around $q = 1$: (12) is not differentiable at $q = 1$, while (13) is smooth. The right figure shows the pinched hysteresis loops of both models with input $i(t) = 5sin(\pi t)$.

Next we use our model to approximate an existing memristive system model. The following window function model is given by [17] to describe the first memristor device found by [6]:

$$v(t) = (R_{ON}x(t) + R_{OFF}(1 - x(t)))i(t), \quad (14a)$$

$$\frac{\mathrm{d}x(t)}{\mathrm{d}t} = k(1 - (2x(t) - 1)^{2p})i(t) \quad (14b)$$

where $x(t)$ is the state variable, k and p are the parameters. This memristive system is a simplified model of the device and widely used. According to [18], there exists a function relation of φ and q of (14). More discussion on window function model can be found in [14], [19], [20], [21]. Here we plot the $\varphi - q$ curve of (14) through simulation, and use our model to approximate this $\varphi - q$ curve, as shown in Fig. 2. Specifically, our model is given by

$$\varphi(q) = a_0 q + b_0 - \ln(1 + \exp\alpha(a_1 q + b_1))/\alpha. \quad (15)$$

Note that for a small q, $\varphi(q) \approx a_0 q + b_0$, for a large q, $\varphi(q) \approx (a_0 - a_1)q + b_0 - b_1$ and $v/i = \mathrm{d}\varphi(q)/\mathrm{d}q$. The parameters are chosen such that $a_0 = R_{ON}$, $a_0 - a_1 = R_{OFF}$, $a_0 q + b_0$ is the approximating line for (q, φ) data point with a small q and $(a_0 - a_1)q + b_0 - b_1$ is the approximating line for (q, φ) data point with a large q. α can adjust the degree of bending of (14) between line $a_0 q + b_0$ and line $(a_0 - a_1)q + b_0 - b_1$.

It is easy to see that the existing piecewise linear model (12) may not properly approximate $\varphi - q$ curve of (14) (Circle points in Fig. 2). Because circle points in Fig. 2 do not have symmetry as (12). Our model can properly approximate the circle points. The pinched hysteresis loops show that our model is a good approximation of memristive system model (14). In this example, only 1 base function of the neural network of smooth hinge functions is used. Using other kinds of neural network instead of the neural network of smooth hinge functions may increase the complexity of the corresponding approximation model, as more base functions and parameters may be needed.

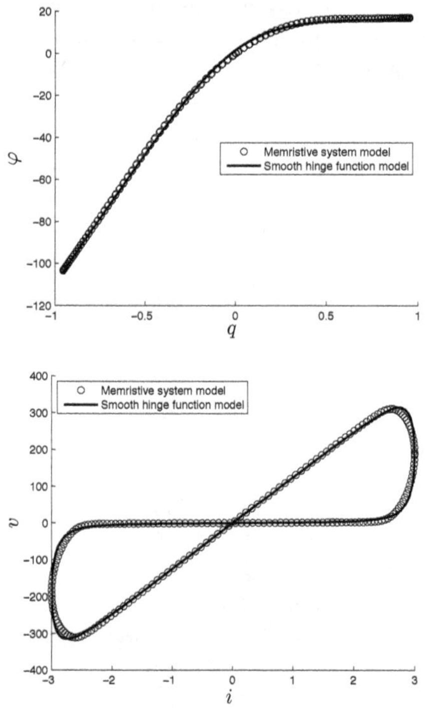

Fig. 2. Approximating existing memristive system model (14) with smooth hinge function model (15). $R_{ON} = 1$, $R_{OFF} = 125$, $k = 1$, $p = 5$, $\alpha = 0.05$, $a_0 = 1$, $a_1 = -124$, $b_0 = 15.5$, $b_1 = 0$. The top figure shows the flux-charge relations of both models. The bottom figure shows the pinched hysteresis loops of both models with input $i(t) = 3\sin(\pi t + \pi/2)$.

In the third example, we show that our generalized model is suitable to model physical memristor devices. Specifically, we model the voltage dependent flux-charge relation of physical AgInSbTe memristor device [9]. The co-existence of extrinsic electrochemical metallization effect and intrinsic memristive characteristics was confirmed in the AgInSbTe memristor [9]. In the gradual resistance tuning of the device, pulses with different voltage amplitudes and 5 μs width were applied to the AgInSbTe memristor device [9]. From the experiment data, we calculate the (φ, q) data for each pulse of different voltage amplitudes. Voltage dependent flux-charge relation is observed, as shown in Fig. 3.

The flux-charge relation of the device varies with different voltage amplitudes. $q = q(\varphi)$ can not describe such voltage dependent flux-charge relation. Our generalized $q = q(\varphi, v)$ model is needed. We use our generalized model in the form of (10) to approximate such voltage dependent flux-charge relation. From Fig. 3 we can see that our model fits the data well. The number of smooth hinge functions m and the parameters a_i, b_i, c_i, $i = 1, \ldots, m$ in (10) are artificially selected. Then parameters a_0, b_0, c_0 and η_i, $i = 1, \ldots, m$ are calculated by least squares method based on the experimental data.

The behavior of physical AgInSbTe memristor device is complicated. Besides the amplitude of the pulse, the pulse width also affects the gradual resistance tuning of the AgInSbTe memristor [9]. Pulses with $-1V$ amplitude and different widths were applied to the AgInSbTe memristor [9]. As shown in Fig. 4, the flux-charge relation of the device varies with the pulse widths. We can analogously model such flux-charge relation by replacing the variable v in (10) with pulse width Δ. Then the base function of $q = q(\varphi, \Delta)$ is in the form of $\ln(1 + \exp(a_k \varphi(t) + b_k \Delta + c_k))$. Fig. 4 shows that our model fits the data properly.

4. Conclusions

In this paper, we propose the generalized flux-charge relation model of memristor considering that the memristor was originally proposed to characterize the flux-charge relation. Such generalization is little studied, but the example of voltage dependent flux-charge relation of the AgInSbTe memristor indicates that such generalization is necessary.

The usage of neural network of smooth hinge functions theoretically guarantees the representation capability of the model. Any functional flux-charge relation, even with multiple parametric variables, can be approximated by our model. With examples, we show that our model is capable of representing the existing memristor models, and approximating voltage dependent flux-charge relation of physical memristor device.

Besides the three examples given in this paper, our model can be applied to model other types of memristors. Because the representation capability of the model is theoretically guaranteed and there is effective identification algorithm for the neural network of smooth hinge functions [13]. Given the (φ, q) data of the specific memristor, our model can be applied to model the memristor.

Acknowledgements

This project is jointly supported by the National Natural Science Foundation of China (61074118, 61134012) and the National Basic Research Program of China (2012CB720505).

The authors would like to thank Y. Li and other authors of [9] for providing the data of the AgInSbTe memristor.

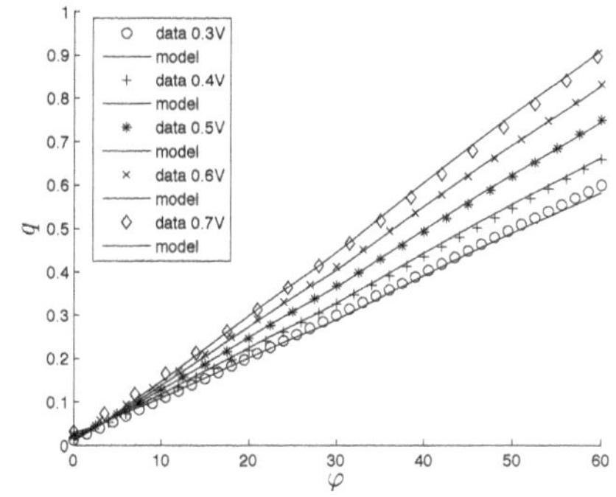

Fig. 3. Approximate the voltage dependent flux-charge relation of the AgInSbTe memristor. The $\varphi(t)$ at the nth pulse is calculated by $\varphi(t) = nA_v\Delta$, where A_v is the amplitude and Δ is the width of the pulse. Similarly, $q(t)$ at the nth pulse is calculated by $q(t) = \Sigma_{k=1}^{n} i_k\Delta$, where i_k is the current measured at the kth pulse. 10 smooth hinge base functions are used in the model.

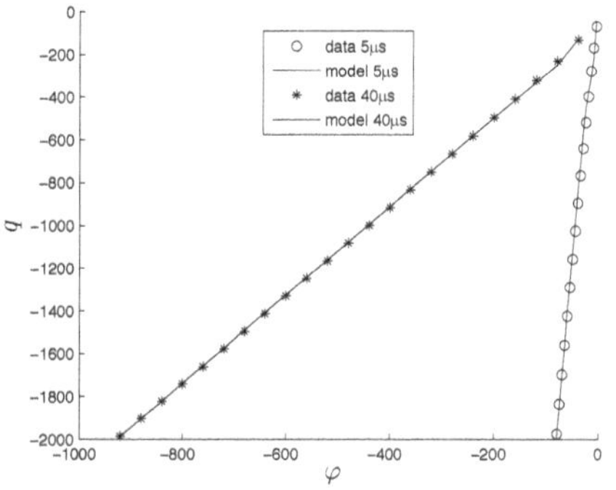

Fig. 4. Approximate the flux-charge relation of the AgInSbTe memristor with different pulse widths. The $\varphi(t)$ at the nth pulse is calculated by $\varphi(t) = nA_v\Delta$, where A_v is the amplitude and Δ is the width of the pulse. Similarly, $q(t)$ at the nth pulse is calculated by $q(t) = \Sigma_{k=1}^{n} i_k\Delta$, where i_k is the current measured at the kth pulse. 4 smooth hinge base functions are used in the model.

References

[1] CHUA, L. O. Memristor – the missing circuit element. *IEEE Transactions on Circuit Theory*, 1971, vol. 18, no. 5, p. 507 - 519.

[2] MAZUMDER, P., KANG S. M., WASER R. Memristors: Devices, models, and applications. *Proceedings of the IEEE*, 2012, vol. 100, no. 6, p. 1911 - 1919.

[3] CHUA, L. O. Resistance switching memories are memristors. *Applied Physics A: Materials Science & Processing*, 2011, vol. 102, no. 4, p. 765 - 783.

[4] CHUA, L. O., KANG, S. M. Memristive devices and systems. *Proceedings of the IEEE*, 1976, vol. 64, no. 2, p. 209 - 223.

[5] VENTRA, M. D., PERSHIN, Y. V., CHUA, L. O. Circuit elements with memory: Memristors, memcapacitors, and meminductors. *Proceedings of the IEEE*, 2009, vol. 97, no. 10, p. 1717 - 1724.

[6] STRUKOV, D. B., SNIDER, G. S., STEWART, D. R., WILLIAMS, R. S. The missing memristor found. *Nature*, 2008, vol. 453, no. 7191, p. 80 - 83.

[7] PERSHIN, Y. V., VENTRA, M. D. Spin memristive systems: Spin memory effects in semiconductor spintronics. *Physical Review B*, 2008, vol. 78, no. 11, p. 113309.

[8] STRACHAN, J. P., TORREZAN, A. C., MIAO, F., PICKETT, M. D., YANG, J. J., YI, W., MEDEIROS-RIBEIRO, G., WILLIAMS, R. S. State dynamics and modeling of tantalum oxide memristors. *IEEE Transactions on Electron Devices*, 2013, vol. 60, no. 7, p. 2194 - 2202.

[9] ZHANG, J., SUN, H., LI, Y., WANG, Q., XU, X., MIAO, X. Aginsbte memristor with gradual resistance tuning. *Applied Physics Letters*, 2013, vol. 102, no. 18, p. 183513.

[10] PICKETT, M. D., STRUKOV, D. B., BORGHETTI, J. L., YANG, J. J., SNIDER, G. S., STEWART, D. R., WILLIAMS, R. S. Switching dynamics in titanium dioxide memristive devices. *Journal of Applied Physics*, 2009, vol. 106, no. 7, p. 074508.

[11] YAKOPCIC, C., TAHA, T. M., SUBRAMANYAM, G., PINO, R. E., ROGERS, S. A memristor device model. *IEEE Electron Device Letters*, 2011, vol. 32, no. 10, p. 1436 - 1438.

[12] KVATINSKY, S., FRIEDMAN, E. G., KOLODNY, A., WEISER, U. C. Team: threshold adaptive memristor model. *IEEE Transactions on Circuits and Systems I: Regular Papers*, 2013, vol. 60, no. 1, p. 211 - 221.

[13] WANG, S., HUANG, X., YAM, Y. A neural network of smooth hinge functions. *IEEE Transactions on Neural Networks*, 2010, vol. 21, no. 9, p. 1381 - 1395.

[14] SHIN, S., KIM, K., KANG, S. Compact models for memristors based on charge-flux constitutive relationships. *IEEE Transactions on Computer-Aided Design of Integrated Circuits and Systems*, 2010, vol. 29, no. 4, p. 590 - 598.

[15] ITOH, M., CHUA, L. O. Memristor oscillators. *International Journal of Bifurcation and Chaos*, 2008, vol. 18, no. 11, p. 3183 - 3206.

[16] BREIMAN, L. Hinging hyperplanes for regression, classification, and function approximation. *IEEE Transactions on Information Theory*, 1993, vol. 39, no. 3, p. 999 - 1013.

[17] JOGLEKAR, Y. N., WOLF, S. J. The elusive memristor: Properties of basic electrical circuits. *European Journal of Physics*, 2009, vol. 30, no. 4, p. 661 - 675.

[18] CORINTO, F., ASCOLI, A. A boundary condition-based approach to the modeling of memristor nanostructures. *IEEE Transactions on Circuits and Systems I: Regular Papers*, 2012, vol. 59, no. 11, p. 2713 - 2726.

[19] BIOLEK, Z., BIOLEK, D., BIOLKOVÁ, V. Analytical solution of circuits employing voltage- and current- excited memristors. *IEEE Transactions on Circuits and Systems I: Regular Papers*, 2012, vol. 59, no. 11, p. 2619 - 2628.

[20] BIOLEK, Z., BIOLEK, D., BIOLKOVÁ, V. Spice model of memristor with nonlinear dopant drift. *Radioengineering*, 2009, vol. 18, no. 2, p. 210 - 214.

[21] YU, J., MU, X., XI, X, WANG, S. A memristor model with piecewise window function. *Radioengineering*, 2013, vol. 22, no. 4, p. 969 - 974.

About Authors ...

Xiaomu MU received the B.S. degree in Control Science and Engineering from Tsinghua University, Beijing, China in 2009. He is currently a Ph.D. candidate in the Department of Automation, Tsinghua University, Beijing, China. His current research areas include nonlinear dynamic system control, piecewise linear identification and memristor device.

Juntang YU received the B.S. degree in Control Science and Engineering from Harbin Institute of Technology, Harbin, China in 2011. He is currently pursuing the Ph.D. degree from the Department of Automation, Tsinghua University, Beijing, China. His current research interests lie in the area of modeling and analysis of memristor circuits and systems.

Shuning WANG received the B.S. degree in Electrical Engineering from Hunan University, China in 1982, the M.S. degree and the Ph.D. degree both in System Engineering from Huazhong University of Science and Technology, China, in 1984 and 1998, respectively. He was an Associate Professor from 1992 to 1993 and a Full Professor from 1994 to 1995, at the Institute of Systems Engineering, Huazhong University of Science and Technology. He joined Tsinghua University, Beijing, China in 1996. Since then, he has been a Full Professor in Department of Automation, Tsinghua University. He was a Visiting Scholar in the College of Engineering, University of California at Riverside in 1994, and a Visiting Fellow in Department of Electrical Engineering, Yale University from 2001 to 2002. His current research interests are mainly in developing practical methods for nonlinear system identification, control and optimization via piecewise-linear approximation.

7

Simple Floating Voltage-Controlled Memductor Emulator for Analog Applications

Mohamed E. FOUDA[1], *Ahmed G. RADWAN*[1,2]

[1]Dept. of Engineering Mathematics, Faculty of Engineering, Cairo University, Giza, Egypt,12613
[2]Nano-electronic Integrated Systems Center (NISC), Nile University, Giza, Egypt

m_elneanaei@ieee.org, agradwan@ieee.org

Abstract. *The topic of memristive circuits is a novel topic in circuit theory that has become of great importance due to its unique behavior which is useful in different applications. But since there is a lack of memristor samples, a memristor emulator is used instead of a solid state memristor. In this paper, a new simple floating voltage-controlled memductor emulator is introduced which is implemented using commercial off the shelf (COTS) realization. The mathematical modeling of the proposed circuit is derived to match the theoretical model. The proposed circuit is tested experimentally using different excitation signals such as sinusoidal, square, and triangular waves showing an excellent matching with previously reported simulations.*

Keywords

Memristor, memductor, mem-element, memristive circuits, emulator, mutator.

1. Introduction

The history of mem-elements was initiated by postulating the existence of the memristor (M) by Chua in his seminal paper in 1971 [1] despite the fact that the memristive behavior was experimentally discovered two centuries ago [2]. Then the basic concept of the memristive devices was introduced by Chua and Kang in 1976 [3]. Later in 2008, a team in HP labs announced that the first solid state memristor was discovered and they introduced the first model of the memristor based on the measurements [4]. The memristor represents the missing relation between the charge $q(t)$ and the flux linkage $\varphi(t)$ that has different characteristics than the well-known elements: resistor (R), capacitor (C) and inductor (L).

Memristive based circuits can participate in very vital applications such as nonvolatile resistive random access memory (ReRAM) [5], [6], analog circuits [7], [8], digital circuits [9], [10], relaxation oscillators [11], [12], [13], chaotic generators [14], and neuromorphic synaptic networks [15].

SPICE models of memristor are essential to design and analyze the complex circuits [16], [17], [18]. New memristive circuit ideas are commonly validated using SPICE models of the memristors first before doing any measurements using solid-state samples [16], [18], [19]. These models study the effect of the boundary of the memristor which models the real solid state memristors.

Due to the lack of commercially available solid state samples, emulator circuits are used to physically mimic the nonlinear dynamics of the mem-elements in various applications. Although, there are many SPICE models which are very useful only for circuit simulation, emulators are needed to be realized in labs in order to enable experimental measurements. Recently, there has been a very intensive research on mem-elements emulators for example, the memristor emulator [20], [21], memcapacitor emulator [22], [23] and meminductor [24]. These emulators were introduced depending on current-, charge- and current-controlled models respectively. The previously published memristor emulators are developed based on current-controlled models but there is no potential research on voltage-controlled models. So, in this work, we are introducing the voltage-controlled memductor emulator for the first time in addition to validation using different excitation signals. In addition, the previous emulators are built using bulky components such as current mirrors, multipliers and dividers. In this work, however, the emulator is built using two simple discrete components and some resistors and capacitors.

This paper is organized as follows: Section 2 discusses the voltage-controlled model of the memductor based on previously reported models as well as the basic building blocks of the memductor. In Section 3, the circuit realization of the memductor is presented with the mathematical modeling of the circuit. In Section 4, the PSPICE simulation and experimental measurements of the proposed circuit are introduced for different voltage excitation signals.

2. Memductor Model

The current-controlled memristor model was discussed in [4] where the constitutive relationship is between the charge q and the flux-linkage φ; and the memristance is a function of the state variable, current and time. Similarly the model of the memductor was discussed in [25] where the transconductance G_m changes depending on the the accumulated flux so it is called memductor (short for memory + conductance). The change rate of memductance G_m is given as follows:

$$\dot{G}_m(\varphi) = \alpha H(\varphi) v_m \tag{1}$$

where α is the proportionality constant ($\Omega^{-1}V^{-1}s^{-1}$) and $H(\varphi)$ is considered as a normalized window function having the non-idealities of the memductance change rate. For the sake of simplicity, let's assume that $H(\varphi) = 1$ representing the linear model of the memductor. By integrating both sides relative to time, the memductance is given by

$$G_m = G_{mo} + \alpha\varphi(t) \tag{2}$$

where G_{mo} is the initial memductance. The memductance is linearly proportional to the accumulated flux.

Besides, in [20], the authors introduced a simple double hysteresis model for the mem-elements where the voltage controlled memductor equation is given by

$$i_m = G_s v_m + \frac{G_s v_m}{T V_{ref}} \varphi(t) \tag{3}$$

where G_s is the initial transconductance, T is the integration factor, and V_{ref} is an arbitrary reference. So the memductance is given by

$$G_m = G_s + \frac{G_s}{T V_{ref}} \varphi(t). \tag{4}$$

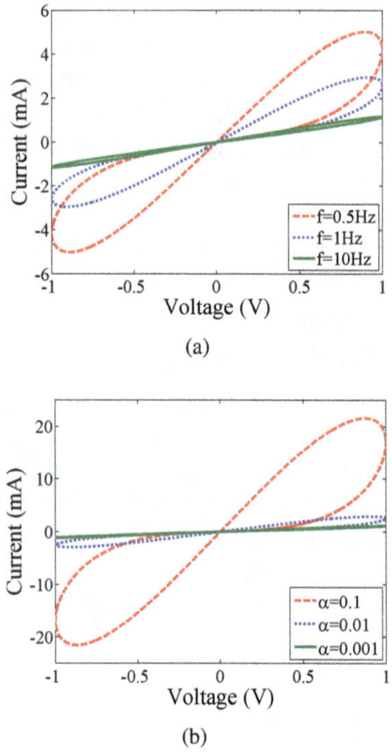

Fig. 1. I-V hysteresis of the memductor for $G_{mo} = 1\ \text{m}\Omega^{-1}$ for different a) frequencies at $\alpha = 0.01$ and b) α at 1 Hz frequency.

Obviously, both models give the same modeling equation for the memductor where $\alpha = G_s/(TV_{ref})$.

Numerical simulations up on changing different parameters of the memductor are shown in Figs. 1 and 2 for sinusoidal input voltage $v(t)$ with 1 V amplitude and $G_{mo} = 10^{-3}\ \Omega^{-1}$. Fig. 1a shows the current-voltage characteristics of the memductor for three different frequencies where the area inside the hysteresis loops decreases by increasing the applied signal frequency. Moreover, Fig. 1b shows the current-voltage characteristics for three different values of α. Also, it is noted that the hysteresis loops size is dependent on the value of α where the hysteresis loops shrink by decreasing α until it tends to zero and the memductance tends to its initial value G_{mo}. Otherwise, the maximum memductance is given by $G_m = G_{mo} + 2\alpha\frac{A}{\omega}$ for sinusoidal input. The maximum value of the memductance G_m is shown in Fig. 2, for a range of α spanning from 0.001 to 0.1 and for the frequency range of the input signal from 0.01 Hz to 100 Hz.

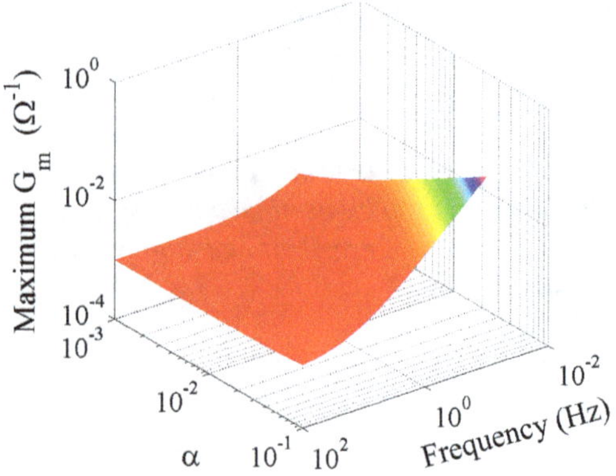

Fig. 2. The maximum memductance for different frequencies and α.

In oder to implement the memductor in which the behavior of memductance is controlled by flux-linkage, a voltage controlled transconductance is needed in addition to a differential voltage integrator to integrate the voltage across the transconductance and generate the flux which controls the transconductance as shown in Fig. 3.

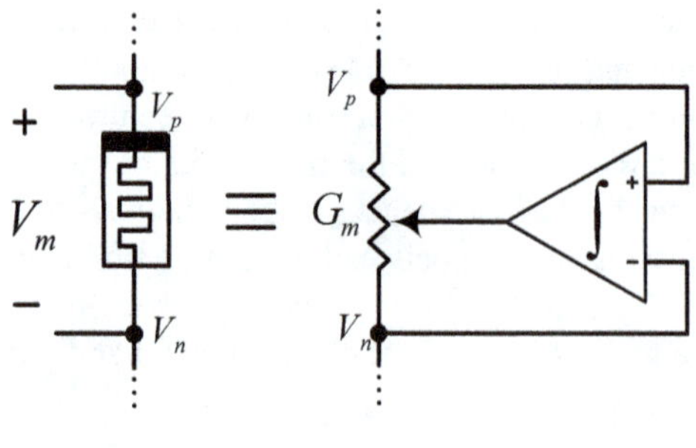

Fig. 3. Behavioral model of linear memductor.

Otherwise, the non-ideal model of the memductor can be implemented by adding a window function $H(\varphi)$ after the integrator to reshape the control voltage and boundary effect of the model.

3. Proposed Emulator Realization

The voltage-controlled transconductance is implemented using LM13700 [26] which is connected as shown in Fig. 4 to implement a floating voltage controlled transconductance G_m where G_m is proportional to the control voltage V_c where its transconductance is given by

$$G_m = 9.6 I_{ABC} \frac{R_A}{R} \tag{5}$$

where I_{ABC} represents the transconductance amplifier bias current. From the PSPICE simulation, it is found that I_{ABC} is linearly proportional to the control voltage ($I_{ABC} = aV_c + I_o$) where a represents the reciprocal of the control resistance R_c and I_o is the initial current. The data sheet of the transconductance [26] states that it is recommended to use $R_c = 15$ kΩ and as a result the corresponding initial current $I_o = 905.057$ μA (These values might vary due to the output offset of the OPAMP). Substituting by I_{ABC} into (5). The transconductance G_m is given by

$$G_m = \frac{0.64 R_A}{R} V_c + 8.6885 \frac{R_A}{R} (\mathrm{m}\Omega^{-1}). \tag{6}$$

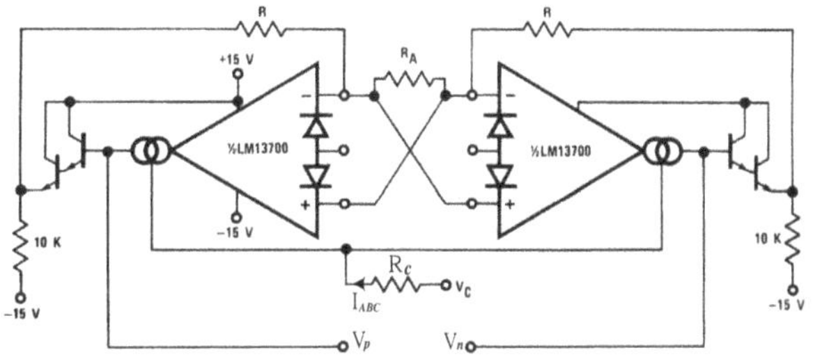

Fig. 4. Floating voltage controlled transconductance implementation using LM13700, adopted from [26].

By comparing (2) and (6), the control voltage V_c should represents the flux linkage of the memductor. The flux linkage can be obtained by integrating the difference voltage of the memductor terminals (V_p, V_n) and is given as follows:

$$V_c = \frac{1}{R_1 C_1} \int_{-\infty}^{t} (V_p - V_n) d\tau = \frac{1}{R_1 C_1} \varphi_{pn}(t). \tag{7}$$

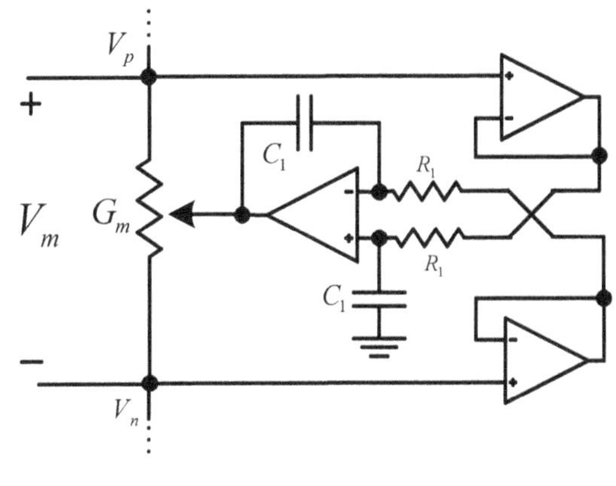

Fig. 5. Generation of the flux control circuit of the memductor.

The integrator circuit is built as shown in Fig. 5, where an inverting integrator is used and two buffer amplifiers to prevent the loading effect of the integrator on the transconductance circuit. By substituting into G_m, the transconductance is given by

$$G_m = 0.64 \frac{R_A}{R R_1 C_1} \varphi(t) + 8.6885 \frac{R_A}{R} (\mathrm{m}\Omega^{-1}). \tag{8}$$

The previous equation emulates the memductor's equation where $G_{mo} = 8.6885 \frac{R_A}{R} (\mathrm{m}\Omega^{-1})$ and $\alpha = 0.64 \frac{R_A}{R R_1 C_1} (\mathrm{m}\Omega^{-1} V^{-1} s^{-1})$.

4. Experimental Results

The circuit is practically assembled on a printed circuit board using LM13700 and TL084 (OPAMP) shown in Fig. 6 where C_1, R_1, R and R_A equal 1 μF, 1 kΩ, 100 kΩ and 10 kΩ respectively. The emulator is tested using NI ELVIS Kit and the voltage results are taken to MATLAB to plot current-voltage hysteresis (as NI ELVIS doesn't plot Lissajous curves). In order to obtain the input current of the emulator, a series resistor is connected where the input current is proportional to the voltage difference across it V_R with gain equals to the inverse of the resistor's value R as shown in Fig. 6(c). In the following measurements, a 1 kΩ resistor is used.

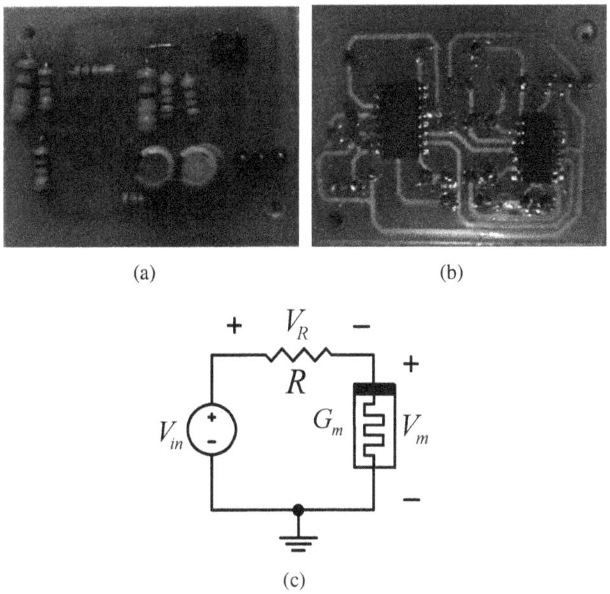

Fig. 6. Printed circuit board of voltage-controlled memristor emulator and circuit setup.

In case of the memductor series with a resistor R, the voltage across the memductor $v_m = V_{in}/(1 + G_m R)$ where v_{in} is the input voltage, and G_m is the memducatance. By substituting into (2), integrating both sides, and simplifying the resulting equation. The memductance is given as follows:

$$G_m = -\frac{1}{R} + \sqrt{\left(\frac{1}{R} + G_{mo}\right)^2 + \frac{2\alpha}{R} \varphi(t)}. \tag{9}$$

The memductance is a function of the time integral of the applied voltage signal , flux, so it is very important to study the effect of the basic analog signals: sinusoidal signal and periodic signal such as square and triangular waveforms. Then, substituting the flux to (9) leads to a closed form expression for the instantaneous memductance. In case of applying a sinusoidal signal with amplitude A_o and frequency ω, the flux is given by

$$\varphi(t) = \frac{2A_o}{\omega}\sin^2(\frac{\omega}{2}t) + \varphi_o. \tag{10}$$

Moreover, in the case of a square signal with A_1 amplitude for positive half cycle and A_2 amplitude for negative half cycle with zero average value, the flux is given by

$$\varphi(t) = \varphi_o + \begin{cases} A_1\tau & \tau \leq T_h, \\ (A_1 + A_2)\tau - A_2 T_h & T_h < \tau \leq T \end{cases} \tag{11}$$

where $\tau = mod(t, T)$. These equations give similar results to the experimental results using appropriate values of G_{mo} and α.

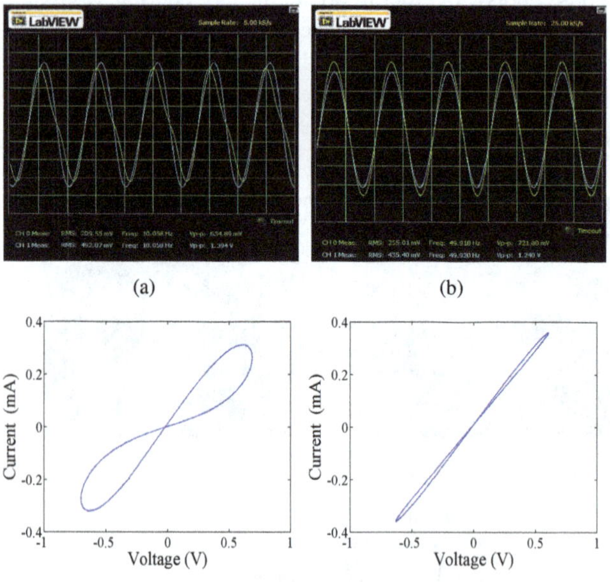

Fig. 7. Emulator response under sinusoidal signal at frequencies 10 Hz and 50 Hz.

The proposed emulator is tested using different voltage excitation signals: sinusoidal signal with frequency 10 Hz and 50 Hz, square wave signal with frequency 10 Hz and triangular wave signal with frequency 10 Hz. Figures 7(a) and 7(b) show the transient voltage V_m (blue line) and current V_R (green line) of the emulator for sinusoidal signal with amplitude 1 volt, also Figs. 7(c) and 7(d) show the corresponding double-loop pinched I-V hysteresis of the memductor which shrinks with increasing the input frequency.

Besides, It is very important to study the effect of periodic signal on the proposed emulator especially square wave signal because of hard switching effect. So a square wave signal with amplitude ± 0.5 Volt with 10 Hz frequency is applied as shown in Fig. 8 where the transient voltage v_m (blue line) and current (V_R/R) (green line) showing a high functionality of the emulator to be used in memristor based relaxation oscillators [12], [13]. The voltage across the memductor increases / decreases for positive / negative pulse where the memductance decreases / increases respectively. Moreover, Fig. 9 shows the response of the triangular wave excitation with amplitude ± 1 Volt with 10 Hz frequency and I-V hysteresis in case of triangular excitation.

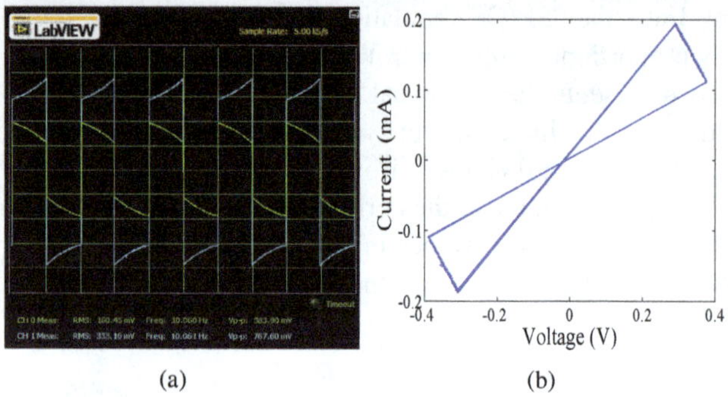

Fig. 8. Emulator response under square wave signal at frequency 10 Hz; a) transient voltage and current, and b) corresponding I-V hysteresis.

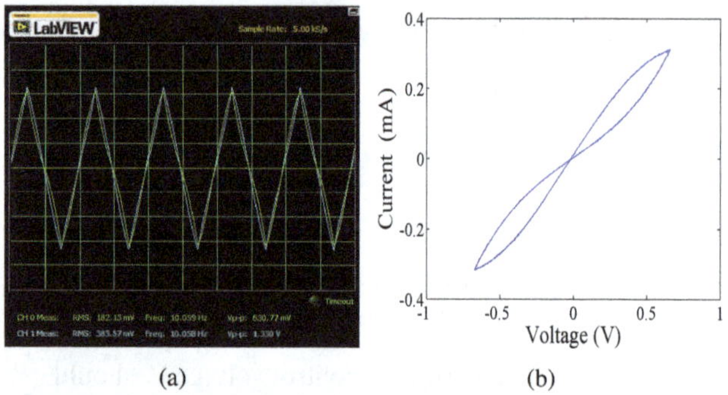

Fig. 9. Emulator response under triangular wave signal at frequency 10 Hz; a) transient voltage and current, and b) corresponding I-V hysteresis.

5. Conclusion

This paper discussed a new simple voltage-controlled memductor emulator circuit where its mathematical model was derived. The effect of changing the emulator parameters was discussed. Moreover, the emulator was assembled experimentally and its functionality was verified for different excitation signals.

References

[1] CHUA, L. Memristor-the missing circuit element. *IEEE Transactions on Circuit Theory*, 1971, vol. 18, no. 5, p. 507 - 519.

[2] PRODROMAKIS, T., TOUMAZOU, C., CHUA, L. Two centuries of memristors. *Nature Materials*, 2012, vol. 11, no. 6, p. 478.

[3] CHUA, L., KANG, L. Memristive devices and systems. *Proceedings of the IEEE*, 1976, vol. 64, no. 2, p. 209 - 223.

[4] STRUKOV, D., SNIDER, G., STEWART, D., WILLIAMS, R. The missing memristor found. *Nature*, 2008, vol. 453, p. 80 - 83.

[5] ESHRAGHIAN, K., CHO, K., KAVEHEI, O., KANG, S., ABBOTT, D., KANG, S. Memristor MOS content addressable memory (MCAM): Hybrid architecture for future high performance. *IEEE Transactions on Very Large Scale Integration (VLSI) Systems*, 2011, vol. 19, no. 8, p. 1407 - 1417.

[6] VONTOBEL, P., ROBINETT, W., KUEKES, P., STEWART, D., STRAZNICKY, J., WILLIAMS, R. Writing to and reading from a nano-scale crossbar memory based on memristors. *Nanotechnology*, 2009, vol. 20, no. 42, p. 425204.

[7] PERSHIN, Y., DI VENTRA, M. Practical approach to programmable analog circuits with memristors. *IEEE Transactions on Circuits and Systems I: Regular Papers*, 2010, vol. 57, no. 8, p. 1857 - 1864.

[8] SHIN, S., KIM, K., KANG, S. Memristor applications for programmable analog ICs. *IEEE Transactions on Nanotechnology*, 2011, vol. 10, no. 2, p. 266 - 274.

[9] XIA, Q., ROBINETT, W., CUMBIE, M., BANERJEE, N., CARDINALI, T., YANG, J., WU, W., LI, X., TONG, W., STRUKOV, D., et al. Memristor CMOS hybrid integrated circuits for reconfigurable logic. *Nano Letters*, 2009, vol. 9, no. 10, p. 3640 - 3645.

[10] SHALTOOT, A., MADIAN, A. Memristor based carry lookahead adder architectures. In *2012 IEEE 55th International Midwest Symposium on Circuits and Systems (MWSCAS)*. 2012, p. 298 - 301.

[11] ZIDAN, M. A., OMRAN, H., RADWAN, A. G., SALAMA, K. N. Memristor-based reactance-less oscillator. *Electronics Letters*, 2011, vol. 47, no. 22, p. 1220 - 1221.

[12] FOUDA, M. E., KHATIB, M., MOSAD, A., RADWAN, A. Generalized analysis of symmetric and asymmetric memristive two-gate relaxation oscillators. *IEEE Transactions on Circuits and Systems I: Regular Papers*, 2013, vol. 60, no. 10, p. 2701 - 2708.

[13] FOUDA, M., RADWAN, A. Memristor-based voltage-controlled relaxation oscillators. *International Journal of Circuit Theory and Applications*, 2013, DOI: 10.1002/cta.1907.

[14] BUSCARINO, A., FORTUNA, L., FRASCA, M., GAMBUZZA, L. V. A gallery of chaotic oscillators based on HP memristor. *International Journal of Bifurcation and Chaos*, 2013, vol. 23, no. 5, p. 1330015.

[15] KOZMA, R., PINO, R. E., PAZIENZA, G. E. *Advances in Neuromorphic Memristor Science and Applications*, vol. 4. Springer, 2012.

[16] BIOLEK, Z., BIOLEK, D., BIOLKOVA, V. SPICE Model of memristor with nonlinear dopant drift. *Radioengineering*, 2009, vol. 18, no. 2, p. 210 - 214.

[17] PERSHIN, Y. V., DI VENTRA, M. SPICE Model of memristive devices with threshold, *Radioengineering*, 2013,vol. 22, no. 2, p. 485 - 489.

[18] BIOLEK, D., DI VENTRA, M., PERSHIN, Y. Reliable SPICE simulations of memristors, memcapacitors and meminductors. *Radioengineering*, 2013, vol. 22, no. 4, p. 945 - 968.

[19] JUNTANG YU, XIAOMU MU, XIANGMING XI, SHUNING WANG A memristor model with piecewise window function. *Radioengineering*, 2013, vol. 22, no. 4, p. 969 - 974.

[20] ELWAKIL, A., FOUDA, M., RADWAN, A. A simple model of double-loop hysteresis behavior in memristive elements. *IEEE Transactions on Circuits and Systems II: Express Briefs*, 2013, vol. 60, no. 8, p. 487 - 491.

[21] KIM, H., SAH, M., YANG, C., CHO, S., CHUA, L. Memristor bridge synapses. *IEEE Transactions on Circuits and Systems I: Regular Papers*, 2012, vol. 59, no. 10, p. 2422 - 2431.

[22] YU, D., LIANG, Y., CHEN, H., IU, H., Design of a practical memcapacitor emulator without grounded restriction. *IEEE Transactions on Circuits and Systems II: Express Briefs*, 2013, vol. 60, no. 4, p. 207 - 211.

[23] FOUDA, M., RADWAN, A. Charge controlled memristor-less memcapacitor emulator. *Electronics Letters*, 2012, vol. 48, no. 23, p. 1454 - 1455.

[24] LIANG, Y., CHEN, H., YU, D. S. A practical implementation of a floating memristor-less meminductor emulator. *IEEE Transactions on Circuits and Systems II: Express Briefs*, 2014, vol. 61, no. 5, p. 299 - 303.

[25] SHIN, S., ZHENG, L., WEICKHARDT, G., CHO, S., KANG, S. Compact circuit model and hardware emulation for floating memristor devices. *IEEE Circuits and Systems Magazine*, 2013, vol. 13, no. 2, p. 42 - 55.

[26] National Semiconductor. *LM13700 Dual Operational Transconductance Amplifiers with Linearizing Diodes and Buffers*, datasheet.

About Authors ...

Mohamed FOUDA received the B.Sc. degree (honors), in Electronics and Communications Engineering in 2011 from Faculty of Engineering, Cairo University, Cairo, Egypt. He is a teaching assistant at Engineering Mathematics and Physics Department, Cairo University, Egypt. He authored and co-authored 13 journal and conference papers. His research interests include Mem-elements based circuits, and analog circuits.

Ahmed RADWAN received the B.S. degree (honors) in Electronics Engineering, and the M.S. and Ph.D. degrees from Cairo University, Cairo, Egypt, in 1997, 2002, and 2006 respectively. His main research interests are in the fields of nonlinear circuit analysis, chaotic systems, fractional order systems, and memristor-based circuits. He is an Associate Professor in the Engineering Mathematics Department, Cairo University, Egypt and since Jan. 2013 he became the Director of the Technical Center for Job Creation (TCJC), Faculty of Engineering, Cairo University. In addition, he is with Nanoelectronics Integrated Systems Center (NISC), Nile University, Egypt. From 2008 to 2009, he was invited as a visiting professor with the Computational Electromagnetics Lab, ECE, McMaster University, Hamilton, ON, Canada. From 2010 to 2012, he was selected to be one form the pioneer research group in the King Abdullah University of Science and Technology (KAUST), Saudi Arabia. He introduced many generalized theorems in the fractional order circuits and electromagnetics. He is the coauthor of more than 100 international papers and six patents. Dr. Radwan received the Egyptian Government first class medal for achievements in the Mathematical Sciences in 2012. He is a senior member of IEEE.

8

Polarization Beam Splitter Based on Self-Collimation of a Hybrid Photonic Crystal

Fulya BAGCI[1], *Sultan CAN*[2], *Baris AKAOGLU*[1], *A. Egemen YILMAZ*[2]

[1] Dept. of Engineering Physics, Ankara University, 06100 Besevler, Ankara, Turkey
[2] Dept. of Electrical-Electronics Engineering, Ankara University, 06830 Golbasi, Ankara, Turkey

fbagci@eng.ankara.edu.tr, sultancan@ankara.edu.tr, akaoglu@eng.ankara.edu.tr, aeyilmaz@eng.ankara.edu.tr

Abstract. *A photonic crystal polarization beam splitter based on photonic band gap and self-collimation effects is designed for optical communication wavelengths. The photonic crystal structure consists of a polarization-insensitive self-collimation region and a splitting region. TM- and TE-polarized waves propagate without diffraction in the self-collimation region, whereas they split by 90 degrees in the splitting region. Efficiency of more than 75% for TM- and TE-polarized light is obtained for a polarization beam splitter size of only 17 μm × 17 μm in a wavelength interval of 60 nm including 1.55 μm.*

Keywords

Photonic crystals, self-collimation, photonic band gap

1. Introduction

Photonic crystals (PCs) have growing reputation due to their large variety of advantages within ultra-compact sizes [1], [2]. One of the most crucial properties of the PCs is having photonic band gap (PBG) property in which electromagnetic waves cannot propagate in any direction. Light of certain frequencies in a PBG can be guided in a photonic crystal waveguide in the lateral direction by the periodicity of the lattice and in the vertical direction by means of total internal reflection [3]. Besides, the interest to these materials led the researchers to discover new ways to control the flow of light through these structures. Of particular interest is the self-collimation effect, by which the electromagnetic waves can propagate without diffraction in a perfect photonic crystal in the absence of a waveguide. First proposed by Kosaka et al. [4] and promoted by Prather et al. [5] and others [6-10], the phenomenon of self-collimation has been employed in many schemes for a large variety of device applications ranging from bends, splitters, interferometers to filters and detectors. In particular, the self-collimation effect is very preferable for device implementations on-chip integrated photonic circuits due to its arbitrary beam routing efficiency without crosstalk.

In this study, a hybrid photonic crystal consisting of a polarization-independent self-collimation region and a splitting region is employed to achieve polarization splitting. The polarization-independent self-collimation (PIS) region is used to direct the TM- and TE-polarized waves simultaneously without diffraction. The splitting region is formed by changing the composition of the $Al_xGa_{1-x}As$ alloy in a rectangular region inside the PIS region, in which the refractive index of that region is changed to be inside the photonic band gap for one of the orthogonal modes. The equi-frequency contours and photonic band diagrams of the PC are obtained with plane-wave expansion (PWE) method, while the field propagation and splitting properties of the device are determined with finite-difference time-domain simulations (FDTD) [11].

2. Design and Method

The polarization beam splitter structure presented in this study is illustrated in Fig. 1. It consists of two regions, namely the PIS region and the splitting region. Both of these structures consist of a square lattice of air holes of the same radius in $Al_xGa_{1-x}As$ alloy; however with higher refractive index introduced in the splitting region by changing the composition (x). When kept in mind that, the special points of the Brillouin zone for a square lattice PC at the center, corner and face are known as Γ, M and X, the splitting region is aligned along the ΓM direction, whereas light is incident along the ΓX direction. When the TM- and TE-polarized light reaches the splitting region, the two orthogonal polarizations split by 90 degrees. The TE-polarized light follows the path 1 by virtue of the self-collimation, while the TM-polarized light makes a turn in the direction of path 2 due to the photonic band gap effect (Fig. 1).

In the preliminary calculations, the TM photonic band gap is found to be increasing up to $r = 0.36a$ and decreasing later as the radius of the holes increases. The radius of air holes is chosen as $r = 0.36a$, where a is the lattice constant so that a better overlap between the self-collimated wavelength region and photonic band gap region is provided. The lattice constant is arranged as 628 nm to render

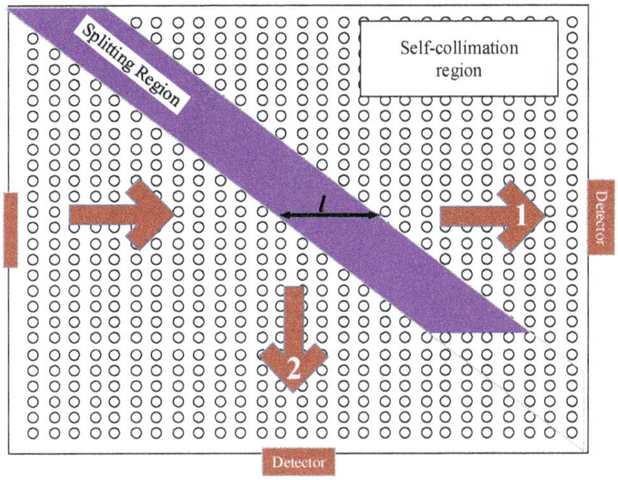

Fig. 1. The geometry of the proposed photonic crystal based polarization beam splitter structure.

the operation wavelength centered at 1.55 µm. Three dimensional PWE and FDTD calculations are approximated in two-dimensions with an effective index value of 2.4 for the PIS region and an effective index value of 2.8 for the splitting region of the PC [12].

3. Simulation and Analysis

The contours of constant frequency along the k_x and k_y wave-vector plane are called equi-frequency contours (EFCs). The group velocity is the velocity of energy transport and defined by $\vartheta_G = \nabla_k \omega(k)$, where k is the Bloch wave-vector. The group velocity is always perpendicular to the EFCs and it is aligned in the increasing $\omega(k)$ direction. Light propagates without diffraction in the flat (not curved) regions of the EFC. A mode can be identified by its unique combination (k, n), where n is called the band number in the band diagram. The EFCs of the second $(n = 2)$ TM (Fig. 2(a)) and TE band (Fig. 2(b)) for the self-collimated PC are square-shaped for the range of normalized frequencies, $\omega = 0.40(c/a) - 0.42(c/a)$, for both TM- and TE-polarizations. In these frequencies, polarization independent self-collimation can be observed along the faces of the square-shaped EFCs which are aligned along the ΓX and XM direction, as seen in Fig. 2.

Full-width half-maximums of the input and output beams are examined for the two orthogonal polarizations by FDTD simulations in order to verify the self-collimation effect that is predicted by the PWE method. As seen in Fig. 3, no remarkable broadening occurs for TM- and TE-polarized waves in the self collimated PC after propagation along 31.4 µm distance (50a).

Once the PIS region of the polarization beam splitter is designed, the properties of the splitting region are investigated in order to achieve the separation of the different polarizations. Fig. 4 shows the photonic band diagram of the splitting region for TM- and TE-modes. As the refractive index of the PC is increased to 2.8 in the

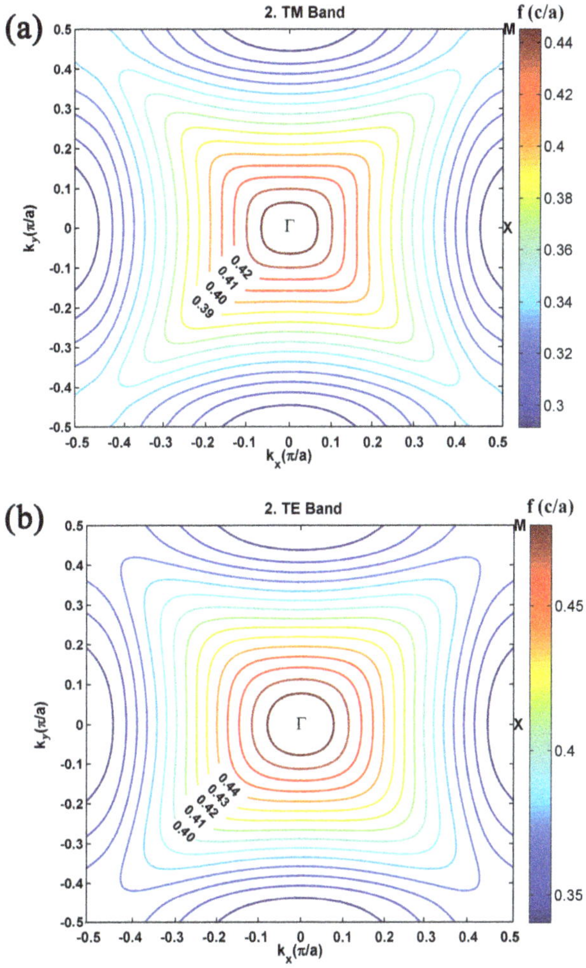

Fig. 2. Equi-frequency contours of the second band for (a) TM modes and (b) TE modes of the self-collimating photonic crystal.

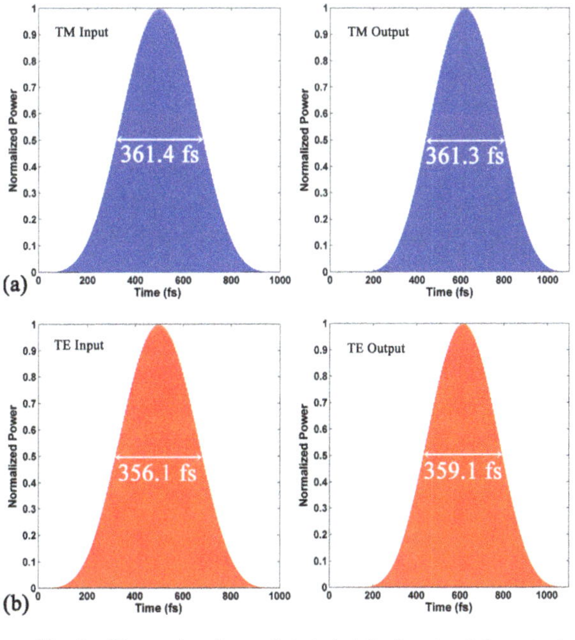

Fig. 3. Time-pulse shapes detected at the input and the output detecting points of the self-collimating photonic crystal for (a) TM modes and (b) TE modes.

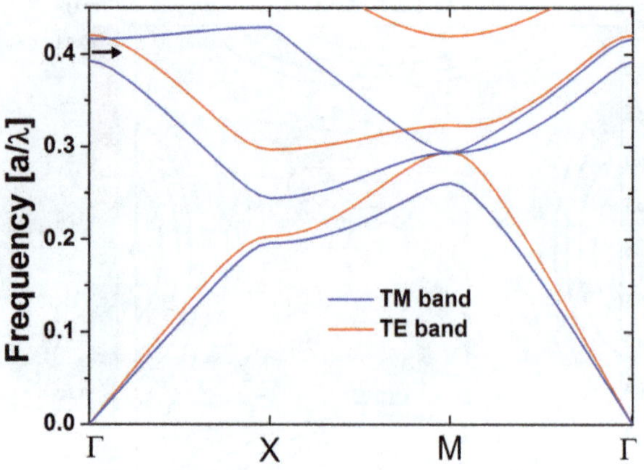

Fig. 4. Band diagram for the splitting region of the photonic crystal with $n = 2.8$. The small arrow points the operation frequency region of the beam splitter.

splitting region, a photonic band gap appears for the TM-polarized light between the normalized frequencies, $\omega = 0.39(c/a) - 0.42(c/a)$. However, since the electric field confines in the air holes for the TE-mode, it is not affected by the refractive index increase as much as the TM-mode. Therefore, there still exists a mode for the TE-polarized waves along the ΓX direction. As the incoming light reaches the splitting region, the TM-polarized waves reflect in the perpendicular direction, whereas the TE-polarized waves propagate directly in the self-collimation regime.

We employ a PC of $27a \times 27a$ size in order to quantify the transmission and reflection efficiency of the polarization beam splitter through FDTD simulations. The perfectly matched layers of $2a$ thickness are imposed on the four edges of the PC layer to prevent reflection. The simulation engine [11] uses uniaxial perfectly matched layers [13] by introducing a special lossy anisotropic material which changes both the dielectric constant and magnetic permittivity. A Gaussian source of frequency $\omega = 0.405(c/a)$ with the full-width half-maximum of 360 fs is launched into the PC along the ΓX direction. Two detectors are placed at the exits of the two ports of the PC as shown in Fig. 1. To obtain the normalized transmission and reflection spectra, the intensity of the transmitted and reflected light beam gathered through these detectors are integrated across the beam cross section and normalized with respect to the incident beam. An abrupt change of refractive index in the splitting region is observed to cause high level of reflection and refraction loss. To minimize the leakage along the boundary between the PIS and the splitting region, a gradual refractive index change is employed through the edges of the splitting region. This is accomplished by inserting three layers of regions with widths of $1.5a$, $2a$ and $1.5a$ inside the PC side by side as the splitting region. From the refractive index of 2.4 to 2.8 the refractive index is increased linearly in the first PC layer, from the refractive index of 2.8 to 2.4 the refractive index is decreased linearly in the third PC layer; whereas the refractive index is fixed to 2.8 in the middle PC layer. Graded refractive index materials are often employed in anti-reflection coatings, light-emitting diodes and solar cells to eliminate Fresnel reflection by fabricating coatings whose refractive index gradually changes from the refractive index of the active semiconductor layer to the refractive index of the surrounding medium [14-16].

It can be seen at Fig. 5 that the reflection of the TM-polarized light is above 75 % between 1.53 μm and 1.59 μm and the transmission of TE-polarized light is above 85 % in the 80 nm operation bandwidth of the polarization beam splitter. A transmission of 93 % for TE-polarized light and reflection of 79.5 % for TM-polarized light are obtained at 1.55 μm. In one of the studies on PC polarization beam splitter, a PC beam splitter is designed by changing the filling fraction of the PC in the splitting region, while maintaining the lattice constant and radius of the holes of the PC [8]. The simulations are realized in 2Ds with an effective index and up to 83 % transmission/reflection for TM/TE polarizations are achieved. Here, we obtain higher reflection and transmission ratios by applying a novel method, i.e. tuning the refractive index of the splitting region locally in the PIS region.

The polarization extinction ratios (PERs) of the transmitted and reflected beams are defined as $-10\log(T_{TM}/T_{TE})$ and $-10\log(R_{TE}/R_{TM})$, respectively. The PERs of the transmitted and reflected beams at 1.55 μm wavelength are calculated as 12.7 dB and 18.3 dB, respectively. Recently, a PC polarization beam splitter based on a self-collimated Michelson interferometer is demonstrated in silicon by using two beam splitters and four mirrors in a large Si PC [17]. The calculations are performed in 2Ds without scaling down the thickness of the Si slab by using an effective index. Although higher PER values are obtained (18.4 dB for TM and 24.3 dB for TE mode) in that study, these values are only valid for the optimized fixed single frequency value and decrease immediately in the surrounding frequencies due to the destructive interference nature of the Michelson interferometer.

As the splitting region, Chen et al. [18] used an air block that is aligned along the ΓM direction inside a Si PC lattice of holes to divide the TE- and TM-polarized waves simultaneously through the two ports. A maximum efficiency of 48 % for TE-polarization at an air block width of $0.18a$ and 49.9 % for TM-polarization at an air block width of $0.33a$ are obtained [18]. Although the production is simple, the width of the air block is critical for achieving maximum efficiency of the splitter in that study. However, since the refractive index of the splitting region is tuned smoothly in our study, the transmission efficiency does not show such an abrupt change. The size of the splitting region can be a little increased to enhance the TM reflection ratio at the cost of a modest decrease in the TE transmission. For example, when the width of the splitting region in the middle PC layer is increased to $3a$, the maximum transmission and reflection efficiencies are 84 % and 87 % for TE- and TM-polarized light, respectively.

The FDTD simulation of the light propagation through the proposed PC based polarization splitting structure is illustrated in Fig. 6. It can be seen that the TE-

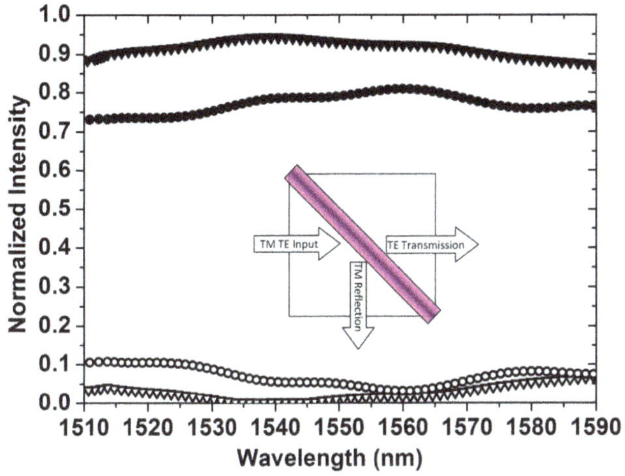

Fig. 5. The transmission of TE (▼), TM (o) and the reflection of TE (∇) and TM (●) polarized light beams against the incident self-collimated light.

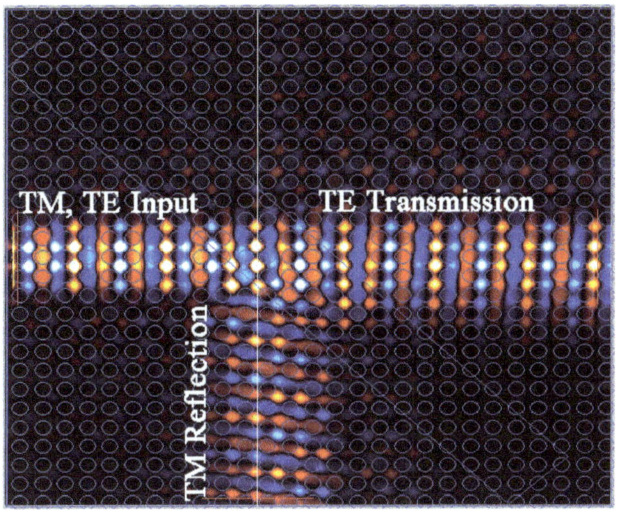

Fig. 6. Simulation of the light propagation through the photonic crystal based polarization beam splitter.

polarized light beams transmit directly and the TM-polarized light beams make a 90 degree turn after passing through the splitting region, with the maintenance of the self-collimation ability.

4. Conclusions

We have proposed and numerically demonstrated a photonic crystal based polarization beam splitter with high splitting efficiency, large separation angle and low crosstalk at the optical communication wavelengths. The transmission and reflection efficiencies of the device for TE and TM polarizations are found to be 93 % and 79.5 %, respectively at 1.55 μm. Moreover the efficiencies stay above 70 % for both polarizations between 1.51 μm and 1.59 μm. The proposed device can be very useful in optical devices and integrated photonic circuits, where polarization insensitivity is needed. For example, it can be used in photonic system-on-chip applications with other photonic components such as light emitters, waveguides, resonators, etc.

Acknowledgements

We gratefully acknowledge the financial support by Scientific Research Projects of Ankara University (BAP) under Grant no. 13B4343015.

References

[1] YABLONOVITCH, E. Inhibited spontaneous emission in solid state physics and electronics. *Physical Review Letters*, 1987, vol. 58, p. 2059–2062.

[2] JOANNOPOULOS, J. D., MEADE, R. D., WINN, J. N. *Photonic Crystals: Molding the Flow of Light*. 2nd ed. NJ: Princeton, 2008.

[3] JOHNSON, S. G., FAN, S. H., VILLENEUVE, P. R., JOANNOPOULOS, J. D., KOLODZIEJSKI, L. Guided modes in photonic crystal slabs. *Physical Review B*, 1999, vol. 60, p. 5751 to 5758.

[4] KOSAKA, H., KAWASHIMA, T., TOMITA, A., NOTOMI, M., TAMAMURA, T., SATO, T., KAWAKAMI, S. Self-collimating phenomena in photonic crystals. *Applied Physics Letters*, 1999, vol. 74, no. 9, p. 1212–1214.

[5] PRATHER, D. W., SHI, S., MURAKOWSKI, J., SCHNEIDER, G. J., SHARKAWY, A., CHEN, C., MIAO, B., MARTIN, R. Self-collimation in photonic crystal structures: a new paradigm for applications and device development. *Journal of Physics D:Applied Physics*, 2007, vol. 40, no. 9, p. 2635–2651.

[6] YU, X., FAN, S. Bends and splitters for self-collimated beams in photonic crystal. *Applied Physics Letters*, 2003, vol. 83, no. 16, p. 3251–3253.

[7] ZHAO, D., ZHANG, J., YAO, P., JIANG, X., CHEN, X. Photonic crystal Mach-Zehnder interferometer based on selfcollimation. *Applied Physics Letters*, 2007, vol. 90, no. 23, p. 231114.

[8] ZABELIN, V., DUNBAR, L. A., THOMAS, N. Le, HOUDRE, R., KOTLYAR, M. V., O'FAOLAIN, L., KRAUSS, T. F. Self-collimating photonic crystal polarization beam splitter. *Optics Letters*, 2009, vol. 17, no. 22, p. 19808–19813.

[9] CHEN, X., QIANG, Z., ZHAO, D., LI, H.,QUI, Y., YANG, W., ZHOU, W. Polarization-independent drop filters based on photonic crystal self-collimation ring resonators. *Optics Express*, 2009, vol. 17, no. 22, p. 19808–19813.

[10] KIM, T. T., LEE, S. G., PARK, H. Y., KIM, J. E., KEE, C. S. Asymmetric Mach-Zehnder filter based on self-collimation phenomenon in two-dimensional photonic crystals. *Optics Express*, 2010, vol. 18, no. 6, p. 5384-5389.

[11] CrystalWave from Photon Design. Available at: http://www.photond.com.

[12] QUI, M. Effective index method for heterostructure-slab-waveguide-based two dimensional photonic crystals. *Applied Physics Letters*, 2002, vol. 81, p. 1163–1165.

[13] GEDNEY, S. D. An anisotropic perfectly matched layer absorbing media for the truncation of FDTD lattices. *IEEE Transactions on Antennas and Propagation,* 1996, vol. 44, no 12, p. 1630–1639.

[14] SOUTHWELL, W. H. Gradient-index antireflection coatings. *Optics Letters*, 1983, vol. 8, no. 11, p. 584–586.

[15] KIM, J. K., CHHAJED, S., SCHUBERT, M. F., SCHUBERT, E. F., FISCHER, A. J., CRAWFORD, M. H., CHO, J., KIM, H., SONE, C. Light-extraction enhancement of GaInN light-emitting diodes by graded-refractive-index indium tin oxide anti-reflection contact. *Advanced Materials*, 2008, vol. 20, no. 4, p. 801–804.

[16] ZHAO, Y., CHEN, F., SHEN, Q., ZHANG, L. Optimal design of light trapping in thin-film solar cells enhanced with graded SiN_x and SiO_xN_y structure. *Optics Express*, 2012, vol. 20, no. 10, p. 11121-11136.

[17] CHEN, X.-Y., LIN, G.-M., LI, J.J., XU, X.F., JIANG, J.Z., QIANG, Z.-X., QUI, Y.S., LI, H. Polarization beam splitter based on a self-collimation Michelson interferometer in a silicon photonic crystal. *Chinese Physics Letters*, 2012, vol. 29, no. 1, p. 014210-1-4.

[18] SHEN, X., HAN, K., YANG, X., SHEN, Y., LI, H., TANG, G., GUO, Z. Polarization-independent self-collimating bends and beam splitters in photonic crystals. *Chinese Optics Letters*, 2007, vol. 5, no. 11, p. 662–664.

About Authors ...

Fulya BAGCI obtained her first degree in Engineering Physics from Ankara University, graduating in 2005. She received her M.Sc. and Ph.D. degrees from the same university in 2008 and 2013, respectively. Dr. Bagci is a research assistant in the Department of Engineering Physics, Ankara University since 2005. Her current research interests include the analysis of photonic crystals and metamaterials for photonic and microwave device applications.

Sultan CAN received her B.Sc. degree in Electrical-Electronics Engineering from the Atilim University in 2008. She received her M. Sc. degree from the same university in 2011. She is now a Ph.D. candidate in Electrical-Electronics Engineering. She is currently with the Dept. of Electrical-Electronics Engineering in Ankara University, where she is a research assistant. Her research interests include electromagnetism, metamaterials and antennas.

Baris AKAOGLU received his B.Sc. degree in Physics from the Middle East Technical University in 1996. He received his M.Sc. and Ph.D. degrees in Physics from the same university in 1998 and 2004, respectively. He is currently with the Dept. of Engineering Physics in Ankara University, where he is an Associated Professor. He did experimental research on semiconductors. His current research interests include photonic crystals and metamaterials.

Asim Egemen YILMAZ received his B.Sc. degrees in Electrical-Electronics Engineering and Mathematics from the Middle East Technical University in 1997. He received his M.Sc. and Ph.D. degrees in Electrical-Electronics Engineering from the same university in 2000 and 2007, respectively. He is currently with the Dept. of Electrical-Electronics Engineering in Ankara University, where he is an Associated Professor. His research interests include computational electromagnetics, nature-inspired optimization algorithms, knowledge-based systems; more generally software development processes and methodologies.

9

Equivalent Circuit Modeling of the Dielectric Loaded Microwave Biosensor

Muhammad Taha JILANI[1], *Wong Peng WEN*[2], *Lee Yen CHEONG*[3], *Mohd Azman ZAKARIYA*, *Muhammad Zaka Ur REHMAN*[4]

[1]Dept. of Electrical and Electronics Engineering, [3]Dept. of Fundamental and Applied Sciences, Universiti Teknologi Petronas, Perak, Malaysia
[4]Dept. of Physics, COMSATS Institute of Information Technology, Islamabad, Pakistan

mtaha.jilani@gmail.com, wong_pengwen@petronas.com.my

Abstract. *This article describes the modeling of biological tissues at microwave frequency using equivalent lumped elements. A biosensor based on microstrip ring resonator (MRR), that has been utilized previously for meat quality evaluation is used for this purpose. For the first time, the ring-resonator loaded with the lossy and high permittivity dielectric material, such as biological tissue, in a partial overlay configuration is analyzed. The equivalent circuit modeling of the structure is then performed to identify the effect of overlay thickness on the resonance frequency. Finally, the relationship of an overlay thickness with the corresponding RC values of the tissue equivalent circuit is established. Simulated, calculated and measured results are then compared for validation. Results agreed well while the observed discrepancy is in an acceptable limit.*

Keywords

Microwave biosensor, microstrip ring resonator, equivalent circuit modeling, effective permittivity, dielectric constant, meat dielectric characterization, overlay thickness.

1. Introduction

Over a decade, electrical properties of materials have gained significant research interest. The measurement of these properties is considered as most promising research area in the field of medical, agriculture and materials engineering [1]. The main advantage of these methods is the non-invasive and non-destructive evaluation, which is desirable for many in-line industrial applications. It is due to the fact that the interaction of electromagnetic fields with a material can be investigated to understand the underlying physical properties of the material. Whereas, this interaction is mainly governed by the relative permittivity, loss factor and conductivity of a material [2]. When a dielectric material is subjected to the electromagnetic field, its behavior can be defined by [3]

$$\varepsilon^* = \varepsilon' - j\,\varepsilon'' \tag{1}$$

where $j = \sqrt{-1}$, ε' is the real part called dielectric constant and the imaginary part is a loss factor. The dielectric constant describes the ability of a material to store energy, whereas, the loss-factor defines energy dissipation of a material. Another, important property of a material to conduct an electric current is known as conductivity σ and it is given as [2]

$$\tan\delta = \sigma / \varpi\varepsilon_0\varepsilon_r \tag{2}$$

where $\tan\delta$ is tangent-loss which is the ratio between loss factor and dielectric constant, ω is the angular frequency, ε_0 is the free space permittivity and ε_r is the dielectric constant of the material. These properties can be utilized effectively, once their relationship can be established with the physical attributes of a material.

In food industry, the application of dielectric properties is evident into various products, such as agriculture, edible-oil, dairy, and meat. Particularly in meat industry, due to increasing demand for high quality products and strict regulations for health and safety. The need for an effective, online and non-destructive dielectric technique is become obvious. Several efforts have been attempted to identify the relationship of dielectric properties with the quality attributes, like: moisture content, composition, freshness, aging, discrimination of frozen & thawed products and microbial activities [4]. Similarly, various new techniques have been proposed and adopted to determine the complex permittivity. Since, biological tissues are heterogeneous, anisotropic and semi-solid in nature, practically only few techniques are preferred for the measurements. The most widely used technique is the coaxial probe, which exhibits high accuracy for broadband measurements [2]. It is based on reflection principle, and it is successfully utilized for determination of meat-aging [5], frozen detection [6], fat analysis [7], and to study the temperature effect [8]. Although, this method has higher accuracy but still is subject to errors. The main source of errors are sample thickness (at least semi-infinite for probe) and the air-gap between the sample and a probe, it is reported that they can introduce errors of up to 20% [9]. The irregularities on a sample surface can reduce accuracy, therefore to decrease error rate MUT should be smoothen. However, sample preparation will become a challenging task for this method. Cable stability during measurements should be considered, since the phase changes cause some additional errors [3]. Repetitive calibration of a measurement system is required to reduce errors.

Recently, a relatively newer method based on microstrip ring-resonator has been used for meat quality evaluation [10], [11]. Obtained results are satisfactory while compared to existing approaches [12]. Since, ring resonator is based on resonance; it offers higher accuracy than other simple planar structures. Additionally, it is simple, cost-effective, fast (measurements in few seconds) and provides higher Q-factor (~250) than conventional microstrip line [13]. Sample loading and unloading is relatively easier than other comparable resonance techniques. This method is based on a fact that the effective permittivity of a planar resonator is strongly dependent on the permittivity of the region or overlay above the ring surface [1]. Hence, any permittivity variations due to physical properties of the material can be evaluated.

In this study equivalent circuit modeling of a biological tissue loaded resonator is performed. The structure of this article is as follows: initially the equivalent lumped elements of the ring structure are derived using transmission line analysis. A thorough study is then carried out to consider the losses, radiation and coupling effects associated to the ring-resonator structure. Similarly, the equivalent circuit of the biological tissue is discussed. The effect of tissue thickness on the resonance frequency and corresponding RC values are studied. This has been used to identify the relationship between RC components & permittivity and then established with overlay thickness. Finally, the resonance frequency of the sample loaded ring-resonator is computed and validated with the measured results.

2. Theoretical Analysis and Modeling of Microwave Biosensor

The microstrip ring resonator is considered as a closed-loop transmission line. To excite the resonator, power is capacitively coupled through feed lines and the gap between them, as shown in Fig. 1(a). The resonance is produced when a mean circumference of the ring is equal to an integral multiple of the guided wavelength [14]

$$2\pi r = n\lambda_g \qquad \text{for } n = 1,2,3\ldots \tag{3}$$

where r is the ring radius, n is mode number and λ_g is the guided wavelength. For resonance frequency, considering (3) with λ_g will give

$$f_0 = \frac{nc}{2\pi r\sqrt{\varepsilon_{eff0}}} \tag{4}$$

where f_0 is the resonance frequency, c is the speed of light in free-space and ε_{eff0} is the effective permittivity of the ring resonator. Once the unloaded resonance frequency is determined, the effective permittivity with the sample loaded ring-resonator can be evaluated by [13]

$$\varepsilon_{eff1} = \varepsilon_{eff0}\left(\frac{f_0}{f_1}\right)^2 \tag{5}$$

where ε_{eff1} and f_1 are the sample-loaded effective permittivity and resonance frequency, respectively.

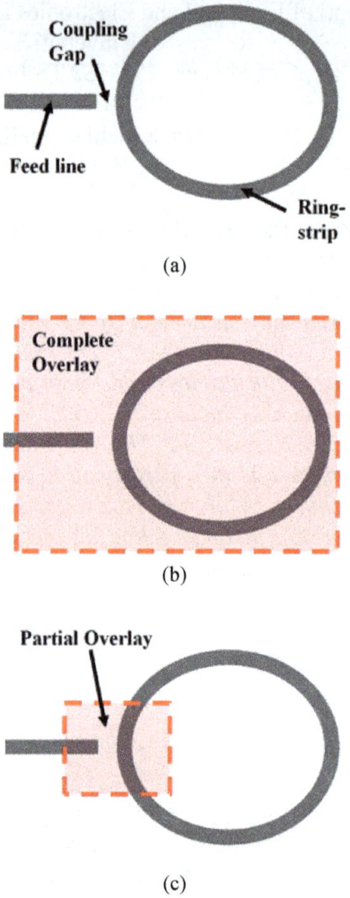

Fig. 1. Microwave biosensor based on one-port microstrip ring-resonator: (a) without a sample, (b) with complete overlay sample, (c) with partial overlay sample.

The structure of microstrip has been widely analyzed, several numerical and dispersion models are used to determine the resonant frequency and effective permittivity. Broadly speaking, these methods of analysis can be classified into two categories; quasi-static analysis and full-wave analysis [15]. In quasi-static approach the propagation mode assumed to be pure TEM and analysis is performed by considering electrostatic capacitance of the structure. In the contrary, for full-wave analysis the propagation mode is hybrid in nature, thus, it is more accurate. Although, the variation of effective permittivity and characteristics impedance can also be studied with respect to the frequency, but altogether make it analytically complex.

Considering quasi-static solutions, the most common techniques are conformal mapping approach (CMA) and the variational method [15]. Although, the results from these two methods are good enough, however, they are limited to only complete dielectric overlay structures, as depicted in Fig. 1(b). In recent study, it is found that due to strong electric fields in the coupling gap, this region has maximum sensitivity to the above surface permittivity variations [16]. Here, field perturbations in this region are linearly proportional to the resonance shifts [17], even this tendency remains same for further higher order modes. Utilizing the partial overlay arrangement, as shown in Fig. 1(c), a small size sample can also be evaluated, while the complete overlay requires relatively a larger sample size. This type of overlay, in fact provides flexibility for the measurement of most of the biomaterials, including grain and pulverized samples [18]. It has been reported that, the said configuration provides an appreciable shift in resonance for the materials having moisture content greater than 15 % [16]. Likewise, for the biological tissues having moisture in 72-80 % range, overlaid on coupling region can produce higher resonance shifts. Although, this arrangement seems more suitable for permittivity measurements, yet, no attempt has been made to analyze the partial overlay arrangement.

Only few reports have discussed the partial overlay configuration, however, they just limited to some specific cases [18], [19]. An empirical expression is obtained, using a ring resonator designed at 10 GHz using Al_2O_3 substrate [18]. Here, the material is partially overlaid on ring-structure to extract the dielectric constant using curve-fitted expression. However, this expression has limited size range and is only applicable to the given resonator.

An alternative solution to numerical and curve-fitted models is an equivalent circuit modeling that can provide a simple and quick analysis, without any time consuming computation [20]. Although, numerical methods give accurate and rigorous results, however, with increased complexity they are difficult to use [14]. Analyzing with these methods often requires large computing time and space for storage. While, circuit analysis can be easily modeled while considering the structure variations and discontinuities [14]. In the following section, we will discuss the modeling scheme to encompass the RLC model of a capacitively-coupled ring-resonator and a tissue structure.

2.1 Equivalent Circuit for a Microstrip Ring Resonator and Biological Tissues

The most straightforward method to analyze the microstrip ring resonator is an equivalent lumped element circuit. It can be described as a simple parallel RLC circuit using the transmission line theory [21], as depicted in Fig. 2(b). This equivalent circuit has parallel connected RLC components, which are associated with the ring losses in terms of resistance R_r and the resonance frequency that corresponded to the particular inductance L_r and capacitance C_r. The series capacitance C_g represents the coupling strength while the additional parallel resistance R_f caters the losses incurred due to feed line. The resonance of the circuit, assuming negligible gap-capacitance, can be calculated as

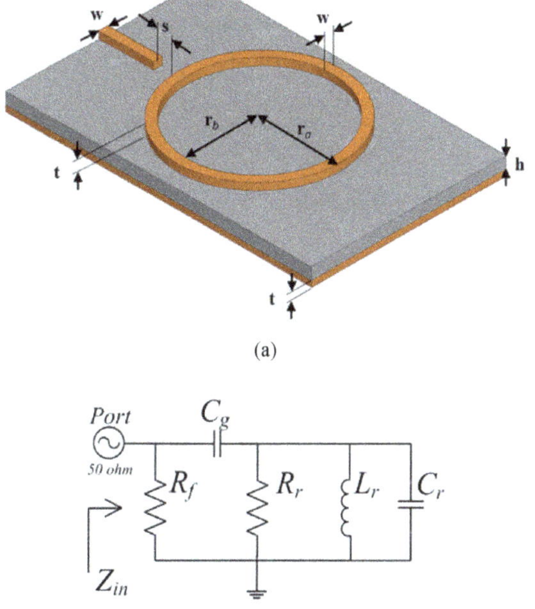

Fig. 2. Capacitively coupled ring resonator: (a) layout, (b) equivalent RLC circuit.

$$f_0 = \frac{1}{2\pi\sqrt{L_r C_r}}. \tag{6}$$

However, to derive the input impedance of the ring same approach has been used as described in [21]. Assuming that the port 2 of a ring-resonator is an open circuit ($i_2 = 0$), where equally divided two-line sections, l_1 and l_2 forms a parallel circuit. Therefore, for a circuit with the individual lengths of transmission lines, the ABCD matrix is given by

$$\begin{bmatrix} A & B \\ C & D \end{bmatrix}_{1,2} = \begin{bmatrix} \cosh(\gamma l_{1,2}) & Z_0 \sinh(\gamma l_{1,2}) \\ Y_0 \sinh(\gamma l_{1,2}) & \cosh(\gamma l_{1,2}) \end{bmatrix}, \gamma = \alpha + j\beta \tag{7}$$

where subscripts 1 and 2 correspond to the transmission lines l_1 and l_2 respectively, Z_0 is the characteristics impedance which is equal to $1/Y_0$, γ is the propagation constant, α is the attenuation constant and β is the phase constant. Similarly, the converted ABCD-matrix into Y-matrix is given as [21]

$$\begin{bmatrix} Y_{11} & Y_{21} \\ Y_{12} & Y_{22} \end{bmatrix} = \begin{bmatrix} Y_0[\coth(\gamma l_1)+\coth(\gamma l_2)] & -Y_0[\operatorname{csch}(\gamma l_1)+\operatorname{csch}(\gamma l_2)] \\ -Y_0[\operatorname{csch}(\gamma l_1)+\operatorname{csch}(\gamma l_2)] & Y_0[\coth(\gamma l_1)+\coth(\gamma l_2)] \end{bmatrix} \tag{8}$$

Letting $i_2 = 0$, the input impedance of the ring (Z_{ring}) can be found by

$$Z_{ring} = \frac{v_1}{i_i}\Big|_{i_2=0} = \frac{Z_0}{2}\frac{\sinh(\gamma l)}{\cosh(\gamma l)-1}. \tag{9}$$

Let $l_g = l/2 = \lambda_g/2$, the Z_{ring} can be re-written as

$$Z_{ring} = \frac{Z_0}{2}\frac{1+j\tanh(\alpha l_g)\tan(\beta l_g)}{\tanh(\alpha l_g)+j\tan(\beta l_g)}. \tag{10}$$

Generally, transmission lines have small loss, so the attenuation can be assumed as $\alpha l_g << 1$ and then $\tanh(\alpha l_g)$

$\approx \alpha l_g$. Considering $\beta = v_p/\omega$, where v_p is the phase velocity and ω is the angular velocity, with some further manipulations [21], Z_{in} can be approximated as

$$Z_{ring} \cong \frac{Z_0}{2} \frac{1+j\alpha l_g \pi\Delta\varpi/\varpi_0}{\alpha l_g + \pi\Delta\varpi/\varpi_0}. \quad (11)$$

Recalling that $\alpha l_g << 1$, similarly $\alpha l_g \pi\Delta\omega/\omega_0 << 1$ thus

$$Z_{ring} \cong \frac{Z_0/2\alpha l_g}{1+j\pi\Delta\varpi/\alpha l_g\varpi_0}. \quad (12)$$

In general, the input impedance given for a parallel GLC resonant circuit is

$$Z_i = \frac{1}{G+2j\Delta\varpi C}. \quad (13)$$

Comparing equation (12) with (13), the obtained expression of Z_{ring} is similar to the parallel GLC circuit. Subsequently, the resistance R_r of the ring will be

$$G_r = \frac{1}{R_r} = \frac{2\alpha l_g}{Z_0} = \frac{\alpha\lambda_g}{Z_0} \quad (14)$$

while the inductance L_r can be derived from $\omega_0 = 1/\sqrt{L_r C_r}$ and is given as

$$L_r = 1/\varpi_0^2 C_r, \quad (15)$$

and finally, the capacitance of the ring circuit is

$$C_r = \frac{\pi}{Z_0\varpi_0}. \quad (16)$$

Although, the ring's equivalent RLC component equations has been derived, however, it should be noted that this circuit analysis does not include the losses pertaining to radiation effects. Henceforth, the total losses R_r along with radiation loss can be described as:

$$R_r = \frac{1}{\left(\frac{1}{R_{rad}} + \frac{1}{R_{cond}} + \frac{1}{R_{dielec}}\right)} \quad (17)$$

where R_{rad} are the losses incured due to radiation, R_{cond} are the losses due to finite conductivity of copper and R_{dielec} are the dielectric losses. The dielectric loss is given as

$$R_d = \frac{Z_0}{\alpha_d l_g} \text{ Np/m} \quad (18)$$

where the attenuation constant for a dielectric loss α_d is described as [22]

$$\alpha_d = 27.3\frac{\varepsilon_r}{\varepsilon_r - 1}\frac{\varepsilon_{eff}-1}{\sqrt{\varepsilon_{eff}}}\frac{\tan\delta}{\lambda_0}. \quad (19)$$

Similarly, for the conductor, the loss is

$$R_c = \frac{Z_0}{\alpha_c l_g} \text{ Np/m} \quad (20)$$

where the attenuation constant for the ohmic loss α_c is defined as

$$\alpha_c = \frac{8.68R_{s1}}{2\pi Z_0 h}\left[\frac{w_{eff}}{h} + \frac{\frac{w_{eff}}{\pi h}}{\frac{w_{eff}}{2h}+0.94}\right]\cdot\left[1+\frac{h}{w_{eff}}+\frac{hV}{\pi w_{eff}}\right]\cdot \quad (21)$$
$$\left[\frac{w_{eff}}{h}+\frac{2}{\pi}\ln\left\{2\pi e\left(\frac{w_{eff}}{2h}+0.94\right)\right\}\right]^{-2}$$

while,

$$V = \ln\left(\frac{2h}{t}\right)+\frac{t}{w},$$
$$R_{s1} = R_s\cdot\left[1+\frac{2}{\pi}\tan^{-1}\left\{1.4\left(\frac{\Delta}{\delta_s}\right)^2\right\}\right], \quad (22)$$
$$R_s = \sqrt{\frac{\pi f\mu_0}{\sigma}}$$

where R_{s1} is the surface-roughness resistance of the conductor, e is the naperian base, R_s is the conductor's surface resistance, Δ is the surface roughness, $\delta_s = 1/R_s\sigma$ is the skin depth, σ is the conductivity, f is the frequency, μ_0 is the free-space permeability, t and w are the line thickness and width, respectively.

As it is mentioned before, transmission line analysis does not account for the radiation loss, whereas, full wave analytical solutions can provide the better insight to this loss mechanism. However, using the cavity resonator model few simple assumptions have been made to reduce unnecessary complexity, as outlined in [23], thus the radiation can be given as

$$R_{rad} = \frac{V_0^2}{2P_r} \quad (23)$$

where V_0 is the voltage produced at the edge of the ring and P_r is the radiated power from the ring. Therefore:

$$V_0 = E_0 h\left[J_n(k_n b_e)Y_n'(k_n a_e) - J'_n(k_n a_e)Y_n(k_n b_e)\right], \quad (24)$$

$$P_r = \frac{2h^2E_0^2}{\pi\eta_0\varepsilon_r}I_1, \quad (25)$$

$$I_1 = \int_0^{\frac{\pi}{2}}\left[\begin{array}{l}\frac{n^2\cos^2\theta}{k_0^2\sin\theta}\left(\frac{J_n(k_0a_e\sin\theta)}{a_e} - \frac{J'_n(k_na_e)}{J'_n(k_nb_e)}\frac{J_n(k_0b_e\sin\theta)}{b_e}\right)^2 \\ +\sin\theta\left(J'_n(k_0a_e\sin\theta) - \frac{J'_n(k_na_e)}{J'_n(k_nb_e)}J'_n(k_0b_e\sin\theta)\right)^2\end{array}\right]d\theta \quad (26)$$

Finally, using (23) the radiation loss can be easily determined. However, in case of an overlay dielectric layer, this loss will be reduced substantially [24].

For a gap-coupled ring-resonator, power coupled to the resonator through feed lines and coupling gaps [14].

Here, with a small gap the coupling will be tight and the gap-capacitance becomes appreciable. This coupling region has maximum electric fields and as reported previously, it will be more sensitive to overlay permittivity variations [16]. Although, this increase in gap-capacitance causes more sensitivity, however, it even deviates the resonator's inherent frequency to the lower frequency, which is known as "pushing effect" [25]. This pushing effect though lowers the insertion-loss but its effect on resonance frequency is more significant [14]. Hence, it cannot be neglected (as in the case of loose coupling) and for more accurate analysis it has to be considered. The gap capacitance can be modeled as π-network, as reported in [26]. The fringing fields are modeled as shunt capacitances C_f which are due to the edges of the strip, while the series capacitance C_g is assumed to be the result of coupling mechanism:

$$C_f = \frac{C_{even}}{2}, \tag{27}$$

$$C_g = \frac{\left(C_{odd} - \frac{C_{even}}{2}\right)}{2}. \tag{28}$$

Closed form-expressions for C_{odd} and C_{even} are described in [27], with their corrected form [24] and they are re-modified as:

$$\left.\begin{aligned} \frac{C_{odd}}{w} &= \left(\frac{s}{w}\right)^{m_0} e^{k_0} \\ \frac{C_{even}}{w} &= 12\left(\frac{s}{w}\right)^{m_e} e^{k_e} \end{aligned}\right\} \text{pF/m} \tag{29}$$

where, for our considered s/w ratio

$$\begin{aligned} m_0 &= \frac{w}{h}(0.619\log w/h - 0.3853) && (\text{for } 0.1 \le \text{s/w} \le 1.0) \\ K_0 &= 4.26 - 1.453\log w/h \\ m_e &= 0.8675 && (\text{for } 0.1 \le \text{s/w} \le 0.3) \\ K_e &= 2.043(w/h)^{0.12} \end{aligned} \tag{30}$$

The modified expressions for low ε_r substrate are given as

$$C_{even}(\varepsilon_r) = 1.167 \cdot C_{even}(9.6) \cdot \left(\frac{\varepsilon_r}{9.6}\right)^{0.9}, \tag{31}$$

$$C_{odd}(\varepsilon_r) = 1.1 \cdot C_{odd}(9.6) \cdot \left(\frac{\varepsilon_r}{9.6}\right)^{0.8}. \tag{32}$$

Now considering the effect of the gap-capacitance, the resonance frequency given in (6), can be modified as

$$f_0 = \frac{1}{2\pi\sqrt{L_r\left(C_r + C_g\right)}}. \tag{33}$$

Finally, a one-port microstrip ring resonator is designed at 1 GHz frequency for measurement purpose. A low loss substrate Roger RT/Duriod 5880 with the dielectric constant ε_r of 2.2 and the loss tanδ of $9 \cdot 10^{-4}$ is used. Strip width and effective permittivity are determined by using LineCalc of Agilent ADS. The corresponding width for 50 Ω characteristics impedance is calculated as 2.4 mm while the ε_{eff} was 1.875. The tight coupling mechanism is incorporated using 250 µm gap size. A feed line with a typical length of a 35 mm is used. The structure is designed and then simulated in FEM based 3D solver HFSS.

Furthermore, using the derived expressions for the equivalent R, L, and C of ring and gap-capacitance, the corresponding values are computed. Since, for the feed loss R_f the exact value is not well defined, due to complex field distribution near the coupling region. Hence, a computer program is used to fit its value to simulation data. The circuit simulation is then performed by using Agilent ADS. Finally, for a fabrication of microstrip ring-resonator a standard photolithography procedure is adapted to print the ring structure on substrate. Design parameters are tabulated in Tab. 1

Parameter	Designed Value
Substrate	Roger RT/Duriod 5880 ε_r 2.2, tanδ $9 \cdot 10^{-4}$
Substrate height	787 µm
Copper thickness	17.5 µm
Coupling gap	250 µm
Line Width (Feed & Ring)	2.4 mm
Feed line length	35 mm
Frequency	1 GHz

Tab. 1. Design parameters of the proposed MRR.

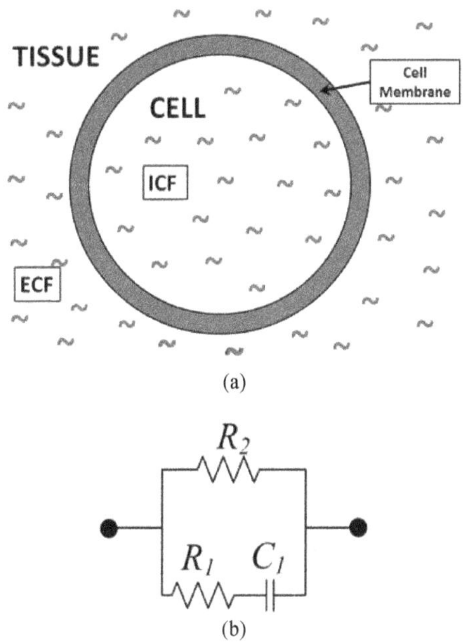

Fig. 3. Biological tissue: (a) structure– ECF is an extracellular fluid and ICF is an intracellular fluid; (b) equivalent circuit.

The structure of biological tissue and its equivalent circuit are shown in Fig. 3. As depicted in Fig. 3(a), the tissue contains extracellular fluid (ECF), this fluid behaves as an electrolyte which mainly comprises Na^+ and Cl^- ions [28]. Cells are immersed into ECF and each cell is basically surrounded by a leaky dielectric membrane that encloses the conductive fluid which is commonly known as intra-cellular fluid (ICF) [29]. This inner plasma suspension is basically called *cytoplasm* and it mainly consists on K^+ ions.

The equivalent circuit of tissues is widely discussed in literature, the typical RC combination, as shown in Fig. 3(b), is presented in many articles [28], [30]. The modeling of a cell structure is complex and extremely difficult, however, various effects related to the fields interaction can be defined by using a simple cell model [31]. At higher frequencies the electric field interacts mainly at cellular level and its behavior can be represented by RC circuit, as shown in Fig. 3(b). The series resistance R_1 basically describes the resistive property of the intracellular fluid. Whereas, the series capacitance C_1 actually represents the capacitive behavior of a cellular-membrane, the parallel resistance R_{12} exhibits the resistive nature of an extracellular fluid. As we have discussed before, at low frequencies, in lower kHz range, the current flows mainly through the extracellular fluid, depicted in Fig. 4(a). This is due to the fact that the cellular membrane is highly resistant to the applied electric field of lower frequency, where the current flow through only ECF. At this point, the impedance is very high, however, as the frequency increases impedance decreases as resistance drops due to its predominant capacitive behavior [32]. In the contrary, at higher frequencies the current starts penetrating into cells, typically from MHz region. This can be observed from Fig. 4(b), where at higher frequencies (MHz range and above) the outer membrane now becomes transparent [31]. Since the membrane is shortened at higher frequencies the impedance becomes zero, hence, the field lines are passed more uniformly at higher frequencies [32].

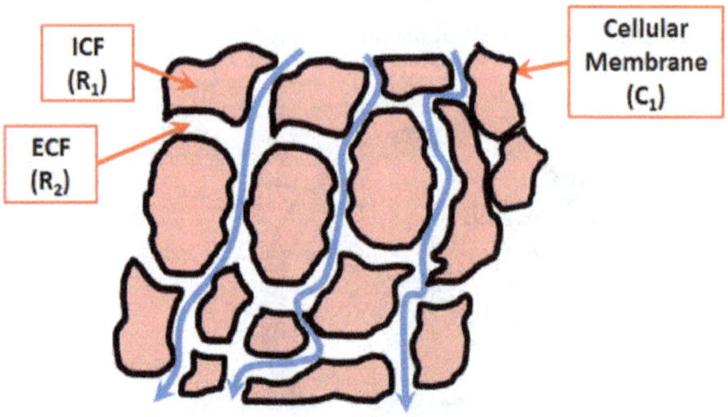

(a)

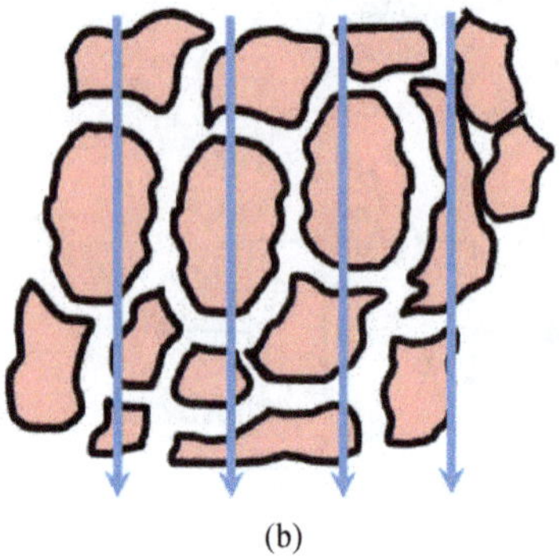

(b)

Fig. 4. The flow of electric current through tissue: (a) at low frequencies; (b) at higher frequencies.

2.2 Modeling of RLC Components of the Capacitively Coupled Dielectric Loaded Ring-Resonator

The expressions for input-impedance, gap-capacitance and lumped elements of a ring-resonator are already derived. Their values corresponding to the parameters that are tabulated in Tab. 1 are then used in Agilent ADS to model the equivalent circuit. Similarly, for the resonator loaded with a meat sample, the equivalent circuit of a tissue (i.e. Fig. 3(b)) will be connected in parallel to the C_g of a ring resonator (i.e. Fig. 2(b)). Using meat equivalent circuit tuning in ADS, the values of R_1, C_1 and R_2 are extracted. These values are based on the data obtained by full-wave simulation on HFSS, for the samples with different thicknesses. To validate the simulation results, the basic circuit theory of a transmission line is used. For this purpose, initially an unloaded capacitively coupled microstrip ring resonator is analyzed. The transfer matrix of a gap capacitance C_g is obtained

$$[T]_{Cg} = \begin{bmatrix} 1 & \dfrac{-j}{\varpi C_g} \\ 0 & 1 \end{bmatrix}. \quad (34)$$

The impedance of a ring Z_{ring} is already calculated using (10). Now, considering ABCD parameters of (34) and Z_{ring} as a load impedance, the total input impedance of a capacitively coupled microstrip ring resonator can be evaluated from:

$$Z_{in} = \frac{AZ_L + B}{CZ_L + D}. \quad (35)$$

Hence, the return loss can be obtained by

$$S_{11} = \frac{Z_{in} - Z_0}{Z_{in} + Z_0}. \quad (36)$$

In a similar manner, the meat loaded ring resonator is analyzed, where the equivalent circuit of a meat sample is connected in parallel to the ring resonator. The analysis begins with a transfer (ABCD) matrix of a meat equivalent circuit. Thus, for a series resistance R_2 $[T]$ is given as:

$$[T]_{R2} = \begin{bmatrix} 1 & R_2 \\ 0 & 1 \end{bmatrix}. \quad (37)$$

Similarly, for a series resistance R_1 and capacitances C_1 the product of the obtained transfer matrices will be

$$[T]_{R1*C1} = \begin{bmatrix} 1 & \dfrac{-j}{\varpi C_1} + R_1 \\ 0 & 1 \end{bmatrix}. \quad (38)$$

Considering the shunt elements, (37) and (38) are converted into Y-parameters respectively, which yields

$$[Y]_{R1} = \begin{bmatrix} \frac{1}{R_2} & -\frac{1}{R_2} \\ -\frac{1}{R_2} & \frac{1}{R_2} \end{bmatrix}$$
$$[Y]_{C1} = \begin{bmatrix} \frac{1}{\frac{-j}{\varpi C_1}+R_1} & -\frac{1}{\frac{-j}{\varpi C_1}+R_1} \\ -\frac{1}{\frac{-j}{\varpi C_1}+R_1} & \frac{1}{\frac{-j}{\varpi C_1}+R_1} \end{bmatrix} \quad (39)$$

Further, for the gap capacitance $[T]_{Cg}$ is converted into Y-matrix, thus

$$[Y]_{Cg} = \begin{bmatrix} \frac{1}{\frac{-j}{\varpi C_g}} & -\frac{1}{\frac{-j}{\varpi C_g}} \\ -\frac{1}{\frac{-j}{\varpi C_g}} & \frac{1}{\frac{-j}{\varpi C_g}} \end{bmatrix} \quad (40)$$

Adding, $[Y]_{R1} + [Y]_{C1} + [Y]_{Cg}$ will give the overall Y-parameters that will transform back into ABCD parameters using

$$\begin{bmatrix} A & B \\ C & D \end{bmatrix} = \begin{bmatrix} \frac{-Y_{22}}{Y_{21}} & -\frac{1}{Y_{21}} \\ \frac{Y_{12}Y_{21}-Y_{11}Y_{22}}{Y_{21}} & \frac{-Y_{11}}{Y_{21}} \end{bmatrix} \quad (41)$$

Thus, again using (35) and (10) Z_{in} can be extracted for the meat sample loaded resonator.

3. Results and Discussion

The equivalent circuit modeling of unloaded and loaded microstrip ring resonator is performed. In Fig. 5, the measured, simulated and calculated results are presented. Obtained results are in good agreement; it can be seen that the resonance is produced over the regular 1 GHz intervals in the given range. However, due to the tight coupling mechanism, the "pushing effect" causes the deviation of a resonator's inherent frequency to a lower frequency. In this case the reflection loss observed from the calculation was 0.981 GHz while from the simulation results it is observed at 0.980 GHz and from measurement it was 0.978 GHz. In measured results the noise figure is higher, that might be the result of limited calibration of portable equipment. But still, the overall difference between these results is not more than 0.3 %.

Further, the simulation result is analyzed for the sample loaded ring resonator over the range from 5 µm to 16 mm. This range is chosen, while considering the results

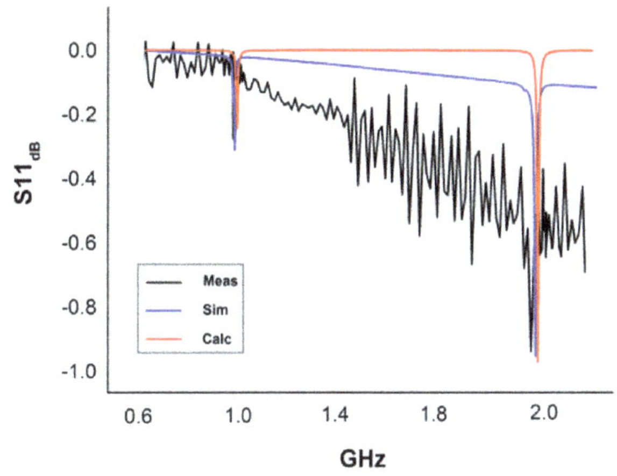

Fig. 5. Reflection loss – Measured, Simulated and Calculated of unloaded microstrip ring resonator.

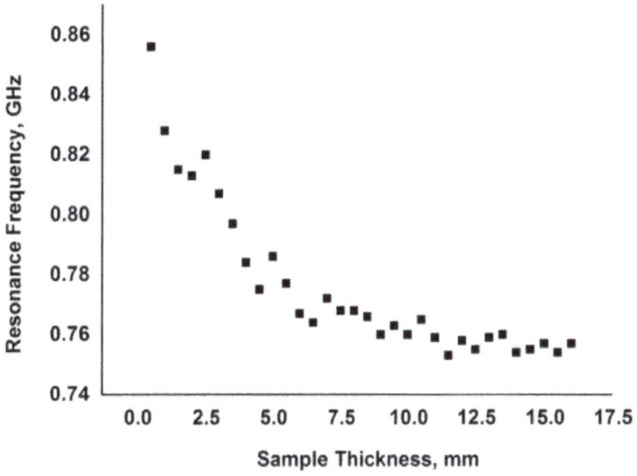

Fig. 6. Resonance frequency as a function of overlay thickness.

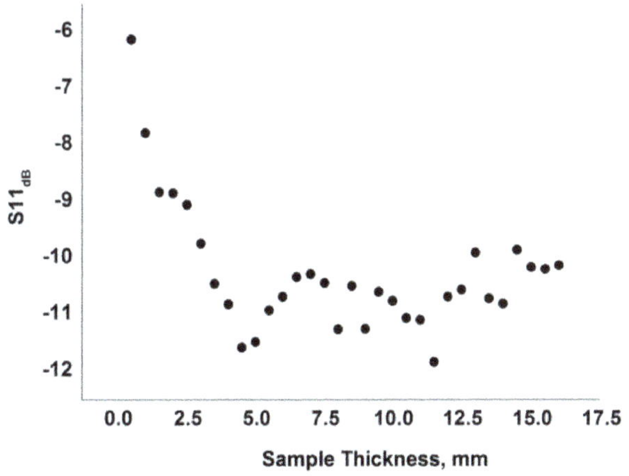

Fig. 7. Return loss as a function of overlay thickness.

of previous study [33], where beyond a certain point, the size of a meat sample will not affect the effective permittivity significantly. Since the thickness effect is more significant than the effect of width, in this study, the width is fixed to 18.3 % of λ_g. Here, beyond this width the effect on resonance frequency is negligible. In Fig. 6 and 7, the resonance frequency and return loss are presented as a function of overlay thickness, respectively. Considering Fig. 6, it can be seen that as the sample thickness increases, the resonance frequency decreases. This is due to the fact that as the sample size increases the field perturbation is also increased, since more fringing fields are getting concentrated into the sample. That results into increasing fringing field capacitance, hence decreases the resonance frequency. Although the shift in resonance frequency is significant but it limited to a lower range only, where beyond a certain thickness, its effect is not significant. Similarly, this happens to the magnitude of the return loss also (Fig. 7), where beyond that point the thickness will not affect the loss effectively. This behavior can be interpreted as the effective permittivity also increases with the overlay height until it reaches the asymptotic value, which indicates that all of the electric field is confined into overlay and substrate [34].

In Fig. 8 and 9, for the given sample thickness the resistive behavior of intracellular fluid and capacitive behavior of cell membrane is presented, respectively. When an electric field is applied to a tissue, the current will pass through the ECF or in some cases, through both ECF and ICF [30]. The path containing ECF is a purely resistive; while for ICF other than resistive it also includes the capacitive effect of a membrane, thus makes it frequency dependent. It is worth to mention that the resistance R_1 and capacitance C_1 of the cell structure are linearly related to

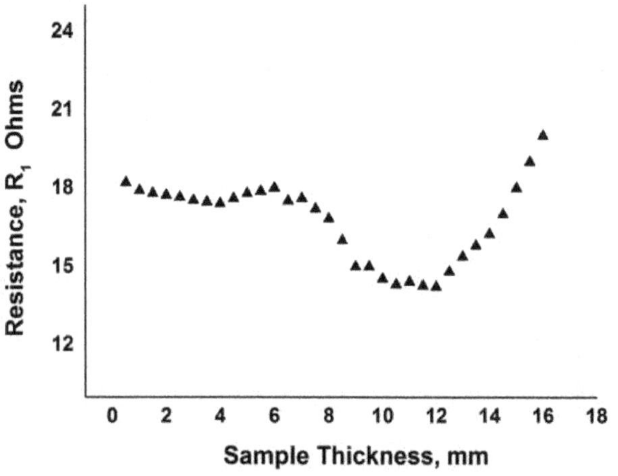

Fig. 8. Resistance of an intracellular fluid as a function of overlay thickness.

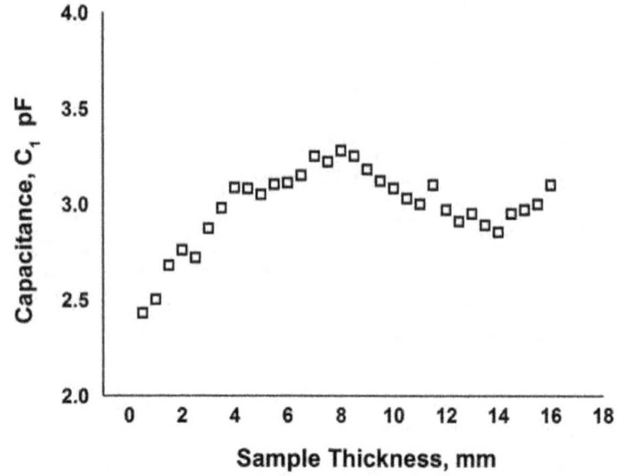

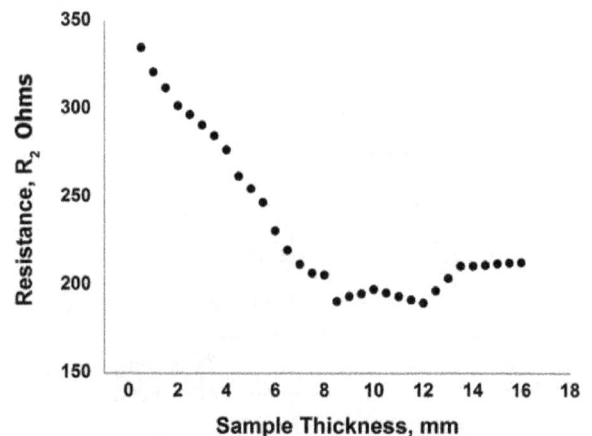

Fig. 9. Capacitance of cell membrane as a function of overlay thickness.

the resistivity of ICF and the dielectric constant of a membrane, respectively [31]. Thus, the effect of resistance is obvious on the magnitude of the resonance and similarly, the capacitance effect is apparent on the resonance shift. The resistance R_1 slightly decreases initially, but the transition is observed at thickness threshold, after that it increases rapidly. This rapid change in R_1 values is more than 27 %, which can be described as the lossy nature of intracellular fluid that contains K^+ ions. The thickness mainly contributes to the increase in resistive losses of the inner suspension.

As it can be seen in Fig. 9, the membrane capacitance C_1 increases up to thickness threshold, after that it seems somehow constant. This behavior of membrane capacitance seems to be inversely proportional to the resonance frequency shift, as f_1 decreases C_1 increases. Initially, the value of C_1 changes about 34.8 % up to the transition range. Beyond that range, there is minimum variation in their values, which is about 6 % only.

In Fig. 10, the resistance R_2 is given as a function of the sample thickness. This resistance exhibits the behavior of extracellular fluid (ECF), where it decreases till the thickness saturation range, but beyond that point the resistance value has minimum variation. The effect on R_2 is significant where it drops approximately from 335 Ω to 200 Ω (gives 40.3 % change) but beyond that transition point the change in its values is not more than 15 %.

Finally, the measurements are carried out using the same procedure as reported in [33]. The measured resonance shift, corresponding to different thicknesses is then

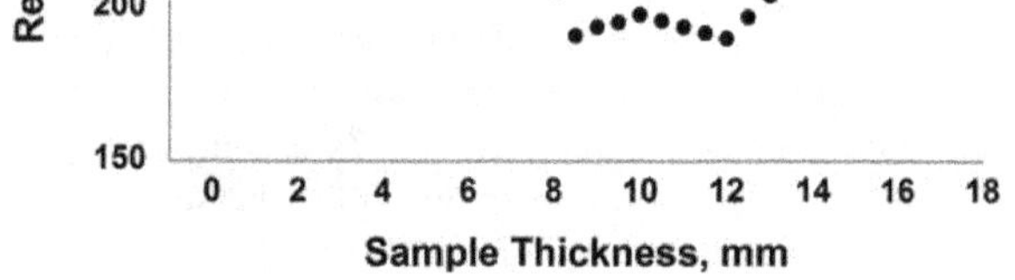

Fig. 10. Resistance of an extracellular fluid as a function of overlay thickness.

Results	Resonance Shift (MHz)	
	10 mm	15 mm
Simulated	213	218
Calculated	153	151
Measured	198	202

Tab. 2. Acquired resonance shift - from simulated, calculated and measured results for the 10 mm and 15 mm thick overlay biological tissues.

compared for validation of computed results. In Tab. 2 the comparison of simulated, calculated and measured results is presented. The observed discrepancy is within the acceptable limit, where the maximum difference observed between the measured and simulated values is about 7.5 %. It is due to actual physical properties of the samples used for measurements, handling and cutting of meat samples can change the amount of moisture, thus having significant effect on resonance frequency. Further, it is really hard to get the exact dimension of meat samples for measurements, due to its malleable nature. This can be also a possible reason for this discrepancy. The values obtained by simulation data are higher than those that are calculated; with a difference of about 24 %. The main reason of this difference is the FEM based full-wave analysis that utilizes the hybrid propagation mode, which involves complex analysis to make it more accurate. Additionally, in full-wave simulation variation of effective permittivity and characteristics impedance with respect to frequency is also performed. Although, transmission line theory can provide sufficient analysis but still it has its own limitations.

4. Conclusion

In this study a thorough analysis of a microwave biosensor based on a dielectric loaded microstrip ring resonator is performed. For the first time, a partial overlay arrangement of a lossy and high permittivity dielectric material, like biological tissue, on ring resonator is analyzed. The values of the lumped components of a 1GHz unloaded resonator are first determined and then, their response is compared with the simulated and measured results. The results are well agreed and the overall difference is not more than 0.3 %. For the loaded sensor, the relationship of overlay thickness with the RC values of the tissue model is established using equivalent circuit modeling. Finally, the resonance frequency of the sample loaded ring resonator is computed and validated with the measured results. The results are in good agreement while the observed discrepancy is in the acceptable limit.

References

[1] VENKATESH, M. S., RAGHAVAN, G. S. V. An overview of dielectric properties measuring techniques. *The Journal of the Canadian Society for Bioengineering (CSBE),* 2005, vol. 47, p. 7.15–7.30.

[2] ZAJÍČEK, R., OPPL, L., VRBA, J. Broadband measurement of complex permittivity using reflection method and coaxial probes. *Radioengineering,* 2008, vol. 17, no. 1, p. 14–19.

[3] AGILENT-TECHNOLOGIES, *Basics of Measuring the Dielectric Properties of Material.* CA, USA, 2006.

[4] DAMEZ, J.-L., CLERJON, S. Meat quality assessment using biophysical methods related to meat structure. *Meat Science,* 2008, vol. 80, no. 9, p. 132–149.

[5] DAMEZ, J.-L., CLERJON, S., ABOUELKARAM, S., LEPETIT, J. Beef meat electrical impedance spectroscopy and anisotropy sensing for non-invasive early assessment of meat ageing. *Journal of Food Engineering,* 2008, vol. 85, p. 116–122.

[6] BASARAN-AKGUL, N., BASARAN, P., RASCO, B. A. Effect of temperature (-5 to 130 degrees C) and fiber direction on the dielectric properties of beef Semitendinosus at radio frequency and microwave frequencies. *Journal of Food Science,* Aug 2008, vol. 73, no. 6, p. E243-9.

[7] SING, K. NG., GIBSON, A., PARKINSON, G., HAIGH, A., AINSWORTH, P., PLUNKETT, A. Bimodal method of determining fat and salt content in beef products by microwave techniques. *IEEE Transactions on Instrumentation and Measurement,* 2009, vol. 58, no. 10, p. 3778–3787.

[8] TRABELSI, S. N., STUART O. Use of dielectric spectroscopy for determining quality attributes of poultry meat. In *Annual International Meeting of the American Society of Agricultural and Biological Engineers (ASABE).* June 21-24, 2009, no. 097035, p. 8.

[9] HAGL, D. M., POPOVIC, D., HAGNESS, S. C., BOOSKE, J. H., OKONIEWSKI, M. Sensing volume of open-ended coaxial probes for dielectric characterization of breast tissue at microwave frequencies. *IEEE Transactions on Microwave Theory and Techniques,* 2003, vol. 51, p. 1194–1206.

[10] JILANI, M. T., WEN, W. P., ZAKARIYA, M. A, CHEONG, L. Y. Dielectric method for determination of fat content at 1 GHz frequency. In *5th International Conference on Intelligent and Advanced Systems.* Kuala Lumpur (Malaysia), June 2014.

[11] JILANI, M. T., WEN, W. P., ZAKARIYA, M. A, CHEONG, L.Y. Microstrip ring resonator based sensing technique for meat quality. In *IEEE Symposium on Wireless Technology and Applications.* Kuching (Malaysia), 2013.

[12] JILANI, M. T., WEN, W. P., ZAKARIYA, M. A, CHEONG, L.Y. A microwave sensor for non-destructive dielectric characterization of biological systems. *Journal of Microwaves, Optoelectronics and Electromagnetic Applications,* 2014 (accepted for publication).

[13] BERNARD, P. A., GAUTRAY, J. M. Measurement of dielectric constant using a microstrip ring resonator. *IEEE Transactions on Microwave Theory and Techniques,* 1991, vol. 39, no. 3 , p. 592 to 595.

[14] CHANG, K., HSIEH, L. H. *Microwave Ring Circuits and Related Structures.* John Wiley & Sons, 2004.

[15] GUPTA, K. C., GARG, R., BAHL, I., BHARTIA, P. *Microstrip Lines and Slotlines.* 2nd ed. Artech House, 1996.

[16] JILANI, M. T., WEN, W. P., ZAKARIYA, M. A, CHEONG, L.Y. Dielectric characterization of meat using enhanced coupled ring-resonator. In *IEEE Asia-Pacific Conference on Applied Electromagnetics,* 2014.

[17] YOGI, R., GANGAL, S., AIYER, R., KAREKAR, R. Split modes in asymmetric microstrip ring resonator by flexible perturbation. *Microwave and Optical Technology Letters,* 1998, vol. 19, no. 2, p. 168–171.

[18] SUMESH SOFIN, R. G., AIYER, R. C. Measurement of dielectric constant using a microwave microstrip ring resonator (MMRR) at 10 GHz irrespective of the type of overlay. *Microwave and Optical Technology Letters,* October 2005, vol. 47, no. 1, p. 11–14.

[19] ABEGAONKAR, M. P., KAREKAR, R., AIYER, R. C. A microwave microstrip ring resonator as a moisture sensor for biomaterials: application to wheat grains. *Measurement Science and Technology,* 1999, vol. 10, no. 3, p. 195.

[20] NAVARRO, J. A., CHANG, K. Varactor-tunable uniplanar ring resonators. *IEEE Transactions on Microwave Theory and Techniques,* 1993, vol. 41, no. 5, p. 760–766.

[21] HSIEH, L. H., CHANG, K. Equivalent lumped elements G, L, C, and unloaded Q's of closed-and open-loop ring resonators. *IEEE*

Transactions on Microwave Theory and Techniques, 2002, vol. 50, no. 2 , p. 453–460.

[22] PUCEL, R. A., MASSE, D. J., HARTWIG, C. P. Losses in microstrip. *IEEE Transactions on Microwave Theory and Techniques,* 1968, vol. 16, no. 6, p. 342–350.

[23] HOPKINS, R., FREE, C. Equivalent circuit for the microstrip ring resonator suitable for broadband materials characterisation. *IET Microwaves, Antennas and Propagation,* 2008, vol. 2, no. 1, p. 66–73.

[24] CHANG, K. MARTIN, S. WANG, KLEIN, J. L. On the study of microstrip ring and varactor-tuned ring circuits. *IEEE Transactions on Microwave Theory and Techniques,* 1987, vol. 35, no. 12 , p. 1288–1295.

[25] BRAY, J. R., ROY, L. Microwave characterization of a microstrip line using a two-port ring resonator with an improved lumped-element model. *IEEE Transactions on Microwave Theory and Techniques,* 2003, vol. 51, no. 5 , p. 1540–1547.

[26] BENEDEK, P., SILVESTER, P. Equivalent capacitances for microstrip gaps and steps. *IEEE Transactions on Microwave Theory and Techniques,* 1972, vol. 20, no. 11 , p. 729–733.

[27] GARG, R., BAHL, I. Microstrip discontinuities. *International Journal of Electronics,* 1978, vol. 45, no. 1, p. 81–87.

[28] DAMEZ, J.-L., CLERJON, S., ABOUELKARAM, S., LEPETIT, J. Dielectric behavior of beef meat in the 1–1500 kHz range: Simulation with the Fricke/Cole–Cole model. *Meat Science,* 2007, vol. 77, no. 12, p. 512–519.

[29] ELLAPPAN, P., SUNDARARAJAN, R. A simulation study of the electrical model of a biological cell. *Journal of Electrostatics,* 2005, vol. 63, p. 297–307.

[30] YANG, Y., WANG, Z.-Y. , DING, Q., HUANG, L., WANG, C., ZHU, D.-Z. Moisture content prediction of porcine meat by bioelectrical impedance spectroscopy. *Mathematical and Computer Modelling,* 2013, vol. 58, no. 3–4 , p. 819–825.

[31] SCHOENBACH, K. H., KATSUKI, S., STARK, R. H., BUESCHER, E. S., BEEBE, S. J. Bioelectrics - new applications for pulsed power technology. *IEEE Transactions on Plasma Science,* 2002, vol. 30, no. 1 , p. 293–300.

[32] DEAN, D., RAMANATHAN, T., MACHADO, D., SUNDARA-RAJAN, R. Electrical impedance spectroscopy study of biological tissues. *Journal of Electrostatics,* 2008, vol. 66, no. 3–4 , p. 165 to 177.

[33] JILANI, M. T., WEN, W. P., ZAKARIYA, M. A, CHEONG, L.Y. Determination of size-independent effective permittivity of an overlay material using microstrip ring resonator. *Microwave and Optical Technology Letters,* 2014 (accepted for publication).

[34] GOUKER, M. A., KUSHNER, L. J. A microstrip phase-trim device using a dielectric overlay. *IEEE Transactions on Microwave Theory and Techniques,* 1994, vol. 42, no. 11, p. 2023–2026.

About Authors ...

Muhammad Taha JILANI received his Bachelor's degree in Electrical Engineering and Master's degree in Communication Technology from Muhammad Ali Jinnah University, Karachi, Pakistan, in 2007 and 2009, respectively. He is currently working towards his PhD degree in Electrical and Electronics Engineering at Universiti Teknologi Petronas, Malaysia. His research interests focus mainly on RF and microwaves spectroscopy for food quality applications.

Wong Peng WEN was born in Perak, Malaysia in 1984. He graduated from the University of Leeds with First Class BEng (Hons.) degree in Electrical and Electronic Engineering and completed his PhD degree in the same university in 2009. He has worked as a Senior Lecturer in Universiti Teknologi Petronas since 2010. His research interests include passive filters, tunable filters and filter miniaturization techniques.

Lee Yen CHEONG received the B.Sc. degree (First Class Hons.) in Mathematics and Physics from the University of Reading, and Ph.D. degree in Mathematics from the University of York, U.K., in 2006 and 2009, respectively. He is a corporate member of the Institute of Physics, U.K. He is currently a senior lecturer at the Dept. of Fundamental and Applied Sciences, Universiti Teknologi Petronas, Malaysia. His main research fields include all aspects of theoretical elementary particle physics and general relativity.

Mohammad Azman Bin ZAKARIYA received the B.Sc. degree in Electrical Engineering from Universiti Teknologi Malaysia, and the M.Sc. degree in Communications and Signal Processing, both from the University of Newcastle upon Tyne, UK. He is a Lecturer at Universiti Teknologi Petronas, Malaysia, and is working towards his PhD at Universiti Sains Malaysia. His research interests include dielectric resonator antennas & defected ground structures.

Muhammad Zaka Ur REHMAN received the B.S. degree in Electronics from COMSATS Inst. of Information Technology, Islamabad, Pakistan in 2007, the MSc. degree in DSP in Communication Systems from Lancaster University, UK, in 2010, and is currently working toward the PhD degree in Electrical Engineering at the Universiti Teknologi Petronas, Perak, Malaysia. His research interests include RF MEMS for microwave applications, substrate integrated waveguide structures and reconfigurable filters design.

An Offset Cancelation Technique for Latch Type Sense Amplifiers

George SOULIOTIS [1], *Costas LAOUDIAS* [1], *Nikolaos TERZOPOULOS* [2]

[1] Dept. of Physics, University of Patras, 26504 Patras, Greece
[2] Oxford Brookes University, Wheatley Campus, Oxford, OX33 1HX, UK

gsoul@upatras.gr, laoudiask@upatras.gr, nterzopoulos@brookes.ac.uk

Abstract. *An offset compensation technique for a latch type sense amplifier is proposed in this paper. The proposed scheme is based on the recalibration of the charging/discharging current of the critical nodes which are affected by the device mismatches. The circuit has been designed in a 65 nm CMOS technology with 1.2 V core transistors. The auto-calibration procedure is fully digital. Simulation results are given verifying the operation for sampling a 5 Gb/s signal dissipating only 360 μW.*

Keywords

Offset cancelation, sense amplifiers, clocked comparators, latch circuits.

1. Introduction

Latch type sense amplifiers are commonly used in integrated circuits for communications and computer peripherals. They are important timing elements on a system that operates with tight timing restrictions, they need to take fast decisions in limited time window and give a fast output. This type of sense amplifiers are clocked circuits with high sensitivity in order to sense low level signal and high gain to be able to generate an amplified output signal. The required high input sensitivity makes them prone to design issues, technology mismatches and process variations.

There are two main types of latches, current mode and voltage mode. The voltage mode latch-type sense amplifiers are usually preferred because of their low static power consumption and high input impedance. One of the most commonly used latch type sense amplifier with several modifications is the so called StrongARM comparator [1], which is shown in Fig. 1 with several variations. Some of variations include complementary versions of this topology, or with the additional transistors, as depicted with the dashed lines. This circuit is a pre-charged differential sense amplifier followed by a pair of cross-coupled NAND gates. The concept behind its operation is based on charging-discharging of the differential pair and the corresponding nodes. The comparison of the discharging speed, between the nodes, gives a decision to the output stage. In that application [1] it is referred that there is no need for a well balanced amplifier because the input signal is in standard CMOS levels. However, many other applications, like 5 Gb/s serializers/deserializers use lower signal amplitude and sometimes this is only a few tens of milivolts. In this case, it is very important for the total differential pair circuitry, including the total capacitance load on the corresponding nodes, to be well balanced. Otherwise, wrong decision may be taken concerning the input signal. The transistor mismatch is an important issue that can affect the balanced capacitive load reducing circuit's input sensitivity. Using transistors with larger dimensions could reduce the mismatch effect [1], but as the bit rate in modern applications is usually high, the transistor size is critical and should be kept low [2], making the mismatch effect even worse. This effect is equivalent to an input-referred offset created by the difference between the two inputs of the differential input and sometimes could be as high as 50 mV. The offset must be compensated in order to ensure the proper sampling of the sense amplifier.

Several techniques have been proposed for offset cancelation. Most of them are based on capacitance recalibration on the charging-discharging nodes [3]-[5] by adding capacitance on the specific node that shows the less capacitive load. The control is digital, selecting capacitor-transistors among an array in order to rebalance the capacitive load. The concept is shown in the circuit of Fig. 2. Although, this is the most common configuration with the capacitors connected on the nodes *A* and *B*, in some other cases the capacitors have been added on the nodes *C* and *D* [6]. The drawback of this approach is that the capacitance of the transistor is varied with the gate voltage [5], reducing the speed or reducing the linearity of the compensation [6]. So the compensation may become inaccurate while the voltage varies between V_{DD} and ground. In [7]-[10] the offset cancelation is realized by the injection of current in nodes *A* and *B* using parallel connected transistors with the input transistors *M1* and *M2*. These extra transistors that have been added at the input can increase the kickback charge and worsen the CMRR [4]. This design requires Digital to Analog Converters (DACs) for switching on/off the parallel transistors making the final topology too complex. A recent work utilizes charge-pump and bulk control [11], while some older techniques used preamplifiers in a unity-gain closed loop, to create input and output offset storage [12].

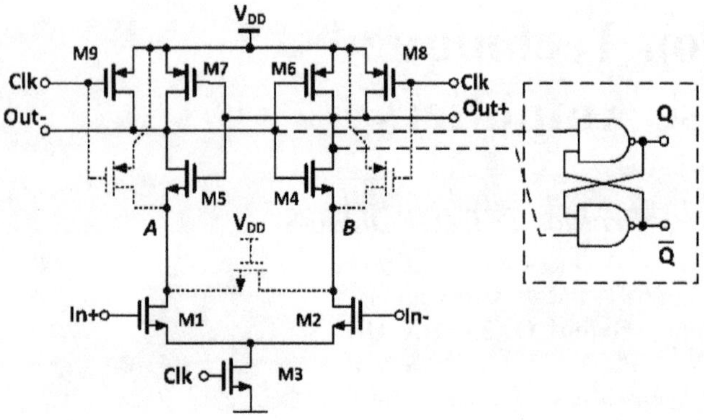

Fig. 1. StrongARM Latch type sense amplifier.

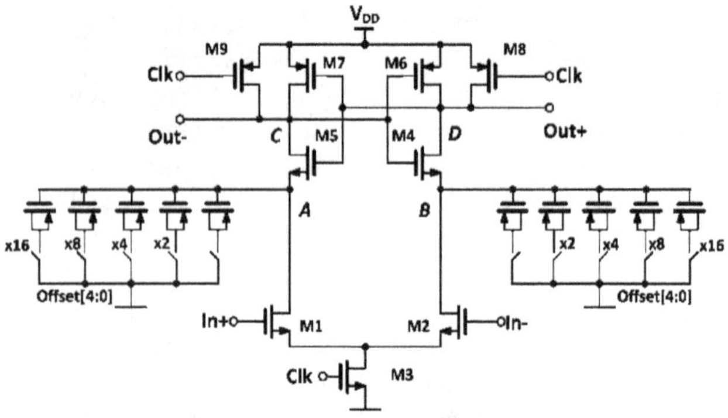

Fig. 2. Offset cancelation scheme based on capacitance control.

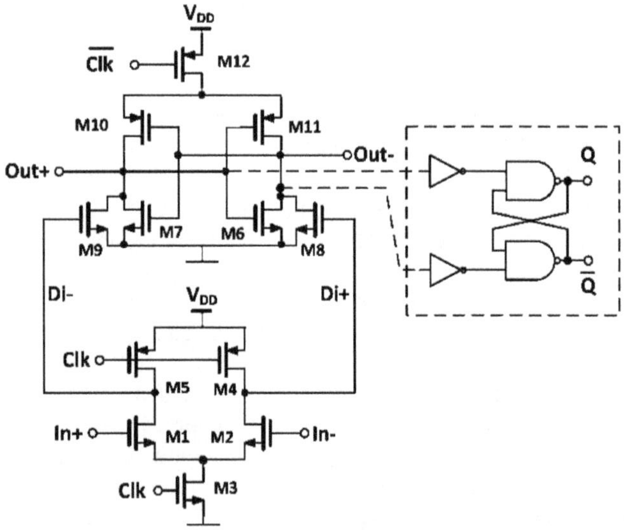

Fig. 3. Improved latch type sense amplifier.

A new accurate technique for the offset cancelation is proposed in this paper. This is based on the current injection in the comparison nodes, without affecting the differential input pair and it is applied on a low voltage latch-type sense amplifier [13]. The result is an accurate offset compensation technique using a low supply voltage, with a very simple and area efficient circuit. In addition, it does not significantly affect the kickback noise and the output delay of the latch-type comparator, while shows smaller power consumption comparing with other proposed topologies.

2. Latch-Type Sense Amplifier

A latch type sense amplifier suitable for a 5 Gb/s deserializer is the double-tail-latch-type sense amplifier, shown in Fig. 3 [13]. Comparing it with the Strong-Arm clocked comparator, this sense amplifier has fewer transistors stacked from rail to rail, has better isolation between input and output and is less sensitive to common mode variation. Though, it needs positive and negative clock.

The Strong-Arm comparator has some drawbacks, as thoroughly analyzed in [14] due to the placement of the cross-coupled inverters in series with the differential pair and because there is only a very short time within one period in which the differential pair has actually gain. This makes it sensitive to common mode input signal, especially when, like in our application, this is close to V_{DD}. Also, the total current is identical for latch and differential pair leaving no space for optimization. The double-tail comparator improves these drawbacks by separating the differential pair with the cross-coupled latch [13]-[16]. The performance under low supply conditions is improved due to fewer transistors stacked. The tail current in the differential pair (through transistor *M3*) is significantly lower than the current of the latching stage (through transistor *M12*) providing high differential gain with fast latching. Then, the decision time can be significantly small while the gain of the operational amplifier is still high, resulting to decreased probability of metastability errors. The fact that the high current flows only in the second stage, which is isolated by the intermediate stage formed by *M8* and *M9*, provides additional shielding to the first stage thus exhibiting less kickback noise. For the aforementioned reasons the double tail sense amplifier has been chosen for our application.

The latch operation of the circuit in Fig. 3 is based on the comparison of the charge between the nodes *Di+* and *Di-* as shown in Fig. 4. It operates in four phases: reset, sample, regeneration and decision. During the reset phase, when *clk* is 0, as shown in Fig. 4a, the input signal goes close to its final logic state (Fig. 4b). The transistors *M4* and *M5* pre-charge the *Di* nodes to V_{DD} as depicted in Fig.4c and transistors *M8* and *M9* discharge the nodes *Out+* and *Out-* to ground, as shown in Fig. 4d. During the sample phase, at the positive edge of the clock, both transistors *M3* and *M12* are turned on. Nodes *Di* start discharging with a rate, approximately, of $I_{M3}/2C_{Di}$. Because of the differential input voltage V_{In}, transistors *M1* and *M2* give different drain current, and the transistor with the higher input voltage help the corresponding drain to discharge slightly faster. A ΔV_{Di} is created, during the sample phase, which is passed by *M8* and *M9* to the cross-coupled inverters, formed by transistors *M6*, *M11* and *M7*, *M10* during the regenerating phase. During this phase, the two inverters start generating the output decision. The role of the transistors *M8* and *M9* is important not only transferring the differential voltage to the next stage but also, minimizing the kickback effect to the input. An SR-latch is connected after *Out+* and *Out-* to create a static output *Q*, as shown in Fig.4e. The SR-latch is always used in latch-type sense amplifiers to keep the final output *Q* constant during the reset phase where both nodes *Out+* and *Out-* return to zero voltage.

Depending on the application, the output stage can be developed by a more sophisticated SR-latch instead of the common SR-latch in order to improve the symmetry in crossing points, rice and fall times, duty cycle, etc [17].

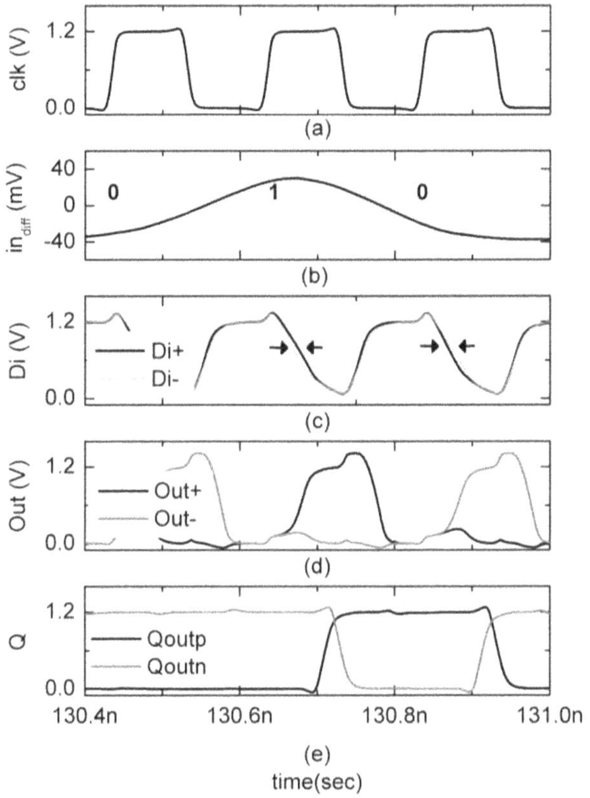

Fig. 4. Signaling of the latch type sense amplifier a) clock, b) differential input, c) charging nodes *Di*, d) pre-output and e) final output Q.

In our design the common SR-latch was selected among other, more advanced topologies that showed, however, high sensitivity on process variations derived from Monte-Carlo simulations. The reason for this high sensitivity is the increased complexity which although improves the operation, at the same time makes the circuit vulnerable to transistor mismatches. The decision phase takes place during the positive edge of the *clk* signal. The drain current I_{D3} of *M3* becomes, then, high and all transistors operate in saturation region. Depending on the input voltage V_{In} the drain currents of the identical transistors *M1* and *M2* are, respectively,

$$I_{D1} = \beta(V_{In+} - V_T)^2, \tag{1}$$

$$I_{D2} = \beta(V_{In-} - V_T)^2 \tag{2}$$

where β is the technology transconductance parameter $\beta = \mu_o C_{ox} W/2L$, $V_T = V_{SI} - V_{th}$, W and L is the width and length of *M1* and *M2*, V_{SI} is the *M1* and *M2* source voltage and V_{th} the threshold voltage of the transistors. The current difference is [18],

$$\Delta I = 2\beta(V_{CM} - V_T)\Delta V_{In} \tag{3}$$

where V_{In} is the differential input voltage and V_{CM} the common mode input voltage, as given in (4) and (5)

$$V_{In} = V_{In+} - V_{In-}, \tag{4}$$

$$V_{CM} = (V_{In+} + V_{In-})/2. \tag{5}$$

From (3) it is obvious that the common mode input must be greater than the threshold voltage depending on the process and the source voltage of *M1*, *M2*. Some important currents and voltages on the internal nodes of the amplifier consisting of the transistors *M1-M5* are shown in Fig. 5. The period of time where the three phases, sample, regeneration and decision take place after reset, is illustrated in the shaded area. During the positive edge of the clock, shown in Fig. 5a, the input signal is already at the logic state of either 0 or 1. Its amplitude may be as low as 40 mV differential, as shown in Fig. 5b. Also, during the positive clock edge the differential pair consisting by transistors M1 and M2 becomes active very fast, as shown in Fig. 5c and Fig. 5d, but they have a small difference between their drain currents, which depends on the logic level of the input signal. The absolute current difference is shown in Fig. 5e. At the same time the nodes *Di* discharge at a slightly different rate, as shown in Fig. 5f and according to the current difference of the differential pair, generating a differential voltage between the nodes Di, as shown in Fig. 5g. The high gain of the differential pair *M1-M2* is important for creating a high differential voltage ΔV_{Di}. The differential amplifier's gain is given by,

$$A_d = g_m r_o \tag{6}$$

where g_m is the transconductance of the transistors *M1, M2* and r_o is the equivalent resistance on the nodes *Di*. Large transistors not only aim to create high transconductance but keep the mismatch at low levels which is important to create proper output decisions, particularly if the input is of an extremely low amplitude as in the case shown in Fig. 5, where the differential input amplitude is less than 40 mV. On the other hand the wide transistors reduce the operating frequency as they add higher capacitive load on the nodes *Di*. This can deteriorate the aperture time and leave a small time window for the reset phase, resulting to a higher number of decision errors.

3. The Proposed Offset Compensation Technique

The mismatch between the symmetric transistors is a significant effect that can lead the latch to completely false decisions, especially if the input voltage level is very small. The mismatch creates an equivalent offset which must be compensated. The offset V_{off} created by the switching differential pair with mismatch transistors is studied in details in [18] and is found equal to

$$V_{off} = \Delta V_{th} - \frac{\Delta\beta}{2\sqrt{2\beta}}\sqrt{\frac{I_{D3}}{\beta}}, \qquad (7)$$

$$\beta = \frac{\beta_1 + \beta_2}{2} = \frac{1}{2}\left(\frac{\mu C_{ox} W_1}{2L_1} + \frac{\mu C_{ox} W_2}{2L_2}\right) \qquad (8)$$

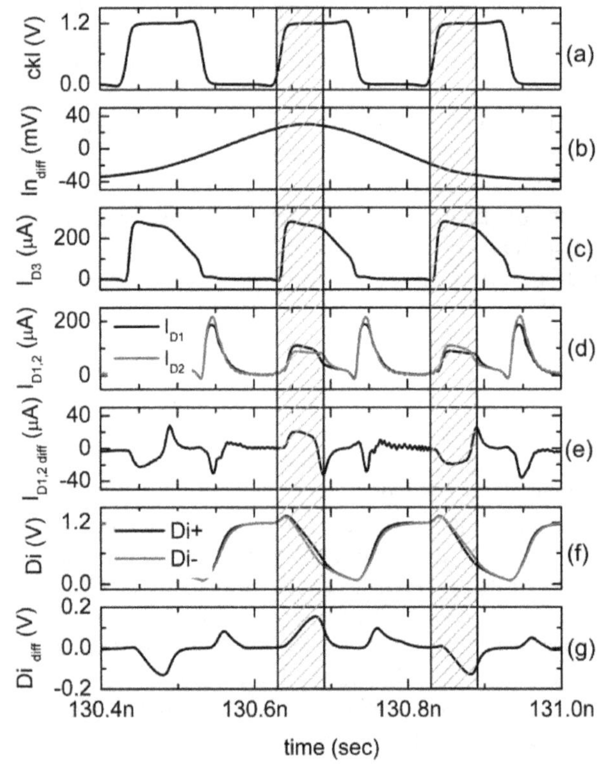

Fig. 5. Current and voltage in transistors M1-M5 a) clock b) differential input c) drain current of M3, d) drain current of M1 and M2, e) difference of drain currents of M1 and M2, f) voltage on charging nodes *Di*, g) differential voltage of *Di+/Di-* nodes.

where $\Delta\beta=\beta_1-\beta_2$ is the difference between the technology transconductance parameters of the two transistors and ΔV_{th} is the difference of the threshold voltage.

The proposed compensation technique was implemented mainly due to the common mode signal sensitivity of the capacitor compensation approach. A detailed analysis including simulation results of that method is given in [5]. Also, the additional capacitors affect the slew rate of the differential pair and may worsen the aperture time, limiting the operation of the clocked comparator. The operating frequency can be estimated by

$$f = \frac{1}{2\pi C_{Di} r_o} \qquad (9)$$

where, C_{Di} is the total capacitance on node *Di* and r_o is the equivalent resistance on those nodes. Above this frequency the gain of the differential pair is gradually reduced.

The discharging rate on each branch of *M1* and *M2* can be controlled not only by the capacitive load but also by the current flowing through these transistors. While some proposed methods use calibrated differential pair to cancel the offset, in our case the discharging current can be controlled by the transistors *M4* and *M5*. Connecting transistors in parallel with *M4* and *M5* the total mismatch influence on the discharging rate can be compensated. The proposed topology is shown in Fig. 6. Although these transistors add load on the clock path, this is not important for two reasons: firstly, because the clock is connected on the transistors through switches and secondly, because the standard high swing CMOS signal is strong enough to drive any block in the system through buffers from the clock tree. In this case, it is preferable to slightly load the clock path over loading any other critical signal nodes coming from the input.

The detection of the mismatches takes place during the training phase of the system. If the pairs *M1-M2* and *M4-M5* are perfectly matched, then the transistors from both sides of the transistor arrays are disconnected. In the case where a mismatch exists, then a proportional number of the weighted transistors are set to on in order to compensate for the mismatch effect. During the training phase there is no data transmission and the two inputs (positive and negative) of the latch type sense amplifier are tied on the same common mode voltage, around 0.9 V, generated by the previous stage. The accurate value of the common mode voltage is not critical for the offset compensation because of the differential nature of the topology, although, it might be critical for the output delay in normal operation [13], depending on the specific topology of the latched comparator. At first, only the transistors of one side are enabled. This enforces the output *Q* at ground potential. A counter starts disabling the transistors from one side to the other. The counting takes place with a clock running at lower speed than the maximum available. This ensures that the compensation procedure will be performed in lower speed, giving all the required time to the counter and to sense amplifier to find accurately the right selection. During this process, whenever the output *Q* changes state from ground to V_{DD}, as shown in Fig. 7b, the counting stops, as shown in Fig. 7a and the selected compensating transistors are locked in their final state using memory latches. The recalibration could be restarted if required by setting the reset signal from 0 to 1, as shown in Fig. 7c. The merit here is that there is no need for recalibration until the next power-on of the system. A possible temperature variation during the operation has

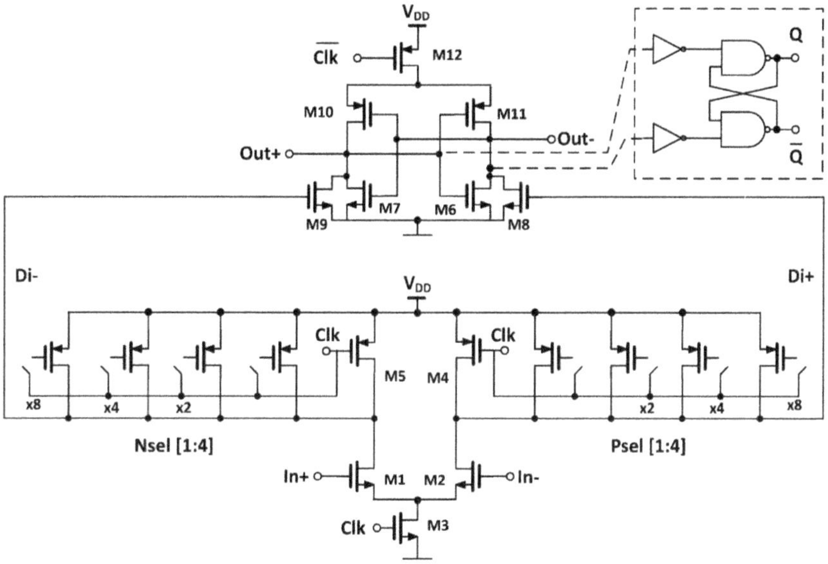

Fig. 6. Improved latch type sense amplifier with offset compensation.

not a potential effect, in first order, due to the differential nature of the circuit.

This compensation technique could be combined with other comparators, as for example with modified versions of the Strong-Arm taking advantage of the transistors with the dashed lines, shown in Fig. 1. These transistors already used to better reset the *Di* nodes and they could be used to control the offset. However, this requires a major redesign of the sense amplifier including the transistors in order to optimize its operation and therefore this design procedure has not been performed.

4. Simulation Results

The double tail latch type sense amplifier with the proposed offset compensation method has been designed in 65 nm CMOS technology and the postlayout simulation results demonstrate the operation before and after the compensation. Only standard threshold transistors (SVT) and not low V_{th} transistors (LVT) were used in this design. The transistors aspect ratio is depicted in Tab. 1. The supply voltage is 1.2 V and the input voltage used for demonstration and verification was less than 20 mV single-ended input. In reality the signal expected at this point on a full system implementation is greater than this value. The data input rate was 5 Gb/s.

Monte-Carlo analysis was performed for the same input with and without compensation. A differential rising ramp voltage is applied to the input in order to test the offset. If the sense amplifier is perfectly matched, then the output changes its state from low to high, when the differential input voltage is zero. In Fig. 8a the switching output Q is shown for multiple Monte-Carlo runs without compensation and in Fig. 8b with compensation, when the rising ramp voltage feeds the input in a corresponding time,

Transistor	W (nm) / L (nm)
M1,M2	3120 / 120
M3, M8, M9	600 / 60
M4, M5	1200 / 120
M6, M7	780 / 60
M10, M11	3120 / 60
M12	6240 / 120

Tab. 1. The transistors' aspect ratio.

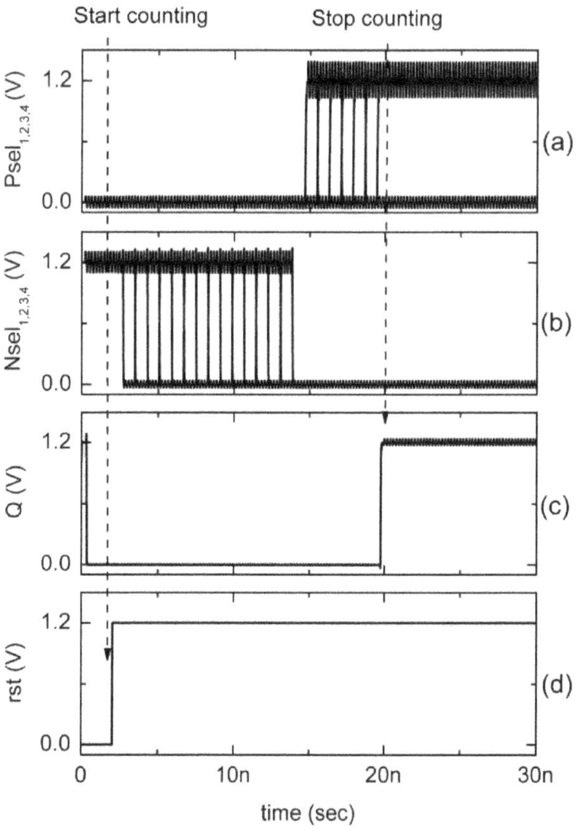

Fig. 7. Counting select bits of a) right side transistors, b) left side transistors and c) reset signal.

as shown in Fig. 8c. Without the compensation, the offset is found to be within the range of ±40 mV, as shown in Fig. 8a. After the compensation the offset is limited into the range of ±2.5 mV. The output *Q* is shown in Fig. 8a and Fig. 8b to change state in quantized values due to clocked operation.

Although the range could be further improved, the obtained° result is good enough for almost any application and therefore no more effort has been given to improve it. However, possible improvement can be accomplished by achieving better resolution of the compensation transistors and increasing the number of control bits. Running 30 Monte-Carlo runs the offset deviations without and with the compensation, respectively, are shown in Fig. 9a and Fig. 9b.

When the offset cancelation process is finalized, some extra transistors stay connected in parallel with *M4* or *M5* in Fig. 6 to create an actual balance in charging/discharging process. After the calibration and under high temperature variation a new small input offset may be regenerated due to varied unbalanced g_m of the correspond transistors. Simulations showed that this is negligible and it is about 2 mV. However, it does not affect the operational ability of the circuit, as the input signal is always much higher than this level. This effect is even smaller in the case of supply voltage variations. The simulation showed negligible offset for supply voltage from 1.1 V to 1.3 V.

The kickback noise in the clocked comparators is important and should not only be kept low, but also to be as invariant from the compensation as possible. The parallel connected transistor in *M4* and *M5* create the current balance between the two branches of *M1* and *M2*, but the total current tail, which finally is responsible for the higher amount of kickback effect, is controlled by *M3*. The current between the two branches has only a small difference, as shown in Fig. 5. The influence of the compensating transistors into the total kickback is small because they only rebalance the available current and they do not increase the total current. For testing purposes, this can be shown by applying common mode signal in the latched-comparator and starting the counting process as doing to find the offset voltage. All the left transistors are initially connected and gradually disconnected. The results are shown in Fig. 10 where in Fig. 10a the counting-down of the switches is depicted and in Fig. 10b the differential input voltage with kickback noise. Fig. 10c and Fig. 10d depict individually the kickback noise in the two inputs *In+* and *In-*, respectively showing the small effect of the compensating on the kickback noise.

The output delay was measured with and without the proposed compensation technique. The results are shown in Fig. 11 where an insignificant additional delay of 5.7°ps is observed (Fig.°11a) due to the connected compensating transistors. This delay is shown when all the compensating transistors are off, but it worsens only 3 ps, when the one side of the transistors are on to compensate an offset of about 25°mV. The delay without the compensation is 39°ps measured from the positive edge of the clock, as shown in Fig. 11b and Fig. 11c, respectively. The active transistors add current helping the slower node to discharge faster. Therefore, there is no significant additional delay. The main difference with the commonly used capacitance-compensation technique is the capacitance added in order to equalize the faster discharging node *Di* with the slower discharging node giving additional total delay to the decision phase.

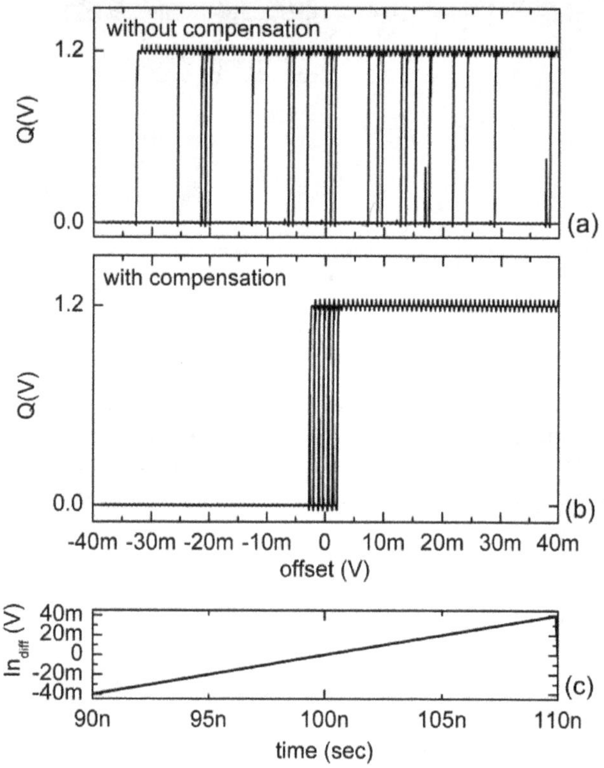

Fig. 8. Offset demonstration a) before and b) after compensation c) with a ramp input signal.

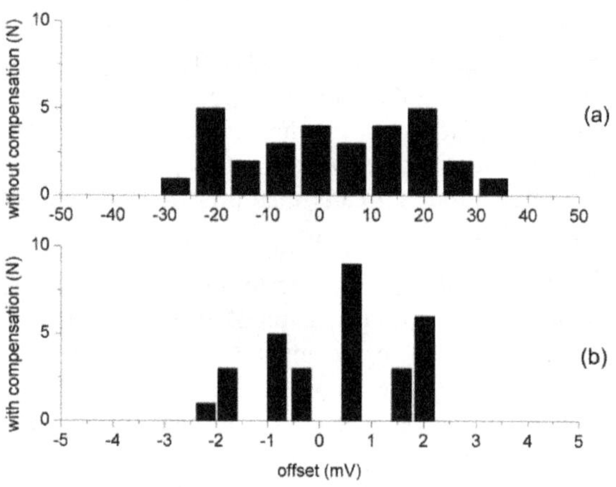

Fig. 9. Offset deviations from Monte-Carlo simulations a) before and b) after compensation.

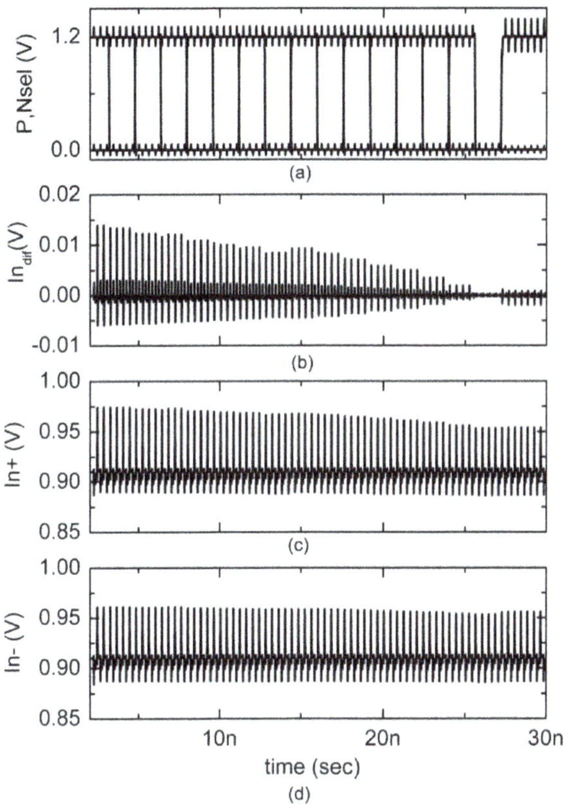

Fig. 10. Kickback noise during the offset compensation process a) right side transistors counting, b) kickback on differential input signal, c) kickback on positive input and d) kickback on negative input.

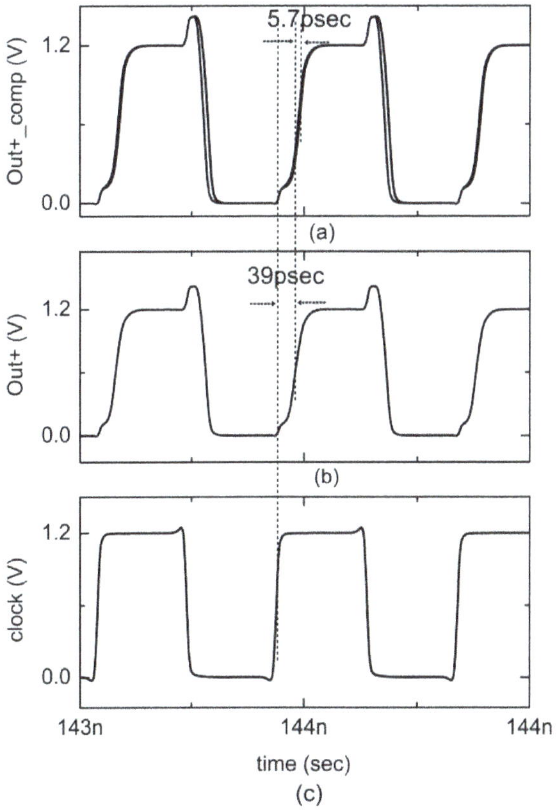

Fig. 11. Output delay a) after the offset compensation, b) before the compensation and c) the clock reference.

The layout of the latch type sense amplifier is shown in Fig. 12. The total area is 31 μm x 33 μm, including the additional SR-latch of the output and the digital counter which is used to select the correct number of the compensating transistors.

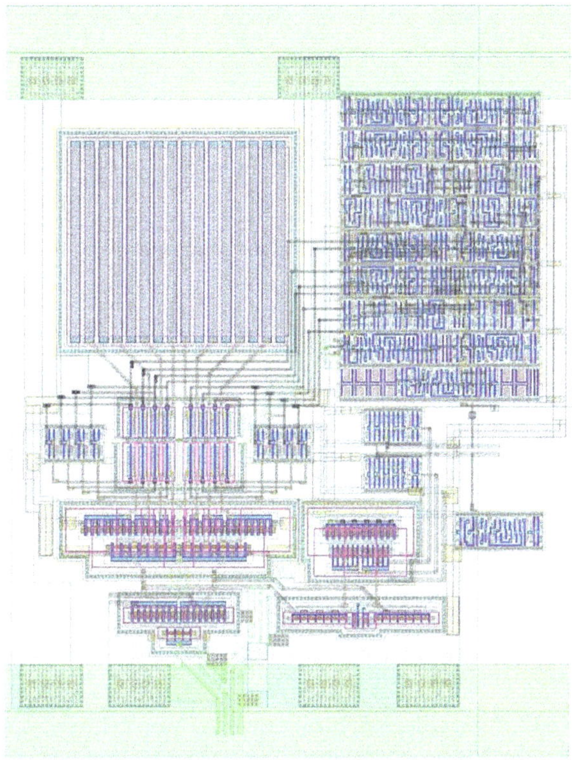

Fig. 12. The layout of the latch type sense amplifier.

The key characteristics of the proposed technique, together with these of some previously published offset compensation topologies are presented in Tab. 2. It should be noted that most of the tabular data in Tab. 2 are measurement results, but in our case the reported findings are based on post-layout simulations and Monte-Carlo analysis. One important point in the proposed technique is that the delay is not affected by the compensation. This is because no capacitive load is added on any of the internal nodes, as in [3], [4], [6] and the current injection in the proposed method is performed in the PMOS clock transistors. Also, that means that the differential pair remains untouched and small in size, without overloading the preceding stage that could be sensitive. This assures that the proposed topology has the minimum power consumption over frequency as shown in Tab. 2. Furthermore, the suggested method is immune to common mode voltage unlike other similar methods, which use mos-capacitors. The layout area is the smallest compared with other techniques and the power consumption is the best compared with the high speed comparators, due to the absence of the compensating capacitors.

Although the residual offset is not the minimum found in the relevant literature, is not far off and is good enough for our application, as the input signal must be

Work	CMOS feature size (nm)	Sample rate (Gb/s) or clock (GHz)	Offset before compensation (mV)	Type of compensation	Area (µm x µm)	Supply Voltage (V)	Power consumption (mW)	P/f (mW/ GHz)	Residual offset (mV)
[3]	250	1.6 GS/s	±70	Capacitive	80 x 50	2.5	3	1.875	±6
[4]	90	3.2 GHz	±70	Capacitive	N.A.	1.0	N.A.	N.A.	±1.5
[6]	90	1.5 Mb/s	18	Capacitive	N.A.	0.5	N.A.	N.A.	±1
[7]	65	1.5 GS/s	19	Current injection in diff. pair	N.A.	1.1	N.A.	N.A.	4.75
[8]	180	1.4 GHz	95	Current injection in diff. pair	N.A.	1.8	0.35	0.25	13
[9]	90	250 MHz	13.7	Current injection in diff. pair	120 x 4.5	1.0	0.04	0.16	1.69
[11]	180	200 MS/s	29.9	Charge pump and bulk driven	N.A.	0.5	0.034	0.17	0.3
This work	65	5 GS/s	40	Current injection in clock transistors	31 x 33	1.2	0.36	0.072	±2.5

Tab. 2. Key characteristics of the different offset compensation techniques.

always greater than 30 mV differential according to the system specifications. The performance in terms of speed and supply voltage can be improved significantly by taking advantage of the low threshold voltage transistors (LVT) offered by the selected process. However this will require additional masking costs when comes to fabrication. For all the reasons described above the proposed technique outperforms all of the previously reported schemes, especially when the output delay performance is important.

5. Conclusions

A new offset compensation technique for latch-type sense amplifier has been presented. The proposed topology controls the bias current of the clocked transistors instead of changing the capacitance on the internal nodes of the sense amplifier. Therefore, it does not need any preamplifier, the linearity does not depend on the voltage of the capacitor transistors which usually are used for the compensation and the calibration process is in a fully digital way. It can operate at 5 Gb/s data rate with 1.2 V voltage supply using typical CMOS transistors (SVT) consuming only 360 µW. The final offset is reduced from ±40 mV to only ±2.5 mV which is much better than the required for most of high speed applications.

References

[1] MONTANARO, J., WITEK, R. T., ANNE, K., BLACK, Z., COOPER, E. M., DOBBERPUHL, D. W. et al. A 160-MHz, 32-b, 0.5-W CMOS RISC microprocessor. *IEEE Journal of Solid-State Circuits*, 1996, vol. 31, no. 11, p. 1703 - 1714.

[2] PELGROM, M., DUINMAIJER, A., WELBERS, A. Matching properties of MOS transistors. *IEEE Journal of Solid-State Circuits*, 1989, vol. 24, no. 5, p. 1433 - 1439.

[3] EMAMI-NEYESTANAK, A., LIU, D., KEELER, G., HELMAN, N., HOROWITZ, M. A 1.6 Gb/s, 3 mW CMOS receiver for optical communication. In *Symposium on VLSI Circuits Digest of Technical Papers*. Honolulu (HI, USA), 2002, p. 84 - 87.

[4] PALERMO, S., EMAMI-NEYESTANAK, A., HOROWITZ, M. A 90 nm CMOS 16 Gb/s transceiver for optical interconnects. *IEEE Journal of Solid-State Circuits*, 2008, vol. 43, no. 5, p. 1235 - 1246.

[5] PALERMO., S. *Design of High-Speed Optical Interconnect Transceivers*, Ph.D. Dissertation. Stanford University, 2007.

[6] GAMBINI, S., RABAEY, J. Low-Power successive approximation converter with 0.5 V supply in 90 nm CMOS. *IEEE Journal of Solid-State Circuits*, 2007, vol. 42, no. 11, p. 2348 - 2356.

[7] EL-CHAMMAS, M., MURMANN, B. A 12-GS/s 81-mW 5-bit time-interleaved flash ADC with background timing skew calibration. *IEEE Journal of Solid-State Circuits*, 2011, vol. 46, no. 4, p. 838 - 847.

[8] WONG, K.-L. J., YANG, C.-K. K. Offset compensation in comparators with minimum input-referred supply noise. *IEEE Journal of Solid-State Circuits*, 2004, vol. 39, no. 5, p. 837 - 840.

[9] MIYAHARA, M., ASADA, Y., DAEHWA, P., MATSUZAWA, A. A low-noise self-calibrating dynamic comparator for high-speed ADCs. In *Proceedings of IEEE Asian Solid-State Circuits Conference*. Fukuoka (Japan), 2008, p. 269 - 272.

[10] ELLERSICK, W., YANG, C.-K. K., HOROWITZ, M. DALLY, W. GAD: A 12-GS/s CMOS 4-bit A/D converter for an equalized multi-level link. In *Proceedings of Symposium on VLSI Circuits. Digest of Technical Papers*. Kyoto (Japan), 1999, p. 49 - 52.

[11] KHANGHAH, M. M., SADEGHIPOUR, K. D. A 0.5 V offset cancelled latch comparator in standard 0.18 µm CMOS process. *Analog Integrated Circuits and Signal Processing*, 2014, vol. 79, no. 1, p. 161 - 169.

[12] RAZAVI, B., WOOLEY, B. A. Design techniques for high-speed, high-resolution comparators. *IEEE Journal of Solid-State Circuits*, 1992, vol. 27, no. 12, p. 1916 - 1926.

[13] SCHINKEL, D., MENSINK, E., KIUMPERINK, E., VAN TUIJL, E., NAUTA, B. A double-tail latch-type voltage sense amplifier with 18 ps setup+hold time. In *Proceedings of IEEE International Solid-State Circuits Conference. Digest of Technical Papers*. San Francisco (USA), 2007, p. 314 - 605.

[14] SCHINKEL, D. *On-Chip Data Communication Analysis, Optimization and Circuit Design*, PhD Dissertation. Twente (Netherlands): University of Twente, 2011. DOI: 10.3990/1.9789036532020.

[15] FIGUEIREDO, P. M., VITAL, J. C. Kickback noise reduction techniques for CMOS latched comparators. *IEEE Transactions on Circuits and Systems – II: Express Briefs*, 2006, vol. 53, no. 7, p. 541 - 545.

[16] KIM, J., LEIBOWITZ, B. S., REN, J., MADDEN C. J. Simulation and analysis of random decision errors in clocked comparators. *IEEE Transactions on Circuits and Systems – I: Regular Papers*, 2009, vol. 56, no. 8, p. 1844 - 1857.

[17] NIKOLIC, B., OKLOBDZIJA, V., STOJANOVIC, V. G., JIA, W., CHIU, J. K.-S., LEUNG, M. M.-T. Improved sense-amplifier-based flip-flop: design and measurements. *IEEE Journal of Solid-State Circuits*, 2000, vol. 35, no. 6, p. 876 - 884.

[18] FILANOVSKY, I. Switching characteristic of MOS differential pair with mismatched transistors. *International Journal of Electronics*, 1988, vol. 65, no. 5, p. 999 - 1001.

About Authors ...

George SOULIOTIS received his B.Sc. in Physics from the University of Ioannina, Greece, in 1992, his M.Sc. in Electronics from the University of Patras, Greece in 1998 and his Ph.D. from the University of Patras, Greece in 2003. He is currently a member of the technical staff of the Department of Physics, University of Patras, Greece. Also, from 2004 he has been as a Post-Doctoral Researcher with the Electronics Laboratory, Department of Physics, University of Patras, Greece. Dr Souliotis serves as a reviewer for many international journals and he is a member in national and international professional organizations. His research interests include integrated analog circuits and filters, current mode circuits and low voltage circuits.

Costas LAOUDIAS was born in Sparta, Greece in 1984. He received the B.Sc. in Physics, the M.Sc. in Electronics and Computers and the Ph.D. in Analog IC design, all from Physics Department, University of Patras, Greece in 2005, 2007 and 2011, respectively. In June 2011, he joined Analogies S. A. as an Analog/Mixed-signal IC Design Engineer working mainly in the circuit design for high-speed serial data communication interfaces. Since April 2013, he is with Cambridge semiconductors where he is working in the area of power management IC design. His recent research interests include design of low-voltage analog integrated circuits for signal processing, ultra low-voltage continuous-time filters for biomedical applications, IQ filters for wireless receivers.

Nikolaos TERZOPOULOS received his B.Eng. from Technical Educational Institution of Thessaloniki, Greece and he holds M.Sc. and Ph.D. in Electronic Engineering, both from Oxford Brookes University, Oxford, UK. In 2011 he also received his MBA degree from Hellenic Management Association. Employed by Analogies S.A. in 2009 as an Analog IC Design Engineer Nikolaos Terzopoulos is the main driver in developing low power high speed analog blocks (such as: Serdes, PLL, equalizers and output drivers) for USB 3.0/PCIe IP Cores. He is also a research fellow at Oxford Brookes University, Oxford, UK. Nikolaos Terzopoulos was previously employed by Texas Instruments as part of the Serdes Analog IC Design group where he has designed a number of silicon proven though mass production low power analog high speed cells such as a 25 Gbps Transmitter for chip to chip communication, a 3.75 GHz PLL and a 5 Gbps Transmitter for Serdes applications both at 45 nm process node. He is also a Research Fellow and a member of the Analog IC design group of Oxford Brookes University, Oxford, UK, supervising a number of Ph.D. candidates in the following research projects: Low Noise Amplifiers (LNA's), Power Amplifiers, Medical instrumentation and PLL's.

Fully CMOS Memristor Based Chaotic Circuit

Şuayb Çağrı YENER [1], H. Hakan KUNTMAN[2]

[1] Dept. of Electrical and Electronics Engineering, Sakarya University, Sakarya, Turkey
[2] Dept. of Electronics and Communication Engineering, Istanbul Technical University, Istanbul, Turkey

syener@sakarya.edu.tr, kuntman@itu.edu.tr

Abstract. *This paper demonstrates the design of a fully CMOS chaotic circuit consisting of only DDCC based memristor and inductance simulator. Our design is composed of these active blocks using CMOS 0.18 μm process technology with symmetric ±1.25 V supply voltages. A new single DDCC+ based topology is used as the inductance simulator. Simulation results verify that the design proposed satisfies both memristor properties and the chaotic behavior of the circuit. Simulations performed illustrate the success of the proposed design for the realization of CMOS based chaotic applications.*

Keywords

Memristor, CMOS design, DDCC, Chua's circuit, chaotic oscillators.

1. Introduction

In 1971, Leon Chua theoretically claimed that the memristor is the fourth circuit element besides the three well-known circuit elements; namely, resistor, capacitor and inductor [1]. For a long time, it remained just as a theoretical element and rarely appeared in the literature because of having no simple and practical realization. In 2008, a group of researchers from HP laboratories announced the fabrication of a physical implementation behaving as a memristor [2]. Its prototype is based on a TiO_2 thin film containing doped and un-doped regions between two metal contacts at nanometer scale. This implementation, realized by HP researchers, has attracted significant attention.

It is expected that memristors can be applied and provide new additional features to analog circuits. Various analog and chaotic applications of memristor to analog, chaotic and synaptic circuits are studied in the literature [3]-[11]. Despite large-scale interest on memristor and emerging many studies, no commercially available memristor exists yet. In this sense, a proper physical implementation representing the memristor behavior is of great importance from the point of view of real-world circuit design.

SPICE macromodels and memristor emulators exhibiting memristor-like behavior are presented in the literature [12]-[22]. SPICE models are useful for modeling characteristics of the memristor, but they have not been an alternative in practical realizations. Emulators can represent the behavior of memristor in a restricted extent and they can be applicable on some real applications.

Some inductorless implementations of Chua's circuit have been presented in the literature [23], [24]. Implementation approach of CMOS memristor employing Differential Difference Current Conveyor (DDCC) based blocks is presented previously [8].

In this paper, we propose a fully CMOS DDCC based scheme to realize memristor. It is pointed out that, this is an appropriate design for circuit. Also physical charge-flux characteristic of memristor. Memristor is constructed to behave convenient to the cubic memristor definition. Beyond our prior work, in this implementation appropriate chaotic behavior is obtained by using fully CMOS DDCC based a new CMOS based inductance simulator is used. Along with that, supply voltages and process technology is selected properly to current CMOS processes.

The paper is structured as follows: Section 2, which follows this introduction, summarizes the nonlinear cubic modeling of memristor. Section 3 includes the memristor-based Chua's circuit along with its dynamics with MATLAB simulations. In Section 4 we present the detailed design steps of our DDCC memristor-based chaotic circuit. Also in this section DDCC based inductance simulator is introduced. In Section 5 we present characteristics of memristor-based Chua's circuit. This section is devoted to the demonstrations of SPICE simulations based on the proposed design. Finally, the conclusions of this work are given in the sixth section.

2. Modeling of Memristor with Cubic Nonlinearity

Memristor can be defined with two types of nonlinear constitute relation between the device voltage and current:

$$v = M(q)i, \tag{1}$$

$$i = W(\varphi)v \tag{2}$$

where $M(q)$ and $W(\varphi)$ are nonlinear functions which are called memristance and memductance respectively and they are defined by:

$$M(q)=\frac{d\varphi(q)}{dq}, \tag{3}$$

$$W(\varphi)=\frac{dq(\varphi)}{d\varphi}. \tag{4}$$

The memristor designed in this work is a flux controlled memristor described by the relation in (2). The relation between the terminal voltage $v(t)$ and the terminal current $i(t)$ of the memristor is obtained by:

$$i(t)=\frac{dq}{dt}=\frac{dq}{d\varphi}\frac{d\varphi}{dt}=\frac{dq}{d\varphi}v(t)=W(\varphi(t))v(t). \tag{5}$$

Nonlinear resistor in Chua's circuit is defined by Zhong with cubic nonlinearity. It has been revealed that, all features of the circuit are captured correctly by this definition [25]. The $q(t)$ - $\varphi(t)$ function of memristor with cubic nonlinearity is used for implementation of chaotic circuits [26]. The cubic polynomial definition of memristor is defined as follows:

$$q(\varphi)=\alpha\varphi+\beta\varphi^3. \tag{6}$$

Thus, the memductance function is given by:

$$W(\varphi)=\frac{dq}{d\varphi}=\alpha+3\beta\varphi^2. \tag{7}$$

Considering (7) and the flux-voltage relation, we get (8). Our definition in this work, is based this relation.

$$i(t)=\left[\alpha+3\beta\left(\int v(t)\,dt\right)^2\right]v(t). \tag{8}$$

The memristor circuit introduced employs only DDCC based sub blocks such as mainly integrator, squarer, multiplier and summer. Designs of these active blocks are given in the following sections.

3. Chaos in Memristor-Based Chua's Circuit

Memristor-based Chua's circuit used in this work is shown in Fig. 1.

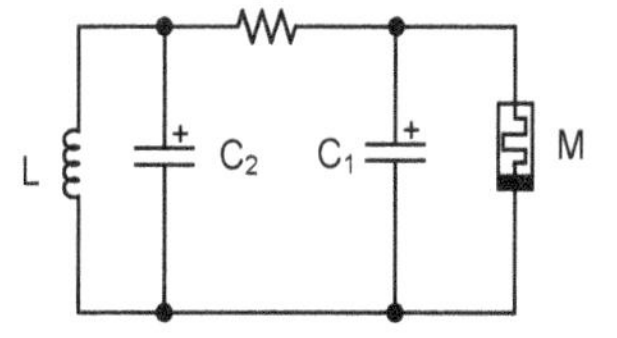

Fig. 1. Memristor-based Chua's circuit.

Note that mathematical background of the memristor-based Chua's circuit illustrated in Fig. 1 is studied earlier in [26]. We apply the DDCC based CMOS circuit proposed in our work to the realization of the Chua's circuit.

The dynamical state equations of memristor-based Chua's circuit can be described by

$$\begin{aligned}
&\frac{d\grave{\varphi}(t)}{dt}=\grave{v}_{C_1}(t)\\
&L\frac{d\grave{i}_L}{dt}=\grave{v}_{C_2}(t)\\
&C_1\frac{d\grave{v}_{C_1}(t)}{dt}=\frac{1}{R}\left(\grave{v}_{C_2}(t)-\grave{v}_{C_1}(t)\right)-W\left(\grave{\varphi}(t)\right)\grave{v}_{C_1}(t)\\
&C_2\frac{d\grave{v}_{C_2}(t)}{dt}=\frac{1}{R}\left(\grave{v}_{C_1}(t)-\grave{v}_{C_2}(t)\right)-\grave{i}_L(t)
\end{aligned} \tag{9}$$

where $W(\varphi(t))$ is memristor memductance and is defined by (7). By considering α, β and other circuit parameter values from [25], [26] this unscaled system gives unrealistic currents (up to hundred amps) and voltages (above kilo volts). In traditional op-amp based implementations currents and voltages are scaled about at mA and ten-volt levels [25], [26]. For our CMOS design we need to reduce much lower than these levels. Rescaled current and voltages are defined as:

$$\begin{aligned}
&\varphi(t)=\frac{1}{\delta}\grave{\varphi}(t)\\
&v_{C_1}(t)=\frac{1}{\delta}\frac{1}{\gamma}\grave{v}_{C_1}(t)\\
&v_{C_2}(t)=\frac{1}{\delta}\frac{1}{\gamma}\grave{v}_{C_2}(t)\\
&i_L(t)=\frac{1}{\delta}\frac{1}{\gamma}\grave{i}_L(t)
\end{aligned} \tag{10}$$

The following set of equations is obtained as our rescaled system.

$$\begin{aligned}
&\frac{d\varphi(t)}{dt}=\frac{v_{C_1}(t)}{\delta}\\
&\frac{di_L(t)}{dt}=\frac{v_{C_2}(t)}{L}\\
&\frac{dv_{C_1}(t)}{dt}=\frac{1}{C_1}\left(\frac{v_{C_2}(t)-v_{C_1}(t)}{R}\right)-\left(\alpha+3\beta\varphi^2(t)\right)v_{C_1}(t)\\
&\frac{dv_{C_2}(t)}{dt}=\frac{1}{C_2}\left(\frac{v_{C_1}(t)-v_{C_2}(t)}{R}\right)-i_L(t)
\end{aligned} \tag{11}$$

where β is also rescaled with

$$\beta=\grave{\beta}\frac{1}{\gamma^2} \tag{12}$$

Let δ = 1.6 kΩ × 47 nF = 7.52×10^{-5}, C_1 = 7.2 nF, C_2 = 70 nF, L = 18 mH, R = 1.97 kΩ. $\alpha = -0.662\times10^{-3}$ and $\beta = 18.75\times10^{-3}$. Initial values are $\varphi(0)=0$, $i_L(0)=0$, $v_{C1}(0)=0.01$, $v_{C2}(0)=0.01$. This parameter set will be used as design basis synthesizing of the CMOS circuit.

The bifurcation diagram of the system with respect to R is shown in Fig. 2. R is very important to create the chaotic behavior. Time domain waveforms and chaotic phase portraits from MATLAB simulations are shown in Fig. 3 and Fig. 4, respectively.

In a recent work, it has been shown that the type of singularities may be an indication of chaos [27]. Following up in this direction, it may be useful to note that the system in (11) has a single singularity on the origin, i.e. $v_{C1} = v_{C2} = i_L = 0$ unlike conventional Chua's system; thus we think that a detailed study of the system from the perspective introduced in [27] would be useful, while being beyond this paper's scope.

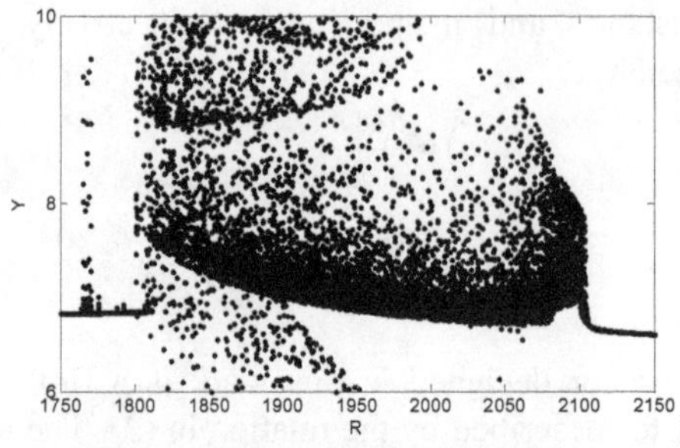

Fig. 2. The bifurcation diagram with respect to R.

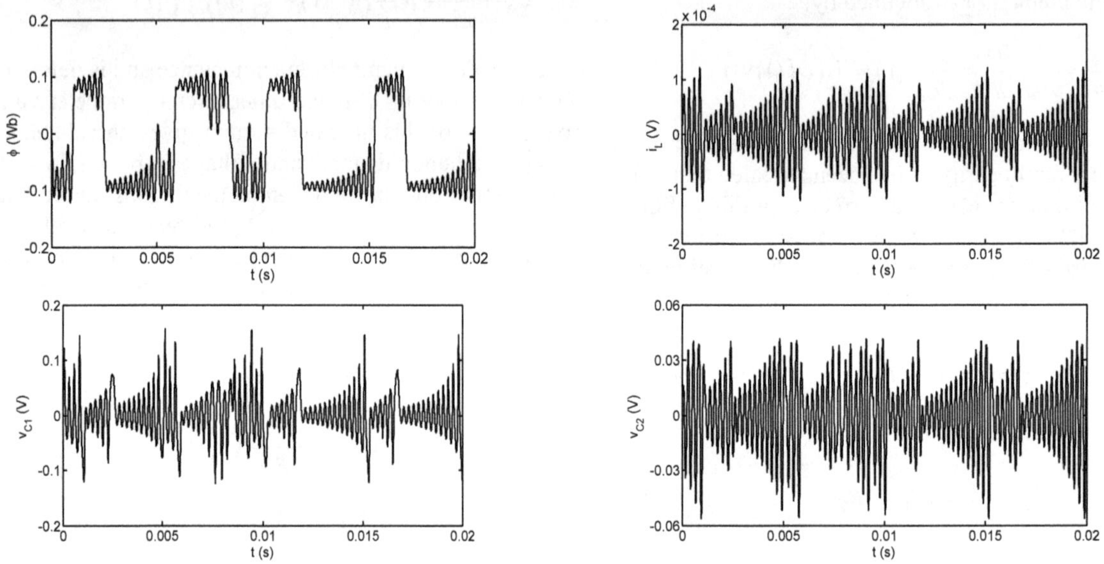

Fig. 3. Time domain waveforms of chaotic signals from MATLAB simulations.

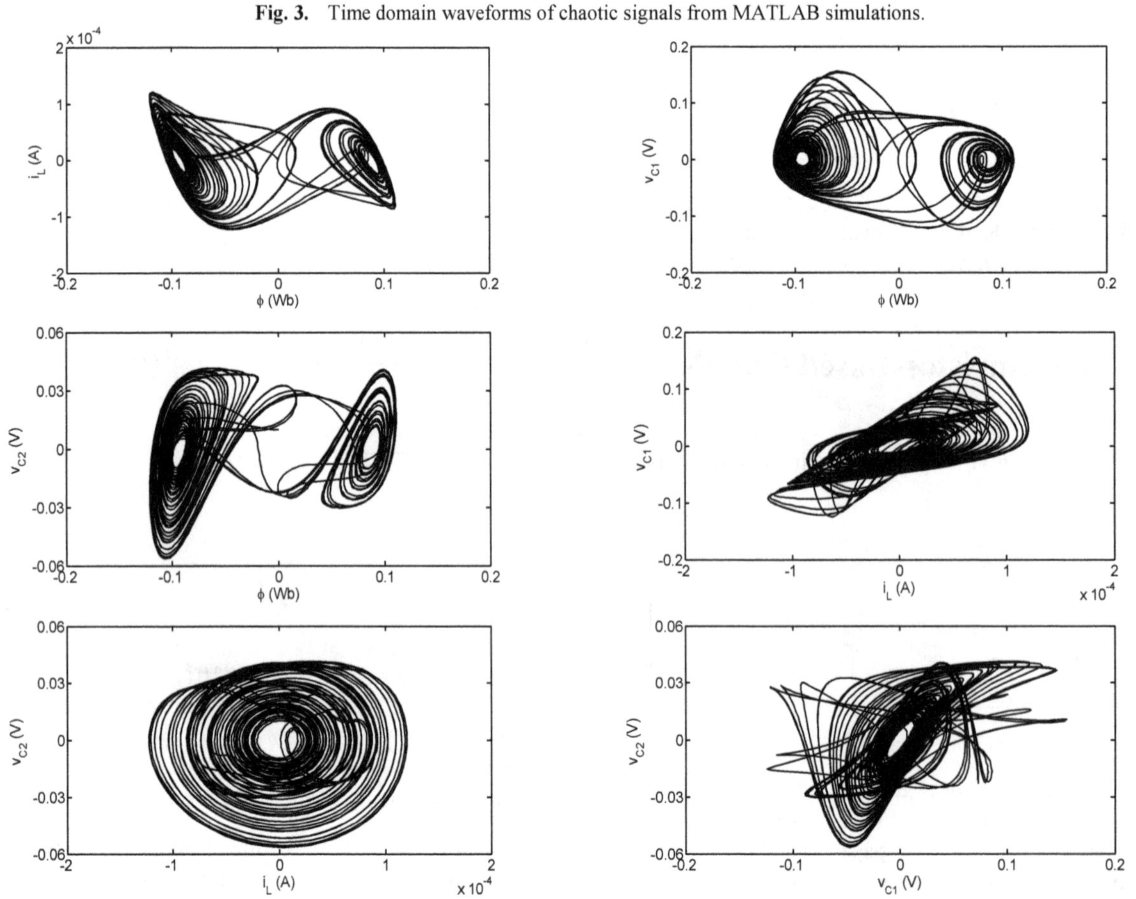

Fig. 4. Phase portraits from the MATLAB simulations.

4. Design of Fully CMOS Memristor-Based Chaotic Circuit

4.1 DDCC

DDCC is a 5-terminal active circuit block. Circuit symbol of DDCC is given in Fig. 5. Its terminal characteristics can be defined by a hybrid matrix giving the output of the five ports in terms of their corresponding inputs as shown in (13) [28].

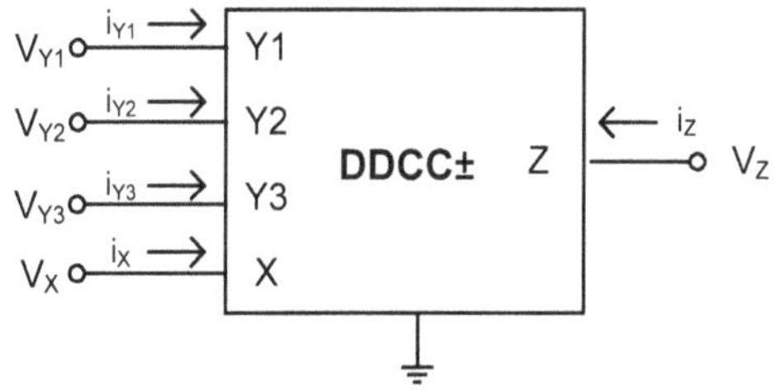

Fig. 5. Circuit symbol of DDCC.

$$\begin{bmatrix} i_{Y1} \\ i_{Y2} \\ i_{Y3} \\ v_X \\ i_Z \end{bmatrix} = \begin{bmatrix} 0 & 0 & 0 & 0 & 0 \\ 0 & 0 & 0 & 0 & 0 \\ 0 & 0 & 0 & 0 & 0 \\ 1 & -1 & 1 & 0 & 0 \\ 0 & 0 & 0 & \pm 1 & 0 \end{bmatrix} \begin{bmatrix} v_{Y1} \\ v_{Y2} \\ v_{Y3} \\ i_X \\ i_Z \end{bmatrix} \quad (13)$$

Here ± (plus or minus) sign indicates whether DDCC is non-inverting or inverting type denoted as DDCC+ or DDCC- respectively.

4.2 Block Diagram of Proposed DDCC Based Memristor

The proposed principle block diagram of memristor constructed employing only CMOS DDCC based sub blocks is shown in Fig. 6.

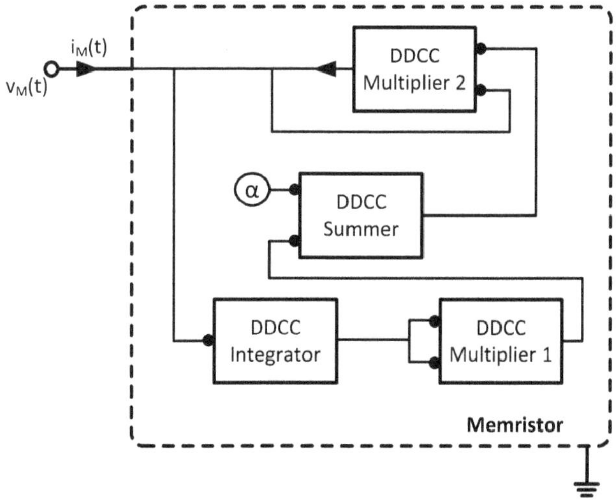

Fig. 6. Principle block diagram of proposed memristor implementation.

Four main CMOS stages are used in our memristor design. The first block works as an integrator. Also this block behaves as a buffer and it is used to avoid loading effect and insulate the current from the other stages. The output of the first block is connected both of the inputs of the multiplier and here it is the square of the voltage. The next block is summing block which also incorporates coefficients defined in (8) to the process. This block also makes current-voltage conversion in order to produce the inputs of the multiplier in voltages form. Finally, the current of our active memristor is generated after the output of this last multiplier block.

4.3 Sub-blocks of Memristor

4.3.1 Integrator

DDCC based integrator is shown in Fig. 7. For $R_X = \infty$ the output characteristics of the circuit can be derived as follows:

$$V_O = \frac{V_1 - V_2 + V_3}{sRC}. \quad (14)$$

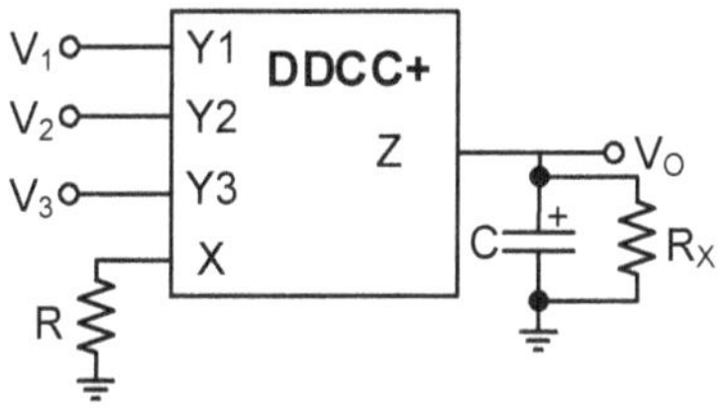

Fig. 7. DDCC based integrator.

Only C is used at the output of the integrator. Its parameters are defined as $R = 1.6$ kΩ and $C = 47$ nF.

4.3.2 Multiplier

In order to obtain DDCC based multiplier, the combination of two DDCC differential squarer is used [28]. The schematic of differential squarer is given in Fig. 8.

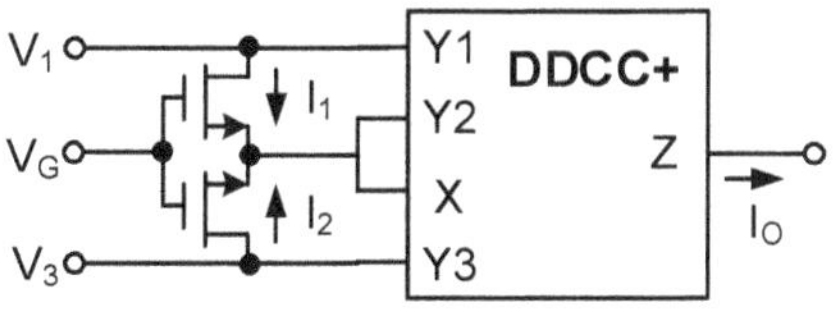

Fig. 8. DDCC based differential squarer.

In triode region, drain current of a MOS transistor can be expressed as follows:

$$I_D = K\left[2(V_{GS} - V_T)V_{DS} - V_{DS}^2\right], \quad (15)$$

$$K = \frac{1}{2}\mu C_{ox}\frac{W}{L}. \quad (16)$$

Considering the voltage relation of DDCC between X-Y inputs and supposing the transistors are well matched, the output current I_O can be derived as

$$
\begin{aligned}
I_O &= -(I_1 - I_2) \\
&= -K\left[2\left(V_G - \frac{V_1+V_3}{2} - V_T\right)\left(V_1 - \frac{V_1+V_3}{2}\right) - \left(V_1 - \frac{V_1+V_3}{2}\right)^2\right] \\
&\quad -K\left[2\left(V_G - \frac{V_1+V_3}{2} - V_T\right)\left(V_3 - \frac{V_1+V_3}{2}\right) - \left(V_3 - \frac{V_1+V_3}{2}\right)^2\right] \\
&= \frac{K}{2}(V_1 - V_3)^2 = K_s(V_1 - V_3)^2
\end{aligned} \tag{17}
$$

If a DDCC- is used here, the output current would be obtained as follows

$$I_O = -\frac{K}{2}(V_1 - V_3)^2 = -K_s(V_1 - V_3)^2 . \tag{18}$$

DDCC based multiplier [28] can be realized employing two squarers as shown in Fig. 9. Summing the outputs of both squarers, the output characteristics of the multiplier is obtained as given in (19).

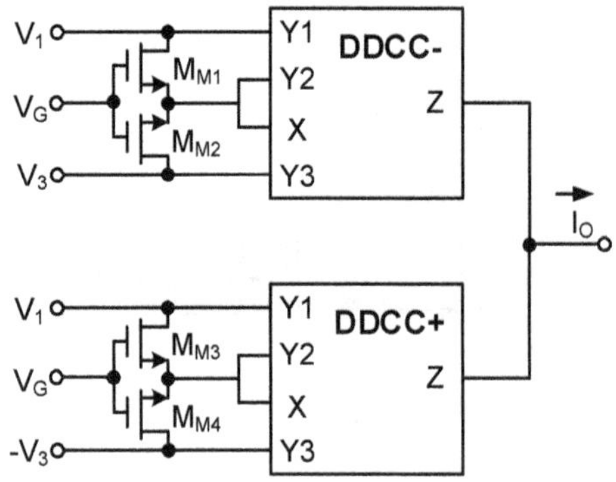

Fig. 9. DDCC multiplier.

$$I_O = \left[\frac{K}{2}(V_1 + V_3)^2 - \frac{K}{2}(V_1 - V_3)^2\right] = 2KV_1V_3 = K_MV_1V_3 \tag{19}$$

where K_M is the multiplying coefficient of the multiplier. Aspect ratios of MOS transistor used at inputs of the multiplier stage are $W/L = 40$ µm / 1 µm and $W/L =$ 10 µm / 1 µm for multiplier 1 and multiplier 2, respectively. Gate voltages of these transistors are defined as $V_G = 0.75$ V and drain currents are decreased as soon as possible preserving required multiplying situation. It is clear from Fig. 9 that the outputs of multipliers are current. The output current of multiplier 2 corresponds current of memristor. However current at the output of memristor 1 will be converted to voltage as stated in description of the summer block.

4.3.3 Buffer

DDCC has buffered inputs with very high gate impedances. Normally there is no need to additional buffer stages. However, DDCC multipliers have additional drain-input MOS transistors. So, DDCC based buffers are used at the input stages of these multipliers as shown in Fig. 10.

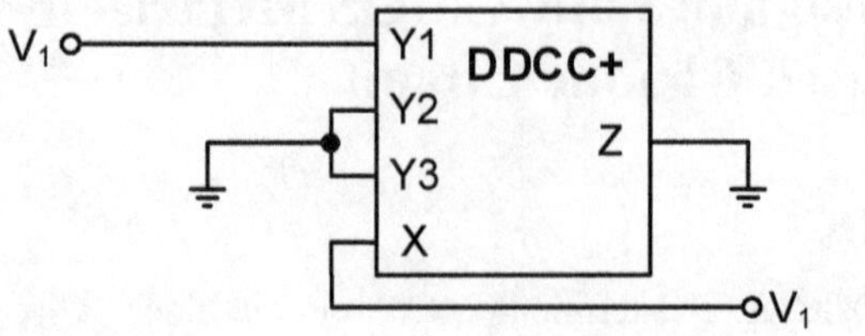

Fig. 10. DDCC based buffer.

4.3.4 Inverter

As can be seen from Fig. 9 we need an inverter at one of the inputs of the each DDCC multiplier. To obtain an inverter employing DDCC, simply it can be obtained by using the Y2 terminal as input and Z terminal as output. However, this topology is proper if the following stage has buffered input(s). Otherwise, an additional buffer will be needed. As a simpler form, DDCC based inverting buffer shown in Fig. 11 is used in our design.

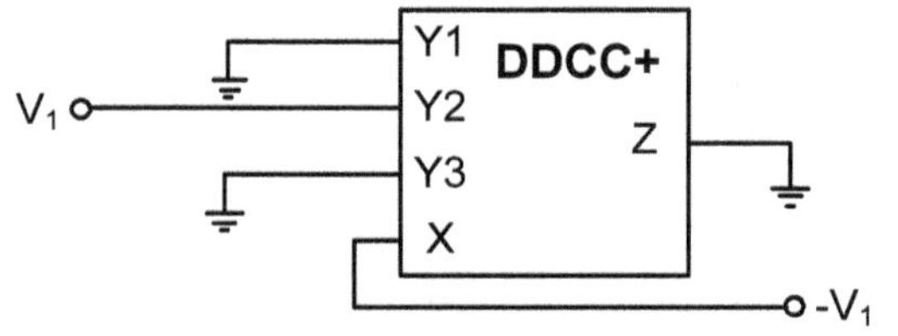

Fig. 11. DDCC based inverting buffer.

4.3.5 Summer

A DDCC is a kind of differential summer circuit due to its characteristic. As the DDCC based summing circuit in Fig. 6, two-input DDCC based topology is used as shown in Fig. 12. In this scheme, by considering differentiated inputs which are applied to Y1 and Y2, the output voltage of the circuit is as follows:

$$V_O = \frac{R_Z}{R_X}(V_1 - V_2). \tag{20}$$

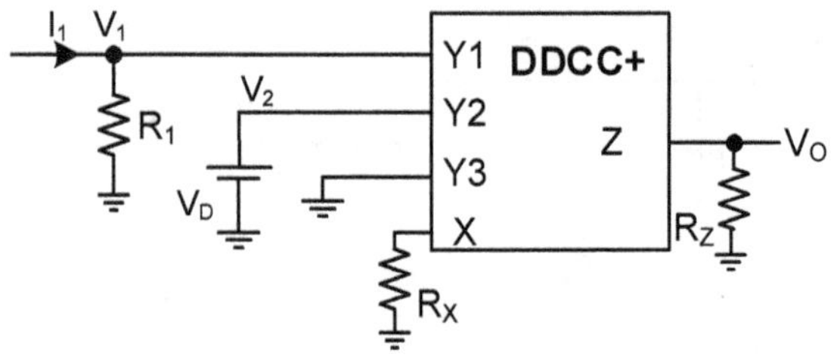

Fig. 12. DDCC based summer topology used in the memristor design.

V_1 corresponds produced voltage from the output current of the previous multiplier with resistor $R_1 = 1.6$ kΩ. V_2 is a DC voltage with the value of $V_D = 135$ mV. It stands for α with scaled a constant. For weighted summation, R_X and R_Z is used with values $R_X = 1$ kΩ and $R_Z = 2.5$ kΩ.

4.4 Single DDCC+ Based Inductance Simulator

Inductors are undesired elements in circuits because of their drawbacks such as in the usage of space, cost and adjustability. They cannot be implemented inside a chip. Actively simulating of the inductors offers appropriate design possibilities and in this sense it is more preferable.

A single CCII+ based inductance simulator was presented by Cicekoglu and Kuntman [30]. Based on this implementation, we have constructed a DDCC+ based new inductance simulator as shown in Fig. 13.

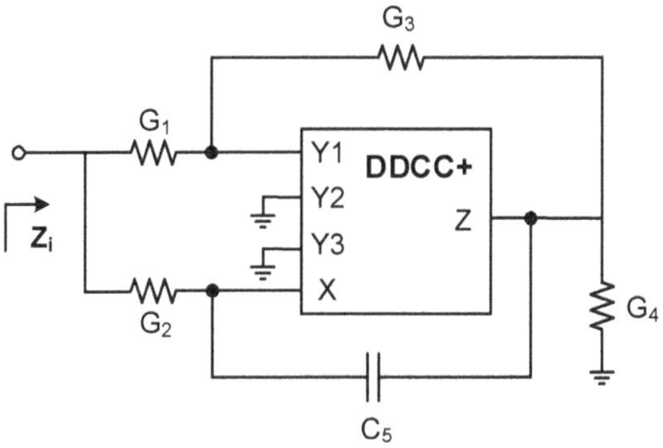

Fig. 13. DDCC based inductance simulator.

By using this topology it is either possible to realize an inductor with series resistance or a lossless inductor. For the grounded inductor with series resistance, equivalent inductance and resistance are defined by

$$L_{eq} = \frac{G_1(G_3+G_4)+G_3(G_4-G_2)+2sG_1C_5}{G_3G_4(G_1+G_8)}, \quad (21)$$

$$R_{eq} = \frac{G_1(G_3+G_4)+G_3(G_4-G_2)}{G_3G_4(G_1+G_8)}. \quad (22)$$

To utilize a lossless inductor, in matching constraint given in (23) the impedance value is obtained as in (24).

$$G_1(G_3+G_4)=G_3(G_2-G_4), \quad (23)$$

$$L_{eq} = \frac{2sG_1C_5}{G_3G_4(G_1+G_2)}. \quad (24)$$

Parameters of inductance simulator are selected as $R_1 = 1.35$ kΩ, $R_3 = 1.35$ kΩ, $R_4 = 1.35$ kΩ, $R_2 = 450$ Ω.

4.5 CMOS Design of DDCC

CMOS design of DDCC± used in this work is shown in Fig. 14. Our DDCC± is constructed based on implementations given in [28], [30]. In the circuit, M1, M2 and M3, M4 are two differential stages. High gain stage is obtained by a current mirror and the differential current is converted to a single-ended output current with the transistors M13, M14, M15. The positive output terminal Z+ is composed with a current source and duplicating elements of the current of the transistor M15 (M8, M9, M16 and M17). At the negative output terminal Z-, the current mirror shaped and the direction of the current is changed by transistors M10, M11, M12 and M18. Offset issue is critical for the multiplier stages in our design. A small DC current has come up at the output of multipliers. It would be solved for multiplier 1 by defining the value of V_D in Fig. 12. However, to systematically get rid of this constant DC current offset, M19 is used.

The supply voltages V_{CC} and V_{SS} are +1.25 V and -1.25 V, respectively. Bias voltages are taken as V_{b1} = -0.34 V and V_{b2} = 0. Aspect ratios of all NMOS and PMOS transistors except M19 are W_N/L_N = 4 μm / 0.8 μm and W_P/L_P = 8 μm / 0.8 μm, respectively. Dimensions of M19 are defined as W_{M19}/L_{M19} = 20 μm / 9 μm.

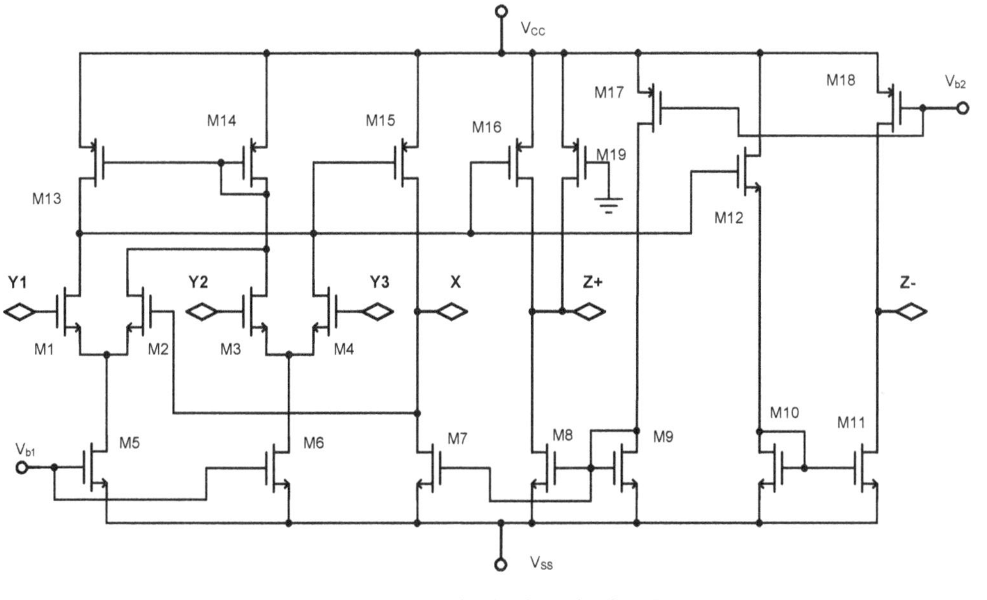

Fig. 14. CMOS circuit schematic of DDCC.

5. Simulation Results

In this section PSPICE simulations of the fully CMOS memristor-based circuit are performed and results are shown. Resistance R in Fig. 1 is still a parameter. It is taken the same as in MATLAB simulations: $R = 1.97\ \text{k}\Omega$. Setting of parameters properly has emerged as a critical issue. In addition to defining of parameters according to the calculations and design constraints, additional adjustments have been carried out by considering characteristics of the whole system.

Time and corresponding frequency domain waveforms and chaotic phase portraits are shown in Fig. 15 and Fig. 16, respectively. Note that the time domain and phase portrait plots obtained from the CMOS circuit exhibit very good visual agreement with corresponding plot in the MATLAB simulations.

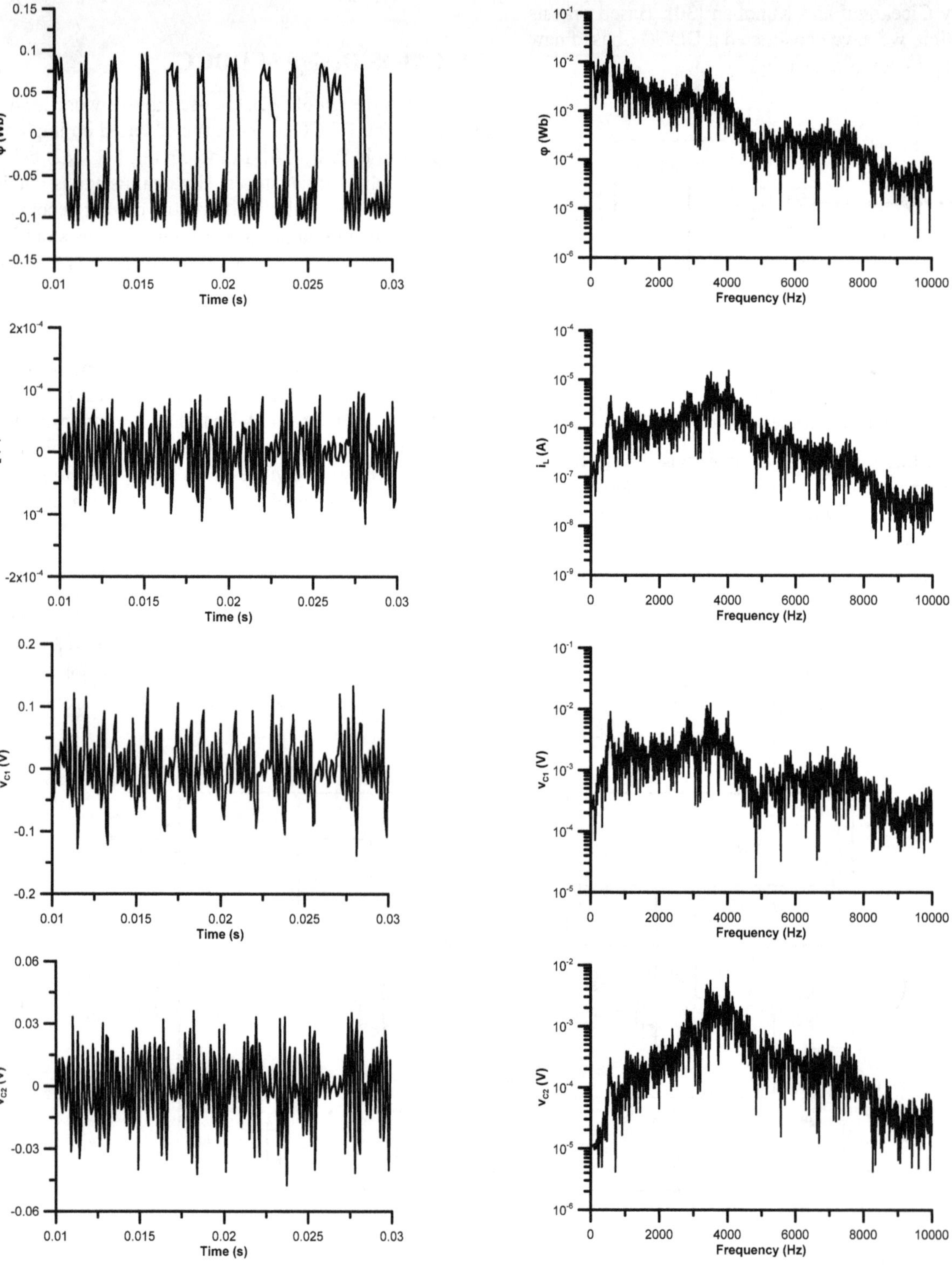

Fig. 15. Time domain waveforms of chaotic signals from SPICE simulations.

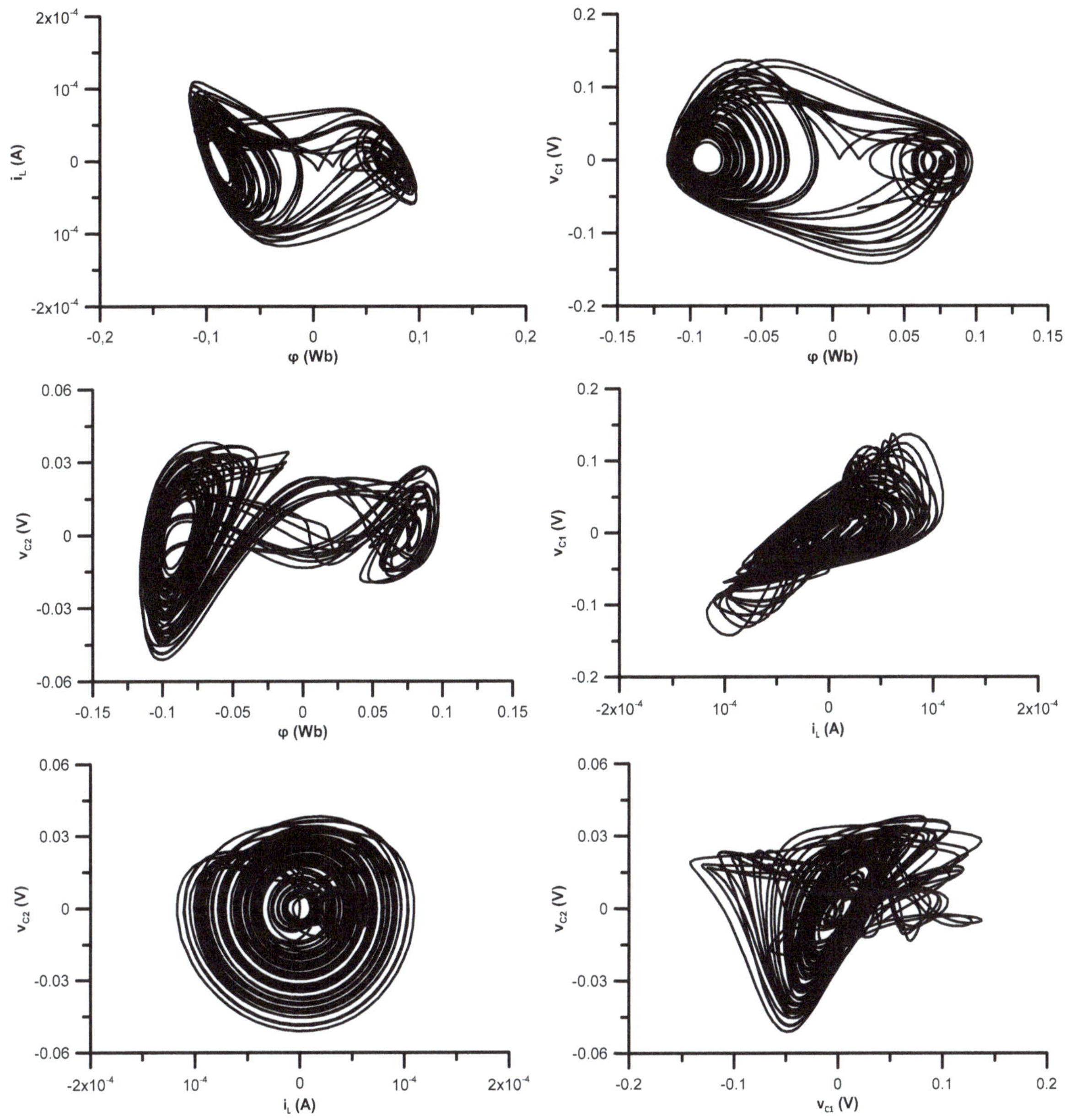

Fig. 16. Phase portraits from the SPICE simulations.

Flux-charge cubic characteristics of the memristor is shown in Fig. 17. Theoretical plot is obtained by using predetermined α and β at the beginning of our design. Experimental characteristics are obtained from CMOS chaotic circuit. From the SPICE simulations, the memductance parameters given in (8), are experimentally redetermined as $\alpha = -0.698 \times 10^{-3}$ and $\beta = 16.987 \times 10^{-3}$.

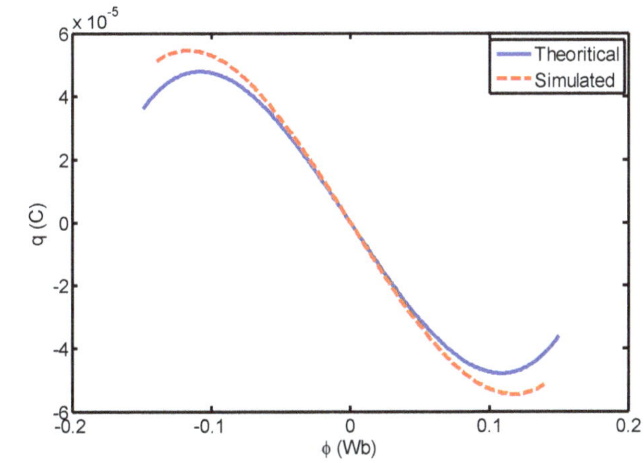

Fig. 17. Theoretical and experimental memristor cubic characteristic curve.

The total power consumption of the system is determined as 3.2 mW. This indicates a very low level comparing of other traditional op-amp based realizations.

Beyond the prior works and traditional op-amp based implementations, our design is entirely CMOS based and implementable only on a chip. From this perspective, it is obvious that it requires much less area. Along with the current process technology, it promises a considerable reduction in terms of the total number of transistors used in the design even except inductor [8], [26]. Unlike the op-amp based implementations, the proposed circuit can operate over a wide range of frequencies from Hz to MHz levels.

6. Conclusions

In this study, a new fully CMOS memristor-based chaotic Chua's circuit is presented. Memristor is implemented employing only CMOS DDCCs based blocks and therefore suitable to current VLSI processes. In the frame of the work performed, besides a systematically realizable DDCC-only realization approach to a CMOS-only memristor, a new CMOS-only chaotic oscillator is also presented. This design consumes less power and requires less area comparing other traditional realizations.

Simulations performed show that with the memristor model proposed, both memristor characteristics and chaotic behavior of the system is provided accurately. Memristor promises some new features such as reduction in the area used for the same function, lower power consumption etc. Furthermore it can be observed clearly that new possibilities and advantages will be added thanks to its natural non-volatile and non-linear structure. Since there is no physical sample yet, physically based correct modeling and design of the memristor is important. Proper implementations can be used in practical applications. It is possible to use a real physical memristor in this system when it is available.

Memristor can be easily applied to chaotic circuits thanks to its natural nonlinear behavior. It seems reasonable to compose memristor-based chaotic synchronization and chaotic communication applications. In addition it is known that memristors can mimic the behavior of biological synapses. We believe that our implementation and its application to chaotic circuits would benefit the memristor-based chaotic circuits, memristor-based analog signal processing applications.

References

[1] CHUA, L. O. Memristor - the missing circuit element. *IEEE Trans. on Circuit Theory*, 1971, vol. 18, no. 5, p. 507–519.

[2] STRUKOV, D.B., SNIDER, G.S., STEWART, D.R., WILLIAMS, R. S. The missing memristor found. *Nature*, 2008, vol. 453, 1 May 2008, p. 80–83.

[3] KAVEHEI, O., IQBAL, A., KIM, Y., ESHRAGHIAN, K., AL-SARAWI, S., ABBOTT, D. The fourth element: characteristics, modelling and electromagnetic theory of the memristor. *Proceedings of the Royal Society A: Mathematical, Physical and Engineering Science*, 2010, 466 (2120), p. 2175–2202.

[4] SAH, M. P.D., YANG, C., KIM, H., CHUA, L.O. A voltage mode memristor bridge synaptic circuit with memristor emulators. *Sensors*, 2012, vol. 12, no. 3, p. 3587–3604.

[5] MUTHUSWAMY, B., CHUA, L. O. Simplest chaotic circuit. *International Journal of Bifurcation and Chaos,* 2010, vol. 20, no. 5, p. 1567–1580.

[6] IU, H.H.C., YU, D.S., FITCH, A.L., SREERAM, V., CHEN, H. Controlling chaos in a memristor based circuit using a twin-T notch filter. *IEEE Transactions on Circuits and Systems I: Regular Papers,* 2011, vol. 58, no. 6, p. 1337–1344.

[7] DRISCOLL, T., PERSHIN, Y. V., BASOV, D. N., DI VENTRA, M. Chaotic memristor. *Applied Physics* A, 2010, vol. 102, no. 4, p. 885-889.

[8] YENER, S.C., KUNTMAN, H. A new CMOS based memristor implementation. In *Proc. of 2012 International Conference on Applied Electronics*. Pilsen (Czech Republic), 2012, p. 345-348.

[9] YENER, Ş. Ç., MUTLU, R., KUNTMAN, H. Performance analysis of a memristor-based biquad filter using its dynamic model. *Informacije MIDEM - Journal of Microelectronics, Electronic Components and Materials*, 2014, vol. 44, no. 2, p. 109–118.

[10] YENER, Ş. Ç., MUTLU, R., KUNTMAN, H. A new memristor-based high-pass filter: Its analytical and dynamical analysis. In *24th International Conference Radioelektronika 2014*. Bratislava, (Slovak Republic), April 15 – 16, 2014, p. 1–4.

[11] YENER, Ş. Ç., MUTLU, R., KUNTMAN, H. Frequency and time domain characteristics of memristor-based filters. In *22nd Signal Processing and Communication Applications Conference (SİU)*. Karadeniz Technical University, Trabzon, 2014, p. 2027–2030.

[12] RAK, A., CSEREY, G. Macromodeling of the memristor in SPICE. *IEEE Transactions on Computer-Aided Design of Integrated Circuits and Systems*, 2010, vol. 29, no. 4, p. 632 – 636.

[13] BIOLEK, Z., BIOLEK, D., BIOLKOVA, V. SPICE model of memristor with nonlinear dopant drift. *Radioengineering*, 2009, vol. 18, no. 2, p. 210–214.

[14] BATAS, D., FIEDLER, H. A memristor SPICE implementation and a new approach for magnetic flux-controlled memristor modeling. *IEEE Transactions on Nanotechnology*, 2011, vol. 10, no. 2, p. 250–255.

[15] BENDERLI, S., WEY, T. A. On SPICE macromodelling of TiO_2 memristors. *Electronics Letters*, 2009, vol. 45, no. 7, p. 377–379.

[16] SHARIFI, M. J., BANAKIDI, Y. M. General SPICE models for memristor and application to circuit simulation of memristor-based synapses and memory cells. *Journal to Circuits, Systems and Computers*, 2010, vol. 19, no. 2, p. 407–424.

[17] PERSHIN, Y. V., DI VENTRA, M. SPICE model of memristive devices with threshold. *Radioengineering,* 2013, vol. 22, no. 2, p. 485–489.

[18] BIOLEK, D., DI VENTRA, M., PERSHIN, Y. V. Reliable SPICE simulations of memristors, memcapacitors and meminductors. *Radioengineering,* 2013, vol. 22, no. 4, p. 945–968.

[19] BIOLEK, Z., BIOLEK, D., BIOLKOVÁ, V. Analytical computation of the area of pinched hysteresis loops of ideal mem-elements. *Radioengineering,* 2013, vol. 22, no. 1, p. 132–135.

[20] PERSHIN, Y. V., DI VENTRA, M. Practical approach to programmable analog circuits with memristors. *IEEE Transactions on Circuits and Systems - I,* 2010, vol. 57, no. 8, p. 1857–1864.

[21] KIM, H., SAH, M. P., YANG, C., CHO, S., CHUA, L. O. Memristor emulator for memristor circuit applications. *IEEE Transactions on Circuit and Systems – I,* 2012, vol. 59, no.10, p. 2422–2431.

[22] MUTLU, R., KARAKULAK, E. Emulator circuit of TiO_2 memristor with linear dopant drift made using analog multiplier. In *National Conference on Electrical, Electronics and Computer Engineering (ELECO) 2010*. Bursa (Turkey), 2010, p. 380–384.

[23] RADWAN, A. G., SOLIMAN, A. M., EL-SEDEEK A. An inductorless CMOS realization of Chua's circuit. *Chaos, Solitons and Fractals,* 2003, vol. 18, p. 149–158.

[24] GOPAKUMAR, K., PREMBLET, B., GOPCHANDRAN, K. G. Implementation of Chua's circuit using simulated inductance. *International Journal of Electronics*, 2011, vol. 98, no. 5, p. 667 to 677.

[25] ZHONG, G. Implementation of Chua's circuit with a cubic nonlinearity. *IEEE Transactions on Circuits and Systems*, 1994, vol. 41, no. 12, p. 934–941.

[26] MUTHUSWAMY, B. Implementing memristor based chaotic circuits. *International Journal of Bifurcation and Chaos*, 2010, vol. 20, no. 5, p. 1335–1350.

[27] ŠPÁNY, V., GALAJDA, P., GUZAN, M., PIVKA, L., OLEJÁR, M. Chua's singularities: great miracle in circuit theory. *International Journal of Bifurcation and Chaos*, 2010, vol. 20, no. 10, p. 2993–3006.

[28] CHIU, W., LIU, S. I., TSAO, H. W., CHEN, J. J. CMOS differential difference current conveyors and their applications. *IEEE Proceedings: Circuits, Devices and Systems*, 1996, vol. 143, p. 91–96.

[29] CICEKOGLU, O., KUNTMAN, H. Single CCII+ based active simulation of grounded inductors. In *ECCTD 2014: European Conference on Circuit Theory and Design*. Budapest (Hungary), 1997, p. 105–109.

[30] CHANG, C. M., LEE, C. N., HOU, C. L., HORNG, J. W., TU, C. K. High-order DDCC-based general mixed-mode universal filter. *IEEE Proceedings: Circuits, Devices and Systems*, 2006, vol. 153, p. 511–516.

About Authors...

Şuayb Çağrı YENER was born in Sakarya, Turkey in 1982. He received B.Sc. degree from Sakarya University in Electrical and Electronics Engineering in 2004, M.Sc. and Ph. D. degree from Istanbul University in Electronics and Communication Engineering in 2007 and 2014, respectively. He is currently a research assistant in Sakarya University. His main research interests are CMOS analog circuits, analog circuit design, circuit and electronic device modeling, modeling of the memristor and memristor based circuits. He is the author or co-author of 26 journal papers published/accepted for publishing and papers presented in national/international conferences.

H. Hakan KUNTMAN received his B.Sc., M.Sc. and Ph.D. degrees from Istanbul Technical University in 1974, 1977 and 1982, respectively. In 1974 he joined the Electronics and Communication Engineering Department of Istanbul Technical University. Since 1993 he is a professor of Electronics in the same department. His research interest includes design of electronic circuits, modeling of electron devices and electronic systems, active filters, design of analog IC topologies. Dr. Kuntman has authored many publications on modeling and simulation of electron devices and electronic circuits for computer-aided design, analog VLSI design and active circuit design. He is the author or co-author of 111 journal papers published or accepted for publishing in international journals, 168 conference papers presented or accepted for presentation in international conferences, 156 Turkish conference papers presented in national conferences and 10 books related to the above mentioned areas. Furthermore he advised and completed the work of 13 Ph.D. students and 41 M.Sc students. Currently, he acts as the advisor of 3 Ph.D. and 4 M.Sc. students. Dr. Kuntman is a member of the Chamber of Turkish Electrical Engineers (EMO).

12

A Novel Stealthy Target Detection Based on Stratospheric Balloon-borne Positional Instability due to Random Wind

Mohamed A. BARBARY[1], *Peng ZONG*[2]

[1] Dept. of Electronic and Information Engineering, Nanjing University of Aeronautics and Astronautics, Nanjing, China
[2] Dept. of Astronautics, Nanjing University of Aeronautics and Astronautics, Nanjing, China

mbarbary300@gmail.com, pengzong@nuaa.edu.cn

Abstract. *A novel detection for stealthy target model F-117A with a higher aspect vision is introduced by using stratospheric balloon -borne bistatic system. The potential problem of the proposed scheme is platform instability impacted on the balloon by external wind force. The flight control system is studied in detail under typical random process, which is defined by Dryden turbulence spectrum. To accurately detect the stealthy target model, a real Radar Cross Section (RCS) based on physical optics (PO) formulation is applied. The sensitivity of the proposed scheme has been improved due to increasing PO-scattering field of stealthy model with higher aspect angle comparing to the conventional ground-based system. Simulations demonstrate that the proposed scheme gives much higher location accuracy and reduces location errors.*

Keywords

Stealthy RCS, bistatic balloon-borne radar, PO method.

1. Introduction

The complexity of stealth target detection is not only related to the target itself, but also influenced by the electromagnetic environment [1]. The countering-stealth technologies are increasingly relevant, and research in this field is ongoing around the world. Stealth technology mostly focuses on defeating conventional ground-based detection radar. Thus, the success of counter stealth endeavors is focused mostly on novel and unique air defense infrastructure configurations. The Radar Cross Section (RCS) is an important evaluation criterion of aircraft's stealthy performance, the envelope of the backscatter from stealthy target varies rapidly with aspect angle. The shaping of stealthy objects to reduce the backscattered energy towards the radar is believed to be less effective when bistatic radar is used [2].

Several researches deal with improving stealthy target detection and tracking based on ground-based bistatic or netted radar system [2–8]. Theses researches didn't evaluate bistatic radar sensitivity and performance of stealthy target with a higher aspect vision. Since the bistatic radar system might be mounted on higher altitude platforms to achieve a larger probability of detecting stealthy target, the bistatic radar sensitivity will be improved due to increasing the scattered field of stealthy target with higher altitude. In this paper, we investigate a novel technique for stealthy target detection based on balloon–borne bistatic radar system. The stations are positioned in the stratosphere about 21 km above the Earth and kept stable in a sphere of radius of 0.5 km [9]. To achieve high location accuracy for stealthy target, the proposed scheme uses a physical optics method (PO) to predict the real RCS of stealthy target. This will better represent the actual situation of the stealthy target detection.

One of the open research issues is whether the platforms positional instability due to sudden gusts of stratospheric winds. In the aerospace field, the study of turbulence effects is of fundamental importance in a lot of different aspects [10], such as improvements of the aerodynamic and structural analysis, prediction of expected behavior of a balloon-borne platform under various levels of turbulence, evaluation of the stability of onboard sensing equipment, and so on. Subject to the extreme complexity of the turbulence phenomena and due to the huge variety of applications, there is not a unique full-comprehensive model for the atmospheric turbulence, but there exist a wide variety of different and simplified models [11–12]. Numerous turbulence models are enumerated and described. The most commonly adopted model to study the impact of the turbulent wind gust on the balloon-borne is the Dryden model. According to this model, the atmospheric turbulence is modeled as a random velocity process added by balloon-borne velocity vector described in a body-fixed Cartesian coordinate system.

The rest of this paper is organized as follows. In Section 2, we present balloon positional instability analysis and random wind mathematical model. In Section 3, we discuss the PO formulation to predict RCS of stealth model. The bistatic range-measurement accuracy adopted for unstable position of the proposed scheme using stealth RCS is discussed in Section 4. Performance of the proposed scheme is evaluated via computer simulation in Section 5, followed by the conclusion in Section 6.

2. Balloon Positional Instability Analysis

The general dynamic equations of a stratospheric balloon platform are derived for flight over flat and non-rotating Earth, considering buoyancy, added mass and relevant conceptual design data of the stratospheric platform. To include the effect of jet stream as a moving wind field, the dynamic equations of motion can be derived in the relative wind-axes, inertial wind-axes, or body-axes coordinate systems [13–14]. The relative wind-axes system is more convenient than other coordinate systems because it expresses the wind-effect terms explicitly, bringing easier understanding. Fig. 1 illustrates the relationship between horizontal wind vector, airspeed velocity vector, and local (Earth-fixed) velocity of the platform. The wind-relative velocity vector is defined by airspeed V, flight path angle γ, and heading ψ.

From Fig. 1, the inertial flight velocity with respect to the local ENU frame is determined as:

$$\begin{aligned} V_I &= \dot{x}_i\, u + \dot{y}_i\, e + \dot{z}_i\, n \\ &= V + W \\ &= V \sin\gamma\, u + (V\cos\gamma \sin\psi + W_E)e + (V\cos\gamma\cos\psi + W_N)n \\ &= V_I \sin\gamma_I\, u + V_I\cos\gamma_I \sin\psi_I\, e + V_I \cos\gamma_I \cos\psi_I\, n\,. \quad (1) \end{aligned}$$

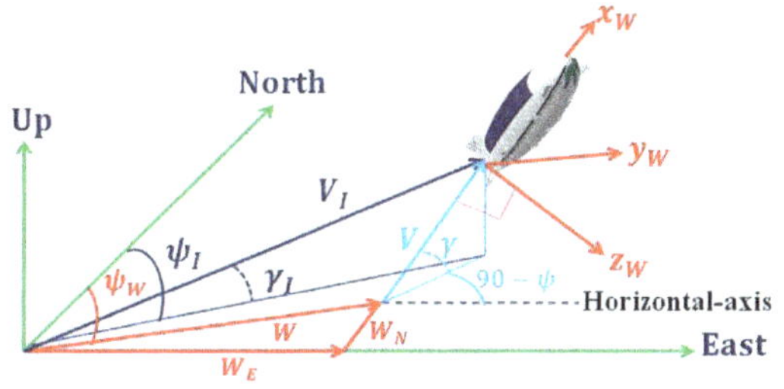

Fig. 1. Balloon –borne velocity in the local and wind-relative frames.

2.1 The Random Wind Mathematical Model

The Dryden model is one of the most useful and tractable models for atmospheric turbulence [13]. To define it we need a body-fixed reference frame attached to the gravity center of balloon-borne which moves with the target. The x-axis is on the position of motion direction, y-axis is on position along the wings, and z-axis is perpendicular to the balloon-borne plane. Then, the turbulence is modeled by adding some random components to balloon-borne velocity defined in the body-fixed coordinate system. An important consideration in this paper is the effect of steady-state horizontal winds. The horizontal wind velocity vector is defined as:

$$W_H = W \sin\psi_W e + W\cos\psi_W n\,, \quad (2)$$

$$= W_E\, e + W_N\, n\,. \quad (3)$$

In Dryden model continuous-time random processes are modeled as zero-mean, Gaussian-distributed processes whose PSD have analytic form given by [10–12]:

$$S_e(\omega) = {\sigma_e}^2 \frac{L_e}{\pi V_0} \frac{1}{1+\left(\frac{L_e}{V_0}\omega\right)^2}\,, \quad (4)$$

$$S_n(\omega) = {\sigma_n}^2 \frac{L_n}{2\pi V_0} \frac{1+3\left(\frac{L_n}{V_0}\omega\right)^2}{\left[1+\left(\frac{L_n}{V_0}\omega\right)^2\right]^2} \quad (5)$$

where V_0 is the gust wind speed in the balloon-borne system. The parameters ${\sigma_e}^2$ and ${\sigma_n}^2$ depend on the level of turbulence to be simulated and are selected accordingly [11]. Parameters L_e and L_n are the scale lengths for the PSDs and depend on the flight altitude. The mean wind velocity at the altitude of 21 km varies between -15 to +15 m/s [9]. Fig. 2 shows the PSDs of (4) and (5) for $\sigma_e = \sigma_n = \sigma_w = 15$ m/s, $L_e = L_n = 533.54$ m, and $V_0 = 15$ m/s. To reflect higher level of turbulence, the curves would be multiplied by the desired values of ${\sigma_e}^2$ and ${\sigma_n}^2$.

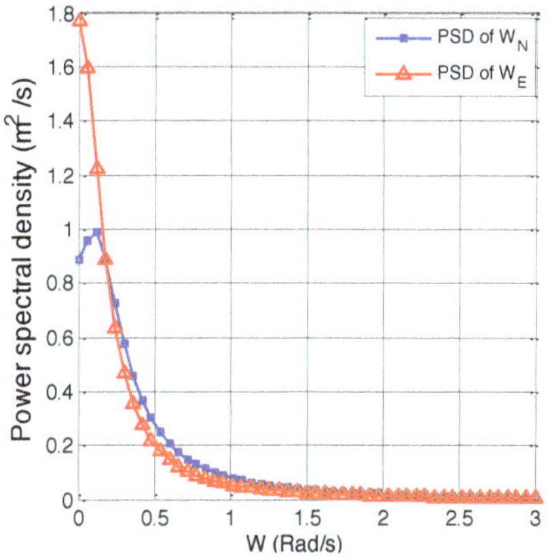

Fig. 2. PSD of Dryden velocity processes.

From (4) and (5), simulation model of wind can be written as:

$$\dot{W}_E + \frac{V_0}{L_e} W_E = \sqrt{\frac{2V_0}{L_e}}\xi_e\,, \quad (6a)$$

$$\dot{W}_1 + \frac{V_0}{L_n} W_1 = \left(\sqrt{3}-1\right)\sqrt{\frac{V_0}{L_n}}\xi_n, \quad (6b)$$

$$\dot{W}_N + \frac{V_0}{L_n} W_1 + \frac{V_0}{L_n} W_N = \sqrt{\frac{3V_0}{L_n}}\xi_n \quad (6c)$$

where W_1 is the transition variable for calculating the wind field model, ξ_e and ξ_n are the random variables subject to normal distribution $N\,(0, {\sigma_w}^2)$.

2.2 External Forces Acting on Balloon Platform

The external forces acting on balloon-borne platform include aerodynamic lift L and drag D, thrust T, weight W and buoyancy B. We also consider a generic lateral force N, which may be generated by any means, such as rolling the lift vector through a small angle, ϕ, or applying lateral thrust. A free-body diagram of the forces in x-z plane is shown in Fig. 3.

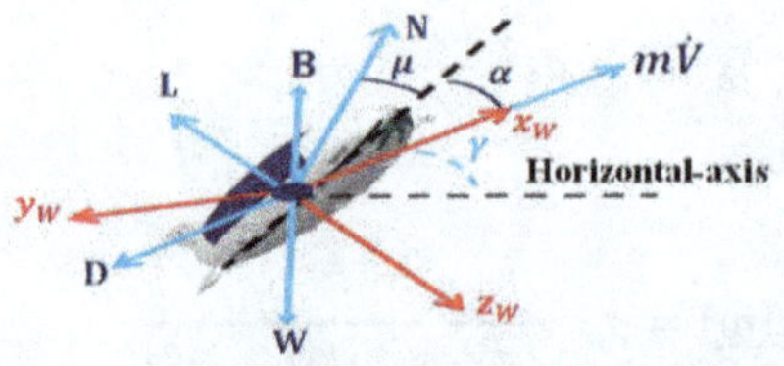

Fig. 3. External forces acting on balloon- borne platform.

The equations of motion are described by equating the time derivative of the momentum vector with the sum of external forces.

$$\Sigma F = \frac{d}{dt}(M\,V_I)$$
$$= [(B-W)\sin\gamma \;+T\cos(\alpha+\mu) - D\,]\hat{x}$$
$$+N\hat{y} + [(W-B)\cos\gamma - L - T\sin(\alpha+\mu)\,]\hat{z} \quad (7)$$

where L and D are the lift and drag forces, respectively. The lift and drag coefficients from Lee [15] are used, and are assumed to vary only with angle of attack. μ represents the tilting angle of the propellers installed on both sides of the airship. The lift and drag force can be calculated by

$$L = q.VOL^{2/3}.C_L(\alpha)\,, \qquad D = q.VOL^{2/3}.C_D(\alpha)\,, \qquad (8a)$$
$$C_L(\alpha) = 0.590\alpha^4 + 1.2231\alpha^3 + 0.3248\alpha^2 + 0.921\alpha + 0.0118,$$
$$C_D(\alpha) = 0.340\,\alpha^4 + 0.0662\alpha^3 + 1.2248\alpha^2 + 0.0334\alpha + 0.04 \qquad (8b)$$

where q and VOL represent the dynamic pressure of free stream and envelope volume of the platform.

The required thrust and power are given by [15]

$$T = q.VOL^{2/3}.C_D, \qquad P = T.V.\frac{1}{\eta_p\eta_m} \qquad (9)$$

where η_p and η_m represent the efficiency of the propellers and electric motor equipped in the airship. The values 0.7 and 0.9 are used for the efficiencies, respectively

The buoyancy force is another typical discriminator between LTA vehicles and conventional aircraft. It plays the role of lifting balloons upward and is equal to the weight of displaced air by its volume immersed in the atmosphere. The net lift that can be available for payload, system, and structure is determined by subtracting the weight of the lift gas and envelope [15]:

$$L_{net} = B - W = VOL\,(\rho_a - \rho_h)g - W_{env}\,,$$
$$B = VOL.\rho_a.g \qquad (10)$$

where ρ_a and ρ_h refer to the density of the surrounding atmosphere and helium, respectively. For the helium that is generally used as the lifting gas, the gross lift per unit volume $(\rho_a - \rho_h)g$ is 10.359 N/m^3.

By Newton's second law, the force equilibrium equation in the inertial frame is expressed as

$$F = m_T a = m_T\left(\frac{dV_I}{dt}\right)_I\,,$$
$$m_T = m + m_{ax} + m_{ay} + m_{az} \qquad (11)$$

where m_T includes the empty mass m and the added masses, m_{ax}, m_{ay}, and m_{az}. When a body moves through fluid, it must push some mass of fluid out on the way. If the body is accelerated, the surrounding fluid must also be accelerated. Under this circumstance, the body behaves as if it were heavier, so that mass is added. The diagonal terms of added mass tensor are the main terms on the body axes of balloon for m_{ax}, m_{ay}, and m_{az} of (11) respectively. Because it is assumed that air density varies in a unit at operating altitude, the values should be multiplied by corresponding density to obtain added mass in the optimization process:

$$M_a = \begin{bmatrix} 2.1391\times10^4 & 1.6502\times10^{-12} & 1.3365\times10^{-11} \\ -2.0890\times10^{-12} & 2.4363\times10^5 & 9.8516\times10^1 \\ -2.2134\times10^{-12} & -9.8516\times10^1 & 2.4363\times10^5 \end{bmatrix} (\mathrm{m}^3) \qquad (12)$$

The total inertial acceleration is acceleration of airship with respect to local ENU frame, plus acceleration of ENU frame in inertial space, plus Coriolis acceleration. Using notation (d/dt) A to denote a derivative taken with respect to frame A, the inertial acceleration expressed in the wind frame is:

$$\left.\frac{dV_I}{dt}\right|_I = \left.\frac{dV}{dt}\right|_I + \left.\frac{dW_I}{dt}\right|_I$$
$$= \left.\frac{dV}{dt}\right|_W + \omega_W \times V|_W + C_l^W \times \left.\frac{dW_I}{dt}\right|_I \qquad (13)$$

where ω_W is angular rate of wind-axes frame regarding to the Earth-fixed frame and it satisfies

$$\omega_W = \begin{bmatrix} 1 & 0 & -\sin\gamma \\ 0 & \cos\phi & \sin\phi\cos\gamma \\ 0 & -\sin\phi & \cos\phi\cos\gamma \end{bmatrix}\begin{bmatrix}\dot{\phi}\\ \dot{\gamma}\\ \dot{\psi}\end{bmatrix}, \qquad (14a)$$

$$C_l^W \times \left.\frac{dW_I}{dt}\right|_I = C_l^W\begin{bmatrix}\dot{W}_N\\ \dot{W}_E\\ 0\end{bmatrix} = \begin{bmatrix}\dot{W}_{wx}\\ \dot{W}_{wy}\\ \dot{W}_{wz}\end{bmatrix}, \qquad (14b)$$

and

$$C_l^W = \begin{bmatrix} C\gamma C\psi & C\gamma S\psi & -S\gamma \\ S\phi S\gamma C\psi - C\phi S\psi & S\phi S\gamma S\psi + C\phi C\psi & S\phi S\gamma \\ C\phi S\gamma C\psi + S\phi S\psi & C\phi S\gamma S\psi - S\phi C\psi & C\phi C\gamma \end{bmatrix} \qquad (14c)$$

where C_l^W is the transformation matrix which transforms both of the local-level and wind-axes frames to each other, {C, S} mean cos and sin respectively. Combining equations (13) and (14) with (7) leads to the final representation of (11) in the wind-axes frame, after several algebraic manipulations. Finally, solving the simultaneous algebraic equations for the derivatives $\dot{V}$, $\dot{\gamma}$, and $\dot{\psi}$, the force equilibrium equations can be represented as

$$\dot{V} = \frac{(T\cos\alpha - D) - (mg - B)\sin\gamma}{m_T} - \dot{W}_{wx}\,, \qquad (15a)$$

$$\dot{\gamma} = \frac{(T\sin\alpha + L)\cos\phi - (mg - B)\cos\gamma}{m_T V} + \frac{\dot{W}_{wz}\cos\phi + \dot{W}_{wy}\sin\phi}{V}\,, \qquad (15b)$$

$$\dot{\psi} = \frac{(T\sin\alpha + L)\sin\phi}{m_T V\cos\gamma} + \frac{\dot{w}_{wz}\sin\phi - \dot{w}_{wy}\cos\phi}{V\cos\gamma} \quad (15c)$$

where

$$\dot{w}_{wx} = \dot{w}_N \cos\gamma\cos\psi + \dot{w}_E\cos\gamma\sin\psi, \quad (16a)$$

$$\dot{w}_{wy} = \dot{w}_N(\sin\phi\sin\gamma\cos\psi - \cos\phi\sin\psi) + \dot{w}_E(\sin\phi\sin\gamma\sin\psi + \cos\phi\cos\psi), \quad (16b)$$

$$\dot{w}_{wz} = \dot{w}_N(\cos\phi\sin\gamma\cos\psi + \sin\phi\sin\psi) + \dot{w}_E(\cos\phi\sin\gamma\sin\psi - \sin\phi\sin\psi). \quad (16c)$$

3. The Physical Optics (PO) Formulation for Stealthy F-117A RCS Model

In the presence of a perfectly conducting surface, the total electromagnetic field of a source may be expressed as a superposition of the incident fields (E_i, H_i) and the fields (E_s, H_s) which are scattered by the surface. The scattered fields can be expressed in terms of the radiation integrals over actual currents induced on the surface of the scatterer. The PO assumes that the induced surface currents on the scatterer surface are given by the geometrical optics (GO) currents over those portions of the surface directly illuminated by the incident magnetic field, $\vec{H}_i$, and zero over the shadowed sections of the surface [16]:

$$\vec{J_s} = \begin{cases} 2\hat{n}\times\vec{H}_i, & \text{illuminated region} \\ 0, & \text{shadow region} \end{cases} \quad (17)$$

where $\hat{n}$ denotes the outward unit normal vector on a surface. The authors in this paper use PO method to predict RCS of a stealth target based on the geometry model of F-117A, which are modeled by the triangular facets. The geometry model of the stealth target based on F-117A is approximated by a model consisting of many triangular facets, in which there are a large number of points on the surface described in terms of Cartesian coordinates. This surface is then approximated by planar triangular facets connecting these points. An arbitrary midpoint p of the triangle surface is assigned coordinates (r_p, θ_p, ϕ_p), the observation point is assigned coordinates (r_s, θ_s, ϕ_s) and unit vectors $(\hat{r}_s, \hat{\theta}_s, \hat{\phi}_s)$. Normal vector $\hat{n}$ is a unit vector with its tip at the midpoint of triangle. Then $\hat{n}$ can be expressed as cross product of vectors $\overrightarrow{AB}$ and $\overrightarrow{AC}$. Once these vectors are found, $\hat{n}$ can directly be found by $\hat{n} = \overrightarrow{AB}\times\overrightarrow{AC} \,/\, |\overrightarrow{AB}||\overrightarrow{AC}|$. These parameters are depicted in Fig. 4.

Thus far, the discussion has involved the calculation of the scattered field from a single facet. Superposition is used to calculate the scattered field from the stealth target. First, the scattered field is computed for each facet. Then, the scattered field from each facet is vector summed to produce the total field in the observation direction. If the source is at a great distance from the target, it will illuminate the target with an incident field which is essentially a plane wave. The incident electric field intensity is given by $\vec{E}_i = (E_{i\theta}\hat{\theta}_i + E_{i\phi}\hat{\phi}_i)e^{-j\vec{k}_i\hat{r}_i\cdot\vec{r}_p}$, where $E_{i\theta}$, $E_{i\phi}$ are the

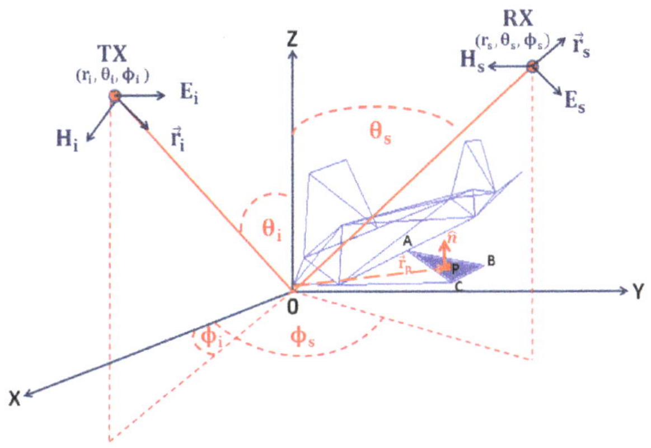

Fig. 4. Vector definitions of an approximation of a stealthy F-117A model using triangular facets on the surface.

orthogonal components in terms of the variables θ and ϕ, (r_i, θ_i, ϕ_i) are the spherical coordinates of the source and $(\hat{r}_i, \hat{\theta}_i, \hat{\phi}_i)$ are the unit vectors, so the magnetic field intensity of the incident field is given by:

$$\vec{H}_i = \frac{\vec{k}_i \times \vec{E}_i}{Z_0} = \frac{1}{Z_0}(E_{i\phi}\hat{\theta}_i - E_{i\theta}\hat{\phi}_i)e^{j\vec{k}_i h} \quad (18)$$

where $(k = \frac{2\pi}{\lambda})$, $\vec{k}_i$ is the propagation vector defined as $\vec{k}_i = -k(\hat{x}\sin\theta_i\cos\phi_i + \hat{y}\sin\theta_i\sin\phi_i + \hat{z}\cos\theta_i)$, Z_0 is the intrinsic impedance of free space and $h = \hat{r}_i \cdot \vec{r}_p = x_p\sin\theta_i\cos\phi_i + y_p\sin\theta_i\sin\phi_i + z_p\cos\theta_i$. Since radiation integral for the scattered field is calculated by employing a GO approximation for the currents induced on the surface, it can be concluded that PO is a high frequency method, which implies that target is assumed to be electrically large. For the scattered field, the vector potential is given by [17]:

$$\vec{A} = \frac{\mu}{4\pi r_s}e^{-jkr_s}\iint_S \vec{J}_s e^{jk\hat{r}_s\cdot\vec{r}_p}\,ds \quad (19)$$

where μ is the permeability of a specific medium. For a far-field observation point, the following approximation holds

$$\begin{aligned}\vec{E}_s(r_s, \theta_s, \phi_s) &= -jw\vec{A} \\ &= -\frac{jw\mu}{2\pi r_s}e^{-jkr_s}\iint_S \hat{n}\times\vec{H}_i e^{jk\hat{r}_s\cdot\vec{r}_p}\,ds \\ &= \frac{e^{-jkr_s}}{r_s}(E_{i\phi}\hat{\theta}_i - E_{i\theta}\hat{\phi}_i)\times\underbrace{\left(\frac{j}{\lambda}\right)\iint_S \hat{n}e^{jk(h+g)}\,ds}_{\vec{S}} \quad (20)\end{aligned}$$

where
$g = \hat{r}_s\cdot\vec{r}_p = x_p\sin\theta_s\cos\phi_s + y_p\sin\theta_s\sin\phi_s + z_p\cos\theta_s$. However, it is not possible to obtain an exact closed form solution for $\vec{S}$ with this integral. Given that the incident wavefront is assumed plane and the incident field is known at the facet vertices, the amplitude and phase at the interior integration points can be found by interpolation. Then, the integrand can be expanded by using Taylor series, and each term integrated to give a closed form result. Usually, a small number of terms in Taylor series (on the order of 5) will give a sufficiently accurate approximation with unit amplitude plane wave ($|Ei| = 1$) [18].

$$\vec{S} = \left(\frac{j}{\lambda}\right) |\overrightarrow{AB} \times \overrightarrow{AC}| \, e^{jD_0} \left\{ \left[\frac{e^{jD_p}}{D_p(D_q - D_p)}\right] - \left[\frac{e^{jD_q}}{D_q(D_q - D_p)}\right] - \frac{1}{D_q D_p} \right\} \quad (21a)$$

where

$$D_p = k[(x_B - x_A)\sin\theta_s \cos\phi_s + (y_B - y_A)\sin\theta_s \sin\phi_s + (z_B - z_A)\cos\theta_s], \quad (21b)$$

$$D_q = k[(x_C - x_A)\sin\theta_s \cos\phi_s + (y_C - y_A)\sin\theta_s \sin\phi_s + (z_C - z_A)\cos\theta_s], \quad (21c)$$

$$D_0 = k[x_A \sin\theta_s \cos\phi_s + y_A \sin\theta_s \sin\phi_s + z_A \cos\theta_s]. \quad (21d)$$

It is now possible to write the formula of PO current as $\vec{J}_s = (J_{sx}\hat{x} + J_{sy}\hat{y} + J_{sz}\hat{z})e^{jkh}$. In the general case, the local facet coordinate system will not be aligned with the global coordinate system. In the local facet coordinate system $(x", y", z")$, the facet lies on the $x"y"$ plane, with $\hat{z}"$ being the normal to the facet surface, hence $\hat{n} = \hat{z}"$. For any arbitrary oriented facet with known global coordinates, its local coordinates can be obtained by a series of two rotations. First, angles α and β, are calculated by $\alpha = \arctan[n_y/n_x]$ and $\beta = \arccos(\hat{z} \cdot \hat{n})$, where $\hat{n} = n_x\hat{x} + n_y\hat{y} + n_z\hat{z}$. The local coordinates can be transformed to global coordinates [19]:

$$\begin{bmatrix} x" \\ y" \\ z" \end{bmatrix} = \begin{bmatrix} \cos\beta & 0 & -\sin\beta \\ 0 & 1 & 0 \\ \sin\beta & 0 & \cos\beta \end{bmatrix} \begin{bmatrix} \cos\alpha & \sin\alpha & 0 \\ -\sin\alpha & \cos\alpha & 0 \\ 0 & 0 & 1 \end{bmatrix} \begin{bmatrix} x \\ y \\ z \end{bmatrix} \quad (22)$$

However, in facet local coordinates, the surface current does not have a $\hat{z}"$component, since the facet lies on the $x"y"$plane. Hence the local surface current is given by $\vec{J}_s = (J''_{sx}\hat{x}" + J''_{sy}\hat{y}")e^{jkh}$, the surface current components are [19]:

$$J''_{sx} = \left[\frac{E"_{i\theta}\cos\phi"\cos\theta"}{2R_s + Z_0\cos\theta"} - \frac{E"_{i\phi}\sin\phi"}{2R_s\cos\theta" + Z_0}\right]\cos\theta" \quad (23a)$$

$$J''_{sy} = \left[\frac{E"_{i\theta}\sin\phi"\cos\theta"}{2R_s + Z_0\cos\theta"} + \frac{E"_{i\phi}\cos\phi"}{2R_s\cos\theta" + Z_0}\right]\cos\theta" \quad (23b)$$

where $E"_{i\theta}, E"_{i\phi}$ are the components of the incident field in the local facet coordinates, $\theta", \phi"$ are the spherical polar angles of the local coordinates and R_s being the surface resistivity of the facet material. When $R_s = 0$, the surface is a perfect electric conductor and assume that surface model is smoothing. To obtain the total scattered field, simply replace (23a) and (23b) in the radiation integral for the triangular facet, which was determined in (21a), the total number of facets (m = 20), so

$$\vec{E}_s(r_s, \theta_s, \phi_s) = \sum_{m=1}^{20} \frac{-jkZ_0 e^{-j(kr_s - D_{0m})}}{4\pi r_s} \left(J''_{smx}\hat{x}" + J''_{smy}\hat{y}"\right) \times |\overrightarrow{AB}_m \times \overrightarrow{AC}_m| \times \left\{ \left[\frac{e^{jD_{Pm}}}{D_{Pm}(D_{qm} - D_{Pm})}\right] - \left[\frac{e^{jD_{qm}}}{D_{qm}(D_{qm} - D_{Pm})}\right] - \frac{1}{D_{qm} D_{Pm}} \right\}. \quad (24)$$

Once the scattered field is known, RCS is computed in terms of the incident and scattered electric field intensities, and given by [20]:

$$RCS(r_s, \theta_s, \phi_s) = \lim_{R\to\infty} 4\pi R^2 \frac{|\vec{E}_s(r_s, \theta_s, \phi_s)|^2}{|\vec{E}_i|^2} \quad (25)$$

where R is the distance between the radar transmitter and target. For most objects, radar cross section is a three-dimensional map of the scattering contributions, which varies as a function of aspect angles (azimuth and elevation) and polarization. The scattering matrix describes the scattering behavior of a target as a function of polarization. Normally it contains four RCS values ($\theta\theta$, $\theta\phi$, $\phi\theta$ and $\phi\phi$), where the first letter denotes the transmission polarization, the second letter is the polarization at receive. Therefore, RCS can be derived at any polarizations:

$$RCS(r_s, \theta_s, \phi_s) = \lim_{R\to\infty} 4\pi R^2 \begin{bmatrix} |S_{\theta\theta}|^2 & |S_{\theta\phi}|^2 \\ |S_{\phi\theta}|^2 & |S_{\phi\phi}|^2 \end{bmatrix}. \quad (26)$$

The s_{pq} denotes the scattering parameter, whose first subscript specifies polarization of the receive antenna and the second one refers to polarization of the incident wave. The elements of scattering matrix are complex quantities and in terms of RCS [20]

$$RCS_{pq} = 4\pi R^2 S_{pq}{}^2 e^{-2j\psi_{pq}},$$

$$\psi_{pq} = \arctan\left\{\frac{Im(\frac{E_{sp}}{E_{iq}})}{Re(\frac{E_{sp}}{E_{iq}})}\right\}. \quad (27)$$

4. Range Accuracy under Positional Instability and Stealthy RCS Data

The bistatic geometry for a stealth target model considered in this paper is shown in Fig. 5, in which stealth target state $X = (p, V)^T = (x, y, z)^T$ is given in 3D Cartesian coordinates. In the same way the localizations of unstable balloon receiver and stationary transmitter are given respectively by $X_R = (P_R, V_R)^T = (x_R, y_R, z_R)^T$ and $X_T = (P_T, 0)^T = (x_T, y_T, z_T)^T$. Here we consider measurements in terms of bistatic range r, azimuth ϕ, elevation θ and bistatic range–rate $\dot{r}$, which is proportional to measured Doppler shift. The measurement equation without a root mean square measurement error (RMSE) $Z = (r, \phi, \theta, \dot{r})^T$ is given by:

$$r = \|P - P_T\| + \|P - P_R\|, \quad (28a)$$

$$\phi = \arctan\left[\frac{y - y_R}{x - x_R}\right], \quad (28b)$$

$$\theta = \arctan\left[\frac{z - z_R}{\sqrt{(x - x_R)^2 + (y - y_R)^2}}\right], \quad (28c)$$

$$\dot{r} = \left[\frac{(P - P_T)\,V}{\|P - P_T\|}\right] + \left[\frac{(P - P_R)\,(V - V_R)^T}{\|P - P_R\|}\right] \quad (28d)$$

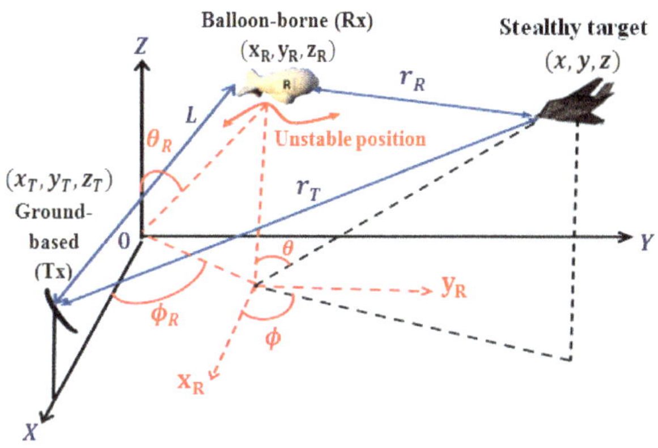

Fig. 5. The stealthy target detection under positional instability of balloon-borne bistatic radar system.

while the measurements at the receiver is characterized by root mean square measurement error (RMSE) as

$$\left.\begin{aligned} r^m &= r + \sigma_R \\ \phi^m &= \phi + \sigma_\phi \\ \theta^m &= \theta + \sigma_\theta \end{aligned}\right\} \quad (29)$$

The measurement accuracy is characterized by RMSE of σ_R, σ_ϕ, σ_θ computed by three error components [21].

$$\sigma_R = ({\sigma_{RN}}^2 + {\sigma_{RF}}^2 + {\sigma_{RB}}^2)^{1/2} \; , \quad (30a)$$

$$\sigma_\phi = ({\sigma_{\phi N}}^2 + {\sigma_{\phi F}}^2 + {\sigma_{\phi B}}^2)^{1/2} \, , \quad (30b)$$

$$\sigma_\theta = ({\sigma_{\theta N}}^2 + {\sigma_{\theta F}}^2 + {\sigma_{\theta B}}^2)^{1/2} \quad (30c)$$

where $\sigma_{RN}, \sigma_{\phi N}, \sigma_{\theta N}$ are SNR dependent random range and angular measurement errors, $\sigma_{RF}, \sigma_{\phi F}, \sigma_{\theta F}$ are range and angular fixed errors, and $\sigma_{RB}, \sigma_{\phi B}, \sigma_{\theta B}$ are range and angular bias errors. The SNR-dependent error usually dominates the radar range angular errors, which are random with a standard deviation and given by:

$$\sigma_{RN} = \frac{\Delta R}{\sqrt{2(SNR)}} = \frac{C}{2B\sqrt{2(SNR)}} \, , \quad (31a)$$

$$\sigma_{\phi N} = \sigma_{\theta N} = \frac{\theta_B / \cos(\phi \; or \; \theta)}{K_M \sqrt{2(SNR)}} \quad (31b)$$

where B is the waveform bandwidth, C is the speed of light, ΔR is range resolution, θ_B is broadside beamwidth in the angular coordinate of the measurement, and K_M is monopulse pattern difference slope, assuming the value of broadside beamwidth is 1° and K_M is typically 1.6 [21]. The bistatic form of radar equation is developed here to evaluate bistatic radar sensitivity properties. The transmitter and the receiver are deployed at two separate locations, either or both of them changing with time. The receiver co-operates with transmitter through a synchronization link. Normally, a co-located Tx and Rx are not described as a bistatic system, even though they don't share a common antenna. Since the separation becomes significant relative to the typical target range, so that bistatic radar systems become relevant. It is also assumed that the target is an isotropic radiator, giving a constant RCS in all directions. Under these assumptions, it is reasonable to calculate bistatic radar sensitivity by summing up the partial signal to noise ratio as [2]:

$$SNR = \frac{P_t G_t G_t RCS_B \lambda^2}{(4\pi)^3 K T_s B_n {R_t}^2 {R_r}^2 L} \quad (32)$$

where P_t is the peak transmitted power, G_t is the transmitter gain, G_r is the receiver gain, RCS_B is bistatic RCS of the target, λ is the transmitted wavelength, B_n is the bandwidth of the transmitted waveform, k is Boltzmann's constant, T_s is the receiving system noise temperature, L is the system loss for transmitter and receiver, R_t is the distance from the transmitter to the target and R_r is the distance from the target to the receiver. Most of the previous researches in bistatic radar systems only considered the simplest case of bistatic radar sensitivity by isotropic radiator giving a constant RCS in all directions except for the distance from the transmitter and receiver to the target [3–8]. But this is not an accurate consideration to calculate the SNR of stealth target because the RCS value varies with elevation angle and azimuth angles according to the position instability of the balloon receiver. Therefore the accurate formula of bistatic balloon-borne radar sensitivity depends on nature RCS of stealth target and the position instability of the balloon receiver, it should be written as:

$$SNR_B(r, \phi, \theta) = M \frac{RCS_B(r, \phi, \theta)}{{R_t}^2 {R_r}^2} \, , \quad (33a)$$

$$M = \frac{P_T G_T G_R \lambda^2}{(4\pi)^3 K T_s B_n L} \, . \quad (33b)$$

From (33) and (31), the accurate form of random range and angular measurement error SNR depend on RCS of stealth target predicted by (PO) method in Section 3 and are given by

$$\sigma_{RN} = \frac{C}{8B\sqrt{\dfrac{M.RCS_B(r, \phi, \theta)}{{R_t}^2 {R_r}^2}}} \, , \quad (34a)$$

$$\sigma_{\phi N} = \sigma_{\theta N} = \frac{\theta_B / \cos(\phi \; or \; \theta)}{K_M \sqrt{2\left(\dfrac{M.RCS_B(r, \phi, \theta)}{{R_t}^2 {R_r}^2}\right)}} \, . \quad (34b)$$

By using the differential (29) and it can be expressed in the matrix form [22]:

$$\begin{bmatrix} dr \\ d\phi \\ d\theta \end{bmatrix} = \begin{bmatrix} C_{R1} + C_{T1} & C_{R2} + C_{T2} & C_{R3} + C_{T3} \\ -\dfrac{\sin^2 \phi}{y - y_R} & \dfrac{\cos^2 \phi}{x - x_R} & 0 \\ \dfrac{C_{R3} \cos \phi}{r_R} & -\dfrac{C_{R3} \sin \phi}{r_R} & \dfrac{\cos \theta}{r_R} \end{bmatrix} \begin{bmatrix} dx \\ dy \\ dz \end{bmatrix} + \begin{bmatrix} k_r \\ k_\phi \\ k_\theta \end{bmatrix} \quad (35a)$$

where

$$C_{i1} = \frac{x - x_i}{r_i}, C_{i2} = \frac{y - y_i}{r_i}, C_{i3} = \frac{z - z_i}{r_i} \, , \quad i = (R, T), \quad (35b)$$

$$k_r = -[C_{R1} dx_R + C_{R2} dy_R + C_{R3} dz_R] \, , \quad (35c)$$

$$k_\phi = \frac{\sin^2 \phi}{y - y_R} dx_R - \frac{\cos^2 \phi}{x - x_R} dy_R \, , \quad (35d)$$

$$k_\theta = \frac{C_{R3}\cos\phi}{r_R}dx_R + \frac{C_{R3}\sin\phi}{r_R}dy_R + \frac{\cos\theta}{r_R}dz_R, \qquad (35e)$$

or

$$dV = \mathbb{C}dX + dS \qquad (36)$$

where $\mathbb{C}$ (3×3) is the matrix of coefficients, dX (3×1) is the vector of stealth target's position error, dV (3×1) is the vector measurement of stealth target's position and dS (3×1) is the vector pertaining to all random measurement error according to position instability of balloon receiver. The solution of (36) is

$$dX = \mathbb{C}^{-1}(dX - dS). \qquad (37)$$

The corresponding covariance matrix of the stealthy target position error is [22]:

$$P_{dx} = \mathbb{C}^{-1}\{E[dVdV^T] + E[dSdS^T]\}\mathbb{C}^{-T}\ . \qquad (38)$$

The expressions of $E[dVdV^T]$ and $E[dSdS^T]$ are given by

$$E[dVdV^T] = \mathrm{diag}\left[\sigma_r^2 \quad \sigma_\phi^2 \quad \sigma_\theta^2\right], \qquad (39a)$$

$$E[dSdS^T] =$$

$$\begin{bmatrix} 2 & 0 & 0 \\ 0 & \frac{1}{r_R^2\cos^2\theta} & \frac{(\sin\phi-\cos\phi)\sin^2\phi\sin\theta}{2r_R} \\ 0 & \frac{(\sin\phi-\cos\phi)\sin 2\phi\sin\theta}{2r_R} & \frac{1}{r_R^2} \end{bmatrix}\begin{bmatrix} \sigma_{xR}^2 \\ \sigma_{yR}^2 \\ \sigma_{zR}^2 \end{bmatrix} \qquad (39b)$$

where $\sigma_{xR}, \sigma_{yR}, \sigma_{zR}$ are the position instability measurement errors of the balloon receiver

The RMSE is used to describe the stealthy target position accuracy, from (38), (39a) and (39b), it is noted that the stealthy position accuracy is related to the position of the considered target and the deployment of the two sites in the bistatic system. So it is called GDOP (Geometrical Dilution Of Position). The expression of the GDOP is [23]:

$$GDOP = \sqrt{tr\,[P_{dx}]}\,. \qquad (40)$$

The radar and target parameters are illustrated in Tab. 1.

Parameter	Value
P_t(kWatt)	250
G_t , G_r (dB)	32
f(MHz)	3000
B_n (MHz)	1
L_t, L_r(dB)	5
σ_{RF}(m)	3
σ_{RB} (m)	10
ΔR (m)	10
$\sigma_{\phi F}$(mrad)	0.2
$\sigma_{\theta F}$(mrad)	0.2
$\sigma_{\phi B}$(mrad)	0.5
$\sigma_{\theta B}$(mrad)	0.5
V(m/s)	400

Tab. 1. The radar and stealthy target parameters.

5. Simulation Results

5.1 Instability of Balloon-borne Receiver according to Wind Speed Results

For all problems considered, the ideal balloon position is fixed at (X = 150 km, Y = 100 km, Z = 21 km) from the ground-based transmitter. The balloon is initialized by flying at 1 m/s airspeed. With solving the optimal control problems, we neglect the contribution from centripetal acceleration, assuming acceleration is constant in the ENU farm, with a magnitude of 0.029 m/s^2. The simulation displays the positional instability of the balloon- borne receiver according to the horizontal wind speed, taking 250 seconds of random wind as an example, σ_w is of 15 m/s shown in Fig. 6. It is clear that the mean wind velocity at altitude of 21 km varies randomly between -15 to +15 m/s. Fig. 7 shows the comparison between the ideal position and unstable balloon position in X-direction and Y-direction. It is clear that the balloon suspends in the stratosphere about 21 km above the Earth and extends in a sphere of 0.5 km radius.

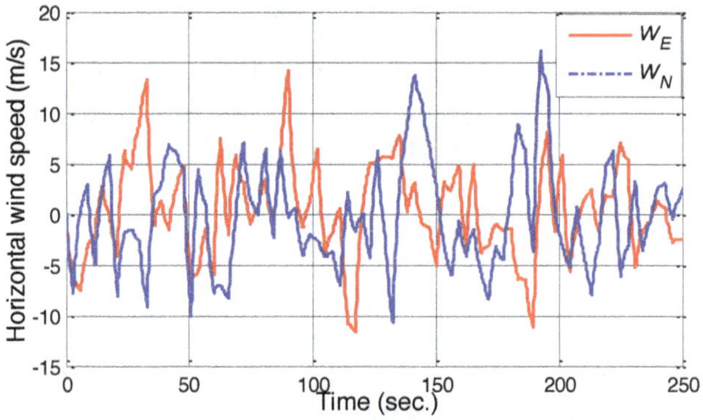

Fig. 6. Simulation of random horizontal wind speed.

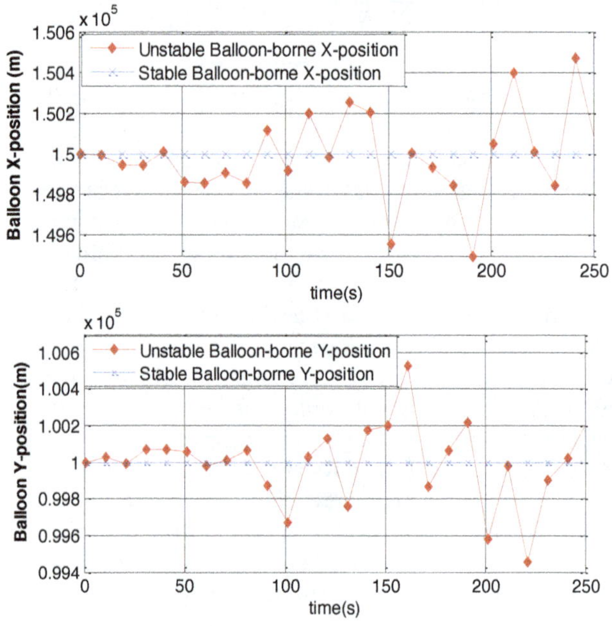

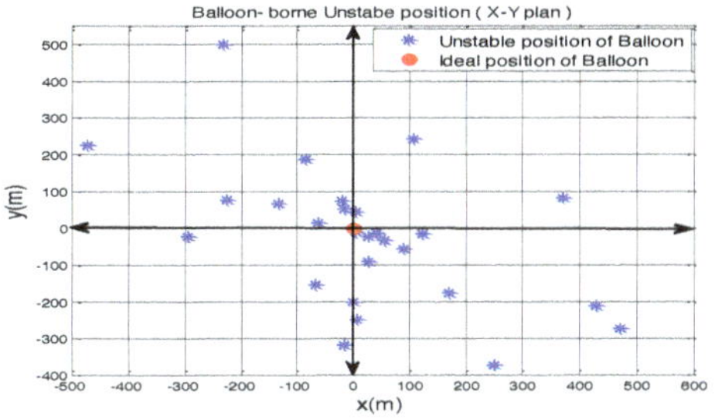

Fig. 7. The comparison between the ideal position and unstable balloon position in X-direction and Y-direction.

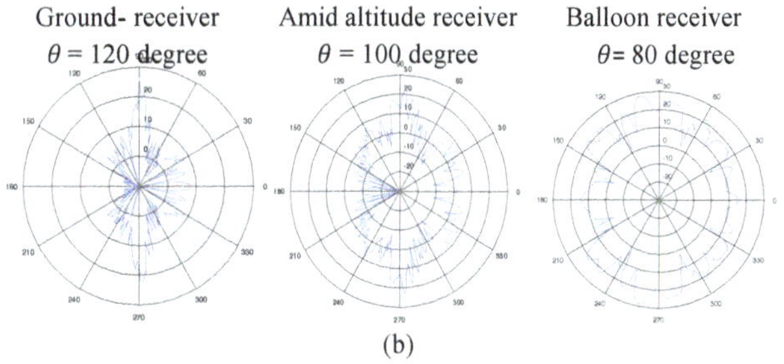

Fig. 9. a) Comparison of RCS within different aspect angle θ according to the altitude of bistatic receiver in 2-D, b) the polar plot using different aspect angles.

5.2 Establishing the Stealthy Target Model and Stealthy RCS Results

Fig. 8(a) shows the geometry model and scatters of stealth target F-117A in the range of ($0 \le \theta \le 360$) and ($0 \le \phi \le 360$). Fig. 8(b) shows 3-D RCS of the stealth target in bistatic system. A comparison of 2-D bistatic RCS within different aspect angle θ according to the altitude of bistatic receiver is demonstrated in Fig. 9(a). We further assume that the incident wave is (θ-polarized), frequency is 3 GHz and elevation angle takes two values ($\theta = 80$ and 120 degree) while azimuth angle ϕ between the horizon and observation direction varies from (0 to 360 degrees). It is clear that the RCS with a higher aspect angle in balloon-borne radar is better than with lower aspect angles in ground-based bistatic system. Fig. 9(b) shows the results in 2-D polar plot.

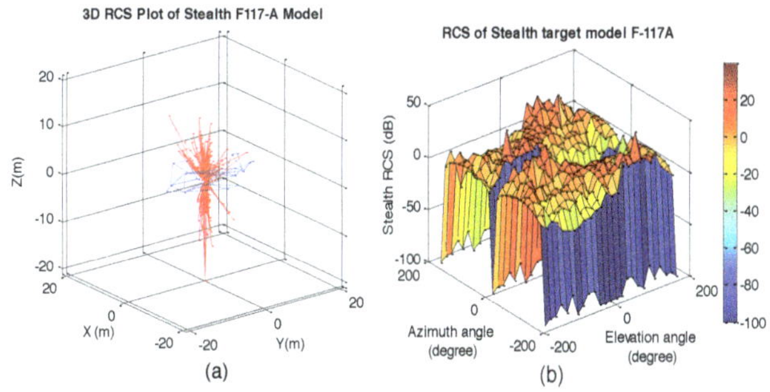

Fig. 8. Bistatic RCS of the stealthy target based on F-117A in 3-D.

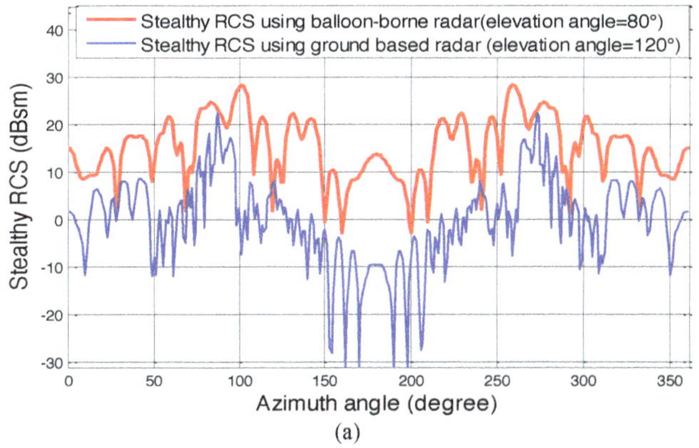

(a)

5.3 SNR Results for Proposed Scheme under Balloon Positional Instability & Stealthy RCS

A comparison of radar sensitivity between the proposed scheme in balloon-borne radar and conventional ground-based bistatic system at X-direction is demonstrated in Fig. 10. To clearly indicate SNR fluctuation of real stealth target RCS in (X-axis) range, we assume that the stealth target is moving at constant altitude 17 km and constant velocity $V = 400$ m/s. Thus, as long as the range changes, the elevation aspect angle changes similarly, say the elevation angle $\theta_s = 180°$ for ground-based radar while for balloon-borne radar $\theta_s = 0°$, so as to satisfy the minimum range between radar and stealth target. The maximum range exists in the far field saturation, as elevation angle $\theta_s \approx 90°$ for ground-based radar and $\theta_s \approx 50°$ for balloon-borne radar. Fig. 10 is a comparison between the SNR of real stealth RCS subject to stable and unstable position of the balloon-borne and flat RCS (0.025 m^2) of conventional ground-based system. It is clear that the sensitivity of the proposed radar scheme has been improved due to increasing scatterer RCS of stealth model with a higher aspect angle comparing to the conventional system. The 3-D bistatic radar sensitivities of taking flat RCS (0.025 m^2) and stealth RCS are shown in Fig. 11.

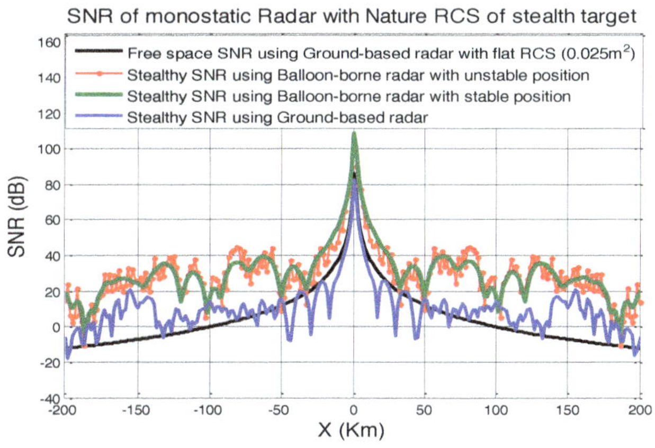

Fig. 10. A comparison between SNR for radar with real RCS of a stealth target under stable and unstable balloon position and ground-based system in X-direction.

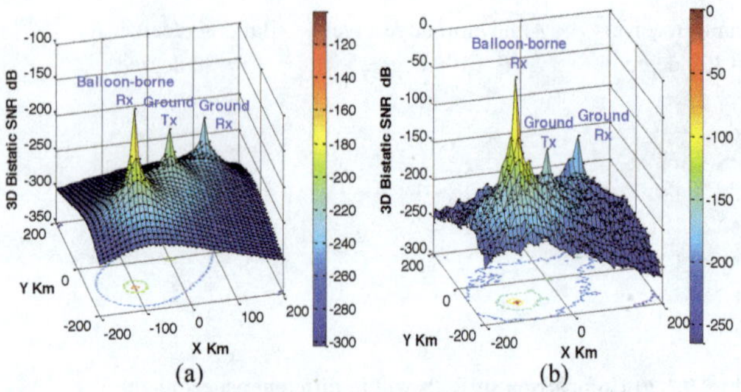

Fig. 11. 3D Bistatic radar sensitivity using (a) flat RCS (0.025 m^2) and (b) real stealth RCS.

5.4 Simulation of Tracking a Stealthy Target

Fig. 12 shows RMSE in range and angle of stealthy target detection. It shows that RMSE of the proposed scheme under instable position has been improved comparing to the conventional ground-based radar by increasing scatterer RCS of stealth model with higher aspect angle. We can find that fluctuation of RMSE value under two cases shows a tendency around the flat (RCS = 0.025 m^2) value along X-axis. From the RMSE plots, it shows that the value is increasing with range enlarging when ground-based system is applied. In balloon-borne system with instable position, RMSE becomes even less due to obtaining real stealth RCS. In ground-based system, nulls (less than 0 dB) of SNR increase along range axis up to the maximum range (200 km), while range RMSE reaches to 300 m, angle RMSE reaches to 4°. In cases of balloon-borne system under instable position, RMSE has been improved with less nulls existing, for the same maximum range, RMSE equals to 50 m, angular RMSE fluctuate around 1°.

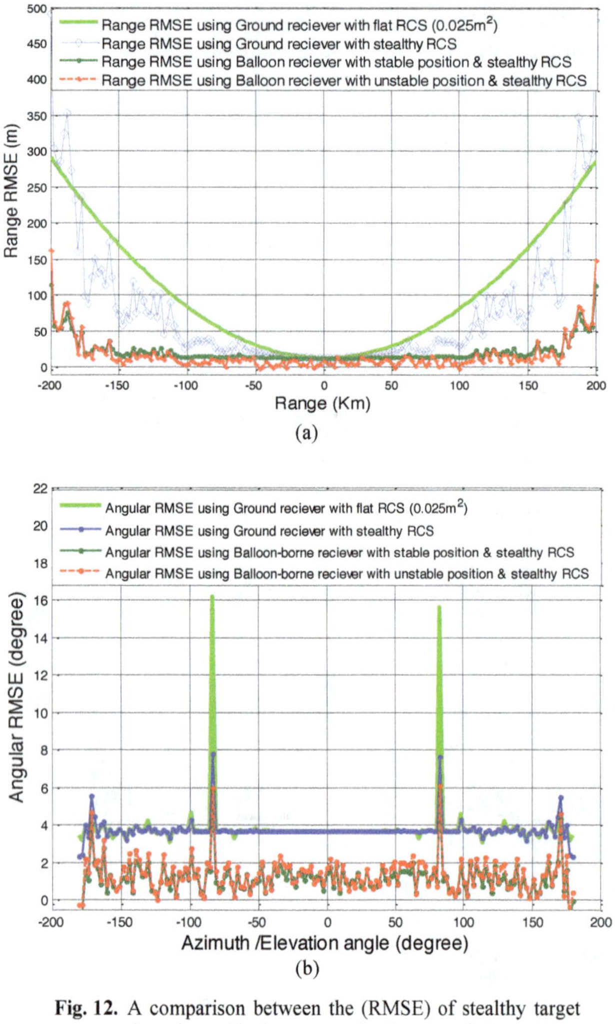

Fig. 12. A comparison between the (RMSE) of stealthy target detection with the proposed scheme and the conventional system.

Fig. 13 presents the contour plots of GDOP values for real stealth RCS data predicted by PO method in different radar geometrical structures. The instable balloon-borne platform is simulated under random wind by using Dryden turbulence model within an area of 400 km × 400 km. The stealth target flies at 17 km altitude. The ground-based transmitter (Tx) is located at (–150,150) km, ground-based receiver (Rx) is located at (150,150) km and the balloon-borne receiver (Rx) is located above (150,–150) km. The altitude of balloon-borne receiver is equal to 21 km. The GDOP of balloon-borne radar has been improved on conventional ground-based radar by higher aspect vision. The worst case is that the transmitter and the receiver both are ground-based. In this case, GDOP around ground-based receiver is poor that the inner contour (30 m) is located at 35 km and the outer contour (180 m) is located at 100 km. The optimal case is that the receiver is put on balloon-borne, it indicates that the GDOP results have been improved and accurately estimated due to decreasing RMSE of the stealth target detection, it is shown that the inner contour (30 m) is located at 150 km and the outer contour (180 m) is located at 350 km from the balloon-borne receiver. The results of Fig. 13 are summarized in Tab. 2.

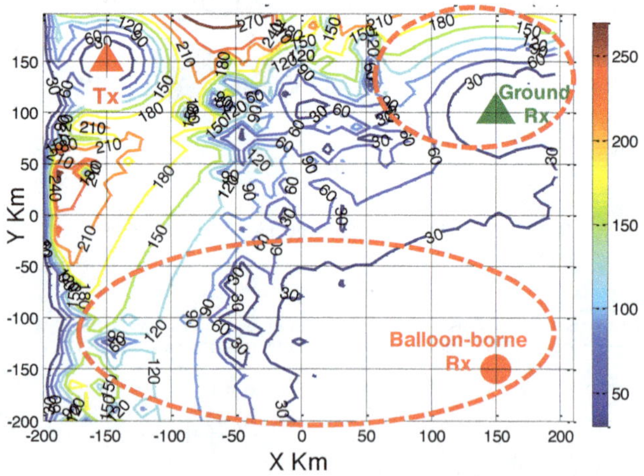

Fig. 13. The GDOP (m) comparison of balloon-borne and conventional ground-based bistatic system.

Radar type	GDOP of receiver (m)			
	Inner contour		Outer contour	
	Value	Range (km)	Value	Range (km)
Ground-based	30	35	180	100
Balloon-borne	30	150	180	350

Tab. 2. The GDOP of different geometrical structures.

It is found that GDOP of the proposed scheme has been improved due to decreasing the range and angular RMSE by increasing scatterer RCS of the stealth model comparing to the conventional ground-based system

In Fig. 14(a), the comparison between tracking of stealth target using the proposed scheme under instable position due to random wind speed and the conventional ground system. It is clear that the position estimate error of stealth target model was reduced by using the proposed scheme at all time interval due to increasing stealth RCS with a higher aspect vision as shown in Fig. 14(b).

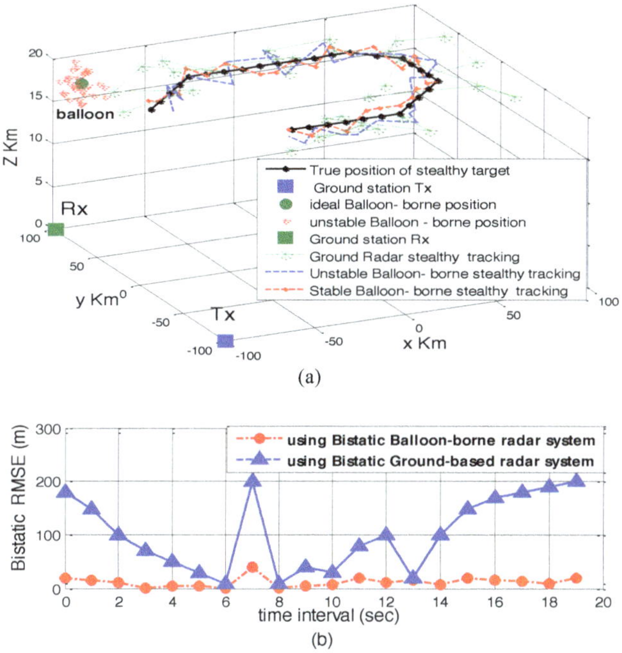

Fig. 14. The comparison between the tracking of stealthy target using balloon-borne bistatic system and conventional ground-based bistatic system.

6. Conclusion

An improvement of stealth RCS detection with higher aspect vision is presented. The stratospheric balloon positional instability due to random wind is considered. The results revealed that the proposed scheme demonstrates higher location accuracy than the conventional ground-based system. It is clear that bistatic radar sensitivity of the proposed scheme has been improved due to increasing scatterer RCS of stealth model with a higher aspect angle as predicted by PO method. The comparison between tracking of stealth target using the proposed scheme and the conventional system is introduced. The GDOP of the proposed scheme has been improved due to decreasing RMSE of the balloon radar system comparing to the conventional system. Finally the proposed system has better performance at almost all time intervals.

Acknowledgments

This work is supported by the China National Found of "863 Project", Ref. 2013AA7010051.

References

[1] CHEN, X., GUAN, J., LIU, N., HE, Y. Maneuvering target detection via Radon-Fractional Fourier transform-based long time coherent integration. *IEEE Trans. Signal Process*, 2014, vol. 62, no. 4, p. 939–953.

[2] TENG, Y., GRIFFITHS, H.D., BAKER, C.J., WOODBRIDGE, K. Netted radar sensitivity and ambiguity. *IET Radar Sonar Navig.*, December 2007, vol. 1, no. 6, p. 479–486.

[3] KUSCHEL, H., HECKENBACH, J., MULIER, ST., APPEL, R. Countering stealth with passive, multi-static, low frequency radars. *IEEE Aerospace and Electronic Systems Magazine*, 2010, vol.25, no. 9, p. 11–17.

[4] KUSCHEL, H., HECKENBACH, J., MULIER, ST., APPEL, R. On the potentials of passive, multistatic, low frequency radars to counter stealth and detect low flying targets. In *IEEE Conference, RADAR '08*, 2008, p. 1-6.

[5] HOWE, D. Introduction to the basic technology of stealthy aircraft: Part 2- Illumination by the enemy (active considerations). *Journal of Engineering for Gas Turbines and Power*, 1991, vol. 113, no. 1, p. 80–86.

[6] EL-KAMCHOUCHY, H., SAADA, K., HAFEZ, A. Optimum stealthy aircraft detection using a multistatic radar. *ICACT Transactions on Advanced Communications Technology (ICACT-TACT)*, 2013, vol. 6, no. 2, p. 337–342.

[7] DENG, H. Orthogonal netted radar systems. *IEEE Aerospace and Electronic Systems Magazine*, 2012, vol. 27, no. 5, p. 28–35.

[8] BEZOUSEK, P., SCHEJBAL, V. Bistatic and multistatic radar systems. *Radioengineering*, 2008, vol. 17, no. 3, p. 53–59.

[9] AXIOTIS, D. I., THEOLOGOU, M. E., SYKAS, E. D. The effect of platform instability on the system level performance of HAPS UMTS. *IEEE Communications Letters*, 2004, vol. 8, no. 2, p. 111 to 113.

[10] BEAL, T. R. Digital simulation of atmospheric turbulence for Dryden and von Karman models. *Journal of Guidance, Control, and Dynamics*, 1993, vol. 16, no. 1, p. 132–138.

[11] FORTUNATI, S., FARINA, A., GINI, F., GRAZIANO, A., GRECO, M. S., GIOMPAPA, S. Impact of flight disturbances on airborne radar tracking. *IEEE Transactions on Aerospace and Electronic Systems*, 2012, vol. 48, no. 3, p. 2698 –2710.

[12] HOGGE, E. B-737 linear autoland simulink model. *NASA, Technical Report* NASA/CR-2004-213021, 2004.

[13] MIELE, A., WANG, T., MELVIN, W. W. Optimal take-off trajectories in the presence of wind shear. *Journal of Optimization Theory and Applications*, 1986, vol. 49, no. 1, p. 1–45.

[14] FELDMAN, M. A. Efficient low-speed flight in a wind field. *Master Thesis*. Blacksburg, VA, Virginia Polytechnic Inst. and State Univ., July 1996.

[15] LEE, S., BANG, H. Three-dimensional ascent trajectory optimization for stratospheric airship platforms in the jet stream. *Journal of Guidance, Control, and Dynamics*, 2007, vol. 30, no. 5, p. 1341 to 1352.

[16] UPENDRA, A., BALAKRISHNAN, J. A novel method for RCS reduction of a complex shaped aircraft using partial RAM coating. *International Journal of Engineering and Innovative Technology (IJEIT)*, 2012, vol. 2, no. 2, p. 52–56.

[17] LI, J., WANG, X., QU, L. Calculation of physical optics integrals over NURBS surface using a delaminating quadrature method. *IEEE Transactions on Antennas and Propagation*, 2012, vol. 60, no. 5, p. 2388– 2397.

[18] MOREIRA, F. J. S., PRATA, A. JR. A self-checking predictor-corrector algorithm for efficient evaluation of reflector antenna radiation integrals. *IEEE Transactions on Antennas and Propagation*, 1994, vol. 42, no. 2, p. 246–254.

[19] CORUCCI, L., GIUSTI, E., MARTORELLA, M., BERIZZI, F. Near field physical optics modelling for concealed weapon detection. *IEEE Transactions on Antennas and Propagation*, 2012, vol. 60, no. 12, p. 6052–6057.

[20] CORBEL, C., BOURLIER, C., PINEL, N., CHAUVEAU, J. Rough surface RCS measurements and simulations using the physical optics approximation. *IEEE Transactions on Antennas and Propagation*, 2013, vol. 61, no. 10, p. 5155–5165.

[21] CURRY, G. R. *Radar System Performance Modeling*. Second Edition, 2005.

[22] WEI, W., HE, L. The location method and accuracy analysis for bistatic systems. In *National Aerospace and Electronics Conference (NAECON)*. Dayton (OH, USA), 1994, p. 62–65.

[23] ZHAO KONGRUI, YU CHANGJUN, ZHOU GONGJIAN, QUAN TAIFAN. Altitude and RCS estimation with echo amplitude in bistatic high frequency surface wave radar. In *16th International Conference on Information Fusion*. Istanbul (Turkey), 2013, p. 1342–1347.

About Authors ...

Mohamed A. BARBARY received the B.Sc degree in Electronics and Communications in 2003 and the M.Sc. in Electrical Engineering in 2012, both from the Faculty of Engineering, Alexandria University, Egypt. He is currently working towards the Ph.D. degree in the College of Electronic and Information Engineering, Nanjing University of Aeronautics and Astronautics (NUAA), Nanjing, China. He received many technical courses in radar, stealth target detection and electronic engineering design and implementation and worked as radar system engineer for more than 4 years.

Peng ZONG received the B.Sc. from Nanjing University of Aeronautics and Astronautics (NUAA), China, and Ph.D. degrees from University of Portsmouth, UK. He has been a professor in College of Astronautics NUAA since 2006. His academic experience is research fellow in the Centre for Communication Systems Research, University of Surrey, England in 2001, following by employee of Hughes Network System (USA) as software engineer and chief designer in Information and Technology Institute of China Aerospace Science & Industry Corp. (CASIC) until 2006. His research fields are mobile communication, satellite networking mostly concerned by routing algorithm of LEO constellation, and stealthy detection of netting radar.

13

3D Capturing with Monoscopic Camera

Miroslav GALABOV

Dept. of Computer Systems and Technologies, University of Veliko Turnovo, street T.Tarnovski 2, 5000 Veliko Turnovo, Bulgaria

lexcom@abv.bg

Abstract. *This article presents a new concept of using the auto-focus function of the monoscopic camera sensor to estimate depth map information, which avoids not only using auxiliary equipment or human interaction, but also the introduced computational complexity of SfM or depth analysis. The system architecture that supports both stereo image and video data capturing, processing and display is discussed. A novel stereo image pair generation algorithm by using Z-buffer-based 3D surface recovery is proposed. Based on the depth map, we are able to calculate the disparity map (the distance in pixels between the image points in both views) for the image. The presented algorithm uses a single image with depth information (e.g. z-buffer) as an input and produces two images for left and right eye.*

Keywords

3D content, multi-view camera, 3D capturing.

1. Introduction

With different types of cameras, the 3D content capturing process is completely different [1]. The stereo camera or depth camera simultaneously captures video and associated per-pixel depth or disparity information; multi-view cameras capture multiple images simultaneously from various angles, then a multi-view matching (or correspondence) process is required to generate the disparity map for each pair of cameras, and then the 3D structure can be estimated from these disparity maps. The most challenging scenario is to capture 3D content from a normal 2D (or monoscopic) camera, which lacks disparity or depth information.

Basically, typical *stereo cameras* use two cameras mounted side by side for the recording, although some variants may build them into one with two lenses.

Depth cameras refer to a class of cameras that have sensors that are able to measure the depth for each of the captured pixels using a principle called time-of-flight. It gets 3D information "by emitting pulses of infrared light to all objects in the scene and sensing the reflected light from the surface of each object." The objects in the scene are then ordered in layers in the z-axis, which gives a grayscale depth map that a game or any software application can use. The depth sensor consists of an infrared laser projector combined with a monochrome CMOS sensor, which captures video data in 3D under any ambient light conditions.

Commercial *multi-view camera* systems are rare in the market, although Honda Motor announced that it had developed a prototype multi-view camera system which displays views from multiple wide-angle CCD cameras on the vehicle's navigation screen to reduce blind spots, support smooth parallel or garage parking, and support comfortable and safe driving at a three-way intersection where there is limited visibility or on narrow roads. There are many multi-view cameras (or camera arrays) set up in labs for research efforts, where the synchronization among these cameras is conducted by gunlock devices.

2. Capturing with Monoscopic Camera

The major difference between a stereo image and a mono image is that the former provides the feel of the third dimension and the distance to objects in the scene. Human vision by nature is stereoscopic due to the binocular views seen by our left and right eyes in different perspective viewpoints. It is our brain that is capable of synthesizing an image with stereoscopic depth. In general, a stereoscopic camera with two sensors is required for producing a stereoscopic image or video. However, most of the current multimedia devices deployed are implemented within the monoscopic infrastructure.

In the past decade, stereoscopic image generation has been actively studied. In [2], the video sequence is analyzed and the 3D scene structure is estimated from the 2D geometry and motion activities (which is also called structure from motion, or SfM). This class of approaches enables the conversion of recorded 2D video clips to 3D; however, its computational complexity is rather high so that it is not feasible to use it for realtime stereo image generation. On the other hand, since SfM is a mathematically ill-posed problem, the result might contain artifacts and cause visual discomfort. Some other approaches first estimate depth information from a single-view image and then generate the stereoscopic views after that. In [3], a method for extracting relative depth information from monoscopic cues, for example retinal sizes of objects, is proposed, which is useful for the auxiliary depth map generation. In [4], a facial-feature-based parametric depth map generation scheme is proposed to convert 2D head-and-shoulders images to 3D. In [5], an unsupervised method for depth-map generation is proposed, but some steps in the approach, for example the image classification in preprocessing, are not trivial and may be very complicated to implement, which undermines the practicality of the proposed algorithm. In [6], a real-time 2D-to-3D image conversion algorithm is proposed using motion detection and region segmentation; however, artifacts are unavoidable due to the inaccuracy of object segmentation and object depth estimation.

Clearly, all the methods mentioned above consider only the captured monoscopic images. Some other approaches use auxiliary sources to help generate the stereo views. In [7], a low-cost auxiliary monochrome or low-resolution camera is used to capture the additional view, and it then uses a disparity estimation model to generate the depth map of the pixels. In [8], a monoscopic high resolution color camera is used to capture the luminosity and chromaticity of a scene, and an inexpensive flanking 3D-stereoscopic pair of low resolution monochrome "outrigger" cameras are used to augment luminosity and chromaticity with depth. The disparity maps generated from the obtained three views are used to synthesis of the stereoscopic pairs. In [9–11], a mixed sets of automatic and manual techniques are used to extract the depth map (sometimes the automatic method is not reliable), and then a simple smoothing filter is used to reduce the visible artifacts of the result image.

In the following text, we introduce the new concept of using the auto-focus function of the monoscopic camera sensor to estimate depth map information [12], which avoids not only using auxiliary equipment or human interaction as mentioned above, but also the introduced computational complexity of SfM or depth analysis. The whole system design is novel, and is generic for both stereo image and video capture and generation. The additional but optional motion estimation module can help to improve the accuracy of the depth map detection for stereo video generation. The approach is feasible for low-power devices due to its two-stage depth map estimation design. That is, in the first stage, a block-level depth map is detected, and an approximated image depth map is generated by using bilinear filtering in the second stage. By contrast, the proposed approach uses statistics from motion estimation, auto-focus processing, and the history data plus some heuristic rules for estimating the depth map.

In Fig. 1, the proposed system architecture that supports both stereo image and video data capturing, processing, and display is shown. In the system, an image is first captured by a monoscopic camera sensor in the video front end (VFE), and then it goes through the auto-focus process, which helps to generate a corresponding approximated depth map. The depth map is further processed either using bilinear filtering for still-image or taking into account the motion information from the video coding process for video. After that, a depth-based image pair generation algorithm is used to generate stereo views. Clearly the 3D effect can be accomplished by choosing different display technologies such as holographic, stereoscopic, volumetric, and so on. In Fig. 2, the system architecture for still images is shown, which is simpler than the generic architecture in Fig. 1.

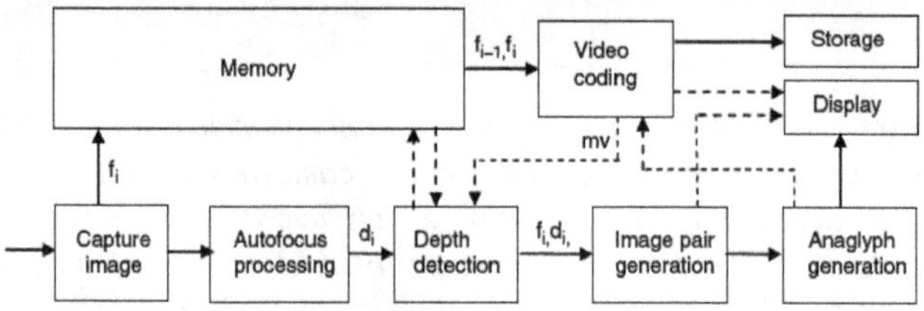

Fig.1. System architecture.

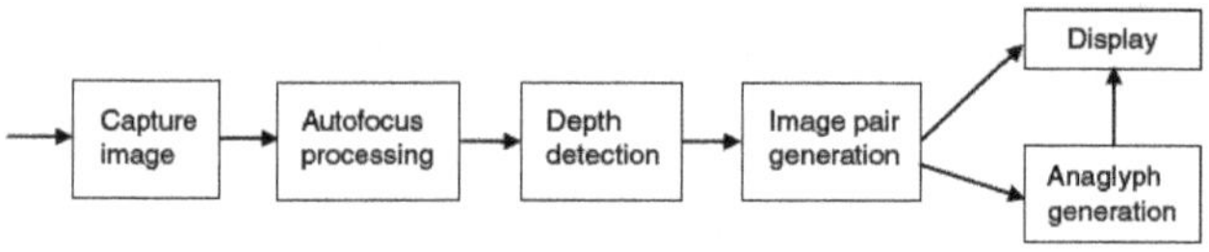

Fig. 2. System architecture for still images.

3. Depth Estimation by Autofocus Processing

In digital cameras, most focusing systems choose the best focus position by evaluating image contrast on the imager plane. Focus value (FV) is a score measured via a focus metric over a specific region of interest, and the autofocusing process of the camera normally chooses the position corresponding to the highest focus value as the best focus position of the lens. In some cameras, the high frequency content of an image is used as the focus value, for example, the high pass filter (HPF)

$$HPF = \begin{bmatrix} -1 & 0 & 0 & 0 & -1 \\ 0 & 0 & 4 & 0 & 0 \\ -1 & 0 & 0 & 0 & -1 \end{bmatrix}, \qquad (1)$$

can be used to capture the high frequency components for determining the focus value.

It is important to know that there is a relationship between the lens position from the focal point and the target distance from the camera (as shown in Fig. 3), and the relationship is fixed for a specific camera sensor. Various camera sensors may have different statistics of such relationships. It means that once the autofocus process locates the best focus position of lens, based on the knowledge of the camera sensor's property, we are able to estimate the actual distance between the target object and the camera, which is also the depth of the object in the scene. Therefore, the proposed depth map detection relies on a sensor-dependent autofocus processing.

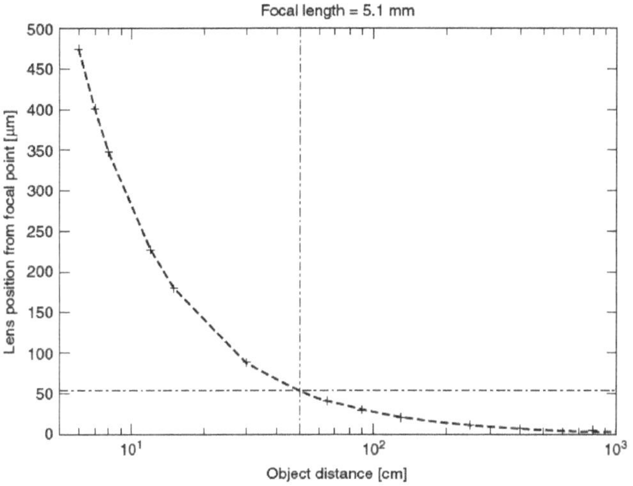

Fig. 3. Relationship between lens position from focal point and object distance.

4. Stereo Image Pair Generation

In this section, we propose a novel stereo image pair generation algorithm by using Z-buffer-based 3D surface recovery. The algorithm flowchart is shown in Fig. 4. We assume that the obtained image is the left view of the stereoscopic system; then, based on the depth map, we are able to calculate the disparity map (the distance in pixels between the image points in both views) for the image. Then a Z-buffer-based 3D interpolation process is called to construct a 3D visible surface for the scene from the right eye. Finally, the right view can be obtained by projecting the 3D surface onto the projection plane.

4.1 Disparity Map Generation

In Fig. 5, the geometry model of binocular vision is shown, where F is the focal length, L(x_L, y_L, 0) is the left eye, R(x_R, y_R, 0) is the right eye, T(x_T, y_T, z) is a 3D point in the scene, and P(x_P, y_P, F) and Q(x_Q, y_Q, F) are the projection points of the T onto the left and right projection planes. Clearly, the horizontal position of P and Q on the projection planes are $(x_P - x_L)$ and $(x_Q - x_R)$, and thus the disparity is $d = [(x_Q - x_R) - (x_P - x_L)]$.

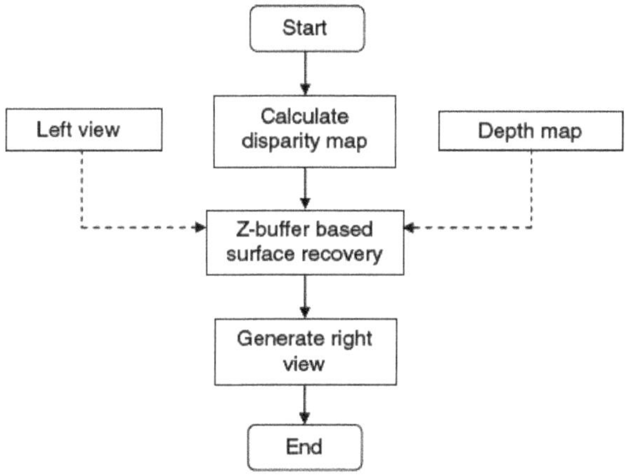

Fig. 4. Flowchart of the stereo image pair generation.

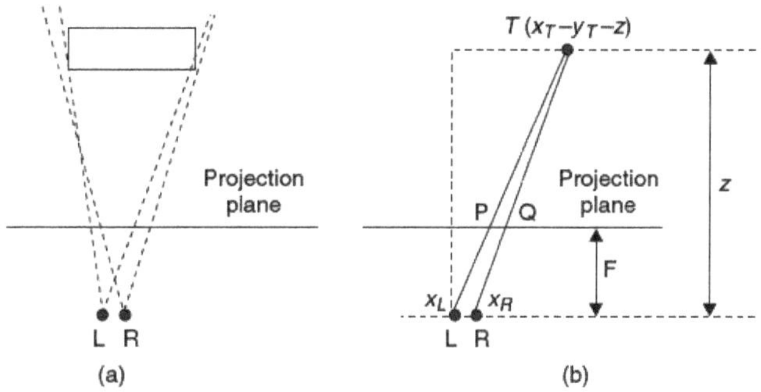

Fig. 5. Geometry model for binocular vision.

As shown in Fig. 5,

$$\frac{F}{z} = \frac{x_P - x_L}{x_T - x_L} = \frac{x_Q - x_R}{x_T - x_R}, \quad (2)$$

so

$$x_P - x_L = \frac{F}{z}(x_T - x_L), \quad (3)$$

$$x_Q - x_R = \frac{F}{z}(x_T - x_R), \quad (4)$$

and thus the disparity can be obtained as:

$$d = \frac{F}{z}(x_L - x_R). \quad (5)$$

Therefore, for every pixel in the left view, its counterpart in the right view is shifted to the left or right side by a distance of the disparity value obtained in (5). However, the mapping from left-view to right-view is not 1-to-1 mapping due to possible occlusions, therefore further process is needed to obtain the right-view image.

4.2 Z-Buffer-Based 3D Surface Recovering

We propose a Z-buffer-based 3D surface recovering algorithm for right-view generation. Since the distance between two eyes compared to the distance from eyes to the objects (as shown in Fig. 5) is very small, we can approximately think that the distance from object to the left eye is equal to the distance from itself to the right eye, which would greatly simplify the calculation.

In this method, we maintain a depth map $Z(x, y)$ for the right view where x, y are pixel positions in the view. Here the purpose is to reconstruct the 3D visible surface for the right view. At the beginning, the depth map is initialized as infinity. Then, for every pixel (x_0, y_0) in the left view with depth z_0 and disparity value d_0, we update the depth map for its corresponding pixel in the right view as:

$$Z(x_0 + d_0, y_0) = \min[Z(x_0 + d_0, y_0), z_0]. \quad (6)$$

After all the pixels in the left view are processed, we check the reconstructed depth map and search for the pixels with values equal to infinity (the pixels without a valid map on the left view). For such pixels, we first calculate its depth by 2D interpolation based on its neighbor pixels with available depth value. After that, we find the disparity value following the computing using (5) and then inversely find its corresponding pixel in the left view. If the corresponding pixel is available, then the corresponding intensity value can be used on the right-view pixel; otherwise, we use interpolation to calculate the intensity value based on its neighbor pixels in the right view with available intensity values. It is important to point out that the benefit of using the proposed algorithm over the direct intensity interpolation method is that it considers the 3D continuity of the object shape which results in better realism for the stereo effect.

Clearly, the problem of recovering the invisible area of the left view is an ill-posed problem. In the solution of [13], [14], the depth of the missing pixel is recovered by using its neighbor pixel in the horizontal direction corresponding to a further surface with an assumption that there are no other visible surfaces behind in the scene. For some cases, the assumption might be invalid. To consider more possible cases, in the proposed solution, the surface recovering considers depths of all the neighbor pixels in all directions, which will reduce the chances of invalid assumption and will result in better 3D continuity of the recovered surface.

5. Experimental Results

In these experiments, we used stereoscopic display to demonstrate the resulting 3D effect. We calculate the stereo image pairs using different kinds of image depth map and generate the corresponding images. As shown in Fig. 6b, this is generated by using the approximated image depth map shown in Fig. 6a, and Fig. 6d is generated by using the accurate image depth map shown in Fig. 6c. The approximate time of an execution of the algorithm for image from Fig. 6 is the following (CPU: Intel Core i3 4130 3.4 GHz 512 kb, resolution of images: 320 × 200): source ray-traced frame 2.5 s, both warped frames 0.15 s.

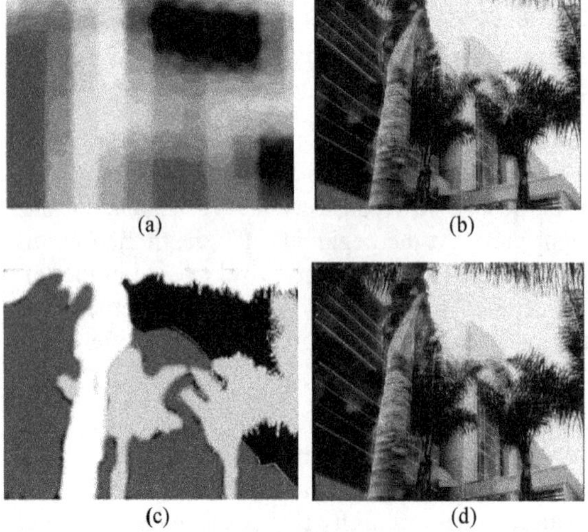

Fig. 6. Examples of the resulting stereoscopic image generated by using the different depth map.

The time of the ray-tracing depends directly on the complexity of the scene and on the contrary the time of the warping algorithm is independent of the scene complexity. Thus in the case of more complex scenes the efficiency gain is bigger. Because the process of generating of the stereo images involves a computation of a source frame with the original algorithm the time gain is at most 50 % of the original computation minus the time of processing of presented algorithm. In the case of ray-tracing the gain is just almost 50 %.

6. Conclusions

We propose a novel stereo image pair generation algorithm by using Z-buffer-based 3D surface recovery. We make experiments, in which we used stereoscopic display to demonstrate the resulting 3D effect. The results indicate that the approximated image depth map results in a similar image quality as using the accurate depth map, which proves the good performance of the proposed algorithm. The advantages of the presented method are the following: simplicity and speed that do not depend on the scene complexity.

All camera parameters need not to be known. It is sufficient to know only the resolution of images, horizontal angle of view and the distance from the camera to the plane of projection.

Acknowledgements

The presented article is part of research work carried out in the "Analysis, research and creation of multimedia tools and scenarios for e-learning" project - Contract No: RD - 09-590-12/10.04.2013, which is financially supported by the St. Cyril and St. Methodius University of Veliko Turnovo, Bulgaria.

References

[1] GUAN-MING SU, YU-CHI LAI, KWASINSKI, A., HAOHONG WANG. *3D Visual Communications.* John Wiley & Sons, 2013.

[2] JEBARA, T., AZARBAYEJANI, A., PENTLAND, A. 3D structure from 2D motion. *IEEE Signal Processing Magazine*, May 1999, vol. 16, no. 3, p. 66–83.

[3] XU, S. B. Qualitative depth from monoscopic cues. In *Proc. of Int. Conf. on Image Processing and its Applications*, Maastricht (The Netherlands), 1992, p. 437–440.

[4] WEERASINGHE, C., OGUNBONA, P., LI, W. 2D to pseudo-3D conversion of head and shoulder images using feature based parametric disparity maps. In *Proc. International Conference on Image Processing*. Thessaloniki (Greece), 2001, p. 963–966.

[5] BATTIATO, S., CURTI, S., CASCIA, M. L., TORTORA, M., SCORDATO, E. Depth map generation by image classification. *Proc. SPIE*, April 2004, vol. 5302, p. 95–104.

[6] CHOI, C., KWON, B., CHOI, M. A real-time field-sequential stereoscopic image converter. *IEEE Trans. Consumer Electronics*, August 2004, vol. 50, no. 3, p. 903–910.

[7] SETHURAMAN, S., SIEGEL, M. W. The video Z-buffer: a concept for facilitating monoscopic image compression by exploiting the 3D stereoscopic depth map. In *Proc. SMPTE International Workshop on HDTV'96*. Los Angeles (USA), 1996, p. 8–9.

[8] KIM, K., SIEGEL, M., SON, J. Y. Synthesis of a high-resolution 3D-stereoscopic image pair from a high-resolution monoscopic image and a low-resolution depth map. In *Proc. SPIE/IS&T Conference*, January 1998, vol. 3295A, p. 76–86.

[9] WANG, H., LI, H., MANJUNATH, S. Real-time capturing and generating stereo images and videos with a monoscopic low power mobile device. *US Patent*, 2012.

[10] KAMENCAY, P., BREZNAN, M., JARINA, R, LUKAC, P., ZACHARIASOVA, M. Improved depth map estimation from stereo images based on hybrid method. *Radioengineering*, 2012, vol. 21, no. 1, p. 70-78.

[11] KALLER, O., BOLEČEK, L., KRATOCHVÍL, T. Profilometry scaning for correction of 3D images depth map estimation. In *Proceedings of the 53rd International Symposium ELMAR- 2011*. Zadar (Croatia), 2011, p. 119-122. ISBN: 978-953-7044-12- 1.

[12] CURTI, S., SIRTORI, D., VELLA, F. 3D effect generation from monocular view. In *Proc. First International Symp. on 3D Data Processing Visualization and Transmission (3DPVT 2002)*. Padua (Italy), 2002, p. 550–553. DOI: 10.1109/TDPVT.2002.1024116.

[13] KOZANKIEWICZ, P. Fast algorithm for creating image-based stereo images. In *Proc. 10th International Conference in Central Europe on Computer Graphics, Visualization and Computer Vision*. Plzen (Czech Republic), 2002.

[14] BATTIATO, S., CAPRA, A., CURTI, S., CASCIA, M. L. 3D stereoscopic image pairs by depth-map generation. In *Proc. 2nd International Symp. on 3D Data Processing, Visualization and Transmission*. Thessaloniki (Greece), 2004, p. 124–131. DOI: 10.1109/TDPVT.2004.1335185.

About Author...

Miroslav GALABOV was born in Veliko Turnovo. He received his M.S.E degree in Radio Television Engineering from the Higher Naval School N. Vapcarov, Varna, Bulgaria, in 1989. After that he worked as a design engineer for the Institute of Radio Electronics, Veliko Turnovo. From 1992 to 2001 he was an assistant professor at the Higher Military University, Veliko Turnovo. He received his Ph.D. degree in Automation Systems for Processing of Information and Control from the Higher Military University, in 1999. Since 2002 he has been an assistant professor and from 2005 he has been an associate professor in the Computer Systems and Technologies Department, St. Cyril and St. Methodius University of Veliko Turnovo. He is the author of ten textbooks, and over 40 papers. His current interests are in signal processing, 3D technologies and multimedia.

Visible Light Communications towards 5G

Stanislav ZVANOVEC [1], *Petr CHVOJKA* [1], *Paul Anthony HAIGH* [2], *Zabih GHASSEMLOOY* [3]

[1] Dept. of Electromagnetic Field, Czech Technical University in Prague, Technicka 2, 166 27 Prague, Czech Republic
[2] Faculty of Engineering, University of Bristol, Bristol, BS8 1TR, UK
[3] Optical Communications Research Group, Faculty of Engineering and Environment, Northumbria University, Newcastle-upon-Tyne NE1 8ST, UK

xzvanove@fel.cvut.cz, petr.chvojka@fel.cvut.cz, paul.anthony.haigh@bristol.ac.uk, z.ghassemlooy@northumbria.ac.uk

Abstract. *5G networks have to offer extremely high capacity for novel streaming applications. One of the most promising approaches is to embed large numbers of co-operating small cells into the macro-cell coverage area. Alternatively, optical wireless based technologies can be adopted as an alternative physical layer offering higher data rates. Visible light communications (VLC) is an emerging technology for future high capacity communication links (it has been accepted to 5GPP) in the visible range of the electromagnetic spectrum (~370–780 nm) utilizing light-emitting diodes (LEDs) simultaneously provide data transmission and room illumination. A major challenge in VLC is the LED modulation bandwidths, which are limited to a few MHz. However, myriad gigabit speed transmission links have already been demonstrated. Non line-of-sight (NLOS) optical wireless is resistant to blocking by people and obstacles and is capable of adapting its' throughput according to the current channel state information. Concurrently, organic polymer LEDs (PLEDs) have become the focus of enormous attention for solid-state lighting applications due to their advantages over conventional white LEDs such as ultra-low costs, low heating temperature, mechanical flexibility and large photoactive areas when produced with wet processing methods. This paper discusses development of such VLC links with a view to implementing ubiquitous broadcasting networks featuring advanced modulation formats such as orthogonal frequency division multiplexing (OFDM) or carrier-less amplitude and phase modulation (CAP) in conjunction with equalization techniques. Finally, this paper will also summarize the results of the European project ICT COST IC1101 OPTICWISE (Optical Wireless Communications - An Emerging Technology) dealing VLC and OLEDs towards 5G networks.*

Keywords

5G networks, light emitting diodes, visible light communications

1. Introduction

In recent years, the worldwide growth in mobile data traffic has led to the development of new technologies for future high capacity communication systems. Every year the number of wireless devices such smartphones, laptops and tablets increases, thus multimedia content becomes the main part of the overall mobile data transferred. This fact results in an increasing throughput requirement from the next generation of communication networks (5G), which are expected to be deployed beyond 2020. Network designers face several critical challenges, all of which need to be addressed, such as optimal spectra allocation, high capacity broadband links, power consumption, quality of services (QoS) and mobility. For instance, approximately one exabyte (EB) of data was transferred across the entire global internet in 2000 (0.083 EB/month) [1]. In contrast, ~30 times more data was carried by the mobile networks per month in 2014, which corresponds to ~2.5 EB/month (refer to Fig. 1). Moreover, the latest projections from Cisco predict that the overall mobile data traffic will reach approximately 24 EB/month by 2019, which is approximately one order of magnitude larger than 2014 (Fig. 1). This corresponds to a compound annual growth rate (CAGR) of 57% for the 2014-2019 period. Following the projection Fig. 1 also shows the fit for this data, which estimates that > 30 EB/month will be transmitted beyond 2020. Most of this mobile data traffic (up to 69%) is expected to consist of video and media by the end of 2018 [1].

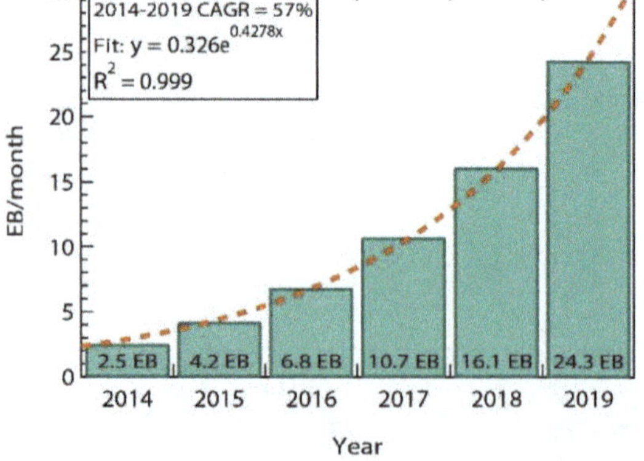

Fig. 1. A prediction of the mobile data traffic per month.

5G networks are expected to meet all the mentioned requirements [2], [3]. The network architecture is expected to be changed dramatically and the limited frequency bands must be used more efficiently. Future systems will be based on heterogeneous networks (HetNets) and advanced radio access technologies (RATs). HetNets include several small cells featuring low transmission power and small coverage area, thus enabling high cell density. Such a system configuration allows spectral reuse, hence improving the capacity of the wireless channel [2], [3], [4]. The cellular architecture should be designed to separate indoor and outdoor scenarios and support massive multiple-input multiple-output (MIMO) technologies with distributed antenna systems [2], [3].

As the number of communication devices increases and the demand for connections grows, power efficiency becomes one of the most important issues for 5G networks. Around 2% of all carbon-dioxide emissions worldwide are produced by communication technologies; this will increase significantly with the aforementioned increasing mobile data traffic demand [5]. Moreover, ~57% of overall wireless network energy consumption is dissipated in radio access nodes [6]. Thus, 'green' and energy efficient systems should reduce the CO_2 emissions and decrease the operating costs.

Indoor communication systems can offer a solution to all the above issues, such as mm-wave systems (3 to 300 GHz) or optical wireless communications such as VLC, which is carried at 400–490 THz. 5G-VLC offers a number of small cells, also referred to as attocells in the literature, in the indoor environment, thus enabling many advantages such as high capacity data transmission, excellent mobility and energy efficient management. Recent experiments demonstrated up to 1.6 Tbit/s for optical wireless backhaul links at 1550 nm [7] and up to 3.4 Gbit/s [8] in VLC based access networks. Moreover, two functionalities are provided with VLC, i.e. the combining of data transmission with illumination of the room using light-emitting diodes (LEDs), which is not available with other network technologies. This paper provides a wide overview of the VLC technology and summarizes its development and the state-of-the-art. The rest of the paper is organized as follows: Section 2 describes the fundamental VLC principles with emphasis on inorganic and organic LEDs, Section 3 is focused on the modulation formats and finally in Section 4, perspective VLC applications are discussed.

2. Visible Light Communications Technology

License-free spectrum, practically unlimited channel bandwidth, high capacity links and energy efficiency are the main features of light fidelity (Li-Fi) networks. Li-Fi is a subset of the VLC domain, which refers to broadcasting access networks with multiple users. The transmitter consists of LEDs either singularly or in an array that are intensity modulated at a rate above which it is not perceivable by the human eye. Since full room illumination is required, the visible range of the electromagnetic spectrum (~370 to 780 nm) is utilized in VLC. Moreover, VLC has been proposed for future 5G networks standards.

There are two types of LEDs used to produce white light: (*i*) inorganic metal alloy semiconductor LEDs (usually a blue emitting gallium nitride (GaN) interface with a cerium doped yttrium aluminum garnet (Ce:YAG) color converting phosphor) and (*ii*) organic LEDs (OLEDs) made from either small molecules or polymers, using epitaxial or wet processing methods, respectively [9], [10]. Inorganic LEDs are commonly used in VLC, while generic OLEDs are attracting significant attention for future VLC networks. When dissolved into solvents and processed with wet methods (i.e. inkjet printing, spray coating) polymer based OLEDs offer several significant advantages over inorganic devices, namely; large, arbitrarily shaped photoactive areas limited only by the size of the printing apparatus, mechanical flexibility, low temperature and ultralow costs.

2.1 Inorganic LEDs

The most common type of inorganic LEDs producing white light are Ce:YAG converted GaN diodes, as mentioned, which are known as white phosphor LEDs (WPLEDs). Alternatively, white light can be produced using a single LED package with on-board red, green and blue (RGB) chips. Considering WPLEDs, the main impediment in achieving high data rates is the modulation bandwidth. GaN diodes can offer modulation bandwidths up to several hundred MHz [11], however the Ce:YAG phosphor layer has a slow transient response, reducing the bandwidth down to the low MHz region, which common values around 4-5 MHz [12]. Thus, increasing the transmission capacity is the key challenge undertaken by researchers. However, the concept of micro-LEDs was introduced in [13], where the photoactive area of the device is reduced significantly to the μm scale, thus relieving the plate capacitance of the device and increasing the bandwidth. Depending on the photoactive area diameter, bandwidths exceeding 400 MHz were demonstrated. Such a device can provide a 3.22 Gb/s throughput when using an adaptive bit- and power-loading technique [14]. On the other hand, reducing the photoactive area means a reduction in optical power, which in turn severely limits the transmission distance. In [13], the optical power ranged from ~0.5 mW to ~5 mW, which is significantly smaller than standard GaN LEDs.

VLC links at Gb/s data rates have already been demonstrated, despite the very limited LED modulation bandwidths. A popular method to increase the capacity of VLC links is to use spectrally efficiency modulation formats such as discrete multi-tone (DMT) [8], [15]. A 3.4 Gb/s transmission speed was achieved in [8] using a single RGB LED and DMT modulation at a distance < 30 cm. Wavelength division multiplexing (WDM) was utilized to transmit independent streams of information on each wavelength. Nevertheless, DMT requires complex signal processing and a feedback channel for bit- and power-loading. A similar approach was adopted in [15] resulting in a transmission speed of 1 Gb/s at a 10 cm distance.

Another possibility for increasing the channel capacity is the implementation of an equalizer [12]. On-off keying (OOK) is the most common modulation format due to simplicity of implementation and compatibility with equalizers. In [12] several equalizers were tested on a commercially available digital signal processing (DSP) board. The LED had low modulation bandwidth (4.5 MHz) and transmission speeds up to 170 Mb/s were recorded using a highly complex artificial neural network (ANN) based equalizer. Real time data processing on field programmable gate arrays (FPGAs) is the next step towards building a fully real time system, as currently most reports in the literature feature offline processing.

2.2 Organic LEDs

There are two types of OLEDs, the first produced using small molecules (SMOLEDs) and the second with polymers (PLEDs). Due to the previously mentioned advantages, organic materials have been focus of the research to be implemented in communication systems. According to [16], the global organic electronic market will approach $80 billion by 2020, thus resulting in (CAGR) of 29.5% between 2014 – 2020.

A possible and generic PLED structure is depicted in Fig. 2. It consists of thin films (total thickness 1-200 nm) of organic emissive and charge transport layers. Polymers such as:

(i) red: poly[2-methoxy-5-(3′,7′-dimethyloctyloxy)-1,4-phenylenevinylene] (MDMO-PPV)

(ii) green: poly[(9,9-di-n-octylfluorenyl-2,7-diyl)-alt-(benzo[2,1,3] thiadiazol-4,8-diyl)] (F8BT) and

(iii) blue: poly(9,9-dioctylfluorene), (poly(9,9-dioctylfluorene-alt-N-(4-butylphenyl)diphenylamine) and poly(9,90-dioctylfluorene-alt-bis-N,N0-(4-butylphenyl)-bis-N,N0-phenyl-1,4-phenylenediamine) (F8:TFB:PFB)

are evaluated as the emissive layer in [17] while a poly(3,4-ethylenedioxythiophene) (PEDOT) doped with poly(styrenesulfonic acid) (PSS) interlayer is used consistently for the charge transport layer in order to minimize the energy gap between the electrode and emissive layer. In Fig. 2 the PLED is a bottom emitter and hence the anode must be transparent (generally indium tin oxide (ITO) is used). When positive and negative charges recombine, the charges combine for a fraction of a second as Frenkel excitons before the energy is emitted as photons. The material can be placed on various substrates including plastic, enabling flexibility.

However, the most limiting factor for OLEDs is very low modulation bandwidth, which is much lower than the bandwidth of conventional inorganic diodes. The reason for this is because the charge transport characteristics of organic materials are approximately three orders of magnitude lower than those of amorphous silicon (a-Si). OLEDs act as a low pass filter with a certain cut-off frequency. The -3 dB frequency is given by:

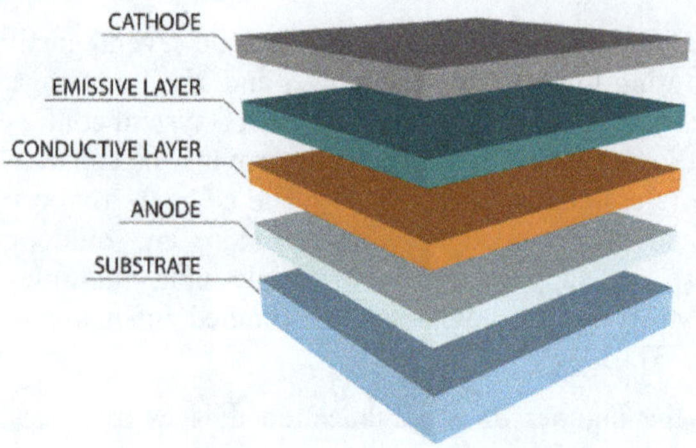

Fig. 2. The structure of OLED.

$$f_c = \frac{1}{2\pi RC} \tag{1}$$

where R is the effective resistance and C is the plate capacitance, given as:

$$C = A\varepsilon_0\varepsilon_r / d \tag{2}$$

where A is the photoactive area and d is thickness of OLED, ε_0 and ε_r are the relative permittivity of free space and the emissive layer, respectively. As d is generally very small (1-200 nm), this tends to a large C, hence a very high capacitance and low bandwidth, which is typically in the order of several hundred kHz [18]. On the other hand, OLEDs with bandwidths up to ~60 MHz were reported in [19], which was achieved by reducing the photoactive area to 0.018 mm^2. Using commercial and custom OLEDs, promising experimental results have been reported in the literature. A SMOLED with a bandwidth 93 kHz is modulated by DMT signal to demonstrate to data rate up to 1.4 Mb/s in [20]. A two-fold improvement of this transmission speed was published in [21] using the same SMOLED as in [20]; a 2.7 Mb/s link is shown using a multi-layer perception (MLP) ANN equalizer offline in MATLAB.

A breakthrough in organic VLC was made in [22], where PLEDs were used to demonstrate transmission speeds in excess of that required for Ethernet connectivity. The data rate of 20 Mb/s was reported in [22], [23] using OOK format and an MLP equalizer. More recently in [17], an aggregated transmission speed of 55 Mb/s was experimentally demonstrated and was achieved using an RGB PLED and WDM, which is a promising improvement for future research activities.

2.3 Other Organic Components

Organic photodetectors (OPD) are also a promising technology and can substitute silicon photodiodes in specific applications. OPDs used in the authors' previous work are based on the bulk heterojunction principle [24]; an interpenetrated and disordered blend of an electron donor and electron acceptor and were fabricated using spray coating [25]. The material costs for the poly(3-hexylthiophene): [6,6]-phenyl C61-butyric acid methylester (P3HT:PCBM) blend are around ~€0.20/cm^2. Furthermore, OPDs can offer superior responsivity in comparison to Si photodetectors in the visible spectrum as shown in Fig. 3.

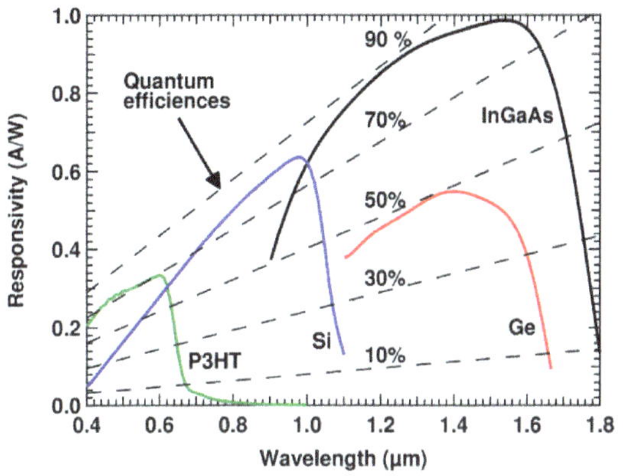

Fig. 3. Photodetector material responsivities.

The OPD with raw 3 dB bandwidth of 30 kHz is tested in [26] with the resulting data rate of 750 kb/s, which is ~20 times increment representing huge potential of such devices. An OPD with ~160 kHz modulation bandwidth is utilized in [27]. Fourth-order pulse position modulation (4-PPM) and an ANN equalizer were used to achieve a transmission speed of 3.75 Mb/s. Fully organic VLC link was introduced in [28], where the low bandwidth components (up to 135 kHz) with OOK modulation and ANN based filtering were used to achieve data rate exceeding 1 Mb/s.

3. Modulation Formats

Besides equalization techniques, spectrally efficient modulation formats such as orthogonal frequency division multiplexing (OFDM) are a popular way to increase the VLC link capacity. For instance, Gb/s transmissions were reported in [8] and [29] by implementing DMT. Power efficient schemes such as low peak-to-average power ratio (PAPR) single carrier frequency-domain equalization (SC-FDE) was proposed in [30]. Here the IFFT block was moved to the receiver side to avoid the generation of complicated waveforms and consequently reduced PAPR. Further spectral efficiency improvements have been achieved by adopting novel OFDM SC-FDE signal formats - polar OFDM and polar SC-FDE in [31].

Recently, carrier-less amplitude and phase modulation (CAP) has appeared as a candidate for optical systems, which outperforms OFDM channel using the same experimental setup [32]. CAP systems have several advantages over OFDM including no Fourier transform as in OFDM or local oscillator, which is utilized in a single carrier modulations such as quadrature amplitude modulation (QAM). The carrier frequencies in CAP are generated by either analogue or digital finite impulse response filters (FIR).

Unlike OFDM, a flat frequency response is required for CAP meaning the low bandwidths available are the major problem once more, due to the 20 dB/decade attenuation for frequencies outside the bandwidth. A possible solution for this was introduced in [33], where the available

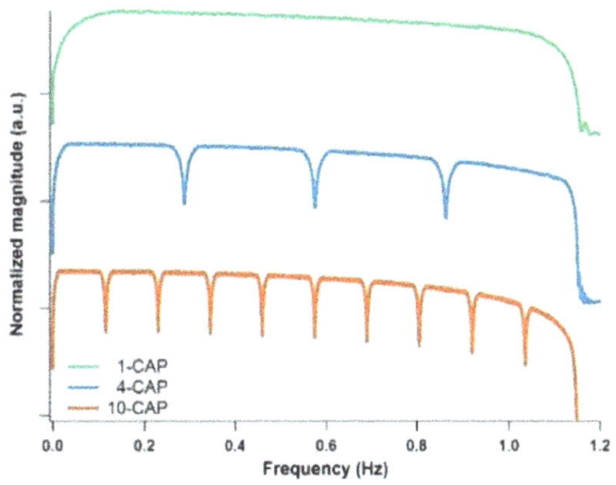

Fig. 4. The concept of m-CAP modulation, where the available system bandwidth is split into 1, 4 and 10 subbands.

bandwidth is divided into a multiband (*m*-CAP) format for an optical fiber channel. The transmission bandwidth was split into 6 sub-bands (subcarriers) and the performance was compared with the traditional CAP (1-CAP) system. The fiber system transmission speed was 102.4 Gb/s and 100 Gb/s for *m*-CAP and 1-CAP, respectively, which is not a significant gain, however a gain in transmission speed was not the focus of the article. The *m*-CAP system noted a significant improvement in dispersion tolerance. Nevertheless, for 1-CAP transmission, two FIR filters are required for signal generation. As *m* increases, the number of FIR filters grows by 2*m*. Thus, one must consider an increment in the system complexity resulting from a multiband approach. On the other hand, such a concept allows optimization of the modulation format used in each subband according to measured signal-to-noise ratio (SNR), i.e. bit- and power-loading in OFDM. The principle of *m*-CAP is illustrated in Fig. 4. The available system bandwidth is split into *m* sub-bands. The more sub-bands are utilized, the less bandwidth is occupied by a single subcarrier, thus resulting in a reduced attenuation caused by the LED low-pass frequency response.

A detailed description of the *m*-CAP modulation format and signal generation can be found in [34], [35]. The first VLC experiment utilizing *m*-CAP was reported in [35], where a data rate of 31.5 Mb/s was demonstrated using 10-CAP; resulting in a spectral efficiency of 4.85 b/s/Hz, which offers huge potential for a future research.

4. Applications

As the technology has evolved over the last decade and with rapid increase of mobile data requirements, there are several challenging areas for specific deployment of VLC systems. Alongside classic indoor communication schemes (i.e., broadcasting networks), VLC can also be utilized for localization or car-to-car and car-to-infrastructure communications. The following sections highlight the main approaches and principles.

4.1 Indoor Communications

Alongside illumination and data communications, VLC systems have been proposed for indoor positioning with a very high accuracy (a few cm). The IEEE 802.15.7 task group [36] has been forming the new standard for VLC, orienting their efforts towards the PHY and MAC standards since 2009. The main focus from the first releases had been given on slow VLC for indoor positioning by means of optical camera communication. More recently high bit rate VLC transmission systems are under standardization, especially based on results produced by the EU COST IC1101 project OPTICWISE consortium.

For indoor applications with a static environment, VLC can offer high QoS even when the movement of people and shadowing by obstacles perturb the beam (so called non-line of sight (NLOS)). This has been proved by several theoretical and experimental investigations [37], [38]. Typically, such NLOS systems use diffuse reflections, where the transmitter illuminates the ceiling, and/or wide field-of-view or receiving apertures instead of direct links. Data rates up to 400 Mbit/s have been reported for a NLOS link in [39]. The first mobile VLC system was reported for typical indoor distances between 2 m and 20 m with data rates decreasing with distance from approx. 500 Mbit/s to 100 Mbit/s in [40].

The typical model of diffuse reflection was described in [41]. The power efficiency for the diffuse signal can be derived as [41]:

$$\mu_{\text{diff}} = \frac{P_{\text{diff}}}{P_{\text{T}}} = \frac{A_{\text{R}} \sin^2(FOV/2)}{A_{\text{room}}} \frac{\langle \rho \rangle}{1-\langle \rho \rangle} \tag{3}$$

where $\langle \rho \rangle$ is mean reflectivity, *FOV* stands for field of view of detector and A_{R} and A_{room} are areas of detector and room, respectively. The decay time then can be given as [41]:

$$\tau = -\frac{\langle t \rangle}{\ln \langle \rho \rangle}. \tag{4}$$

The average time $\langle t \rangle$ between two diffuse reflections for a rectangular room with dimensions $l \times w \times h$ (length × width × height) is expressed by [41]:

$$\langle t \rangle = \frac{2lwh}{c(lw+lh+wh)}. \tag{5}$$

The DC gain of the reflected path can afterwards be determined as [10]:

$$H_{\text{ref}}(O) = \begin{cases} \dfrac{(m+1)}{2(\pi d_1 d_2)^2} \rho A_{\text{det}} dA_{\text{w}} \cos^m(\theta_r)\cos(\alpha) & \\ \cdot\cos(\beta) T_{\text{S}}(\vartheta) g(\vartheta) \cos(\vartheta) & \text{for } 0 \le \vartheta \le \vartheta_{\text{FOV}} \\ 0 & \text{for } \vartheta > \vartheta_{\text{FOV}} \end{cases} \tag{6}$$

where d_1 is the distance between the transmitter and the reflective point, d_2 is the distance between the reflective point and the receiver, ρ is the reflectance coefficient, dA_{w}

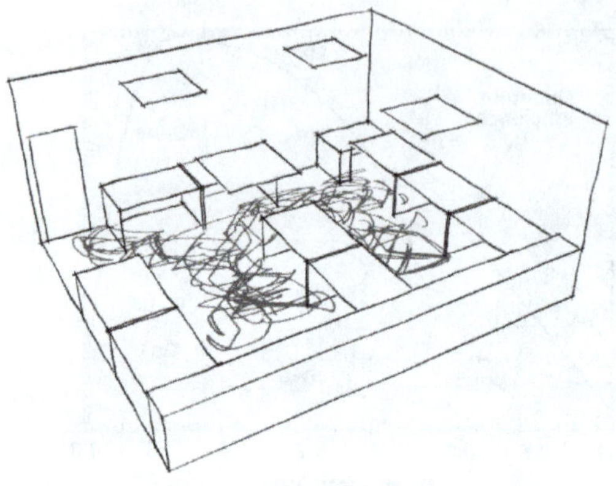

Fig. 5. The simulation of people's movement within an office using VLC with 4-LED.

is a small reflective area on the wall, α is the angle of incidence from the transmitter and β is the angle of irradiance from a reflected point.

Fig. 5 illustrates people's movement within a typical office environment utilizing VLC by employing 4 LEDs mounted on the ceiling.

In such a case sometimes the direct paths from transmitter to receiver are blocked or temporally shadowed by a human. Several studies derived the percentage of shadowing. For instance [42] reported shadowing with a probability of < 2% for a multiple-input multiple-output (MIMO) system covering a typical office. Similar results were derived in [43] where time division multiple access was investigated. Higher order reflections induce significant influence on the temporal dispersion according to [37], [38]. The reflection component always appears respective to

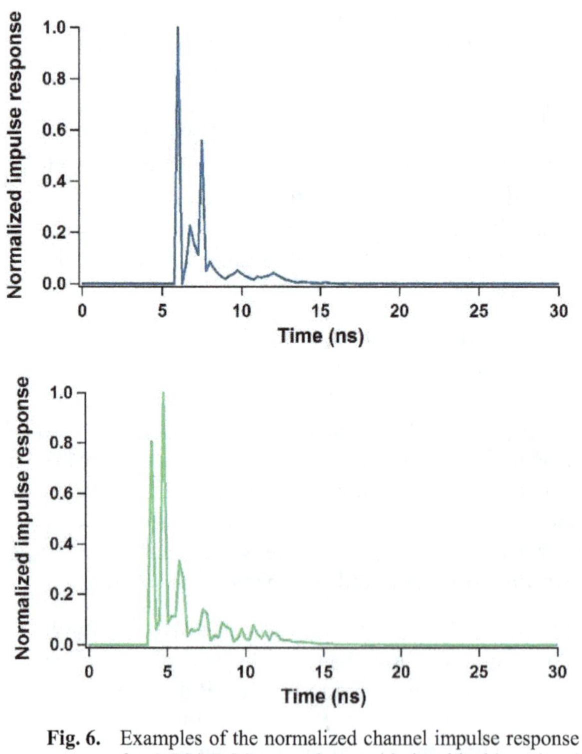

Fig. 6. Examples of the normalized channel impulse response for LOS/NLOS scenario considering blocking signal path by people in an office with: a) 4 LEDs, b) 18 LEDs.

LOS incident paths [38]. This influence can be easily observed in the transmission bandwidth. Two examples of a normalized channel response for the scenario shown in Fig. 5 are illustrated in Fig. 6.

The measured probability density function of the normalized received optical power considering people's movement as well as the analysis of the RMS delay spread for different indoor scenarios and people densities in rooms was carried out in [44]. Based on the measurement campaign the normalized received power showed a Rayleigh distribution with the scale parameter varying from 0.98 to 1.79 for an empty to a crowded room. The RMS delay spread statistics have been derived for three different indoor scenarios. For the case of furnished office environment (people density > 0.16 people/m^2), the cumulative distribution function (CDF) of the received power differs in the worst case by up to 7% contrary to an RMS delay of 2% that was experienced under the same people density in the corridor [44].

4.2 Positioning and Localization

Several positioning systems have been tested over the last few years. In [45] a digital camera was used as a receiver to capture a sequence of images of the LED positioning beacon transmitter. By using image-processing algorithms, the system was able to decode the location information encoded in the visual patterns transmitted by LEDs. The system demonstrated that improved performance can be attained even at low values of SNR. A system for localization of vehicles using the global positioning system (GPS) together with a light beacon device mounted on the vehicle to receive information from transmitter positioned at the road intersections is developed by Honda motors Co., Ltd. in 2010 [46].

Another challenge represents the utilization of the localization within the indoor scenario, where standard GPS signals cannot be received (see illustrative deployment in Fig. 7). In recent years, we have seen research and development in optical based indoor positioning schemes (IPS) offers multitude of advantageous including smaller transceiver size, immunity to electromagnetic interference, and inherent security [47]. VLC based IPS offers the advantage of LED and VLC technologies such as ubiquitous coverage, static channel, multiple lighting elements etc. [48]. Using VLC with synchronization between the transmitter

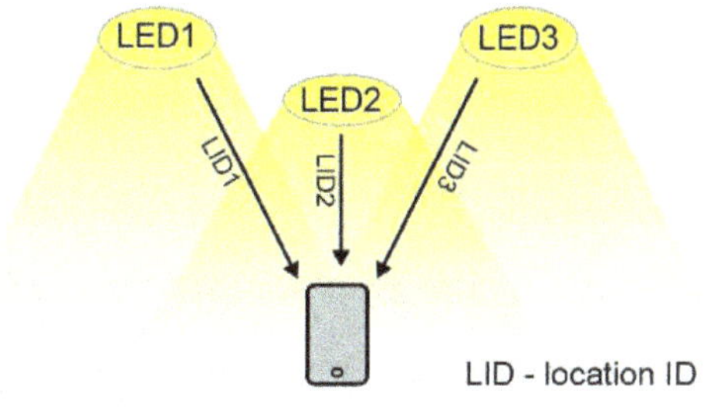

Fig. 7. Concept of VLC positioning system (multi-access mechanism among single terminal and multiple lights + position estimation).

and receiver, the bounds on position estimation accuracy are typically in the order of millimeters or centimeters depending on the geometry of the room, the frequency and power of the transmitted signal and the properties of the LED and the photoreceiver [49]. A VLC-based IPS employing 3-LED, the dual-tone multi-frequency technique and a dedicated algorithm was reported in [48]. Unlike the time-division multiplexing based VLC-IPS, this scheme does not require synchronization between the transmitter and receiver, thus makes it simple, robust and cost effective. VLC-IPS is highly accurate offering an average positioning error of about 1.6 cm, which is much less than many existing IPS.

Typically several transmitters serve as beacons and the RMS delay spread can be used to quantify the amount of multipath distortion that can occur at a particular point within a room [50]. Parameter $D_{\mathrm{rms\_Max}}$ corresponds to the largest multipath distortion, which limits the maximum transmission data rate $R_{\max}$ of the system. $R_{\max}$ for the indoor VLC channel can be calculated following [51], which is given by:

$$R_{\max} \le \frac{1}{10 D_{\mathrm{rms}_{\mathrm{Max}}}} . \tag{7}$$

In order to evaluate the positioning accuracy, we have to consider the Cramer-Rao bound (CRB) [52] as a performance reference, which is the lower bound on the mean square estimation error in the set of unbiased estimates. Typical CRB ranges within the rooms for 4-, 6- and 9-cell configurations were derived for an optimized Lambertian order (OLO) LED case for an indoor cellular optical wireless communication system in [53]. Simulations of particular scenario revealed reached values of CRB from 12.8 cm, from 8.6 cm and from 5.8 cm, respectively, for above mentioned cells' deployment in for rooms of $5 \times 5 \times 3$ m^3, $4 \times 6 \times 3$ m^3, and $5 \times 5 \times 3$ m^3 [53]. This has shown VLC as very useful tool for developers of indoor positioning systems.

4.3 Car-to-Car Communications

VLC can also be utilized for outdoor applications such as the public transport. Note how the infrastructure of public lights has changed over the last 5 years. Typical incandescent lamps have been replaced by LED lighting across whole cities. For example the Los Angeles LED Streetlight Replacement Program has replaced over 140,000 existing streetlights in the city with LED units which has brought energy saving 68 GWh/year and money saving $ 10M/year [54].

Several test use-cases and experimental results have been published for a vehicular VLC network consisting of on-board units, vehicles, and road side units, i.e., traffic lights, street lamps, digital signage etc. Cars fitted with LED-based front and back lights can communicate with each other and with the road side units (RSUs) through the VLC technology. Furthermore, LED-based RSUs can be used for both signaling and broadcasting safety-related

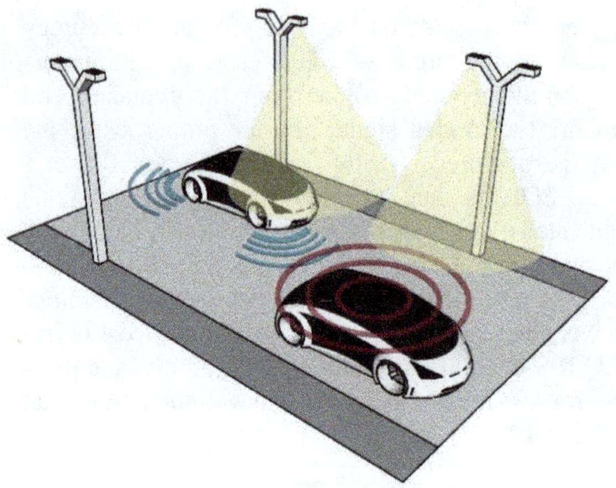

Fig. 8. Concept of VLC for car to car and car to infrastructure.

information to vehicles on the road. Such a network is illustrated in Fig. 8.

An analytical performance analysis of VLC based car-to-car communications for a range of communication geometries consider both the LOS and NLOS paths over a link span of 20 m at a data rate of 2 Mbps was outlined in [55]. Optical wireless communications systems based on an LED transmitter and camera receiver was proposed for automotive applications in [56]. The signal reception experiment has been performed for static and moving camera receivers. Up to 15 Mb/s error-free throughput under fixed conditions was sustained. This represents very good performance for optical wireless communication systems, since the experiment did not involve further correction methods like coding and equalization. In [57] it was shown that the receiver in the driving situation can detect and accurately track an LED transmitter array with error-free communication over distances of 25–80 m. Further tests and experiments are however needed to prove these concepts.

5. Conclusion

The research and development in VLC at a global level has increased more than ten times for last two years. This paper gave an overview of VLC and its use in a number of applications. In outdoor environment VLC can provide internet hot spots using street lighting and mobile access as part of the 5G technology in highly congested areas and within indoor environment it can be used for localization and small cells coverage networks. VLC has been accepted as part of the 802.15.7 task group and is being proposed as a supplement technology in 5G networks. We assume further increase of research efforts within selected applications.

Acknowledgements

This joint research is supported by the EU COST ICT Action IC1101 Optical Wireless Communications: An Emerging Technology (OPTICWISE) and by the Grant Agency of the Czech Technical University in Prague, grant no. SGS14/190/OHK3/3T/13.

References

[1] C. V. N. Index, *Global Mobile Data Traffic Forecast Update*, 2014-2019, White paper, ed. 2013.

[2] BANGERTER, B., TALWAR, S., AREFI, R., STEWART, K. Networks and devices for the 5G era. *IEEE Communications Magazine*, 2014, vol. 52, p. 90–96. DOI: 10.1109/MCOM.2014.6736748

[3] WANG CHENG-XIANG, HAIDER, F., XIQI GAO, XIAO-HU YOU, YANG YANG, DONGFENG YUAN, et al. Cellular architecture and key technologies for 5G wireless communication networks. *IEEE Communications Magazine*, 2014, vol. 52, p. 122 to 130. DOI: 10.1109/MCOM.2014.6736752

[4] LI, Q. C., HUANING NIU, PAPATHANASSIOU, A., GENG WU. 5G network capacity: Key elements and technologies. *IEEE Vehicular Technology Magazine*, 2014, vol. 9, no. 1, p. 71–78. DOI: 10.1109/MVT.2013.2295070

[5] YONG SHENG SOH, QUEK, T. Q. S., KOUNTOURIS, M., HYUNDONG SHIN. Energy efficient heterogeneous cellular networks. *IEEE Journal on Selected Areas in Communications*, 2013, vol. 31, p. 840–850. DOI: 10.1109/JSAC.2013.130503 .

[6] HU, R. Q., YI QUIAN. An energy efficient and spectrum efficient wireless heterogeneous network framework for 5G systems. *IEEE Communications Magazine*, 2014, vol. 52, p. 94–101. DOI: 10.1109/MCOM.2014.6815898

[7] PARCA, G., SHAHPARI, A., CARROZZO, V., TOSI BELEFFI, G. M., TEIXEIRA, A. L. J. Optical wireless transmission at 1.6-Tbit/s (16×100 Gbit/s) for next-generation convergent urban infrastructures. *Optical Engineering*, 2013, vol. 52, no. 11, p. 116102. DOI: 10.1117/1.OE.52.11.116102

[8] COSSU, G., KHALID, A. M., CHOUDHURY, P., CORSINI, R., CIARAMELLA, E. 3.4 Gbit/s visible optical wireless transmission based on RGB LED. *Optics Express*, 2012, vol. 20, no. 26, p. B501–B506. DOI: 10.1364/OE.20.00B501

[9] BURROUGHES, J. H., BRADLEY, D. D. C., BROWN, A. R., MARKS, R. N., MACKAY, K., FRIEND, R. H., et al. Light-emitting diodes based on conjugated polymers. *Nature*, 1990, vol. 347, p. 539–541. DOI:10.1038/347539a0

[10] TANG, C. W., VANSLYKE, S. A. Organic electroluminescent diodes. *Applied Physics Letters*, 1987, vol. 51, p. 913–915. DOI: 10.1063/1.98799

[11] MCKENDRY, J. J. D., GREEN, R. P., KELLY, A. E., ZHENG, G., GUILHABERT, B., MASSOUBRE, D., et al. High-speed visible light communications using individual pixels in a micro light-emitting diode array. *IEEE Photonics Technology Lett.*, 2010, vol. 22, no. 18, p. 1346–1348. DOI: 10.1109/LPT.2010.2056360

[12] HAIGH, P. A., GHASSEMLOOY, Z., RAJBHANDARI, S., PAPAKONSTANTINOU, I., POPOOLA, W. Visible light communications: 170 Mb/s using an artificial neural network equalizer in a low bandwidth white light configuration. *Journal of Lightwave Technology*, 2014, vol. 32, no. 9, p. 1807–1813. DOI: 10.1109/JLT.2014.2314635

[13] MCKENDRY, J. J. D., MASSOUBRE, D., ZHANG, S., RAE, B. R., GREEN, R. P., GU, E., et al. Visible-light communications using a CMOS-controlled micro-light-emitting-diode array. *Journal of Lightwave Technology*, 2012, vol. 30, no. 1, p. 61–67. DOI: 10.1109/JLT.2011.2175090

[14] TSONEV, D., HYUNCHAE, C., RAJBHANDARI, S., MCKENDRY, J. J. D., VIDEV, S., GU, E., et al. A 3-Gb/s single-LED

OFDM-based wireless VLC link using a gallium nitride μLED. *IEEE Photonics Technology Letters,* 2014, vol. 26, p. 637–640. DOI: 10.1109/LPT.2013.2297621

[15] KHALID, A. M., COSSU, G., CORSINI, R., CHOUDHURY, P., CIARAMELLA, E. 1-Gb/s transmission over a phosphorescent white LED by using rate-adaptive discrete multitone modulation. *IEEE Photonics Journal,* 2012, vol. 4, p. 1465–1473. DOI: 10.1109/JPHOT.2012.2210397

[16] SINGH, R. *Global Organic Electronics Market (Application and Geography) - Size, Share, Global Trends, Company Profiles, Demand, Insights, Analysis, Research, Report, Opportunities, Segmentation and Forecast, 2013 – 2020.* 2014.

[17] HAIGH, P. A., BAUSI, F., LE MINH, H., PAPAKONSTANTINOU, I., POPOOLA, W., BURTON, A., et al. Wavelength-multiplexed polymer LEDs: Towards 55 Mb/s organic visible light communications. *IEEE Journal on Selected Areas in Communications.* Accepted, 2014.

[18] HAIGH, P. A., GHASSEMLOOY, Z., RAJBHANDARI, S., PAPAKONSTANTINOU, I. Visible light communications using organic light emitting diodes. *IEEE Communications Magazine,* 2013, vol. 51, p. 148–154. DOI: 10.1109/MCOM.2013.6576353

[19] BARLOW, I. A., KREOUZIS, T., LIDZEY, D. G. High-speed electroluminescence modulation of a conjugated-polymer light emitting diode. *Applied Physics Letters,* 2009, vol. 94, p. 243301–3. DOI: 10.1063/1.3147208

[20] HAIGH, P. A., GHASSEMLOOY, Z., PAPAKONSTANTINOU, I. 1.4-Mb/s white organic LED transmission system using discrete multitone modulation. *IEEE Photonics Technology Letters,* 2013, vol. 25, no. 6, p. 615–618. DOI: 10.1109/LPT.2013.2244879

[21] HAIGH, P. A., GHASSEMLOOY, Z., PAPAKONSTANTINOU, I. HOA LE MINH. 2.7 Mb/s with a 93-kHz white organic light emitting diode and real time ANN equalizer. *IEEE Photonics Technology Letters,* 2013, vol. 25, no. 17, p. 1687–1690. DOI: 10.1109/LPT.2013.2273850

[22] HAIGH, P. A., BAUSI, F., KANESAN, T., LE, S. T., RAJBHANDARI, S., GHASSEMLOOY, Z., et al. A 20-Mb/s VLC link with a polymer LED and a multilayer perceptron equalizer. *IEEE Photonics Technology Letters,* 2014, vol. 26, p. 1975–1978. DOI: 10.1109/LPT.2014.2343692

[23] LE, S. T., KANESAN, T., BAUSI, F., HAIGH, P. A., RAJBHANDARI, S., GHASSEMLOOY, Z., et al. 10 Mb/s visible light transmission system using a polymer light-emitting diode with orthogonal frequency division multiplexing. *Optics Letters,* 2014, vol. 39, p. 3876–3879. DOI: 10.1364/OL.39.003876

[24] BRABEC, C. J., SARICIFTCI, N. S., HUMMELEN, J. C. Plastic solar cells. *Advanced Functional Materials,* 2001, vol. 11, no. 1, p. 15–26.

[25] TEDDE, S. F., KERN, J., STERZL, T., FURST, J., LUGLI, P., HAYDEN, O. Fully spray coated organic photodiodes. *Nano Letters,* 2009, vol. 9, p. 980-3. DOI: 10.1021/nl803386y

[26] HAIGH, P. A., GHASSEMLOOY, Z., HOA LE MINH, RAJBHANDARI, S., ARCA, F., TEDDE, S. F., et al. Exploiting equalization techniques for improving data rates in organic optoelectronic devices for visible light communications. *Journal of Lightwave Technology,* 2012, vol. 30, no. 19, p. 3081–3088. DOI: 10.1109/JLT.2012.2210028

[27] GHASSEMLOOY, Z., HAIGH, P. A., ARCA, F., TEDDE, S. F., HAYDEN, O., PAPAKONSTANTINOU, I., et al. Visible light communications: 3.75 Mbits/s data rate with a 160 kHz bandwidth organic photodetector and artificial neural network equalization. [Invited] *Photonics Research,* 2013, vol. 1, no. 2, p. 65–68. DOI: 10.1364/PRJ.1.000065

[28] HAIGH, P. A., GHASSEMLOOY, Z., PAPAKONSTANTINOU, I., ARCA, F., TEDDE, S. F., HAYDEN, O., et al. A 1-Mb/s visible light communications link with low bandwidth organic components. *IEEE Photonics Technology Letters,* 2014, vol. 26, no. 13, p. 1295–1298. DOI: 10.1109/LPT.2014.2321412

[29] AZHAR, A. H., TRAN, T., O'BRIEN, D. A Gigabit/s indoor wireless transmission using MIMO-OFDM visible-light communications. *IEEE Photonics Technology Letters,* 2013, vol. 25, no. 2, p. 171–174. DOI: 10.1109/LPT.2012.2231857

[30] TEICHMANN, V. S. C., BARRETO, A. N., PHAM, T. T., RODES, R., MONROY, I. T., MELLO, D. A. A. SC-FDE for MMF short reach optical interconnects using directly modulated 850 nm VCSELs. *Optics Express,* Nov 5 2012, vol. 20, no. 23, p. 25369–25377. DOI: 10.1364/OE.20.025369

[31] ELGALA, H., LITTLE, T. D. C. Polar-based OFDM and SC-FDE links toward energy-efficient Gbps transmission under IM-DD optical system constraints [Invited]. *Journal of Optical Communications and Networking,* 2015/02/01, vol. 7, no. 2, p. A277–A284. DOI: 10.1364/JOCN.7.00A277

[32] WU, F. M., LIN, C. T., WEI, C. C., CHEN, C. W., CHEN, Z. Y., HUANG, H. T., et al. Performance comparison of OFDM signal and CAP signal over high capacity RGB-LED-based WDM visible light communication. *IEEE Photonics Journal,* 2013, vol. 5, article no. 7901507. DOI: 10.1109/JPHOT.2013.2271637

[33] OLMEDO, M. I., TIANJIAN ZUO, JENSEN, J. B., QIWEN ZHONG, XIAOGENG XU, POPOV, S., et al. Multiband carrierless amplitude phase modulation for high capacity optical data links. *Journal of Lightwave Technology,* 2014, vol. 32, no. 4, p. 798–804. DOI: 10.1109/JLT.2013.2284926

[34] HAIGH, P. A., LE, S. T., ZVANOVEC, S., GHASSEMLOOY, Z., LUO, P., XU, T., et al. Multi-band carrier-less amplitude and phase modulation for bandlimited visible light communications systems. *IEEE Wireless Communication Magazine.* In print, 2015.

[35] HAIGH, P. A., BURTON, A., WERFLI, K., HOA LE MINH, BENTLEY, E., CHVOJKA, P., et al. A multi-CAP visible light communications system with 4.85 b/s/Hz spectral efficiency. *IEEE Journal on Selected Areas in Communications.* Accepted, 2015.

[36] *IEEE 802.15 WPAN™ Task Group 7 (TG7) Visible Light Communication.* [Online] Cited 2015-03-05. Available at: http://www.ieee802.org/15/pub/TG7.html

[37] LEE, K., PARK, H., BARRY, J. R. Indoor channel characteristics for visible light communications. *IEEE Communications Letters,* 2011, vol. 15, no. 2, p. 217–219. DOI: 10.1109/LCOMM.2011.010411.101945

[38] BARRY, J. R., KAHN, J. M., KRAUSE, W. J., LEE, E. A., MESSERSCHMITT, D. G. Simulation of multipath impulse response for indoor wireless optical channels. *IEEE Journal on Selected Areas in Communications,* 1993, vol. 11, p. 367–379. DOI: 10.1109/49.219552

[39] LANGER, K.-D., HILT, J., SHULZ, D., LASSAK, F., HARTLIEB, F., KOTTKE, C., et al. Rate-adaptive visible light communication at 500Mb/s arrives at plug and play. *Optoelectronics & Communications, SPIE Newsroom,* 2013. DOI: 10.1117/2.1201311.005196

[40] GROBE, L., PARASKEVOPOULOS, A., HILT, J., SCHULZ, D., LASSAK, F., HARTLIEB, F., et al. High-speed visible light communication systems. *IEEE Communications Magazine,* 2013, vol. 51, no. 12, p. 60–66. DOI: 10.1109/MCOM.2013.6685758

[41] JUNGNICKEL, V., POHL, V., NONNIG, S., VON HELMOLT, C. A physical model of the wireless infrared communication channel. *IEEE Journal on Selected Areas in Communications,* 2002, vol. 20, p. 631–640. DOI: 10.1109/49.995522

[42] JIVKOVA, S., KAVEHRAD, M. Shadowing and blockage in indoor optical wireless communications. In *IEEE Global Telecommunications Conference GLOBECOM '03.* 2003, vol. 6, p. 3269–3273. DOI: 10.1109/GLOCOM.2003.1258840

[43] KOMINE, T., NAKAGAWA, M. A study of shadowing on indoor visible-light wireless communication utilizing plural white LED

lightings. In 1st *International Symposium on Wireless Communication Systems*. 2004, p. 36–40. DOI: 10.1109/ISWCS.2004.1407204

[44] CHVOJKA, P., ZVANOVEC, S., HAIGH, P. A., GHASSEMLOOY, Z. Channel characteristics of visible light communications within dynamic indoor environment. *Journal of Lightwave Technology,* 2015, vol. 33, no. 9, p. 1719–1725. DOI: 10.1109/JLT.2015.2398894

[45] LIU, H. S., PANG, G. Positioning beacon system using digital camera and LEDs. *IEEE Transactions on Vehicular Technology,* 2003, vol. 52, no. 2, p. 406–419. DOI: 10.1109/TVT.2002.808800

[46] KATAYAMA, MUTSUMI, M. KAZUYUKI, K. KAZUMITSU, *Vehicle position detection system*. JP Patent, 2010.

[47] ARAFA, A., XIAN JIN, KLUKAS, R. Wireless indoor optical positioning with a differential photosensor. *IEEE Photonics Technology Letters,* 2012, vol. 24, no. 12, p. 1027–1029. DOI: 10.1109/LPT.2012.2194140

[48] PENGFEI LUO, GHASSEMLOOY, Z., HOA LE MINH, KHALIGHI, A., XIANG ZHANG, MIN ZHANG, et al. Experimental demonstration of an indoor visible light communication positioning system using dual-tone multi-frequency technique. In *2014 3rd International Workshop in Optical Wireless Communications (IWOW)*. 2014, p. 55–59. DOI: 10.1109/IWOW.2014.6950776

[49] WANG, T. Q., SEKERCIOGLU, Y. A., NEILD, A., ARMSTRONG, J. Position accuracy of time-of-arrival based ranging using visible light with application in indoor localization systems. *Journal of Lightwave Technology,* 2013, vol. 31, no. 20, p. 3302–3308. DOI: 10.1109/JLT.2013.2281592

[50] CARRUTHERS, J. B., CAROLL, S. M., KANNAN, P. Propagation modelling for indoor optical wireless communications using fast multi-receiver channel estimation. *IEE Proceedings-Optoelectronics,* 2003, vol. 150, no. 5, p. 473–481. DOI: 10.1049/ip-opt:20030527

[51] RAPPAPORT, T. S. *Wireless Communications*. Prentice-Hall, 2002.

[52] MCDONOUGH, R. N., WHALEN, A. D. *Detection of Signals in Noise*. San Diego (CA, USA): Wiley, 1995.

[53] WU, D., GHASSEMLOOY, Z., ZHONG, W.-D., KHALIGHI, M.-A., HOA LE MINH, CHEN, C., et al. Effect of optimal Lambertian order on the performance of cellular indoor optical wireless communications and positioning. *Journal of Lightwave Technologies,* 2015, submitted.

[54] *The LED Streetlight Replacement Program.* [Online] Cited 2014-02-10 Available at: http://bsl.lacity.org/led.html

[55] PENGFEI LUO, GHASSEMLOOY, Z., HOA LE MINH, BENTLEY, E., BURTON, A., TANG, X. Performance analysis of a car-to-car visible light communication system. *Applied Optics,* 2015, vol. 54, no. 7, p. 1696–1706. DOI: 10.1364/AO.54.001696

[56] TAKAI, I., ITO, S., YASUTOMI, K., KAGAWA, K., ANDOH, M., KAWAHITO, S. LED and CMOS image sensor based optical wireless communication system for automotive applications. *IEEE Photonics Journal,* 2013, vol. 5, article no. 6801418. DOI: 10.1109/JPHOT.2013.2277881

[57] NAGURA, T., YAMAZATO, T., KATAYAMA, M., YENDO, T., FUJII, T., OKADA, H. Tracking an LED array transmitter for visible light communications in the driving situation. In *2010 7th International Symposium on Wireless Communication Systems (ISWCS)*. 2010, p. 765–769. DOI: 10.1109/ISWCS.2010.5624361

About the Authors ...

Stanislav ZVANOVEC received his M.Sc. and Ph.D. from the Czech Technical University in Prague, in 2002 and 2006, respectively. Now he is a full professor and vice-head of the Department of Electromagnetic Field at the Faculty of Electrical Engineering, Czech Technical University in Prague. He leads a Free-space and Fiber Optics team from the Faculty of Electrical Engineering, CTU and several research projects. His current research interests include wireless optical communications, visible light communications, remote sensing and optical fiber sensors.

Petr CHVOJKA was born in 1987. He received his M.Sc. from the Czech Technical University in Prague in 2013. Now he is a postgraduate student and a researcher at the Department of Electromagnetic Fields, Czech Technical University in Prague, where he is a member of a Free-space and Fiber Optics team. His research area includes visible light communications, OLED technologies and wireless optical communications.

Paul Anthony HAIGH received the PhD degree in Visible Light Communications from Northumbria University in 2014, publishing 13 articles in high ranking journals. Between 2010 and 2011 he held the prestigious Marie Curie Fellowship at CERN where he worked on optoelectronic links for large hadron collider experiments. In 2010 Paul received the BEng (Hons) degree in Communications Engineering from Northumbria University. Currently, he is a Research Associate within the High Performance Networks group at the University of Bristol working on the EPSRC TOUCAN project. His research is focused on real time seamless, transparent and adaptive and programmable interfaces between wireless and wired multi-technology networks.

Zabih GHASSEMLOOY, CEng, Fellow of IET, Senior Member of IEEE received his BSc (Hons) from the Manchester Metropolitan University in 1981, and MSc and PhD from the University of Manchester, Institute of Science and Technology (UMIST), in 1984 and 1987, respectively. 1986-87 worked in UMIST and from 1987 to 1988 was a Post-doctoral Research Fellow at the City University, London. 1988 joined Sheffield Hallam University as a Lecturer, becoming a Professor in Optical Communications in 1997. 2004-2012 was an Associate Dean for Research in the School of Computing, Engineering and from 2012-2014 Associate Dean for Research and Innovation in the Faculty of Engineering and Environment, Northumbria University at Newcastle, UK. He currently heads the Northumbria Communications Research Laboratories within the Faculty. He is the Editor-in-Chief of the International Journal of Optics and Applications, and British Journal of Applied Science & Technology. His researches interests are on optical wireless communications, visible light communications and radio over fiber/free space optics. He has supervised over 48 PhD students and published more than 550 papers (195 in journals + 4 books). He is a co-author of a CRC book on "Optical Wireless Communications – Systems and Channel Modelling with Matlab (2012)". From 2004-06 he was the IEEE UK/IR Communications Chapter Secretary, the Vice-Chairman (2004-2008), the Chairman (2008-2011), and Chairman of the IET Northumbria Network (Oct 2011-).

Design of High Performance Microstrip Dual-Band Bandpass Filter

Nafiseh KHAJAVI[1], *Seyed Vahab AL-Din MAKKI*[2], *Sohrab MAJIDIFAR*[3]

[1]Dept. of Electrical Engineering, Dezful Branch, Islamic Azad University, Dezful, Iran
[2]Dept. of Electrical Engineering, Razi University, Kermanshah, Iran
[3]Dept. of Electrical Engineering, Kermanshah University of Technology, Kermanshah, Iran

n_khajavi89@yahoo.com, v.makki@razi.ac.ir, s.majidifar@razi.ac.ir

Abstract. *This paper presents a new design of dual-band bandpass filters using coupled stepped-impedance resonators for wireless systems. This architecture uses multiple couple stubs to tune the passband frequencies and the filter characteristics are improved using defected ground structure (DGS) technique. Measurement results show insertion losses of 0.93 dB and 1.13 dB for the central frequencies of 2.35 GHz and 3.61 GHz, respectively. This filter is designed, fabricated and measured and the results of the simulation and measurement are in good agreement.*

Keywords

Dual-band bandpass filters, stepped-impedance resonators, defected ground structure (DGS), insertion loss

1. Introduction

Nowadays microwave systems have different applications in the society including satellite televisions, mobile phones, and civilian and military satellite systems. Given the growth in daily demand of communication systems like satellite communication and mobile phones, a contemporary trend in microwave technology can be seen towards compact and small-sized circuits and lower prices. To achieve this objective, active and passive microwave circuits are designed in a compact, multi-band and frequency-adjustable manner. After developing portable communication systems, multi-band systems have attracted great attention. Using this characteristic, results in smaller sizes and lower prices. Designing dual-band filters using loaded-stub open-loop resonators is a common practice [1]. This structure has return loss of more than 3 dB in two bands. Dual-band bandpass filters have also been provided using spiral stepped-impedance resonators [2]. In [3] dual-band filters have been designed using defected ground structure resonator and a dual-mode open-stub loaded stepped impedance resonator. A dual mode microstrip fractal resonator is proposed in [4] and optimized perimeter of the proposed resonator by using fourth iteration T-square fractal shape. L. C. Tsai [5] combined a wide-band bandpass filter and a band-stop filter and designed a dual-band filter. In another work, X. Guan designed dual-band filters using transmission lines and open-ended stubs [6]. A non-degenerate dual-band microstrip filter has been developed using non-degenerate resonator loads and open-ended stubs in [7]. A dual-band bandpass microstrip filter has been designed in [8] using microstrip periodic stepped-impedance ring resonators. For better description, design process of a LC model is presented in [9]. High selectivity and good suppression is obtained in [10] using embedded scheme resonator. The parallel LC resonant circuit can be represented for equivalent circuit of the DGS [11]. In other case a bandpass filter is designed using coupled DGS open loop resonators [12]. By applying the fractal theory and defected ground structure, filter dimensions have been reduced in [13].

In this paper a microstrip dual-band bandpass filter is designed using coupled stepped-impedance resonators. Designing filter is performed in three stages. To adjust the frequency of pass bands, we add a low impedance section to the middle of the basic structure. For fixed frequencies according to WLAN standards, multiple coupling structures is added to the proposed resonator and filter characteristics like return loss, insertion loss and bandwidth were enhanced using defected ground structure technique. In design process of the proposed filter, the exact combination of several known methods (resonator design base on the $\lambda/2$-stepped impedance structure, multiple coupling- defected ground structure) is used instead of a complex dual band structure. In this method each parts of filter response is influenced by a part of the design process and complexity of the final structure is replaced with the multiplicity of design process.

2. Filter Design

Figure 1 shows basic structure of open-ended stepped-impedance resonator with the half-wave of $\lambda/2$. This structure consists of a high impedance section Z_1 with the electric length of θ_1 and two low impedance sections Z_2 with the electric length of θ_2 as in [14].

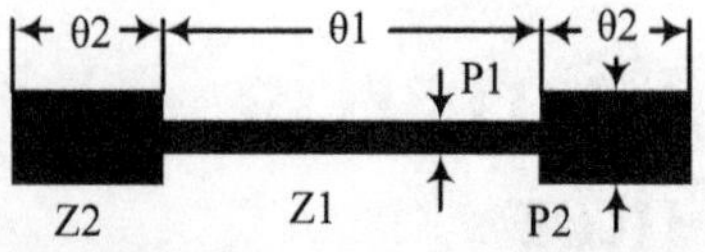

Fig. 1. Basic structure stepped-impedance resonator with the half-wave of λ/2.

The resonance condition of the resonator can be derived by:

$$R_Z = Z_2 / Z_1 = \tan\theta_1 \tan\theta_2 \quad (1)$$

where R_Z is the impedance ratio of the stepped impedance resonator (SIR). The fundamental frequency f_0 and first spurious frequency f_{sb1} of the resonator are related by:

$$\frac{f_{sb1}}{f_0} = \frac{\pi}{2\tan^{-1}\sqrt{R_Z}}. \quad (2)$$

Figure 2 shows the normalization of the first neutral frequency as a function of R_Z. By selecting a lower impedance ratio, the first neutralization mode shifts towards the upper frequency range.

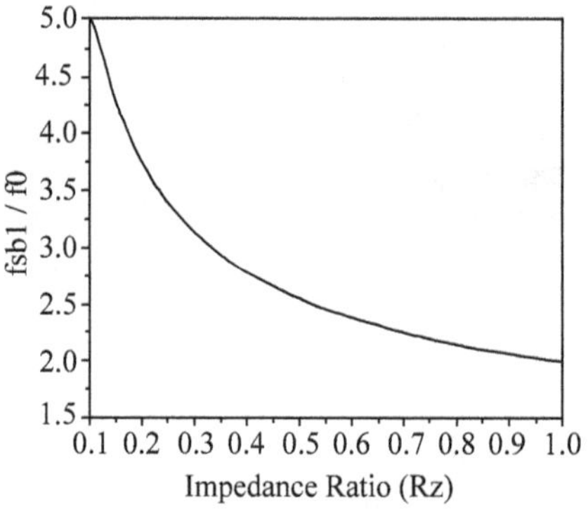

Fig. 2. Normalization of the first neutral frequency as a function of R_Z.

Designing of the filter is performed in three stages:

- Resonator design.
- Adding multiple couplings to the proposed resonator for neutralize neutral harmonics in stop band.
- Using defected ground structure (DGS) technique to improve the filter characteristics.

2.1 Resonator Design

2.1.1 Basic Resonator

According to (2) and f_0 / f_{sb1} = 1.85, we will have R_Z =1.3. Now, if Z_1 = 100 Ω, given (1), Z_2 = 130 Ω. Given the values of Z_1 and Z_2, P1 = 0.289 mm and L2 = 0.13 mm are calculated. As shown in Fig. 3(a), the basic resonator is designed using structure of Fig. 1. Dimension of Fig. 3(a) are as follow: L1 = 30.4 mm, P1 = 0.298 mm, L2 = 0.13 mm, P2 = 2.16 mm and G = 0.1 mm.

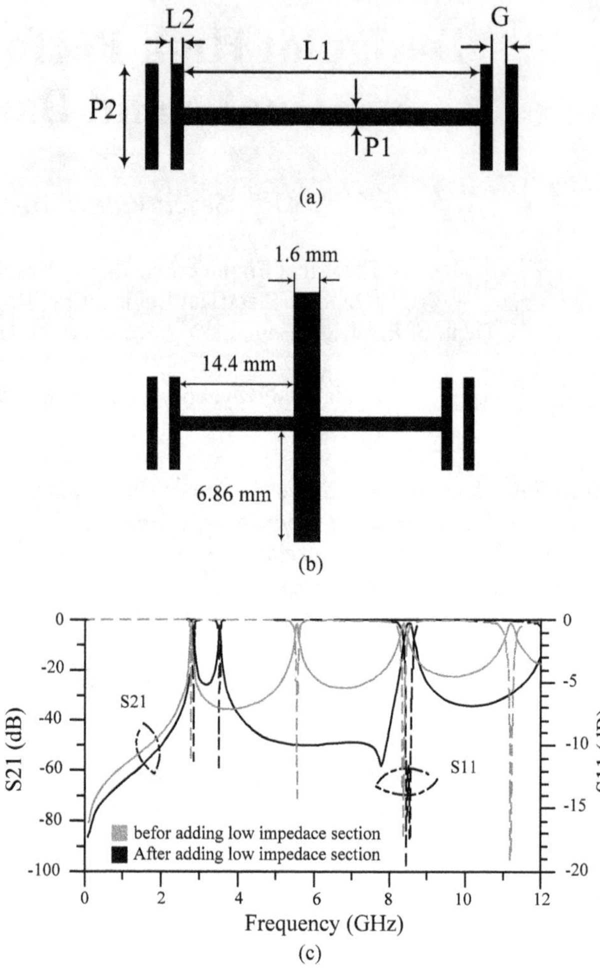

Fig. 3. Basic resonator. (a) The layout of the basic resonator. (b) The layout of the basic resonator with added low impedance section. (c) Simulation result of the part (a) and (b).

2.1.2 Tuning of the Passband Frequencies

In the basic resonator, the first passband frequency reaches 2.7 GHz and the second one reaches 5.5 GHz. With adding a low impedance stub to the basic resonator (Fig. 3b), the second passband frequency shifts to the first one. Passband frequencies can be adjusted in two stages:

- By changing L1, first passband frequency is placed at a desired value.
- By changing the dimensions of the low impedance stub, while the first pass band frequency is fix, the second one tunes at the desired value.

2.1.3 Physical Size Optimizations

To optimize the dimensions of the filter, we change the 30.4 mm long middle transmission line as shown in Fig. 4(a). Dimensions of the proposed resonator are as follow: W1 = 0.13 mm, L1 = 0.16 mm, W2 = 0.13 mm, L2 = 17.57 mm, W3 = 0.186 mm, L3 = 1.86 mm, L4 = 7.852 mm, L5 = 14.412 mm, W4 = 0.4 mm, L6 = 2.4 mm, W5 = 0.43 mm, L7=4.065 mm, W6 = 0.289 mm, G = 0.2 mm.

2.1.4 Proposed LC Model of the Resonator

In order to better describe the designed resonator, its LC model, as it is shown in Fig. 4(b) is presented. In this model, Cg represents the coupling capacitance between the open stubs and the transmission lines are modeled by a T-junction circuit including two series inductances and a central ground ended capacitor. Lt1 and Ct1 are the inductance and capacitance of L7 and Cop1, Cop2 are the capacitances of the open stubs with respect to ground. Lt2 and Ct2 are the inductance and capacitance of T-junction part of the middle loop. The values of circuit parameters are calculated using equations (3)-(8) [15].

$$\varepsilon_{ref} = \frac{\varepsilon_r + 1}{2} + \left\{ \left(1 + 12\frac{H}{W}\right)^{-0.5} + 0.04\left(1 - \frac{W}{H}\right)^2 \right\} \qquad \frac{W}{H} \le 1 \tag{3}$$

$$\varepsilon_{ref} = \frac{\varepsilon_r + 1}{2} + \frac{\varepsilon_r - 1}{2}\left(1 + 12\frac{H}{W}\right)^{-0.5} \qquad \frac{W}{H} \ge 1$$

$$Z_C = \frac{120\pi}{\frac{C_a}{\varepsilon_r}\sqrt{\varepsilon_{ref}}} \tag{4}$$

where W is the microstrip line width, ε_r is the relative dielectric constant of the substrate, H is the substrate thickness, Z_C is the characteristic impedance, ε_{ref} is the effective dielectric constant, C_a is the capacitance per unit length with the dielectric substrate replaced by air and c is the velocity of electromagnetic waves in free space (c = 3.0×10^8 m/s). C_a can be determined by (5).

$$C_a = \frac{2\pi\varepsilon_r}{l_n\left(\frac{8H}{W} + \frac{W}{4H}\right)} \qquad \frac{W}{H} \le 1 \tag{5}$$

$$C_a = \varepsilon_r\left(\frac{W}{H} + 1.393 + 0.66 l_n\left(\frac{W}{H} + 1.444\right)\right) \qquad \frac{W}{H} \ge 1$$

While l is the length of the microstrip line, L and C as inductance and capacitance of microstrip line can be determined by (6) and (7).

$$L = \frac{Z_C l}{V_P}, \tag{6}$$

$$C = \frac{l}{Z_C V_P} \tag{7}$$

V_p is the phase velocity and can be determined by (8).

$$V_p = \frac{c}{\sqrt{\varepsilon_{ref}}} \tag{8}$$

The calculated values of the circuit parameters are as follow: L = 0.08 nH, Cop1 = 0.346 pF, Cg = 0.3 pF, Cop2 = 0.36 pF, Lt1 = 1.09 nH, Ct1 = 0.17 pF, Lt2 = 3.916 nH and Ct2 = 0.67 pF. The electromagnetic (EM) and LC simulated results of this structure are shown in Fig. 4(c). As it is shown, good agreement between the EM and LC simulated results is achieved. This resonator is designed on Ro 4003 substrate with 3.38 dielectric constant, 20 mil heights and 0.0021 loss tangent.

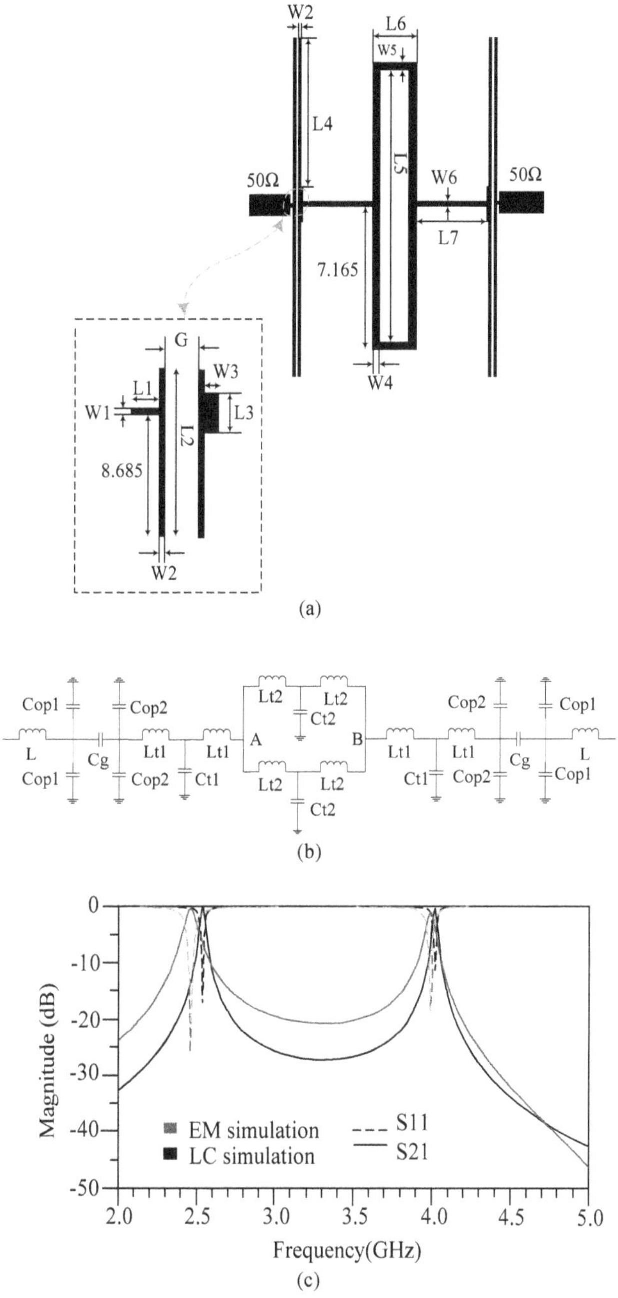

Fig. 4. Proposed resonator. (a) The layout of the proposed resonator. (b) LC equivalent circuit of the proposed resonator. (c) EM and LC circuit simulated results.

The frequency response of the proposed resonator in a wide span is shown in Fig. 5. As can be seen in this figure, this resonator has two bands with the central frequencies of f_1= 2.40 GHz and f_2= 4.0 GHz. The objective is to design a dual-band bandpass filter with the same central frequencies as those of WLAN. Central frequency of the first band is fixed on 2.4 GHz, while in the second band our goal is to achieve central frequency of 3.7 GHz. These bands have insertion losses of 0.478 dB and 1.137 dB, return losses of 26.05 dB and 18.549 dB and bandwidths of 76 MHz and 49 MHz, respectively. In order to achieve the better response characteristics of the resonator, the following changes should be considered: balancing the insertion loss of two bands, decreasing the return loss of two bands and increasing the passbands' bandwidth. Furthermore, the neutral harmonic at the 6.866 GHz decreases the stopband bandwidth.

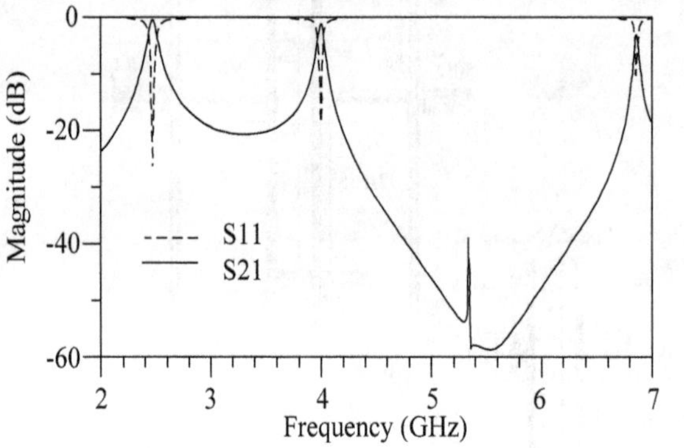

Fig. 5. Frequency response of the proposed resonator in a wide span.

2.2 Adding Multiple Couplings to the Proposed Resonator for Neutralizing Neutral Harmonics in the Stopband

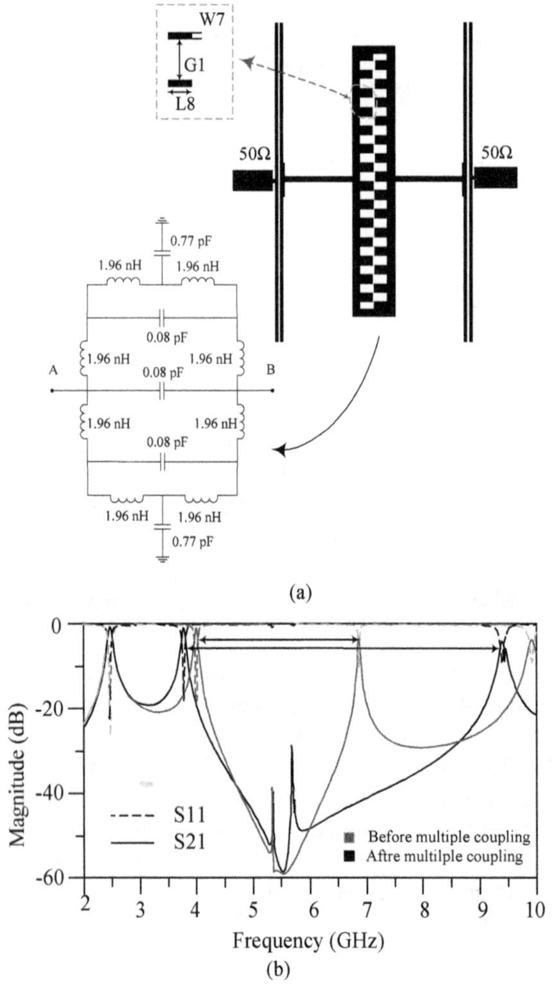

Fig. 6. The proposed filter (a) Layout and LC model (b) Compares between frequency responses of the proposed resonator and the proposed filter.

One of the problems in the last stage is the limited stopband bandwidth. This problem can be solved using multiple couplings. Figure 6(a) shows the proposed filter. In this filter the added multiple coupling are located inside the middle lope of the resonator. The dimensions of these coupling stubs are as follow: W7 = 0.524 mm, L8 = 0.762 mm, G1 = 0.524 mm. The size and number of the coupling stubs and its LC model are depicted in Fig. 6(a). In this model the Lt2 is divided into two parts and the effects of the multiple couplings are depicted with added coupling capacitors and increased grounded capacitors. Simulation results of the proposed filter are shown in Fig. 6(b).

As it is shown in Fig. 6(b), this method suppresses the harmonics from 4.2 GHz to 9.377 GHz, increases the stopband bandwidth and sets the second passband at the 3.7 GHz. In the proposed filter, the first and second passbands have insertion losses of 0.521 dB and 0.842 dB, return losses of 25.3 dB and 17.914 dB and bandwidths of 77 MHz and 61 MHz, respectively. This results show that return loss is increased but passband bandwidth is improved in two bands. In the next stage we try to increase the passbands bandwidth and achieve modified return loss in two bands.

2.3 Using Defected Ground Structure (DGS) Technique to Improve the Filter Characteristics

Figure 7 shows the applied defected ground structure and its LC model to improve the filter characteristics. In this filter, improvement of the passband performance and also increasing the bandwidth of each band are the reasons of DGS utilization. Dumbbell and rectangular shape were used in defected ground structure technique. As it is shown in Fig. 7(a) the dumbbell shaped DGS is introduced by a parallel LC resonator and this part is added to the LC model of the multiple couplings. The rectangular shape DGS is applied to moderate the effect of the bends (on both sides of the central line).

The dimensions of the dumbbell parts and rectangular planes are as follow: W8 = 0.9 mm, L9 = 2.79 mm, W9 = 0.198 mm, L10 = 3.33 mm, L11 = 0.89 mm, L12 = 1.69 mm, L13 = 0.34 mm. A picture of the fabricated filter and the results of the simulation and measurement of the final filter are depicted in Fig. 7(b) and Fig. 7(c), respectively. EM simulations are performed in ADS and measurement performed using the Agilent network analyzer N5230A.

Figure 8 shows the current density distribution at the surface of the filter, with/without DGS. As it is shown in Fig. 8(a) and Fig. 8(b), dumbbell shaped DGS is utilized because of the balanced increasing at the current density distribution of the filter and this has improved the filter response in the passbands (in terms of insertion loss, bandwidth and return loss).From Fig. 8(c) it is apparent that the

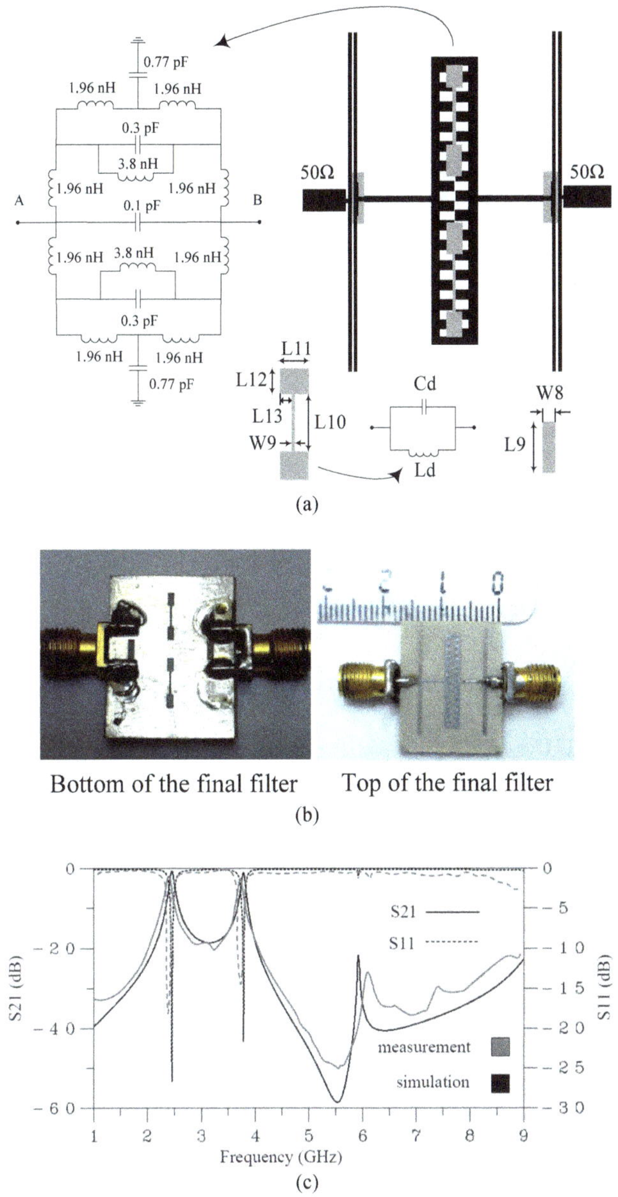

Fig. 7. Final filter. (a) Layout and LC model of the proposed filter. (b) A picture of the fabricated filter. (c) Simulation and measurement results of the filter.

Characteristics of proposed resonator and proposed filter	S_{21} in the first and second passband respectively (dB)	S_{11} in the first and second passband respectively (dB)	Bandwidth in the first and second passband respectively (MHz)
Proposed resonator (without multiple coupling)	- 0.478 & - 1.137	- 26.05 & - 18.549	76 & 49
Proposed filter (using multiple coupling)	- 0.521 & - 0.842	- 25.3 & - 17.914	77 & 61
Final structure of the proposed filter (using DGS technique)	- 0.457 & - 0.682	- 23.57 & - 17.2	95 & 87

Tab. 1. Compares between response characteristics of proposed structures in passbands.

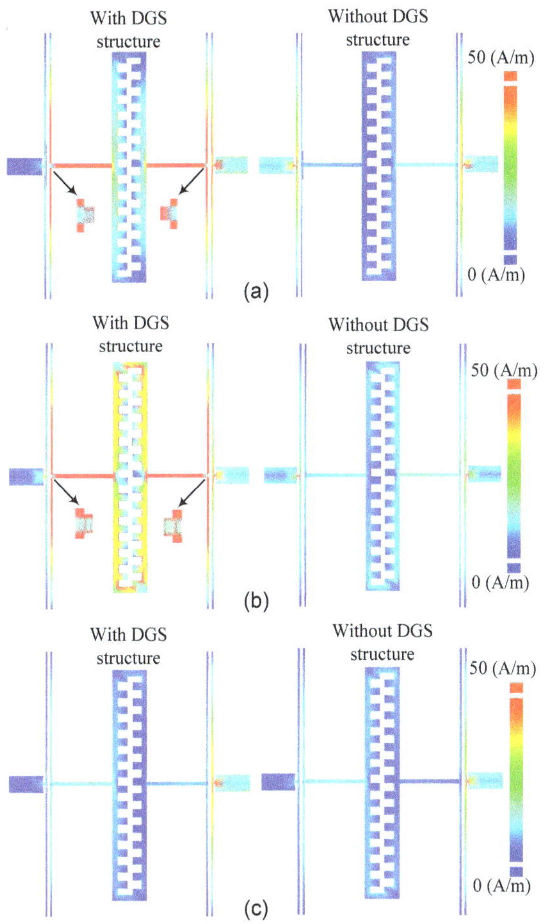

Fig. 8. Current density distribution at the surface of the filter: (a) at 2.45 GHz; (b) at 3.7 GHz; (c) at 3.07 GHz.

current density distribution of the filter with/ without DGS has no obvious change at 3.07 GHz (as the isolation frequency between two bands). So dumbbell shaped DGS has improved the passband performance of the filter. Also, the isolation between passbands and stopband characteristics has not been changed. The bends affect on the high frequency harmonics. In order to decrease the effects of these bends, the rectangular shaped DGS is utilized. As Figures 8(a) and 8(b) indicate the current density distribution of the bends is fixed using rectangular shaped DGS (while the current density distribution is increased in other lines) and the effects of them are limited. A comparison between the response characteristics of the proposed structures (in term of passbands) is shown in Tab. 1.

3. Conclusion

In this paper, at the first, using specified relations, a basic structure is designed. In order to shift the closest harmonic as the second passband towards the first passband, and also, to attenuate other harmonics, we added a low impedance stub to the middle of the basic structure. Using this structure, the proposed resonator for optimization of dimensions is provided. In order to further suppress the harmonics, and also to fix frequencies according to WLAN standards, multiple coupling structures are added to the proposed resonator. Finally, DGS technique was applied to improve filter response specifications. In final structure of the filter, frequency of the first and second passbands are 2.35 GHz and 3.61 GHz respectively, which have return losses of 0.93 dB and 1.13 dB and insertion losses of 17.56 dB and 14.08 dB and bandwidths of 89 MHz and 81 MHz, respectively.

References

[1] MONDAL, P., MANDAL, M. K. Design of dual-band bandpass filters using stub-loaded open-loop resonators. *IEEE Transactions on Microwave Theory and Techniques*, 2008, vol. 56, no. 1, p. 150–155. DOI: 10.1109/TMTT.2007.912204

[2] GUO, L., YU, Z., ZHANG, Y. A dual-band band-pass filter using stepped impedance resonator. *Microwave and Optical Technology Letters*, 2011, vol. 53, no. 1, p. 123–125. DOI: 10.1002/mop.25648

[3] WANG, L., GUAN, R. Novel compact and high selectivity dual-band BPF with wide stopband. *Radioengineering*, 2012, vol. 21, no. 1, p. 492–495.

[4] AHMED, E. S. Dual-mode dual-band microstrip bandpass filter based on fourth iteration T-square fractal and shorting pin. *Radioengineering*, 2012, vol. 21, no. 2, p. 617–623.

[5] YUN, T. S., NOH, S. K., KIM, OH, E. K., SON, H. M., LEE, J. C. Compact dualband bandpass filter with two transmission zeros using dual-mode microstrip resonator and tapped-line geometry. *Microwave and Optical Technology Letters*, 2011, vol. 53, no. 1, p. 108–111. DOI: 10.1002/mop.25627

[6] TSAI, L. C., HSUE, C. W. Dual-band bandpass filters using equal-length coupled-serial-shunted line and Z-transform technique. *IEEE Transactions on Microwave Theory and Techniques*, 2004, vol. 52, no. 4, p. 1111–1117. DOI: 10.1109/TMTT.2004.825680

[7] GUAN, X., MA, Z., CAI, P., KOBAYASHI, Y., ANADA, T., HAGIWARA, G. Synthesis of dual-band bandpass filters using successive frequency transformation and circuit conversions. *IEEE Transactions on Microwave and Wireless Components Letters*, 2006, vol. 16, no. 3, p. 110–112. DOI: 10.1109/LMWC.2006.869868

[8] SHETA, A. F. Narrow band compact non-degenerate dual-mode microstrip filter. In *Proceedings of the 25th National Radio Science Conference*. Egypt, 2008.

[9] MAKKI, S. V. AL-DIN, AHMADI, A., MAJIDIFAR, S., SARIRI, H., RAHMANI, Z. Sharp resonator microstrip LPF using folded stepped impedance open stubs. *Radioengineering*, 2013, vol. 22, no. 1, p. 328–329.

[10] ZONG, B. F., WANG, G. M., ZENG, H. Y., WANG, Y. W. Compact and high performance dual-band bandpass filter using resonator-embedded scheme for WLANs. *Radioengineering*, 2012, vol. 21, no. 4, p. 1050–1053.

[11] PARUI, S. K., DAS S. A new defected ground structure for different microstrip circuit applications. *Radioengineering*, 2007, vol. 16, no. 1, p. 16–22.

[12] VAGNER, P., KASAL, M. A novel bandpass filter using a combination of open-loop defected ground structure and half-wavelength microstrip resonators. *Radioengineering*, 2010, vol. 19, no. 3, p. 392–396.

[13] KUFA, M., RAIDA, Z. Comparison of planar fractal filters on defected ground substrate. *Radioengineering*, 2012, vol. 21, no. 4, p. 1019–1024.

[14] SARKAR, P., GHATAK, R., PODDAR, D. R. A dual-band bandpass filter using SIR suitable for WiMAX band. In *International Conference on Information and Electronics Engineering IPCSIT*. (Malaysia), 2011.

[15] HONG, J. S., LANCASTER, M. J. *Microstrip Filters for RF/ Microwave Applications*. A Wiley-Interscience Publication. John Wiley & Sons, 2001. ISBN 0-471-38877-7

About the Authors...

Nafiseh KHAJAVI received the B.Sc. degree in Medical Engineering from Islamic Azad University Dezful Branch, 2008, and the M.Sc. degree in Electrical Engineering from Kermanshah Science and Research Branch, Islamic Azad University, Kermanshah, 2012. She has been with the Department of Engineering, Islamic Azad University Dezful Branch. Her research interests include microstrip filter, the analysis and design of high-frequency electronics and microwave passive circuits.

Seyed Vahab Al-Din MAKKI was born in Kermanshah. He received his PhD in Electrical Engineering-Waves from Khaje Nasir Toosi University in 2008. He is with the Electrical Engineering Department of Razi University in Kermanshah and Islamic Azad University, Kermanshah Branch, since 2008. His current research interests include modern digital radio propagation systems, microwave devices and radio transmitters.

Sohrab MAJIDIFAR received his B.Sc. and M.Sc. in Electrical Engineering from Razi University in 2009 and 2011, respectively. He joined Kermanshah University of Technology in 2011 as a lecturer. His research interests include microwave passive circuits and RFIC.

16

A Wideband Direct Data Domain Genetic Algorithm Beamforming

Hassan ELKAMCHOUCHI, Mohamed HASSAN

Dept. of Electrical Engineering, University of Alexandria, Alhuria Street, Alexandria, Egypt

helkamchouchi@hotmail.com, engmohamedmokhtar1@yahoo.com

Abstract. *In this paper, a wideband direct data-domain genetic algorithm beamforming is presented. Received wideband signals are decomposed to a set of narrow sub-bands using fast Fourier transform. Each sub-band is transformed to a reference frequency using the steering vector transformation. So, narrowband approaches could be used for any of these sub-bands. Hence, the direct data-domain genetic algorithm beamforming can be used to form a single 'hybrid' beam pattern with sufficiently deep nulls in order to separate and reconstruct frequency components of the signal of interest efficiently. The proposed approach avoids most of drawbacks of already-existing statistical and gradient-based approaches since formation of a covariance matrix is not needed, and a genetic algorithm is used to solve the beamforming problem.*

Keywords

Wideband array, direct data-domain, genetic algorithm, signal of interest, matrix pencil method

1. Introduction

An adaptive array is able to electronically steer its main lobe to any desired direction and place deep pattern nulls to directions of interference sources. That way, the antenna could adaptively minimize the interference power while maintaining the array gain in the direction of the target signal [1–3].

Statistical methods of adaptive antennas are computationally intensive processes and require stationary data to construct a covariance matrix. Direct data-domain (D^3) methods could overcome drawbacks of statistical techniques by processing data on snapshot-by-snapshot basis without constructing a covariance matrix. Hence, D^3 methods could handle non-stationary environments and coherent interferers [4].

Genetic algorithms (GAs) might be more efficient than gradient-based methods for nulling a linear antenna array since the gradient-based methods have following disadvantages:

- The methods are highly sensitive to starting points when the number of variables, and hence the size of the solution space, increases.
- The methods frequently converge to local suboptimum solutions.
- The methods require a continuous and differentiable objective function.
- The methods require piecewise linear cost approximation (for linear programming).
- The methods have problems with convergence and algorithm complexity (for non-linear programming).

Future wireless systems have to utilize wideband smart antennas to meet high speed data transmission while avoiding undesired interference [15]. Beamforming techniques used in narrowband systems are inappropriate for wideband systems due to the limited ability of tracking a desired user or forming nulls in directions of interfering sources over a large frequency band [5], [6]. Some earlier work has been done to solve such problems. Existing concepts of wideband beamforming exhibit disadvantages which could be summarized as follows:

- Different array patterns are used for different frequencies [12–14]. Obviously, such an approach is quite cumbersome.
- Gradient-based wideband beamforming [1], [7], [16] has the already described disadvantages.
- Statistical methods depend on the formation of a covariance matrix [17]. Therefore, stationary data are expected to estimate the covariance matrix. In case of non-stationary data, resulting errors in the covariance matrix reduce the ability to handle coherent interferers [4].
- In [7] it is mentioned that "although the main beam is directed to the signal of interest (SOI) direction and the jammers are nulled correctly but the nulls are not deep enough"; same drawback appears in [1]. Such disadvantage affects the accuracy of SOIs reconstruction. In addition, in [1] and [7], nulls' depths and DOAs estimations' accuracy are found to be frequency dependent. Therefore, selecting one beam pattern corresponding to single sub-band does not assure sufficient nulls' depths to cancel interferers in all other sub-bands.

The proposed wideband direct data-domain genetic algorithm (WD^3GA) beamforming relies on decomposing the received wideband signals into a set of narrow sub-bands by using fast Fourier transform (FFT) and transform all sub-bands to a reference frequency by using the steering vector transformation [5]. Any of these sub-bands could be used by narrowband techniques for direction of arrival (DOA) estimation and beamforming. The investigated narrowband direct data-domain adaptive nulls genetic algorithm (D^3ANGA) beamformer [9] is used to cancel jammers' frequency components in each sub-band. Finally, inverse fast Fourier transform could be used to retrieve the SOI in time domain.

WD^3GA beamforming has the following *unique* set of characteristics:

- WD^3GA beamforming uses only a single hybrid array beam pattern; thus the method complexity is reduced.
- GA is used to solve the beamforming problem. Hence, the gradient-based methods' drawbacks are avoided.
- Both FFT and covariance-matrix-based techniques require recorded samples of the received signals [4]. However, WD^3GA beamforming doesn't make any assumption about the statistics of the environment. Therefore, data non-stationarity has a little effect on the method performance.
- The problem of frequency dependent estimations of DOAs and nulls' depths [1], [7] could be solved by taking into consideration that in D^3ANGA beamforming [9], the determination of the nulls' depths and the DOAs estimation are done before and independent on the beamforming algorithm. Hence, for wideband beamforming, the nulls' depths could be selected prior to the beamforming algorithm to be proportional to strongest interferers' frequency components and the most accurate estimated DOAs could be selected priory. Hence, one 'hybrid' reconstruction array beam pattern combines both sufficient deep nulls and accurate estimated DOAs could be formed.

2. Steering Vector Transformation

For a wideband antenna array of N elements and d spacing between adjacent elements, consider $q+1$ uniformly spaced directions covering a pre-specified angular azimuth region $\mathbf{\Phi}_q$ [5], [7] where,

$$\mathbf{\Phi}_q = \left[\phi_0, \phi_1, \ldots, \phi_q\right] \quad (1)$$

In this paper, all coming signals are considered to be in the azimuth plane ($\theta = 90°$). The steering vector transformation is based on transforming the array steering matrix at the kth frequency f_k to another array steering matrix at a pre-specified reference frequency f_o using a transformation matrix $\boldsymbol{T}_q(k)$ for the angular region $\mathbf{\Phi}_q$ such that:

$$\boldsymbol{A}\left(\mathbf{\Phi}_q, f_o\right) = \boldsymbol{T}_q(k)\boldsymbol{A}\left(\mathbf{\Phi}_q, f_k\right) \quad (3)$$

where $\boldsymbol{A}\left(\mathbf{\Phi}_q, f_o\right)$ and $\boldsymbol{A}\left(\mathbf{\Phi}_q, f_k\right)$ are the array steering matrices for the angular region $\mathbf{\Phi}_q$ at the kth frequency f_k and at the reference frequency f_o respectively. The array steering matrix could be computed as follows:

$$\boldsymbol{A}\left(\mathbf{\Phi}_q, f\right) = \left[\boldsymbol{a}\left(\phi_0, f\right), \boldsymbol{a}\left(\phi_1, f\right), \ldots, \boldsymbol{a}\left(\phi_q, f\right)\right] \quad (4)$$

where $\boldsymbol{a}(\phi, f)$ is the steering vector defined by:

$$\boldsymbol{a}(\phi, f) = \left[1,\ e^{\frac{2\pi fd}{c}\cos(\phi)}, \ldots,\ e^{\frac{2\pi f(N-1)d}{c}\cos(\phi)}\right]^T \quad (5)$$

where T denotes the transpose of the vector. Equation (3) could be solved for $\boldsymbol{T}_q(k)$ using the least squares method which yields the solution:

$$\boldsymbol{T}_q(k) = \boldsymbol{A}\left(\Phi_q, f_o\right)\boldsymbol{A}\left(\Phi_q, f_k\right)^H\left(\boldsymbol{A}\left(\Phi_q, f_k\right)\boldsymbol{A}\left(\Phi_q, f_k\right)^H\right)^{-1} \quad (6)$$

where the H superscript represents the conjugate transpose of a complex matrix. The processed input voltage vector at the kth frequency which has been transformed to the reference frequency f_o, $\boldsymbol{x}(f_o)$, could be written as:

$$\boldsymbol{x}\left(f_o\right) = \boldsymbol{T}_q(k)\boldsymbol{x}\left(f_k\right) \quad (7)$$

where $\boldsymbol{x}(f_k)$ is the input voltage vector at the kth frequency. Using this transformation, a single narrowband beamformer tuned at the reference frequency f_o could be used.

3. Matrix Pencil Method

The MP method is a narrowband D^3 method to estimate the DOA of various signals impinging on an antenna array; the signals' complex amplitudes could be estimated as well [4], [8].

For a uniformly linear array composed of $N+1$ element, the voltage induced in the array n^{th} element, x_n, could be written as:

$$x_n = \sum_{k=1}^{P} S_k e^{\left(\frac{j2\pi nd\cos(\phi_k)}{\lambda}\right)} + \xi_n = \sum_{k=1}^{P} S_k a_k^n + \xi_n \quad (8)$$

where ξ_n is the noise at the nth array element, P is the number of incident signals, S_k is the complex amplitude of the kth incident signal, λ is the wave length, d is the distance between two adjacent elements, ϕ_k is the DOA of the kth signal, and a_k are the poles to be estimated.

The poles a_k could be estimated by constructing and processing a Hankel matrix as illustrated in [4], [8], then the DOAs of various signals could be obtained as follows:

$$\phi_k = \cos^{-1}\left[\frac{\lambda \ln\left(a_{es\,k}\right)}{j2\pi d}\right] \quad (9)$$

where $a_{es\,k}$ is the kth estimated pole. The complex amplitudes vector of the P signals, ***AMP***, could be obtained by:

$$AMP = \left(P_0^H P_0\right)^{-1} P_0^H x \tag{10}$$

where P_0 is the matrix containing the pole of each incident signal at each antenna element and x is a vector containing the induced voltages at the array elements.

4. D^{3}ANGA Beamforming

D^{3}ANGA is a narrowband beamformer [9]. Consider a linear antenna array with uniformly spaced N elements. Hence, N complex weights are used for beamforming. Given the DOAs and the strengths of all coming signals, a genetic algorithm (GA) is used to find the optimal values of these weights in order to fulfill the algorithm objectives [10], [11] which are as follows:

- Minimizing beam the pattern average value to minimize the pattern side lobes level.
- Maximizing the pattern value in the direction of the SOI (P_S) to radiate maximum possible power in this direction.
- Placing deep nulls in directions of the interferers. In addition, nulls' depths are selected to be proportional to interference incident signals' intensities.

Hence, the fitness function could be written as:

$$Fit = w\sum_{i=1}^{i=J}\left|\left|\frac{P_i}{P_s}\right| - N_i\right| + \left|\frac{P_{av}}{P_s}\right| \tag{11}$$

where J is the number of the interferer (jammer) signals, P_i is the array beam pattern complex value in the direction of the ith jammer, P_S is the array beam pattern complex value in the direction of the SOI, N_i is the ith normalized pattern null value corresponding to the ith jammer, w is the weighting factor used to balance GA optimization between the two terms of the fitness function. | | denotes the absolute (magnitude) of the complex quantities and P_{av} is the pattern average value which could be calculated by:

$$P_{av} = \frac{\sum_\theta \sum_\phi |P(\theta,\phi)|}{N_P} \tag{12}$$

where N_P is the number of points at which the pattern values are calculated. The pattern value at any direction, $P(\theta,\phi)$, could be computed by

$$P(\theta,\phi) = W^{\mathrm{T}} A(\theta,\phi) \tag{13}$$

where W is the complex weights' vector to be estimated by the GA and $A(\theta,\phi)$ is the steering vector in the direction of (θ,ϕ). Assuming all coming signals are in the azimuth plane ($\theta = 90°$), $A(\theta,\phi)$ could be expressed as:

$$A = \left[1,\ e^{j2\pi\frac{d}{\lambda}cos\phi}, e^{j2\pi\frac{2d}{\lambda}cos\phi}, \ldots, e^{j2\pi\frac{(N-1)d}{\lambda}cos\phi}\right]^{\mathrm{T}} \tag{14}$$

where d is the space between array adjacent elements and λ is the wavelength corresponding to the operating frequency. The ith normalized pattern null value, N_i, corresponding to the ith jammer could be computed by:

$$N_i = (|S_i| * C)^{-1} \tag{15}$$

where S_i is the ith jammer intensity and C is the cancelling factor (CF) used to make the nulls' depths sufficient to cancel interference signals efficiently.

In this paper, all the pattern values and all the obtained complex weights are normalized with respect to P_s. The GA-estimated complex weights could be used to separate and reconstruct the SOI using:

$$RSOI = W_n^T x \tag{16}$$

where $RSOI$ is the reconstructed SOI, W_n is the normalized weights vector, and x is the received signals' vector.

5. Genetic Algorithm Components

GA is a powerful optimization technique based on the concept of natural selection and natural genetics [10]. GA repeatedly modifies a population of individual solutions. At each step the GA selects individuals at random from the current population to be parents and uses them to produce the children of the next population. Over successive generations, the population "evolves" toward an optimal solution which is considered to be the solution which gives the minimum of the fitness function [10], [11]. In this paper, GAs are implemented based on the built-in genetic algorithm of R2013a MATLAB software package. The basic GA components, used in this paper, are reviewed briefly as follows [10]:

- Genetic representation of solution: Real number encoding is used to represent individual solutions or chromosomes.
- Population initialization: uniform random initialization is used and the population size is selected to be ten times the number of the antenna array elements taking into consideration that two chromosomes are used to represent each array element complex weight one for the real part and the other for the imaginary part [11].
- Evaluation of the fitness function: The GA should find the global minimum of the fitness function.
- Fitness scaling: The ranking method in which the scaling of raw scores is based on the rank of each individual instead of its score is used [11].
- Selection methods: Stochastic uniform selection method is used. Stochastic uniform selection method lays out a line in which each parent corresponds to a section of the line of length proportional to its scaled value. The algorithm moves along the line in steps of equal size. At each step, the algorithm allocates a parent from the section it lands on. The first

step is a uniform random number less than the step size. Also, Elitism forces GA to retain some number of the best individuals at each generation [10], [11]. In this paper, the most fit two chromosomes survive directly to the next generation as elite chromosomes.

- Genetic operators: genetic operators are used to produce new individuals. Crossover and mutation are the most frequently used genetic operators and are described as follows:

1- The crossover operator is the exchange of genes between parent's chromosomes to produce offspring. The scattered crossover method is used. In this method, crossover is done by creating a random binary vector and selecting the genes where the vector's elements are ones from the first parent, and the genes where the vector's elements are zeros from the second parent, and combines the genes to form the child [10], [11]. The fraction of each population, other than elite children, that are made up of crossover children is set to 0.8. The remaining chromosomes are mutation children.

2- Mutation is done by the addition of a random number which is chosen from a Gaussian distribution to each entry of the parent vector.

- Termination condition: a maximum number of 500 generations is used to terminate GA.

6. The Proposed WD^{3}GA

WD^{3}GA beamformer is implemented by a narrowband decomposition structure whereby each signal received at each array element is transformed into its frequency domain components using FFT. Each narrow frequency sub-band is transformed to a reference frequency using the steering vector transformation. The transformed sub-bands could be processed by narrowband techniques. Hence, they are sent to the MP stage, as shown in Fig. 1. In the MP stage, the DOAs as well as the complex magnitudes of all incident signals' frequency components are estimated.

One array beam pattern is used for reconstructing all SOI frequency components from all sub-bands. Two parameters determine the accuracy of reconstruction of the SOI frequency components: 1- the DOA estimation accuracy of all coming signals; 2- the pre-specified pattern nulls' depths. The estimated DOAs should be as accurate as possible, so it is recommended to select the DOAs estimated using the nearest sub-band to the reference frequency f_o since it is observed that the nearest sub-band to the reference frequency has the minimum error in the steering vector transformation [5]. The error in the steering vector transformation could be calculated as follows:

$$\boldsymbol{Error} = \boldsymbol{A}(\boldsymbol{\Phi}_q, f_o) - \boldsymbol{T}_q(k)\boldsymbol{A}(\boldsymbol{\Phi}_q, f_k). \quad (17)$$

When the reference frequency is selected to be one of the considered signals' frequencies, the steering vector transformation is not required for the corresponding sub-band. Hence, the corresponding steering vector transformation error is zero. For the array beam pattern nulls' depths, it is recommended to select the deepest nulls corresponding to the strongest estimated interferers' frequency components by the MP method in order to cancel all the interferers' frequency components in all sub-bands efficiently. So, one hybrid reconstruction beam pattern combines both accurate estimated DOAs and deep nulls and could be used for the estimation of all SOI frequency components then inverse FFT could be used to retrieve the SOI in time domain.

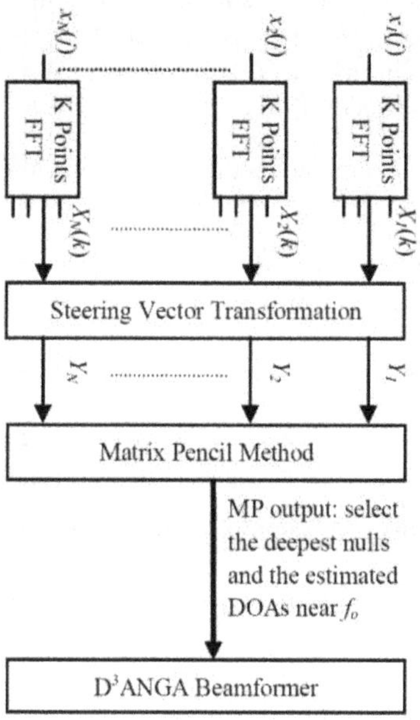

Fig. 1. The proposed WD^{3}GA beamforming structure.

7. Numerical Examples

Consider a uniform linear array of 6 elements. The received wideband signals are decomposed into their narrowband components at 6 frequencies within the design frequency band [3 GHz – 4 GHz], these frequencies are: 3, 3.2, 3.4, 3.6, 3.8, and 4 GHz. The inter-element spacing used equals half wavelength corresponding to the maximum frequency within the design frequency band in order to avoid spatial aliasing [5], [6]. Consider one SOI and one jammer, with arrival angles of $\phi = 120°$ and $\phi = 100°$ respectively. All SOI frequency components' amplitudes are equal to 1 V/m. The signal to noise ratio is 30 dB and the reference frequency is chosen to be 4 GHz. This configuration and values are used in the following two examples; the difference is in the jammer strength.

7.1 Example 1: Constant Jammers Frequency Components Magnitudes

In the 1st example, all jammer frequency components' magnitudes are 40 dB over the corresponding SOI frequency components. Fig. 2 and Fig. 3 show the estimated

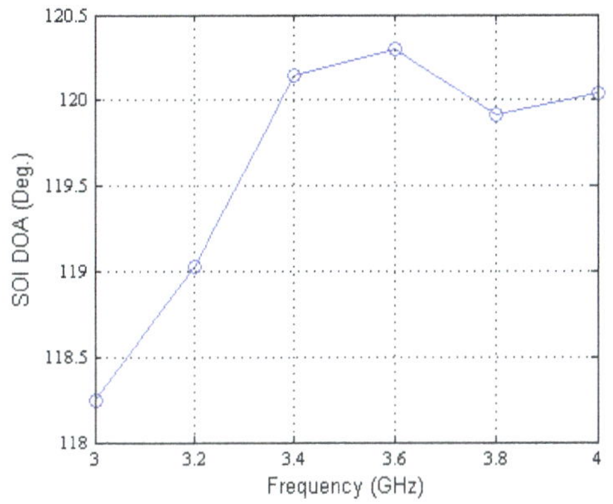

Fig. 2. SOI estimated DOA.

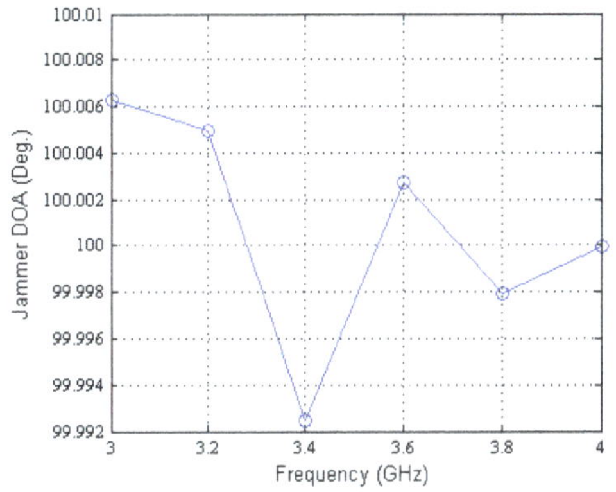

Fig. 3. Jammer estimated DOA.

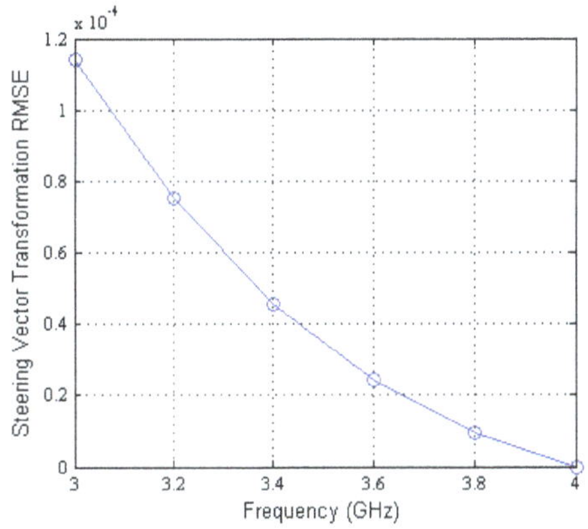

Fig. 4. Steering vector transformation RMSE.

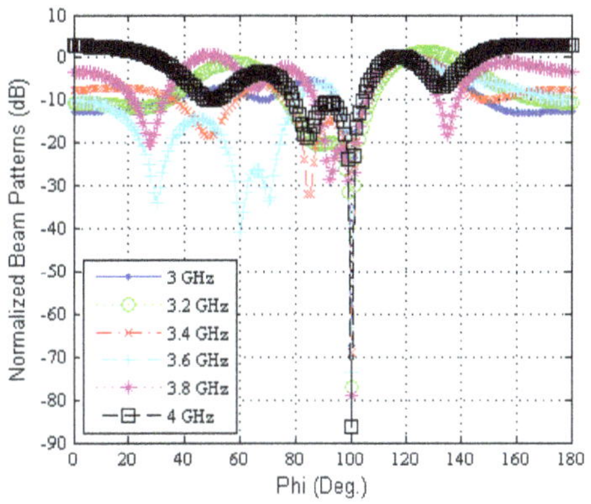

Fig. 5. Normalized array beam patterns.

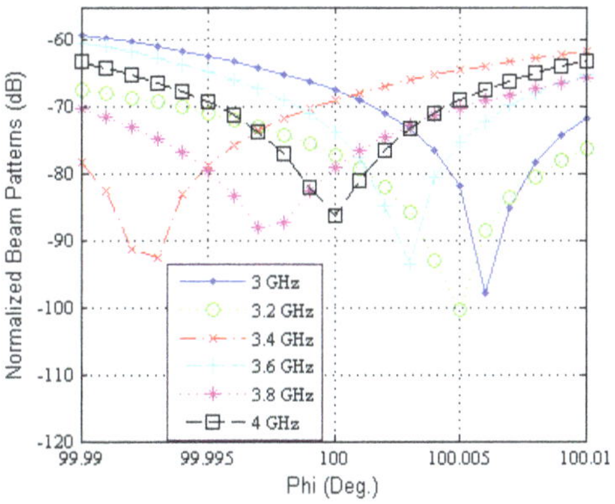

Fig. 6. Normalized detailed nulls' patterns.

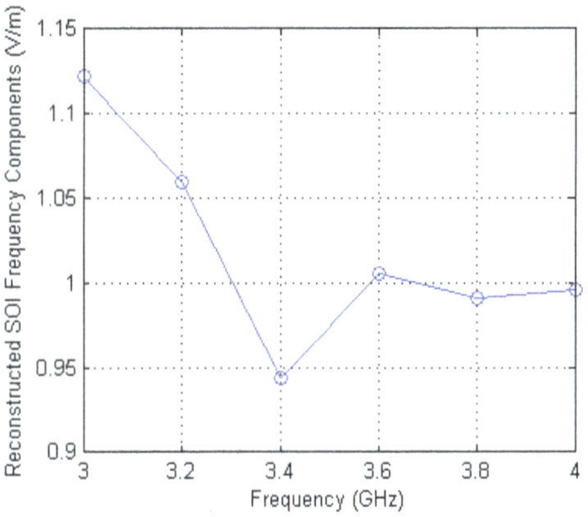

Fig. 7. Reconstructed SOI frequency components (V/m).

DOA for both the SOI and the jammer signal, respectively, corresponding to each frequency sub-band. Obviously, the estimated DOAs become more accurate as the frequency is approaching the reference value, 4 GHz. This could be explained by noticing the decreasing steering vector transformation RMSE values shown in Fig. 4.

Fig. 5 shows six generated beam patterns corresponding to the six sub-bands. It is clear that the array beam patterns are maintained at 0 dB in the direction of the SOI and the nulls are placed correctly in the direction of the jammer. Fig. 6 shows the detailed 'zoomed' nulls' patterns. Since nulls' patterns are very sensitive to the errors in the estimated jammer DOA, the nulls' pattern values in the jammer direction become deeper as the frequency becomes closer to the reference frequency.

Fig. 7 shows the estimated magnitudes of the SOI frequency components using each corresponding beam pattern. Note that the estimation error is large at the lower frequencies and decreases as the reference frequency is approached. To improve the SOI frequency components' reconstruction accuracy, a hybrid reconstruction beam pattern is used. The hybrid pattern combines the most accurate estimated DOAs, i.e. DOAs estimated at the reference frequency, and the deepest nulls corresponding to the strongest estimated interferer frequency component since all the jammer frequency components magnitudes are equal, the already formed beam pattern corresponding to the highest frequency, f_o, is used as a single reconstruction beam pattern by which all the SOI frequency components' magnitudes are accurately estimated as shown in Fig. 8.

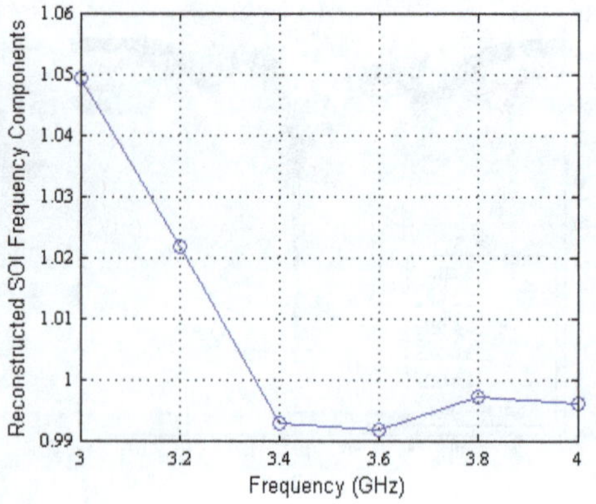

Fig. 8. Reconstructed SOI frequency components (V/m).

7.2 Example 2: Varying Jammers' Frequency Components Magnitudes

In the 2nd example, all jammer frequency components' magnitudes are 60 dB over the corresponding SOI frequency components except the last two jammer frequency components, i.e. at 3.8 GHz and 4 GHz, which are only 40 dB over the corresponding SOI frequency components' strengths. In this case, the beam pattern corresponding to the highest frequency sub-band has a null in the jammer direction but its depth is not sufficient to cancel the high jammer frequency components' strengths at the other sub-bands. Fig. 9 and Fig. 10 show the beam pattern corresponding to the highest frequency sub-band and the corresponding estimated SOI frequency components respectively. In Fig. 10, it is obvious that the estimations at the last two frequencies are acceptable but the estimations at the other frequencies, whereby the jammer frequency components' strengths are high, are not acceptable. These bad estimations could be improved by using a hybrid reconstruction beam pattern which combines the accurate DOA estimations at the reference frequency and the null depth corresponding to the strongest estimated jammer frequency component. The hybrid reconstruction pattern is shown in Fig. 11 and the corresponding estimated SOI frequency components are shown in Fig. 12. The improvement in the SOI frequency components estimations is clear as the null shown in Fig. 11 is deep enough to eliminate the strong jammer frequency components efficiently.

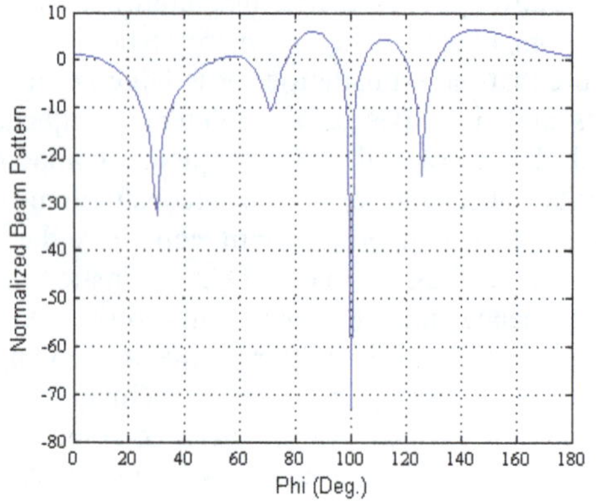

Fig. 9. Normalized beam pattern.

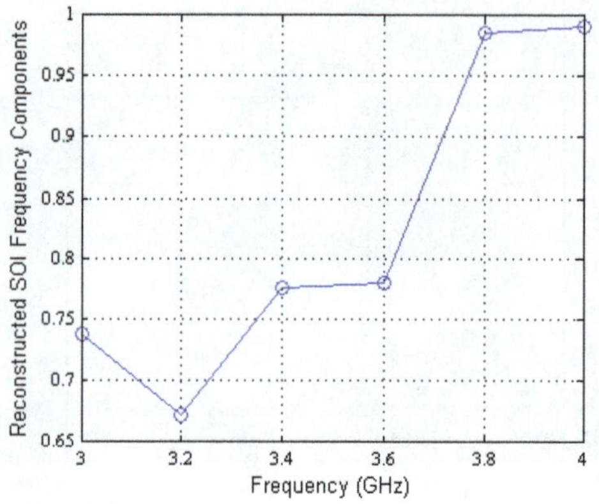

Fig. 10. Reconstructed SOI frequency components (V/m).

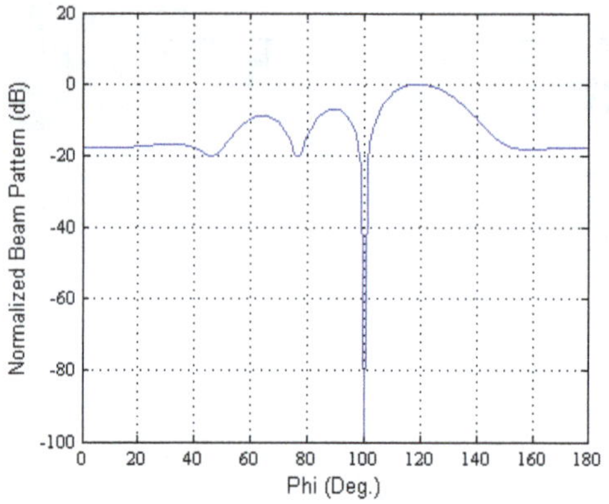

Fig. 11. Hybrid reconstruction beam pattern.

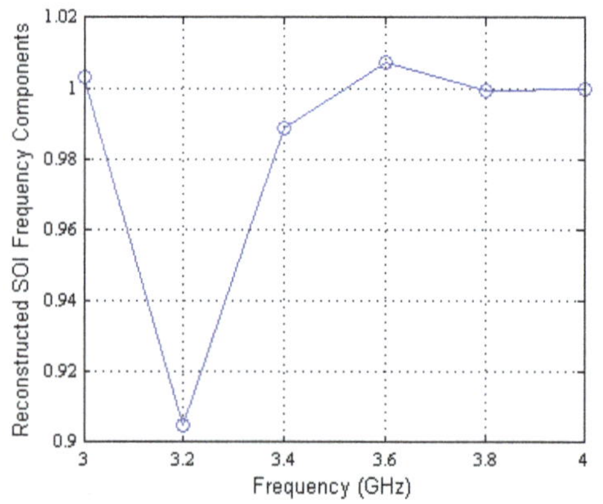

Fig. 12. Reconstructed SOI frequency components (V/m).

8. Conclusion

This paper presents a new wideband direct data domain genetic algorithm (WD^{3}GA) beamforming. The proposed beamformer is based on decomposing the received wideband signal into its narrow sub-bands using FFT. The beamformer is using a single hybrid reconstruction beam pattern which combines both sufficient deep nulls and accurate DOA estimations.

References

[1] ELKAMCHOUCHI, H., MOHAMED, D., ALI, W. D^{3}LS STAP approach on wideband signals using uniformly spaced real

elements. *International Journal of Computer Applications,* 2011, vol. 22, no. 4, p. 42–47. DOI: 10.5120/2568-3530

[2] SVENDSEN, A., GUPTA, I. The effect of mutual coupling on the nulling performance of adaptive antennas. *IEEE Antennas and Propagation Magazine,* June 2012, vol. 54, no. 3, p. 17–38. DOI: 10.1109/MAP.2012.6293947

[3] ADVE, R., SARKAR, T. Compensation for the effects of mutual coupling on direct data domain adaptive algorithms. *IEEE Transactions on Antennas and Propagation*, January 2000, vol. 48, no. 1, p. 86–94. DOI: 10.1109/8.827389

[4] SARKAR, T., WICKS, M., SALAZAR-PALMA, M., BONNEAU, R. *Smart Antennas*. Wiley-IEEE Press, 2003.

[5] SHABAN, M., KISHK, S. Steering vector transformation technique for the design of wideband beamformer. In *The Proceeding of 27th National Radio Science Conference*. Menouf (Egypt), 2010.

[6] LIU, W., WEISS, S. *Wideband Beamforming Concepts and Techniques.* Wiley (Wiley Series on Wireless Communication and Mobile Computing), 2010. 302 p. ISBN: 978-0-470-71392-1

[7] ELLATIF, W. *Smart Antennas: Space Time Adaptive Processing Based on Direct Data Domain Least Squares Using Real Elements. Doctoral Dissertation.* Electrical Engineering, Alexandria University, 2011.

[8] SARKAR, T., PEREIRA, O. Using the matrix pencil method to estimate the parameters of a sum of complex exponentials. *IEEE Antennas and Propagation Magazine*, May 1995, vol. 37, no. 1, p. 48–55. DOI: 10.1109/74.370583

[9] ELKAMCHOUCHI, H., HASSAN, M. Space time adaptive processing using real array elements based on direct data domain adaptive nulls genetic algorithm beam forming. In *Proceedings of the International Conference on Electronics and Communication System*. Coimbatore (India), 2014, vol. 2, p. 183–187. DOI: 10.1109/ECS.2014.6892563

[10] HASSAN, M. A *Proposed Fault Identification Scheme in Systems Using Soft Computing Methodologies. M.Sc. Dissertation.* Electrical Engineering, Alexandria University, 2008.

[11] *Genetic Algorithm and Direct Search Tool Box User's Guide*. The MathWorks, Inc., 2006.

[12] LIU, W., WEISS, S., HANZO, L. A generalized side lobe canceller employing two-dimensional frequency invariant filters. *IEEE Transactions on Antennas and Propagation,* 2005, vol. 53, no. 7, p. 2339–2343. DOI: 10.1109/TAP.2005.850759

[13] LIU, W., WEISS, S. A new class of broad arrays with frequency invariant beam patterns. In *Proceeding of International Conference on Acoustics, Speech, and Signal Processing ICASSP 2004.* Montreal (Canada), 2004, vol. 2, p. 185–188. DOI: 10.1109/ICASSP.2004.1326225

[14] MOGHADDAM, P., AMINDAVAR, H. Direction of arrival estimation: a new approach. In *Proceeding of Signal Processing Conference.* Nordic (Sweden), 2000.

[15] MONTHIPPA, U., BIALKOWSKI, M. E. A wideband smart antenna employing spatial signal processing. *Journal of Telecommunication and Information Technology*, 2007, no. 1, p. 13–17.

[16] REN, Y., HE, H., ZHANG, Y., ZHANG, K. An amplitude-only direct data domain least square algorithm of wideband signals based on the uniform circular array. In *Proceeding of the 5th International Conference on Wireless Communications, Networking and Mobile Computing (WiCom 2009).* Beijing (China), 2009, p. 1–4. DOI: 10.1109/WICOM.2009.5302753

[17] MANI, V.V.; BOSE, R. Genetic algorithm based smart antenna design for UWB beamforming. In *IEEE International Conference on Ultra-Wideband ICUWB 2007.* Singapore, 2007, p. 442–446. DOI: 10.1109/ICUWB.2007.4380985

About the Authors ...

Hassan ELKAMCHOUCHI was born in Egypt, 1943. He received his Ph.D. degree in Communication Engineering in 1972 from Alexandria University. He is now a professor for Alexandria University, Egypt. His research interests include adaptive antennas, optimization algorithms, radars, electromagnetic theory, cryptography and digital communication.

Mohamed HASSAN was born in Egypt. He received his M.Sc. from Alexandria University in 2008. His research interests include adaptive antennas, optimization algorithms, radars, electromagnetic theory and modeling.

17

Electromagnetic Scattering and Statistic Analysis of Clutter from Oil Contaminated Sea Surface

Cong-hui QI, Zhi-qin ZHAO

School of Electronic Engineering, University of Electronic Science and Technology of China, Xiyuan Ave 2006, Chengdu, Sichuan, China

qiconghui0826@163.com, zqzhao@uestc.edu.cn

Abstract. *In order to investigate the electromagnetic (EM) scattering characteristics of the three dimensional sea surface contaminated by oil, a rigorous numerical method multilevel fast multipole algorithm (MLFMA) is developed to preciously calculate the electromagnetic backscatter from the two-layered oil contaminated sea surface. Illumination window and resistive window are combined together to depress the edge current induced by artificial truncation of the sea surface. By using this combination, the numerical method can get a high efficiency at a less computation cost. The differences between backscatters from clean sea and oil contaminated sea are investigated with respect to various incident angles and sea states. Also, the distribution of the sea clutter is examined for the oil-spilled cases in this paper.*

Keywords

Oil contaminated sea, sea clutter, MLFMA, resistive loading

1. Introduction

Detection of oil spills is a long-term objective in sea remote sensing [1], [2]. The oil contaminated sea water is a threat to the marine environment. The vertical polarization of high frequency surface wave radar (HFSWR) has the feature of small energy attenuation, all weather operating and over-horizon. Most importantly, targets can be detected beyond the range of visibility. Thus HFSWR can be used to detect oil spills in far distance. When the sea water is contaminated by oil, the sea dynamics as well as the surface tension are changed [3]. Accordingly, the EM scattering characteristics from the oil contaminated sea surface are changed, such as radar echo and sea clutter distributions. The difference between backscatters from clean sea and oil contaminated sea is a useful feature and desirable in Synthetic Aperture Radar (SAR) imagery simulation of ocean scene which is often used to detect oil pollution on the ocean surface. Therefore, how to quantify the scattering differences is an open problem for the radar engineering.

Many researches have been made to examine the EM backscatter from the oil contaminated sea surface, but most of them are limited to two dimensional (2D) sea surface or focus on approximate method [3], [4]. However, the hypothetical 2D model lames their applications to three dimensional (3D) cases in practical engineering. Meanwhile, the oil contaminated sea surface is a two-layered dielectric problem. Therefore, the simulation in computational electromagnetics (CEM) becomes much complicate than the homogenous dielectric medium case, and needs further researches. It is our motivation.

In this paper, the electric field integral equation (EFIE) on the interfaces is derived using the wave equation combining the boundary condition. In order to exactly simulate the EM backscatter from the sea surface, the rigorous numerical multilevel fast multipole algorithm (MLFMA) [5], [6] is applied to obtain reliable results. In order to accelerate the EM simulation process when using MLFMA at low-grazing-angle (LGA) cases, a combination model of Thorsos illumination window [7] and resistive loading window [8] is applied to avoid the edge effect induced by artificial truncation of the sea surface in range and azimuth directions. These treatments can greatly improve the computation efficiency thus the simulation of the sea clutter is possible using relatively less computation resources. It is noted that this proposed method is more universal than these existing analytical methods, and unlimited by the 3D sea states or larger incident angles.

This rigorous numerical method and its treatments make the Monte Carlo simulation at LGA incidence possible. Then, the electromagnetic backscatter from the oil contaminated sea surfaces is computed with the variation of incident angles and sea states. As a comparison, the backscatter reduction and the statistic characteristics of the two kinds of sea clutter are analyzed compared with the clean sea surface. Some interesting conclusions are summarized in our experiments.

2. Sea Model

The sea surface can be generated using a spectral method which considers it as a superposition of harmonics.

The amplitudes of the harmonics are proportion to a certain wind-dependent surface-roughness spectrum $W(K, \varphi)$. Then the sea surface can be obtained by inverse fast Fourier transform (IFFT). In this paper, the 2D Pierson-Moskowitz (PM) spectrum is adopted to generate the 3D sea surface [9]. The PM spectrum is defined by

$$W(K_w, \varphi) = \frac{\alpha}{2K_w^4}\exp\left\{-\frac{\beta g^2}{K_w^2 U^4}\right\}\cos^4\left(\frac{\varphi-\varphi_w}{2}\right) \quad (1)$$

where K_w is the spatial wavenumber, U is the wind speed at the height of 19.5 m above the sea surface, two constants $\alpha = 8.1 \times 10^{-3}$ and $\beta = 0.74$. The angle φ is measured in the horizontal x-y plane with respect to the x-axis and φ_w is the wind direction. In our simulation, the wind direction is towards positive x-axis.

When the sea surface is contaminated by oil, the sea surface tension as well as the sea movement is changed. In the research of Lombardini et al. [10], it is demonstrated that the height of oil contaminated sea surface is damped comparing with the clean sea surface. This damping effect can be expressed by an attenuation coefficient, called Marngoni viscous damping coefficient [11]. In this paper, the oil layer thickness of the contaminated sea is 1 mm.

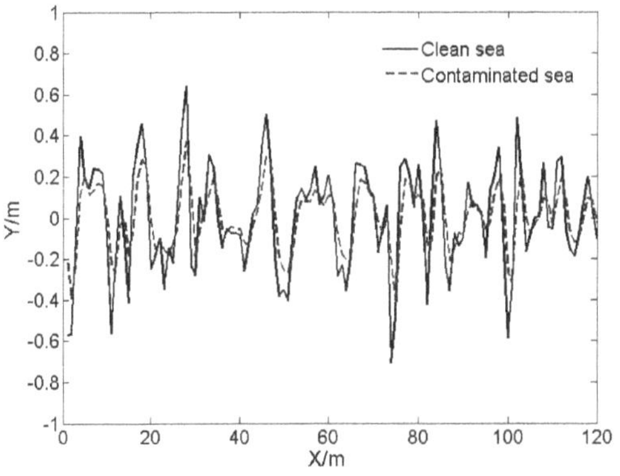

Fig. 1. Comparison of clean sea surface and contaminated sea surface.

Figure 1 shows the comparison of the geometrical profile between the clean sea surface and the contaminated sea surface. The simulation is performed at the wind speed of $U = 7$ m/s which is considered as the average wind speed over the ocean. It is observed that the surface heights of the contaminated sea are indeed damped comparatively to the clean sea surface. Moreover, the small capillarity-scale variations are strongly damped, which implies a strong damping of the surface slopes.

3. EM Modeling

Clean sea water is homogeneous dielectric medium with a relative permittivity ε_1 and permeability μ_1. However, when the sea water is contaminated by oil whose relative permittivity is ε_2 and permeability is μ_2 (generally, $\mu_1 = \mu_2 = 1$), the contaminated sea surface becomes a two-layered dielectric material with its thickness d, as shown in Fig. 2.

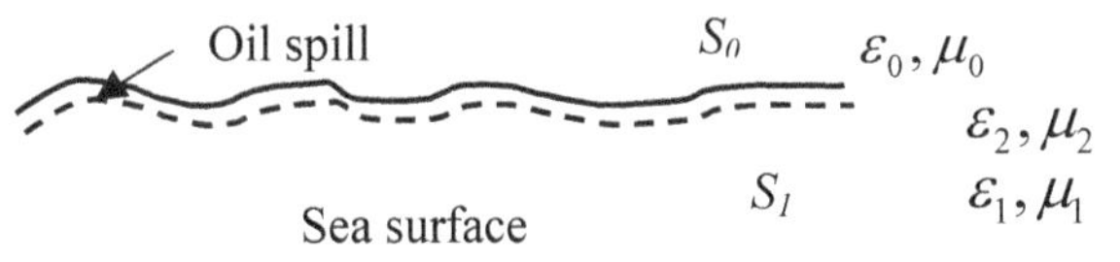

Fig. 2. Geometry of two-layered medium.

Generally, the HFSW can work at LGA for some extreme cases. Therefore, the multipath effect is complicated for EM simulation. In this section, the rigorous numerical method MLFMA and its treatments are proposed to get the backscatters from the two-layered medium problem. The computation efficiency can be greatly improved, thus makes the Monte Carlo simulation of the sea surface possible.

3.1 MLFMA

Using the wave equation and the boundary conditions, EFIE can be derived as functions of the electric equivalent electric current $\boldsymbol{J}$ and equivalent magnetic current $\boldsymbol{M}$ which exist on the interface S_0 between the oil and air as well as the interface S_1 between oil and water. The two sets of surface currents on the boundaries S_0 and S_1 ($\boldsymbol{J}_0$, $\boldsymbol{M}_0$, $\boldsymbol{J}_1$, $\boldsymbol{M}_1$) can be determined using the following numerical simulation procedure.

$$-ik_0 Z_0\int_{S_0}\left\{\boldsymbol{J}_0 G_0 + \frac{1}{k_0^2}\nabla'\cdot\boldsymbol{J}_0\nabla G_0\right\}ds' + \oint_{S_0}\boldsymbol{M}_0\times\nabla G_0 ds' - \frac{1}{2}\boldsymbol{M}_0 = -E^i, \bar{r}\in S_0 \quad (2.a)$$

$$ik_1 Z_1\int_{S_0}\left\{\boldsymbol{J}_0 G_1 + \frac{1}{k_1^2}\nabla'\cdot\boldsymbol{J}_0\nabla G_1\right\}ds' - \oint_{S_0}\boldsymbol{M}_0\times\nabla G_1 ds' - ik_1 Z_1\int_{S_1}\left\{\boldsymbol{J}_1 G_1 + \frac{1}{k_1^2}\nabla'\cdot\boldsymbol{J}_1\nabla G_1\right\}ds' + \oint_{S_1}\boldsymbol{M}_1\times\nabla G_1 ds' - \frac{1}{2}\boldsymbol{M}_0 = 0, \bar{r}\in S_0 \quad (2.b)$$

$$ik_1 Z_1\int_{S_0}\left\{\boldsymbol{J}_0 G_1 + \frac{1}{k_1^2}\nabla'\cdot\boldsymbol{J}_0\nabla G_1\right\}ds' - \oint_{S_0}\boldsymbol{M}_0\times\nabla G_1 ds' - ik_1 Z_1\int_{S_1}\left\{\boldsymbol{J}_1 G_1 + \frac{1}{k_1^2}\nabla'\cdot\boldsymbol{J}_1\nabla G_1\right\}ds' + \oint_{S_1}\boldsymbol{M}_1\times\nabla G_1 ds' - \frac{1}{2}\boldsymbol{M}_1 = 0, \bar{r}\in S_1 \quad (2.c)$$

$$-ik_2 Z_2\int_{S_0}\left\{\boldsymbol{J}_1 G_2 + \frac{1}{k_2^2}\nabla'\cdot\boldsymbol{J}_1\nabla G_2\right\}ds' + \oint_{S1}\boldsymbol{M}_1\times\nabla G_2 ds' - \frac{1}{2}\boldsymbol{M}_1 = 0, \bar{r}\in S_1 \quad (2.d)$$

where $\oint_{S_0} ds'$ denotes principle integration, Z_n is the characteristic impedance, k_n is the wave number and

$G_n = g_n(r,r') = 4\pi e^{ik_n|r-r'|}/|r-r'|$ is the Green function. The subscript n = 1, 2, 3 denotes the three cases in free space, sea water and oil medium, respectively.

Here the equivalent impedance boundary condition is used to deal with the boundary condition scattering problem [12]. It means the electric current $\boldsymbol{J}$ and equivalent magnetic current $\boldsymbol{M}$ satisfy $n' \times \boldsymbol{M}_0 = Z_1 \boldsymbol{J}_0$ and $n' \times \boldsymbol{M}_1 = Z_2 \boldsymbol{J}_1$, where n' is the outer normal vector at each field point. Therefore, the unknown number in (2) will be halved. The unknown equivalent electric current can be obtained by the Galerkin's method in method of moment (MoM), as following.

$$\begin{bmatrix} [Z_{11}^{nm}] & [Z_{12}^{nm}] & 0 & 0 \\ [Z_{21}^{nm}] & [Z_{22}^{nm}] & [Z_{23}^{nm}] & [Z_{24}^{nm}] \\ [Z_{31}^{nm}] & [Z_{32}^{nm}] & [Z_{33}^{nm}] & [Z_{34}^{nm}] \\ 0 & 0 & [Z_{43}^{nm}] & [Z_{44}^{nm}] \end{bmatrix} \bullet \begin{bmatrix} [I_1^n] \\ [I_2^n] \\ [I_3^n] \\ [I_4^n] \end{bmatrix} = \begin{bmatrix} [V_1^m] \\ [0] \\ [0] \\ [0] \end{bmatrix} . \quad (3)$$

To investigate the statistic characteristics of the backscatter of the sea surface, Monte Carlo simulation is necessary. However, it is rather computationally expensive and time-consuming. In order to accelerate the computing process, MLFMA is used to solve the above equation. In the process of MLFMA, the interactions between the elements are classified as near-region and the far-region. The near-region matrix elements are calculated directly using MoM, while the far-region elements are acquired by using MLFMA.

$$\sum_{l} Z^{nm} I^n = \sum_{l \in NR} Z^{nm} I^n + \sum_{l \in FR} Z^{nm} I^n \quad (4)$$

where NR represents the near-region and FR represents the far-region, l means the basis function, I^n are the expansion coefficients needed to solve using MoM, and Z^{nm} are the elements of the impedance matrix. More details can be referenced in [5].

3.2 Treatments

Numerical methods can guarantee sufficient precision when solving EM computing problems even at LGA. However, it will bring edge diffraction which is induced by the artificial truncation to the sea surface. The electric current $\boldsymbol{J}$ will change suddenly at the edge of the truncated surface. This phenomenon is called edge effect. In the research of EM scattering in those numerical methods, how to avoid the edge effect is an important issue. The most popular way to suppress the edge effect is using the illumination tapered window (Thorsos window) [7] which is got by modulating a tapered function to incident electric field, as following.

$$p(x,z) = e^{-j\frac{2(x - z\tan\theta_i)^2/g^2 - 1}{(kg\cos\theta_i)^2}\hat{k}\hat{k}\cdot r'} e^{-\frac{(x - z\tan\theta_i)^2}{g^2}} \quad (5)$$

where g is a constant that controls the width of the illumination beam, θ_i is the incident angle.

However, according to the studies by Jin [13], the length L of rough surface has to satisfy the condition: $L \geq 24\lambda/(\cos\theta_i)^{1.5}$, where λ is the EM wavelength. Accordingly, at LGA the scale of the computing sea surface is very large. For example, if $\theta_i = 85°$, then $L \geq 932\lambda$. And it leads to a large amount of unknown number on the sea surface. Therefore, instead of the illumination window, the resistive loading window is applied to suppress the edge effect under the plane wave illumination. It can be accomplished by simply adding $\boldsymbol{RJ}$ to the EFIE formula, taking (2a) as an example, as following

$$\overline{\overline{L}}(\boldsymbol{J}) + R\boldsymbol{J} = -E^i, \bar{r} \in S_0 \quad (6)$$

where R is the resistance, and $\overline{\overline{L}}(\cdot)$ is the matrix-vector multiply operator and denotes the procedure in left side of (2a). The resistance is loaded within the resistive loading area, which was discussed in detail in reference [14]. Once the resistive loading method is used, the length L of sea surface only needs to be as large as 10 times of the correlation length of the rough surface. It is much smaller than the condition of Thorsos window case.

However, it is found that the resistive loading will bring the convergence problem because that the impedance matrix Z^{nm} is a non-diagonally dominant matrix for this two-layered dielectric medium scattering problem in our simulations. Since the resistance R will attenuate the value of self-impedance element, the condition number of the final discretized impedance matrix will become worse. Therefore, more computational CPU time will be taken in the impedance matrix iteration processes.

A method combining the two different windows is proposed to accelerate its convergence. Resistive loading window is applied in range direction while a Gaussian illumination window is used in azimuth direction. In azimuth direction, the incident plane wave is modulated by a Gaussian function $p(y) = e^{-y^2/g^2}$ instead of the Thorsos window. Meanwhile, an important work in range direction is to extend the original sea surface with smoothly curved sections (for example, we used radius of 10λ) that join to planar sections which are angled 30° down from the horizon. So that all the points on the extension surfaces are shadowed by the original rough surface. The resistive loading of the edges has the advantage of not requiring the surface to be modified at LGA.

As a comparison, a 600 m (L)×600 m (W) PM sea surface realization at U = 3 m/s was generated and simulated under those system parameters: radar frequency 15 MHz, incident angle 60°, vertical polarization. As shown in Fig. 4, the convergence iteration is 228 steps for the traditional resistive window [8] in range and azimuth directions whereas only 50 steps are needed for the proposed combination of the two different windows.

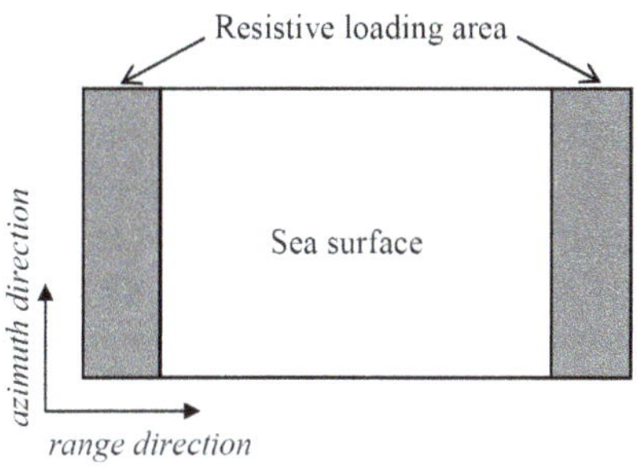

Fig. 3. The scheme of two windows.

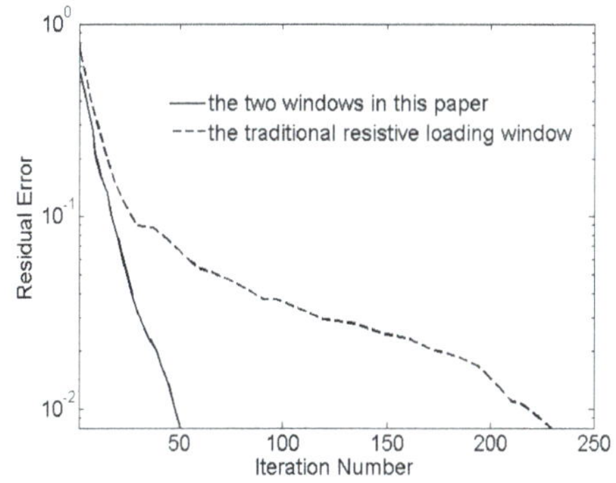

Fig. 4. Convergence comparison.

Therefore, the proposed method to suppress the edge effect is more effective and can greatly reduce the CPU time, which is a big improvement in Monte Carlo simulation.

4. Numerical Simulations

4.1 Comparison of Backscatters

By using this proposed method, some experiments are presented in this section. The HFSWR operates at the frequency of 15 MHz, vertical polarization. Monte Carlo simulation is applied to get average radar backscatter. A total number of 1000 sea surfaces are generated for each sea state. The simulation sea area is 600 m long with a width of 600 m. The simulation is performed when wind speed is 3 m/s and 9 m/s, corresponding to Douglas sea state 2 and 4.

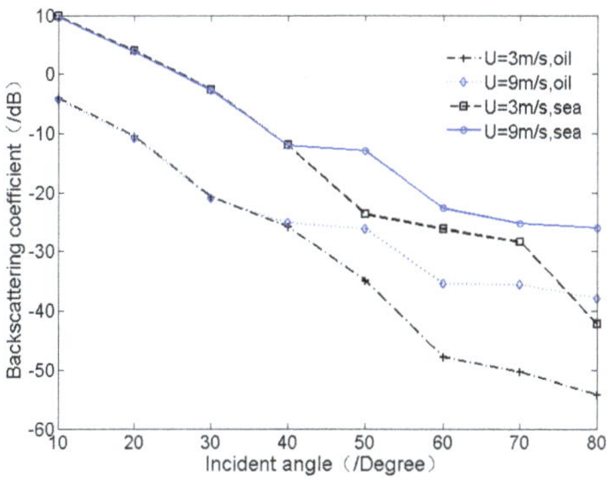

Fig. 5. Backscatter coefficients with various wind speeds.

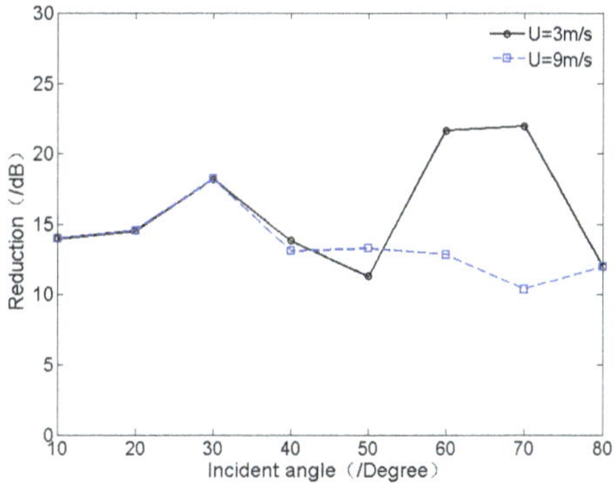

Fig. 6. Backscatter coefficient reduction.

As shown in Fig. 5, the radar backscatters of the contaminated sea surface is much lower than that of the clean sea surface. The reduction of the backscatter coefficient corresponding to different sea states is shown in Fig. 6. A reduction of more than 10 dB at moderate incident angles is found through our simulation, and even 15 dB at LGA angle. In SAR systems, the oil polluted area is seen as "dark" sea. Capillary waves on the sea surface reflect electromagnetic energy. However, when the oil layer on the sea surface, the capillary waves are dampened. As a consequence, the backscatter is much weaker than clean sea surface.

4.2 Sea Clutter Distribution

By using the radar echo data acquired in the above experiments, the distribution of the sea clutter is also examined. For radar system, statistical models (Rayleigh, Lognormal, Weibull and K-distributions) are often used to describe the sea clutter. In this paper, the probability density functions (PDF) of the backscatter of the sea clutter for clean sea and oil-contaminated sea are studied. The parameters of Rayleigh, Lognormal and Weibull distribution are estimated by the maximum likelihood (ML) method. It is a standard approach to parameter estimation and provides optimum estimates in the sense that these estimates are the most probable parameter values. However, ML is computationally expensive to estimate the parameters of K distribution. A new method based on higher order and fractional moments is proposed in these references [15], [16]. It can significantly reduce the computational requirement. Details about the formulas of these four distributions and the estimation methods are summarized in reference [16]. Take this sea state of $U = 9$ m/s as an example, the sea clutter distribution is shown for different incident angles in Fig. 7.

Kolmogorov-Smirnov test (K-S test) is used to find the best fitting distribution of the sea clutter among the four distributions. K-S test has the advantage of making no assumption about the distribution of the sea data. It is nonparametric and distribution free. The K-S test statistic quantifies a distance between the empirical distribution function of the sample and the cumulative distribution function of the reference distribution, or between the

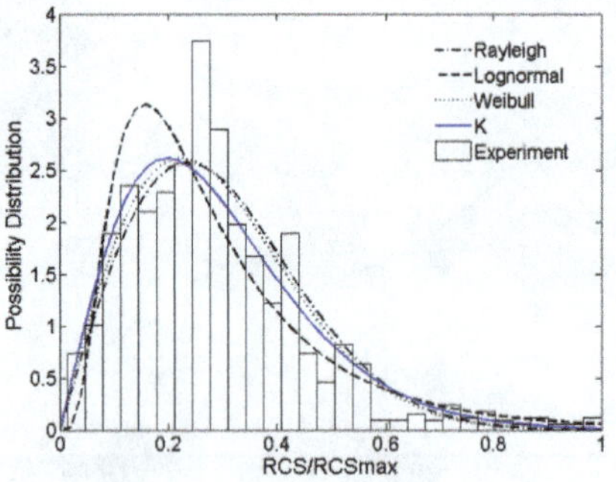

(a) Clean sea surface at incident angle of 30°

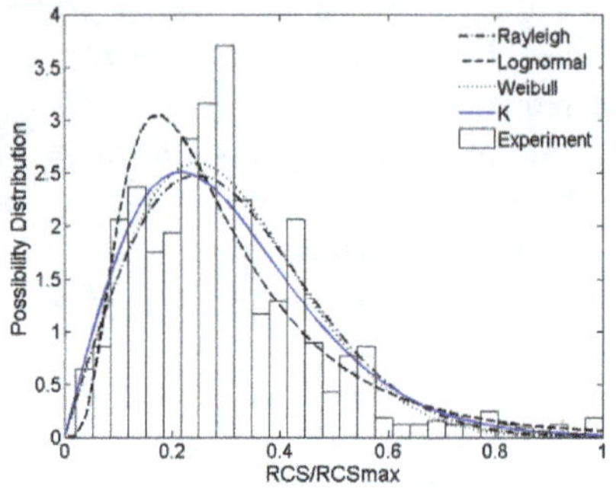

(b) Oil contaminated sea at incident angle of 30°

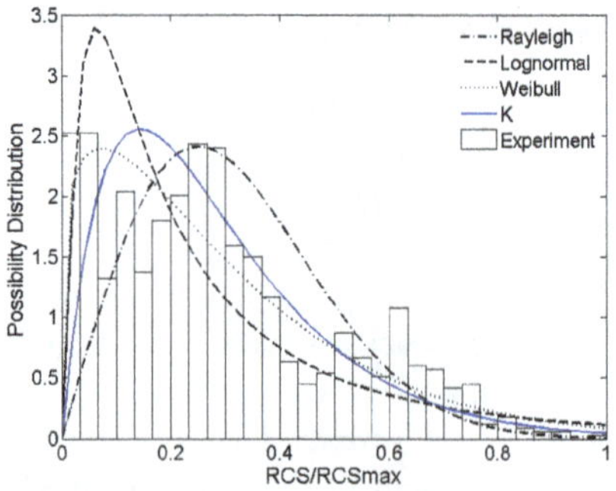

(c) Clean sea at incident angle of 80°

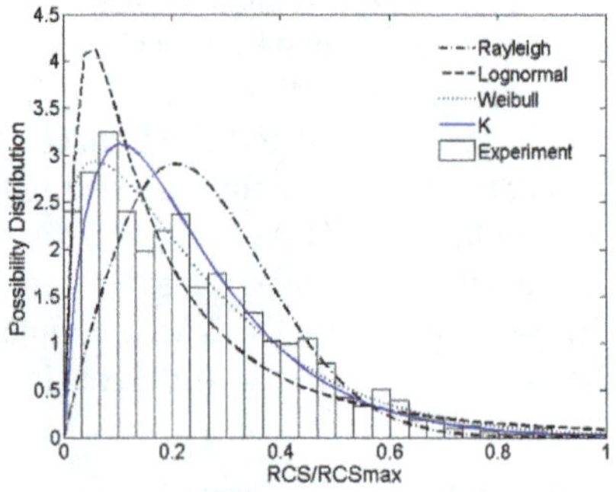

(d) Oil contaminated sea at incident angle of 80°

Fig. 7. The backscatter distribution of sea clutter.

empirical distribution functions of two samples. The K-S test statistic is defined as *De* which is shown

$$De = \max_{1\le i\le N} \left| F(x_i) - F_t(x_i) \right| = \max_{1\le i\le N} \left| n/N - F_t(x_i) \right|. \quad (7)$$

$F_t(x_i)$ is the cumulative distribution of the distribution being tested and $F(x_i)$ is the cumulative distribution of the simulated sea clutter. *De* means the maximum deviation between the cumulative distribution of the tested distribution and the statistic distribution of the simulation result. The best fit distribution is with the smallest De. Tab. 1 is the K-S test of the sea clutter distribution corresponding to Fig. 7. And the sea clutter distributions are also examined under different sea states in Tab. 2 and Tab. 3.

De	Rayleigh	Lognormal	Weibull	*K*
Sea /30°	**0.0419**	0.1140	0.0451	0.0597
Sea /80°	0.1281	0.1446	**0.0614**	0.0914
Oil /30°	0.0493	0.1156	**0.0444**	0.0609
Oil/ 80°	0.1530	0.1106	**0.0397**	0.0566

Tab. 1. K-S test.

Incident angle	10°	20°	30°	40°	50°	60°	70°	80°
State 2 (3 m/s)	L	K	K	L	W	W	L	W
State 4 (9 m/s)	K	R	W	W	W	W	W	W

Tab. 2. The distribution of sea clutter for contaminated sea.

Incident angle	10°	20°	30°	40°	50°	60°	70°	80°
State 2 (3 m/s)	K	K	K	L	W	L	L	W
State 4 (9 m/s)	K	K	R	W	W	W	W	W

Tab. 3. The distribution of sea clutter for clean sea.

As can be seen from Tab. 2 and Tab. 3, the distribution of clutter from oil contaminated sea surface is different from that of clean sea surface. For most cases, the sea clutter satisfies K distribution and Weibull distribution. When the wind speed above the sea surface is moderate (3 m/s), K distribution describes the sea clutter quite well at small incident angle. As the incident angle getting large, it is Weibull distribution for the contaminated sea whereas it is Lognormal distribution for the clean sea. When the wind speed gets higher (9 m/s), Weibull distribution will be dominant at LGA. The distribution of clutter from oil contaminated sea is first mentioned in this paper. Rigorous numerical EM simulation method can ensure sufficient accuracy. Moreover, with the treatments this method can obtain a high efficiency. Consequently, Monte Carlo simulation can be performed on the electric large problem. The real radar data from HFWSR system is usually very expensive. Instead of the real data from the radar, the simulation results by using rigorous numerical computational electromagnetic method as shown in this paper is workable as well. Moreover, it is more cost-effective and available.

5. Conclusions

In this paper, the sea clutter distribution is examined when there is oil pollution on the sea surface. In HFSWR system, the observation area is very large; accordingly the accurate EM simulation of the large sea surface is rather difficult. The integral equations both on the clean sea surface and the oil contaminated sea surface are presented. Numerical method with treatments on it can ensure sufficient accuracy to get reliable results of the backscatter from sea surface. Meanwhile, the computational resources are saved a lot, and then Monte Carlo simulation can be performed. It is found that the backscatter from oil contaminated sea surface is much weaker than that from clean sea surface. That is because the oil damped the capillary waves. Accordingly, the reflected electromagnetic energy is weaker than before. This effect can be seen in SAR image that the oil contaminated area is dark. The distribution of the sea clutter which is an important statistic characteristic is investigated by using the simulated results. Differences of the sea clutter distribution are found between the oil contaminated sea surface and the clean sea surface.

Acknowledgements

This work was supported by the National Natural Science Foundation of China under grant 61171044; the Fundamental Research Funds for the Central Universities under grant ZYGX2012YB006; and Research Fund for the Doctoral Program of Higher Education of China under grant 20120185110024.

References

[1] SOLBERG, A. H. S. Remote sensing of ocean oil-spill pollution. *Proceedings of the IEEE*, 2012, vol. 100, no. 10, p. 2931–2945. DOI: 10.1109/JPROC.2012.2196250

[2] FINGAS, M., BROWN, C. Review of oil spill remote sensing. *Marine Pollution Bulletin*, 2014. DOI: 10.1016/S1353-2561(98)00023-1

[3] PINEL, N., DECHAMPS, N., BOURLIER, C. Modelling of the bistatic electromagnetic scattering from sea surfaces covered in oil for microwave applications. *IEEE Transactions on Geoscience and Remote Sensing*, 2008, vol. 46, no. 2, p. 385–392. DOI: 10.1109/TGRS.2007.902412

[4] GHANMI, H., KHENCHAF, A., COMBLET, F. Numerical modeling of electromagnetic scattering from sea surface covered by oil. *Journal of Electromagnetic Analysis and Applications*, 2014, vol. 6, p. 15–24. DOI: 10.4236/jemaa.2014.61003

[5] SONG, J., LU, C.-C., CHEW, W. C. Multilevel fast multipole algorithm for electromagnetic scattering by large complex objects. *IEEE Transactions on Antennas and Propagation*, 1997, vol. 45, no. 10, p. 1488–1493. DOI: 10.1109/8.633855

[6] YANG, W., ZHAO, Z., QI, C., NIE, Z. Electromagnetic modeling of breaking waves at low grazing angles with adaptive higher order hierarchical Legendre basis functions. *IEEE Transactions on Geoscience and Remote Sensing*, 2011, vol. 49, no. 1, p. 346–352. DOI: 10.1109/TGRS.2010.2052817

[7] THORSOS, E. I. The validity of the Kirchhoff approximation for rough surface scattering using a Gaussian roughness spectrum. *Journal of the Acoustical Society of America*, 1988, vol. 83, p. 78 to 92. DOI: DOI: 10.1121/1.396188

[8] ZHAO, Z., WEST, J. C. Resistive suppression of edge effects in MLFMA scattering from finite conductivity surfaces. *IEEE Transactions on Antennas and Propagation*, 2005, vol. 53, p. 1848–1852. DOI: 10.1109/TAP.2005.846810

[9] LI, X., XU, X. Scattering and Doppler spectral analysis for two-dimensional linear and nonlinear sea surfaces. *IEEE Transactions on Geoscience and Remote Sensing*, 2011, vol. 49, p. 603–611. DOI: 10.1109/TGRS.2010.2060204

[10] LOMBARDINI, P., FISCELLA, B., TRIVERO, P., CAPPA, C., GARRETT, W. Modulation of the spectra of short gravity waves by sea surface films: slick detection and characterization with a microwave probe. *Journal of Atmospheric and Oceanic Technology*, 1989, vol. 6, p. 882–890. DOI: 10.1175/1520-0426(1989)006<0882:MOTSOS>2.0.CO;2

[11] GADE, M., ALPERS, W., HUHERFUSS, H., WISMANN, V. R., LANGE, P. A, On the reduction of the radar backscatter by oceanic surface films: Scatterometer measurements and their theoretical interpretation. *Remote Sensing of Environment*, 1998, vol. 66, p. 52–70. DOI:10.1016/S0034-4257(98)00034-0

[12] GLISSON A. W. Electromagnetic scattering by arbitrarily shaped surface with impedance boundary conditions. *Radio Science*, 1992, vol. 27, no. 6, p. 935–943. DOI: 10.1029/92RS01782

[13] LIU, P., JIN, Y. Q. Numerical simulation of bistatic scattering from a target at low altitude above rough sea surface under an EM-wave incidence at low grazing angle by using the finite element method. *IEEE Transactions on Antennas and Propagation*, 2004, vol. 52, no. 5, p. 1205–1210. DOI: 10.1109/TAP.2004.827497

[14] ZHAO, Z., WEST, J. C. Resistive treatment of edges in MLFMA LGA scattering from finite conductivity 2D surfaces. In *IEEE Antennas and Propagation Society International Symposium*, 2002, p. 264–267. DOI: 10.1109/APS.2002.1016974

[15] ISKANDER, D. R., ZOUBIR, A. M. Estimation of the parameters of the K-distribution using higher order and fractional moments [radar clutter]. *IEEE Transactions on Aerospace and Electronic Systems*, 1999, vol. 35, p. 1453–1457. DOI: 10.1109/7.805463

[16] FARINA, A., GINI, F., GRECO, M., VERRAZZANI, L. High resolution sea clutter data: statistical analysis of recorded live data. *IEE Proceedings-in Radar, Sonar and Navigation*, 1997, vol. 144, no. 3, p. 121–130. DOI: 10.1049/ip-rsn:19971107

About the Authors ...

Cong-hui QI was born in Hebei, China. She received her M.Sc. from University of Electronic Science and Technology of China in 2009. Her research interests include computational electromagnetic, backscatter from rough surface and SAR image.

Zhi-qin ZHAO was born in Hunan, China. He received B.S. and M.S. degrees in Electronic Engineering from the University of Electronic Science and Technology of China, Sichuan, and the Ph.D. degree in Electrical Engineering from Oklahoma State University, Stillwater, in 1990, 1993, and 2002, respectively. His research interests include radar signal processing and computational electromagnetics.

A Simplified Scheme of Estimation and Cancellation of Companding Noise for Companded Multicarrier Transmission Systems

Siming PENG[1], *Zhigang YUAN*[1], *Jun YOU*[2], *Yuehong SHEN*[1], *Wei JIAN*[1]

[1]Dept. of Wireless Communications, PLA University of Science and Technology, 210014 Nanjing, China
[2]Dept. of Command Information System, PLA University of Science and Technology, 210014 Nanjing, China

lgdxpsm@gmail.com, yzhigang_cn@163.com, chunfeng22259@126.com

Abstract. *Nonlinear companding transform is an efficient method to reduce the high peak-to-average power ratio (PAPR) of multicarrier transmission systems. However, the introduced companding noise greatly degrades the bit-error-rate (BER) performance of the companded multicarrier systems. In this paper, a simplified but effective scheme of estimation and cancellation of companding noise for the companded multicarrier transmission system is proposed. By expressing the companded signals as the summation of original signals added with a companding noise component, and subtracting this estimated companding noise from the received signals, the BER performance of the overall system can be significantly improved. Simulation results well confirm the great advantages of the proposed scheme over other conventional decompanding or no decompanding schemes under various situations.*

Keywords

Multicarrier transmission systems, peak-to-average power ratio (PAPR), nonlinear companding transform (NCT), companding noise cancellation

1. Introduction

Multicarrier transmission is a promising technique in future communication systems. However, one of the major drawbacks of multicarrier systems is the inherent high peak-to-average power ratio (PAPR) of the transmitted signals. It's known that the efficiency of the high power amplifier (HPA) is directly related to the PAPR of the input multicarrier signals especially in the orthogonal frequency division multiplexing (OFDM) systems, which is applied in many important wireless communication standards such as the Third Generation Partnership Project (3GPP) Long-Term Evolution Advanced (LTE-A) standard [1], [2]. The PAPR problem still prevents OFDM from being adopted in the uplink of wireless communication standards [1].

Up to now, many works have been conducted to deal with this high PAPR problem [3]. Such as the iterative clipping and filtering [4], [5], [6], coding [7], Partial Transmission Sequence (PTS) [8], [9], [10], Selective Mapping (SLM) [11], Tone Reservation (TR) [12], companding transform (CT) [13]–[20] and so on. Among them, the clipping and companding may be the simplest two methods, since they can be employed directly to the multicarrier systems without any restrictions on the number of subcarriers and frame format and so on. However, due to the clipping often introduces serious in-band distortions as well as out-of-band radiation, consequently, the bit-error-rate (BER) performance of the system is greatly degraded. On the contrary, the companding transform can not only achieve more effective PAPR reduction but also better BER performance than the clipping method. Hence, it attracts more and more researchers' attention in recent years.

Since companding transform is an extra predistortion process employed on the original signals, hence, the introduced companding noise may also degrade the system performance to some extend. Conventional decompanding operation can approximately remove the companding noise at the receiver [13], [14], [15], [16], but the channel noise will be amplified by the decompanding function simultaneously, and consequently, the system performance will not be so optimistic especially under low signal-to-noise ratio (SNR) region. In order to avoid amplifying the channel noise caused by the decompanding operations, in [17], the authors proposed to abandon the decompanding operation at the receiver, and although a great BER performance improvement can be achieved, however, there is still a relative large gap of BER performance away from the performance bound. In [19], the authors proposed an iterative receiver to estimate and cancel the companding noise. By referring to the Bussgang theorem, the companding signals are regarded as the summation of a useful attenuated input replica and an uncorrected nonlinear distortion noise, and then removed from the receiver. Although the BER performance can be greatly improved in this scheme, however, due to the inherent complex expression of the companding function, the accurate attenuation coefficient of the input signals will difficultly to be determined, and consequently, greatly restrains the effectiveness for general companding transforms.

In this paper, a simplified scheme of estimation and cancellation of nonlinear companding noise is proposed. By expressing the companded signals as the summation of the companding noise added to the original signals, and then, subtracting the estimated companding noise from the receiver, the above difficulties of analysis and calculation of the attenuation coefficient of the original signals can be well resolved. Moreover, a significant BER performance improvement than conventional operations, such as the decompanding or no decompanding, at the receiver can be achieved simultaneously. It also shows that the presented scheme is robust in various practical situations.

The rest of this paper is organized as follows. In Sec. 2, a typical multicarrier system model is described and the PAPR problem is formulated briefly. The theoretical analysis of the proposed scheme is presented in Sec. 3. In the next section, the overall BER performance of a typical exponential companding (EC) equipped with the proposed companding noise cancellation scheme is evaluated and followed by the conclusion summarized in Sec. 5.

2. System Model

Fig. 1 shows the block diagram of a typical multicarrier transmission system with companding transform. Let us denote the data symbols $X_k, k = 0, 1, \ldots, N-1$, as a vector $\mathbf{X} = [X_0, X_1, \ldots, X_{N-1}]^T$ with N subcarriers, where $(\cdot)^T$ is the matrix transpose operation. The complex baseband representation of a multicarrier signal is given by

$$x(t) = \frac{1}{\sqrt{N}} \sum_{k=0}^{N-1} X_k e^{j2\pi k \Delta f t}, 0 \leqslant t \leqslant NT_s \quad (1)$$

where $j = \sqrt{-1}$, Δf is the subcarrier interval, and NT_s is the useful data block period. In general, the subcarriers are chosen to be orthogonal (i.e. $\Delta f = 1/NT_s$).

Generally, the PAPR of multicarrier signals $x(t)$ is defined as the ratio between the maximum instantaneous power and its average power, i.e.

$$PAPR = \frac{\max\limits_{0 \leqslant t \leqslant NT} \left[|x(t)|^2 \right]}{1/NT \int\limits_0^{NT} |x(t)|^2 dt}. \quad (2)$$

To better approximate the PAPR of continuous-time OFDM signals, the OFDM signals samples are obtained by L times oversampling. L-times oversampled time-domain samples can be achieved by performing a LN-point IFFT of the data block with $(L-1)N$ zero-padding, i.e.

$$\mathbf{X}_p = \left[X_0, \ldots, X_{\frac{N}{2}-1}, \underbrace{0, \ldots, 0}_{(L-1)N}, X_{\frac{N}{2}}, \ldots, X_{N-1} \right]^T. \quad (3)$$

Therefore, the oversampled IFFT output can be expressed as

$$x(n) = \frac{1}{\sqrt{N}} \sum_{k=0}^{N-1} X_k e^{j\frac{2\pi nk}{LN}}, 0 \leq n \leq LN-1. \quad (4)$$

The corresponding PAPR computed from the L-times oversampled time domain OFDM signal sample is defined as

$$PAPR = \frac{\max\limits_{0 \leq n \leq LN-1} \left[|x(n)|^2 \right]}{\mathbb{E}\left[|x(n)|^2 \right]} \quad (5)$$

where $\mathbb{E}[\cdot]$ denotes the expectation operator.

Assume that the input information symbols are statistically independent and identically distributed. Based on the central limit theory, $x(n)$ can be approximated as a complex Gaussian process with zeros mean and variance σ^2 when the number of sub-carriers N is large enough (e.g. $N \geqslant 64$). Thus, the amplitude $|x(n)|$ follows a Rayleigh distribution with the probability distribution function (PDF) as

$$f_{|x_n|} = \frac{2x}{\sigma^2} \exp(-\frac{x^2}{\sigma^2}), x \geqslant 0. \quad (6)$$

From (6), we can see that the peak power of $|x_n|$ can take a value much large than its average power. That is to say, the multicarrier signal has a large PAPR, which leads to a negative impact on the system performance. While, by reallocating the power or the statistics of multicarrier signal reasonable, the companding transform can well resolve this high PAPR problem.

The fundamental principle of nonlinear companding transform can be described as follows [14]. The original signal $x(n)$ is companded before converted into analog waveform and amplified by the HPA. The companded signal is denoted as

$$y_n = \zeta(x_n) \quad (7)$$

where $\zeta(\cdot)$ is the companding function which only changes the amplitude of x_n. When passing through the AWGN channel, the transmitted signals can be recovered by the corresponding decompanding function $\zeta^{-1}(\cdot)$, i.e.

$$\tilde{x}_n = \zeta^{-1}(y_n + w_n) \approx x_n + \zeta^{-1}(w_n) \quad (8)$$

where w_n is the channel noise.

It has been pointed in [17] that the BER performance of companded multicarrier systems can be improved by carefully design the companding function $\zeta(\cdot)$ [13]–[20], however, due to the amplified channel noise (by decompanding operation) or unprocessed companding noise (by no decompanding operation), this modification may be limited in practice. Hence, there is considered to be a trade-off between PAPR reduction and BER performance for companding transform. However, in the subsequent works, we will show that this problem can be well resolved by the proposed companding noise cancellation scheme.

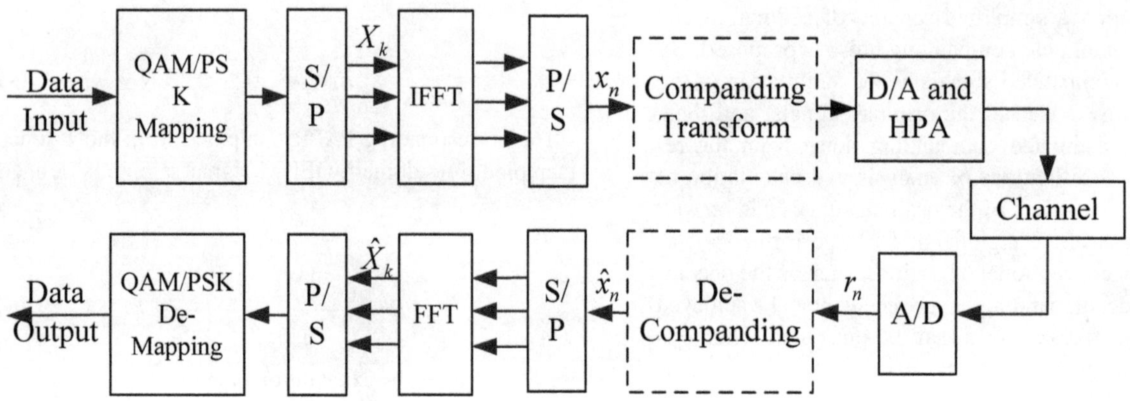

Fig. 1. Block diagram of typical multicarrier transmission system with companding transform.

3. Algorithm Formulation

In this section, we will review the basic concepts of conventional Bussgang based companding noise cancellation scheme briefly, and then, the proposed scheme will be presented subsequently.

3.1 Bussgang Theorem Based Scheme

According to the Bussgang theorem [19], the companded multicarrier signal $y(n)$ can be modeled as the aggregate of an attenuated signal component and companding noise d_n, i.e.

$$y(n) = \alpha x(n) + d_n, n = 0, 1, \ldots, LN-1 \tag{9}$$

where α is the attenuate coefficient, and which is a time invariant for stationary input processes.

From (9), the attenuate coefficient α can be calculated as

$$\alpha = \frac{\mathbb{E}\{y(n)x^*(n)\}}{\mathbb{E}\{x(n)x^*(n)\}} = \frac{1}{\sigma^2}\int_0^\infty x \cdot \zeta(x) \cdot f_{|x|}(x)dx \tag{10}$$

where $f_{|x|}(x)$ is as defined in (6) and $x^*(t)$ is the complex conjugate of $x(t)$.

In the receiver, by making full use of the received signal $y(n)$ and reconstructing the companding process as that of the transmitter, the estimated companding noise component can be calculated as

$$\hat{d}_n = \hat{y}(n) - \alpha\hat{x}(n), n = 0, 1, \ldots, LN-1 \tag{11}$$

where $\hat{y}(n), \hat{x}(n)$ are the reconstructed companded signal and the detected signal at the receiver, respectively.

However, it should be noted that the companding function $\zeta(\cdot)$ in (10) often has a complex expression (for example, see the companding functions in [14], [15] and [17]), which make the coefficient α difficult to be accurately calculated for general companding transforms. Moreover, the precision of α will have a great impact on the ultimate system performance. Hence, the effectiveness of conventional estimation and cancellation of companding noise will be greatly restrained in practice.

3.2 Proposed Scheme

Different from (9), if the attenuation of original signals is regarded as caused by the companding noise, then the output of the compander can be simply expressed as the original signals added with an extra companding noise component, i.e.

$$y(n) = x(n) + \hat{d}_n, n = 0, 1, \ldots, LN-1 \tag{12}$$

where $\hat{d}_n$ is the equivalent companding noise.

Hence, the received signal can be expressed as

$$\begin{aligned} r(n) &= h(n) * y(n) + w_n \\ &= h(n) * x(n) + h(n) * \hat{d}_n + w_n, n = 0, 1, \ldots, LN-1 \end{aligned} \tag{13}$$

where '$*$' is the convolution operation and $h(n)$ is the impulse response of the transmitting channel.

Assume the detected symbol is $\tilde{X}_k$(or equivalently $\tilde{x}(n)$), and the output of the compander at the receiver is $\tilde{y}(n)$, i.e.

$$\tilde{y}(n) = \tilde{x}(n) + \tilde{d}_n, n = 0, 1, \ldots, LN-1 \tag{14}$$

where $\tilde{d}_n$ is the companding noise regenerated at the receiver.

Since $\tilde{y}(n)$, $\tilde{x}(n)$ are all achievable at the receiver, hence, the companding noise can be estimated as

$$\tilde{d}_n = \tilde{y}(n) - \tilde{x}(n), n = 0, 1, \ldots, LN-1. \tag{15}$$

According to current channel estimation $\tilde{h}(n)$, then, $\tilde{h}(n) * \tilde{d}_n$ is subtracted from the current channel observation $r(n)$ to obtain the refined channel observation $\hat{r}(n)$, i.e.

$$\begin{aligned}\hat{r}(n) &= r(n) - \tilde{h}(n) * \tilde{d}_n \quad n = 0,1,\ldots,LN-1 \\ &= h(n) * x(n) + (h(n) * \hat{d}_n - \tilde{h}(n) * \tilde{d}_n) + w_n.\end{aligned} \tag{16}$$

It can be seen from (16) that the channel estimation error between $\tilde{h}(n)$ and $h(n)$ may affect the ultimate result of $\hat{r}(n)$, however, this error may also affect the estimation precision of $\tilde{d}_n$ (due to the decision error of $\tilde{X}_k$). On the other hand, when $\tilde{h}(n)$ severe deviates from $h(n)$, then, the channel estimation error will become the dominate interference component. Hence, we will mainly consider the approximate ideal channel estimation, i.e. $\tilde{h}(n) \approx h(n)$ in the subsequent works.

When $\tilde{d}_n \rightarrow \hat{d}_n$, then we will have the asymptotic ideal[1] result, i.e.

$$\hat{r}_n \rightarrow h(n) * x(n) + w_n, n = 0,1,\ldots,LN-1. \tag{17}$$

The above processes can be summarized in Fig. 2.

Comparing (9) and (12), we can have the conclusions that

- The proposed scheme is more simple than that of the conventional estimation and cancellation of companding noise scheme which bases on the Bussgang theorem [19]. Moreover, since the calculation of the attenuated coefficient α can be avoided in the presented scheme, hence, it can be easily adopted to general companding schemes.
- The novel scheme is more robust than that of the conventional scheme. This is due to the estimated companding noise in the presented scheme will not be affected by the precision of α as that in the conventional scheme.

4. Simulation Results

To evaluate the overall system performance, computer simulations are performed based on an OFDM system with $N = 256$ subcarriers and the input bit stream is modulated by Quaternary Phase Shift Keying (QPSK) and 16 Quadrature Amplitude Modulation (16QAM). The AWGN channel as well as frequency-selective multi-path fading channel [15] is considered in the simulations. Moreover, the oversampling factor $L = 4$ and a cycle prefix with the length of $1/4$ symbol is employed to mitigate the inter-symbol interference. Furthermore, for the convenience of comparison, a typical nonlinear companding function, i.e. the EC scheme proposed in [13] with companding function[2]

$$\zeta(x) = \mathrm{sgn}(x)\sqrt[d]{\nu[1-\exp(-\frac{|x|^2}{\sigma^2})]} \tag{18}$$

where

$$\nu = \left(\frac{\mathbb{E}\left[|x|^2\right]}{\mathbb{E}\left[\sqrt[d]{\left[1-\exp(-\frac{|x|^2}{\sigma^2})\right]}^2\right]}\right)^{\frac{d}{2}} \tag{19}$$

and d is the companding degree (the PAPR performance of EC according to [13] with different companding degree d is summarized in Tab. 1), is equipped with several conventional operations at the receiver, including conventional decompanding (DC) [13], [14], [15], [16], [20], no decompanding (NDC) [17], [18] as well as the proposed companding noise cancellation (CNC) operations are all considered in the simulations.

	EC		Original signals
	$d=1$	$d=2$	
PAPR(dB), CCDF=10^{-3}	4.98	3.21	11.38

Tab. 1. PAPR performance of EC with different companding degrees.

4.1 BER in AWGN Channel

Figure 3 shows the BER performance of OFDM system with QPSK modulation employing EC and equipped with different operations at the receiver over AWGN channel. In this figure, the curve of "Performance bound" is the signals transmitted without living through any nonlinear distortions. Given that the BER at $P_e = 10^{-5}$, we can find that the companded signals are suffering from a serious nonlinear distortions especially when with conventional decompanding operations at the receiver, and there is a 3.23 dB ε_b/N_0 gap between EC-DC with $d = 1$ and the performance bound and more than 6 dB ε_b/N_0 gap for EC-DC with $d = 2$. While, although there is a significant BER improvement when with no decompanding operation at the receiver, however, there is still a gap of 0.78 dB ε_b/N_0 at EC-NDC with $d = 1$ and 3.03 dB ε_b/N_0 at EC-NDC with $d = 2$ away from the performance bound. However, when with the proposed companding noise cancellation scheme, the BER gap between EC-CNC and the performance bound can be effectively restored to no more than 0.29 dB regardless of the companding degree d.

Figure 4 shows the BER performance of EC with 16QAM modulation and different operations at the receiver over AWGN channel. It's obvious that the overall BER performance has suffering a more serious degradation than that of with QPSK modulation. Furthermore, when with EC-DC and $d = 1$, there is a 5.58 dB ε_b/N_0 gap away from the performance bound. While, we can also find that when with high ε_b/N_0, the conventional decompanding operation will trend to outperform that with no decompanding operation at

[1] As we will show later, this is due to the inherent estimated errors of the detected signals $\tilde{x}(n)$, and this error is unavoidable but can be improved by some other techniques, such as the channel coding.

[2] It's obvious that (10) is not appropriate for this companding function, since the coefficient ν in (18) can only be calculated numerically, and this phenomenon is widely exits in some other companding functions (see [14]–[20]).

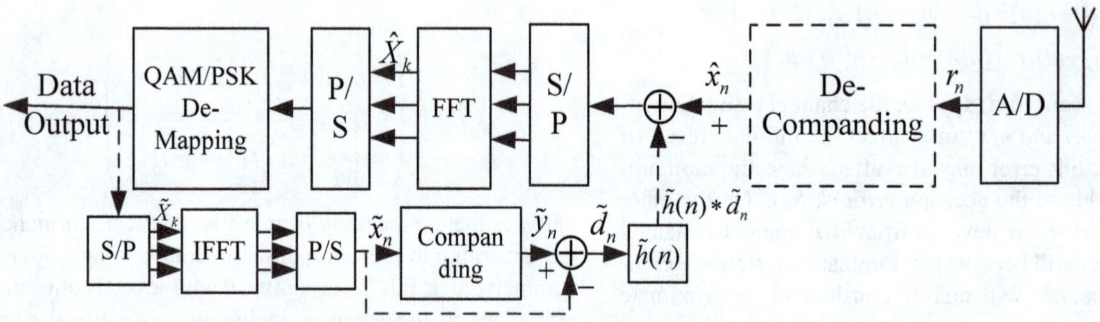

Fig. 2. Block diagram of the proposed scheme for companding transform. Note that when the SNR is low, it's prefer to with no decompanding at the receiver, and then $\hat{x}(n) = r(n)$.

the receiver. This is due to that although the decompanding operation may amplifier the channel noise, while, when ε_b/N_0 is high enough (e.g. $\varepsilon_b/N_0 > 18$ dB), the companding noise will become the dominant interference component and must be removed from the received signals. When with the proposed companding noise cancellation scheme, there is a 1.09 dB ε_b/N_0 gap between EC-CNC with $d = 2$ and the performance bound, while the relative gap is only 0.16 dB for EC-CNC with $d = 1$, which further demonstrates the great advantages of the proposed scheme over other conventional operations.

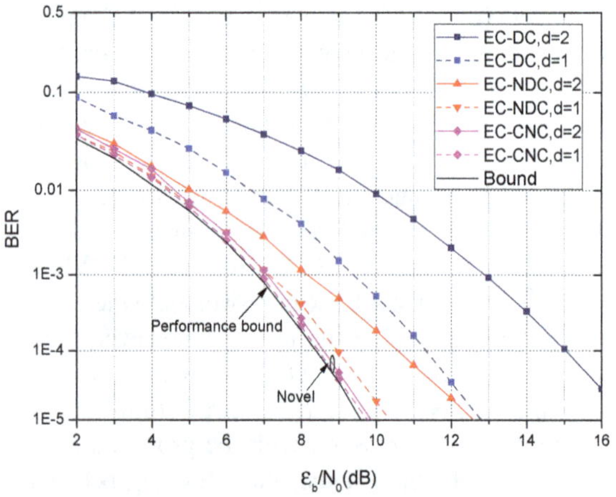

Fig. 3. BER performance of QPSK modulation and EC equipped with different operations over AWGN channel.

4.2 BER with HPA over AWGN Channel

For most wireless communication systems, the HPA is widely used to provide adequate transmit power. In this paper, the solid state power amplifier (SSPA) model described in [21] is considered in the subsequent works. Fig. 5 shows the BER performance of different operations at the receiver with SSPA and $IBO = 0$ dB over AWGN channel. It is seen that after passing through the SSPA, the BER performance has been degraded to some extent. In this figure, the "Original signals" is the signals without companding transform and directly transmitted to the SSPA, consequently, result in 3.33 dB ε_b/N_0 degradation compared to the performance bound as shown in Fig. 3. Moreover, we can also find that the EC-CNC with $d = 2$ outperforms the original signals 2.55 dB and 0.16 dB better than EC-CNC with $d = 1$.

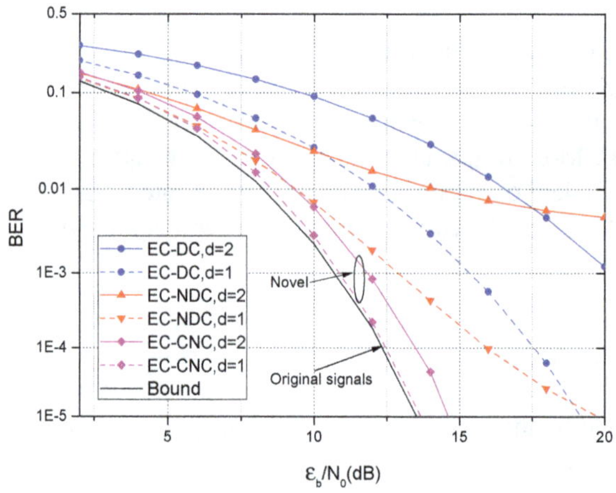

Fig. 4. BER performance 16QAM modulation of EC equipped with different operations at the receiver over AWGN channel.

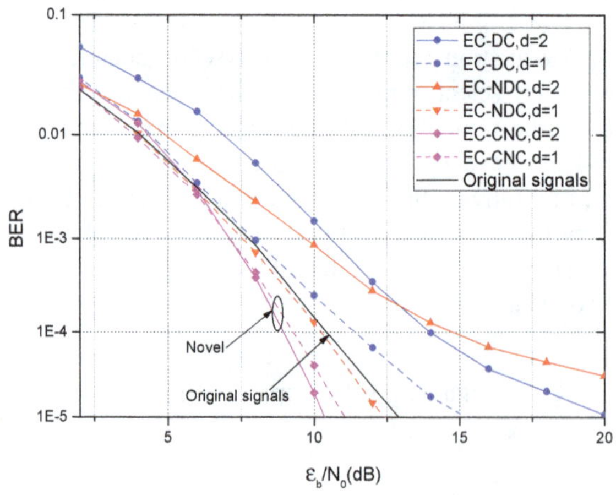

Fig. 5. BER performance of EC with HPA over AWGN channel with different operations at the receiver.

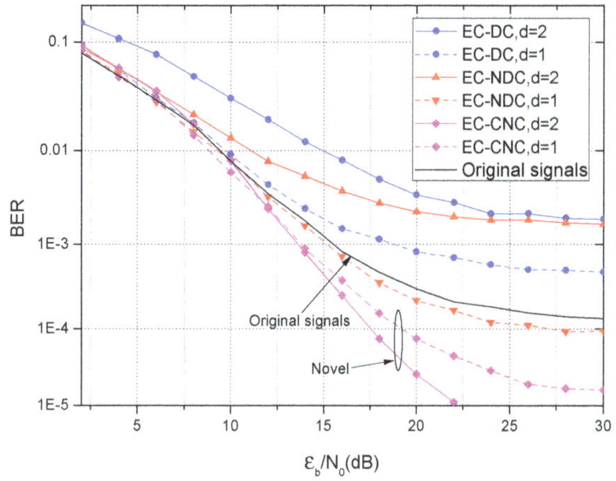

Fig. 6. BER performance of the receiver with HPA and equipped with various operations over fading channel.

4.3 BER with HPA over Fading Channel

Figure 6 shows the BER performance of EC with SSPA and equipped with different operations at the receiver over fading channel. In this figure, the IEEE 802.16 fading channel model described in [15] is considered in the simulations. From it we can see that different with other conventional operations at the receiver, the EC equipped with CNC greatly outperforms the original signals. Furthermore, comparing Fig. 6 and Fig. 5, we can also find that the advantages of the proposed scheme are more significant in practical wireless communication systems than that of in the AWGN channel.

4.4 BER with Iterative Filtering

Since the companding operation is a type of nonlinear process that may lead to out-of-band radiation, hence, an iterative companding and filtering technique may need to remove the out-of-band radiation and peak regrowth. However, it's known that the filtering may lead to the degradation of the BER performance, hence, it's necessary to estimate the practical BER performance of companding transform when considering the iterative filtering technique.

In this subsection, we mainly employ filtering in the baseband signals in the frequency domain [4], and the filter is based on the rectangular window, and can be defined as

$$\mathbf{Y}_f = \mathbf{H}_f \mathbf{Y}_c \tag{20}$$

where $\mathbf{Y}_c$ is the companded signal in frequency domain, $\mathbf{Y}_f$ is the filtered signals, and $\mathbf{H}_f$ is the frequency response of the filter defined by

$$H_f(k) = \begin{cases} 1, 0 \leqslant k \leqslant N/2-1, L-N/2 \leqslant k \leqslant L-1 \\ 0, \text{otherwise.} \end{cases} \tag{21}$$

Figure 7 shows the BER performance of EC over AWGN channel with iterative filtering at the transmitter and equipped different operations at the receiver with 2 iterations. Note that the receiver also performs the same iterative companding and filtering process as that at the transmitter. From it we can see that although the SNR gap between EC-CNC and the performance bound is slightly increased to 0.2 dB at $d = 1$ and 0.98 dB at $d = 2$ compared with Fig. 3, however, it's obvious that the proposed scheme still greatly outperforms other conventional schemes, which well verifies the robustness of the proposed scheme under various situations.

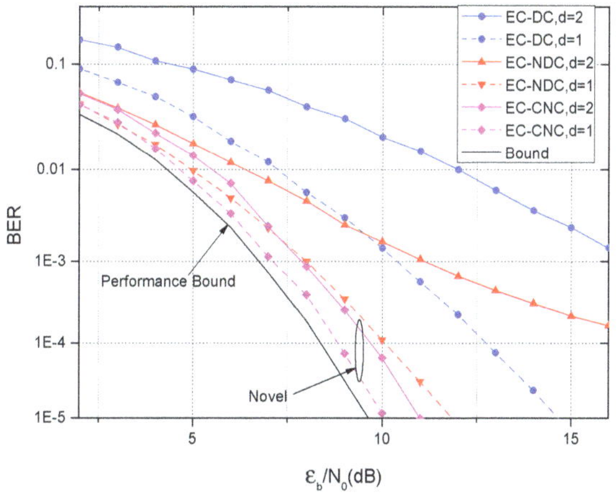

Fig. 7. BER performance of EC with iterative filtering and equipped with various operations at the receiver over AWGN channel.

5. Conclusion

Nonlinear companding noise in companded multicarrier systems greatly degrades the BER performance of the overall system. In this paper, a simplified but effective scheme of estimation and cancellation of companding noise is proposed to enhance the BER performance of the companded multicarrier systems. By expressing the companding process as the original signals added with a companding noise component and removing the estimated companding noise from the receiver, a great BER performance improvement can be achieved compared to conventional operations. Although with a slightly increased complexity (due to additional IFFT/FFT and companding operations at the receiver), we believe that the proposed scheme may well resolve the embarrassment, i.e. the trade-off problem between PAPR reduction and BER performance, for the conventional companding schemes.

Acknowledgements

This work was supported by the National Natural Science Foundation of China (grant nos 60772083 and 61201242); The Natural Science Foundation of Jiangsu Province (grant no. BK2012057) and the PLA University Preresearch Foundation (KYTYZLXY1208).

References

[1] WUNDER, G., FISHCHER, R. F. H., BOCHE, H., ET AL. The PAPR problem in OFDM transmission: new direction for long-lasting problem. *IEEE Signal Processing Magazine*, 2013, vol. 51, no. 11, p. 130–144. DOI: 10.1109/MSP.2012.2218138

[2] POLAK, L., KRATOCHVIL, T. Exploring of the DVB-T/T2 performance in advanced mobile TV fading channels. In *Proc. 36th International Conference on Telecommunictaions and Signal Processing (TSP2013)*, Rome (Italy), 2013, p. 768–772. DOI: 10.1109/TSP.2013.6614042

[3] HAN, S. H., LEE, J. H. An overview: peak-to-average power ratio reduction techniques for multicarrier transmission. *IEEE Wireless Communications*, 2004, vol. 12, no. 2, p. 56–65. DOI: 10.1109/MWC.2005.1421929

[4] ZHU, X., PAN, W., LI, H., ET AL. Simplified approach to optimized iterative clipping and filtering for PAPR reduction of OFDM signals. *IEEE Transactions on Communications*, 2013, vol. 61, no. 5, p. 1891–1901. DOI: 10.1109/TCOMM.2013.021913.110867

[5] CHEN, H., HAIMOVICH, A., M. Iterative estimation and cancellation of clipping noise for OFDM signals. *IEEE Communications Letters*, 2003, vol. 7, no. 7, p. 305–307. DOI: 10.1109/LCOMM.2003.814720

[6] XIA, L., LI, Z., YOUXI, T., ET AL. Analysis of the performance of iterative estimation and cancellation of clipping non-linear distortion in OFDM. In *Proceeding of Future Generation of Communications and Networking*, 2007, p. 1–5. DOI: 10.1109/FGCN.2007.68

[7] SLMANE, S., B. Reduction the peak-to-average power ratio of OFDM signals through precoding. *IEEE Transactions on Vehicular Technology*, 2007, vol. 56, no. 2, p. 686–695. DOI: 10.1109/TVT.2007.891409

[8] QI, X., LI, Y, HUANG., H. A low complexity PTS scheme based on tree for PAPR reduction. *IEEE Communications Letters*, 2012, vol. 16, no. 9, p. 1486–1488. DOI: 10.1109/LCOMM.2012.072012.121228

[9] VARAHRAM, P., ALI, M., B. A low complexity partial transmit sequence for peak to average power ratio reduction in OFDM systems. *Radioengineering*, 2011, vol. 20, no. 3, p.677–682.

[10] MARSALEK, R. On the reduced complexity interleaving method for OFDM PAPR reduction. *Radioengineering*, 2006, vol. 15, no. 3, p. 49–53.

[11] JIANG, T., NI, C., GUAN, L. A novel phase offset SLM scheme for PAPR reduction in Alamouti MIMO-OFDM systems without side information. *IEEE Signal Processing Letters*, 2013, vol 20, No. 4, p. 383–386. DOI: 10.1109/LSP.2013.2245119

[12] WANG, L., TELLAMBURE, C. Analysis of clipping noise and tone reservation algorithms for peak reduction in OFDM systems. *IEEE Transactions on Vehicular Technology*, 2008, vol. 57, no. 3, p. 1675–1694. DOI: 10.1109/TVT.2007.907282

[13] JIANG, T., YANG, Y., SONG, Y., H. Exponential companding technique for PAPR reduction in OFDM systems. *IEEE Transactions on Broadcasting*, 2005, vol. 51, no. 2, p. 244–248. DOI: 10.1109/TBC.2005.847626

[14] WANG, Y., GE, J., WANG, L., ET AL. Nonlinear companding transform for reduction of peak-to-average power ratio in OFDM systems. *IEEE Transactions on Broadcasting*, 2013, vol. 59, no. 2, p. 369–375. DOI: 10.1109/TBC.2012.2219252

[15] JENG, S., S., CHEN, J., M. Effective PAPR reduction in OFDM systems based on a companding technique with trapezium distribution. *IEEE Transactions on Broadcasting*, 2010, vol. 56, no. 2, p. 258–262. DOI: 10.1109/TBC.2011.2112237

[16] PENG, S., SHEN, S., YUAN, Z., ET AL. A novel nonlinear companding transform for PAPR reduction in lattice-OFDM systems. *Frequenz*, 2014, vol. 69, no. 2, p. 461–469. DOI: 10.1515/freq-2013-0169

[17] HOU, J., GE, J., ZHAI, D.,ET AL. Peak-to-average power ratio reduction of OFDM signals with nonlinear companding scheme. *IEEE Transactions on Broadcasting*, 2010, vol. 56, no. 2, p. 258–262. DOI: 10.1109/TBC.2010.2046970

[18] PENG, S., SHEN, Y., YUAN, Z., ET AL. PAPR reduction of LOFDM signals with an efficient nonlinear companding transform. In *Proceeding of International Conference on Wireless Communications and Signal Processing*, WCSP'13, Hangzhou, 2013, p. 1–6. DOI: 10.1109/WCSP.2013.6677217

[19] JIANG, T., YAO, W., SONG, Y., ET AL. Two novel nonlinear companding schemes with iterative receiver to reduce PAPR in multicarrier modulation systems. *IEEE Transactions on Broadcasting*, 2006, vol. 52, no. 2, p. 268–273. DOI: 10.1109/TBC.2006.872992

[20] PENG, S., SHEN, Y., YUAN, Z. PAPR reduction of multi-carrier systems with simple nonlinear companding transform. *Electronic Letters*, 2014, vol. 50, no. 6, p. 473–475. DOI: 10.1049/el.2013.4216

[21] COSTA, E., MIDRIO, M., PUPOLIN, S. Impact of amplifier nonlinearities on OFDM transmission systems performance. *IEEE Communications Letters*, 1999, vol. 3, no. 2, p. 37–39. DOI: 10.1109/4234.749355

About the Authors...

SIMING PENG was born in Hubei, China Republic in 1990. He received his bachelor's degree from Wuhan University of Science and Technology in 2012. He is currently a M.Sc. candidate at the Department of Wireless Communications, College of Communications Engineering, PLA University of Science and Technology. His research interests include multicarrier communications and signal processing.

ZHIGANG YUAN was born in Hebei, China Republic in 1980. He received his Ph.D. degree in Communication and Information System from PLA University of Science and Technology in 2008. He is currently a lecture at the same university. His interests are signal processing and lattice multi-carrier communication theory.

JUN YOU was born in Jiangsu Province. He received the M.S. and Ph.D. degree from Institute of Communications Engineering, PLA University of Science and Technology, Nanjing, China, in 2001 and 2005, respectively. He is currently an associate professor in the College of Command Information System, PLAUST. His interest is fast signal processing.

YUEHONG SHEN was born in Hubei, China Republic in 1959. He received his Ph.D. degree in Communication Engineering from Nanjing University of Science and Technique in 1999. And now, he is a professor and doctor advisor of Wireless Department at the Institute of Communication Engineering, PLA University of Science and Technology,

China. His interests include blind and source separation in communication systems.

WEI JIAN was born in Henan, China Republic in 1976. He received his Ph.D. degree in Communication and Information Engineering from PLA University of Science and Technology in 2006. His interests cover communication theory and high speed wireless communication systems.

Enhanced Model of Nonlinear Spiral High Voltage Divider

Václav PAŇKO [1,3], *Stanislav BANÁŠ* [1,3], *Richard BURTON* [4],
Karel PTÁČEK [2,3], *Jan DIVÍN* [1,3], *Josef DOBEŠ* [1]

[1]Dept. of Radio Engineering, Czech Technical University in Prague, Technická 2, 166 27 Praha 6, Czech Republic
[2]Dept. of Microelectronics, Brno University of Technology, Technicka 3058/10, 61600 Brno, Czech Republic
[3]ON Semiconductor, SCG Czech Design Center, 1. maje 2594, 75661 Roznov p. R., Czech Republic
[4]ON Semiconductor, 5005 East McDowell Road, Phoenix, AZ 85008, USA

vaclav.panko@onsemi.com, stanislav.banas@onsemi.com, richard.burton@onsemi.com,
karel.ptacek@onsemi.com, jan.divin@onsemi.com, dobes@fel.cvut.cz

Abstract. *This paper deals with the enhanced accurate DC and RF model of nonlinear spiral polysilicon voltage divider. The high resistance polysilicon divider is a sensing part of the high voltage start-up MOSFET transistor that can operate up to 700 V. This paper presents the structure of a proposed model, implemented voltage, frequency and temperature dependency, and scalability. A special attention is paid to the ability of the created model to cover the mismatch and influence of a variation of process parameters on the device characteristics. Finally, the comparison of measured data vs. simulation is presented in order to confirm the model validity and a typical application is demonstrated.*

Keywords

High voltage start-up MOSFET, pinch-off, high voltage spiral divider, statistical modeling

1. Introduction

Nowadays, the power consumption is one of the most important integrated circuit parameters. High voltage power start-up MOSFET transistor described in this paper is used to minimize the power consumption [1, 2]. It is designed to provide initial current directly from the high voltage source. This MOSFET transistor charges up the regulator voltage on an external capacitor to about 14 V. The main goal is to minimize power consumption of the circuit that is directly connected to the rectified DC high voltage source. This high voltage can be up to 400 V for a 230 V AC supply and 700 V for switcher applications using power factor correction.

The HV start-up MOSFET is fabricated in an analog 1 μm CMOS technology. The simplified structure of this MOSFET is depicted in Fig. 2. The source and drain are formed from a low-doped Nwell and are contacted by N+ diffusion. The drain drift area contains a floating P doped resurf diffusion (ptop) fabricated before field oxide. The MOSFET channel is created from Pwell not isolated from the P-substrate and it is covered by polysilicon gate. This drain-gate-source structure is rotary symmetrical around vertical axis in the center of the drain. It means that the drain is created in the shape of a circle and the gate and the source in the shape of an annulus.

The drain is located in the center of the device and contains rounded bonding pad. A drain bonding wire is connected directly to this bonding pad and this is only one possible way how the drain can be connected. The oxide breakdown is much lower (about 100 V) than maximum allowed drain voltage. The drain can be biased up to 700 V and this makes integrated direct sensing of the high drain voltage impossible. Hence, the high resistance polysilicon spiral voltage divider is used for sensing of high drain voltage. The spiral is connected to the drain and continues spirally toward the gate. How the polysilicon spiral divider is connected to other device components is depicted in the schematic symbol of HV MOSFET in the Fig. 1 (terminals d, tap1, tap2).

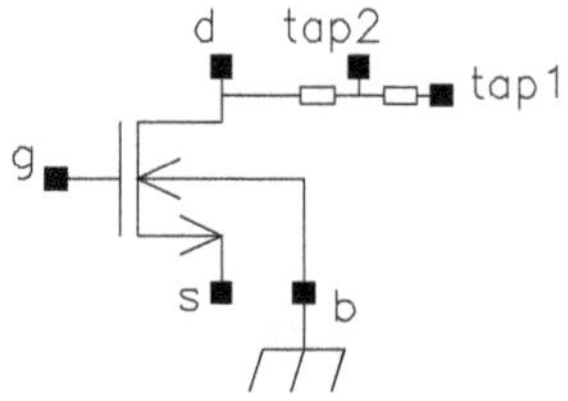

Fig. 1. Schematic symbol of HV start-up MOSFET.

The spiral divider is designed to have the electric field distribution as much similar as possible as the drain drift area under it. This ensures the voltage between divider and silicon does not exceed oxide breakdown voltage. The polysilicon spiral divider has a big impact on a distribution of electric field in low doped drain drift area. And on the contrary, the strong electric field in low doped drain drift area causes a lot of parasitic effects that have a big influence on DC and RF device characteristics. These attributes make the modeling of this start-up MOSFET complicated, especially the divider ratio voltage and frequency dependency. The divider is usually modeled by the simple RC network, but there exist the operation areas where such simple model is not sufficient.

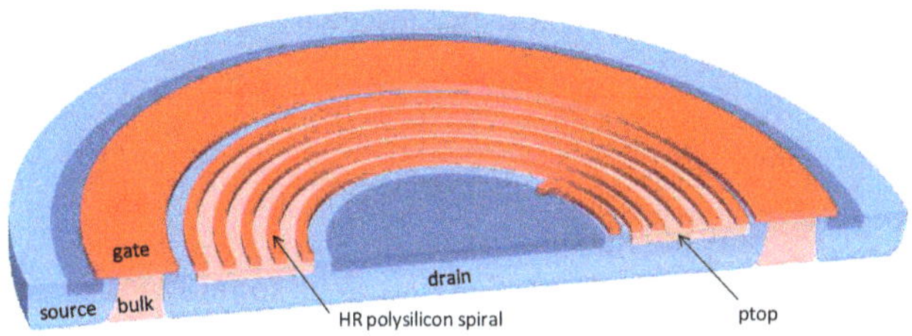

Fig. 2. The simplified 3D structure of HV start-up MOSFET transistor.

2. Spiral Divider Modeling

For the purpose of the equivalent lumped element circuit creation the polysilicon spiral is divided into several separate spiral elements. This division is shown in Fig. 3(a) where each spiral element has a different color. For better lucidity only the first four turns are depicted in this figure. The equivalent 3D circuit in Fig. 3(b) is obtained if these spiral elements are uncoiled to parallel plains. The 3D equivalent circuit in Fig. 3(b) can be redrawn for better lucidity to the 2D equivalent circuit, which is depicted in Fig. 4.

2.1 Spiral Element Length

The spiral divider of the HV MOSFET transistor is a special case of the Archimedes spiral [3]. The radius r of the spiral is increased in one turn by a radius increment Δr. The basic equation defined in polar coordinates for the radius is

$$r = r_0 + \varphi \frac{\Delta r}{2\pi} \tag{1}$$

where r_0 is an initial radius of the spiral and φ is an actual angle circumscribed by the spiral.

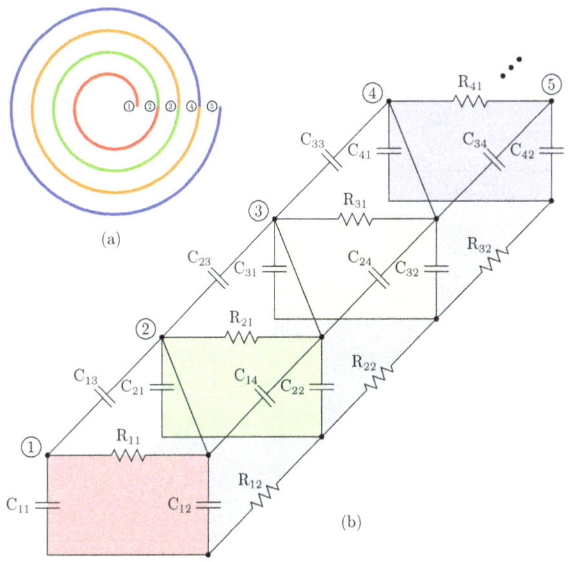

Fig. 3. Equivalent lumped 3D circuit of first four spiral poly subsegments and ptop: (a) colored spiral subsegment, (b) equivalent circuit. Colors from (a) match (b).

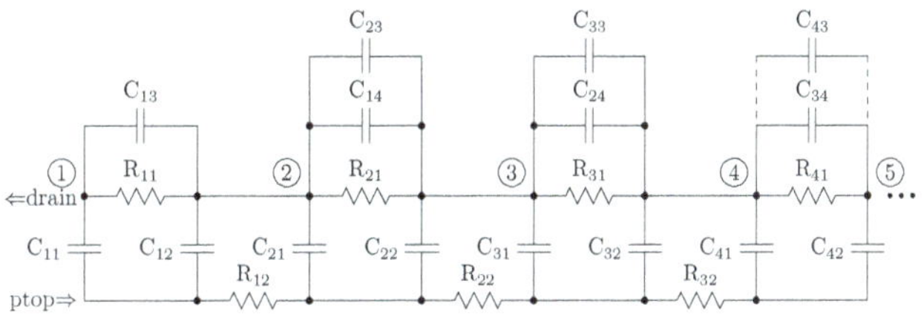

Fig. 4. Equivalent lumped 2D circuit of first four spiral poly subsegments and ptop.

The curve length can be calculated in the following way. If $f(\phi)$ is the function of the curve in polar coordinates (ϕ is an angle) then the length L of the curve is defined as

$$L = \int_0^{\varphi} \sqrt{[f(\phi)]^2 + \left(\frac{\mathrm{d}f(\phi)}{\mathrm{d}\phi}\right)^2}\,\mathrm{d}\phi. \tag{2}$$

For the spiral defined in polar coordinates by (1) the spiral length L is obtained by substituting the equation (1) into (2), and by solving this integral the spiral length is

$$\begin{aligned} L = {} & \frac{\Delta r}{4\pi} \ln \left(\frac{r_0 + \varphi\frac{\Delta r}{2\pi} + \sqrt{\left(r_0 + \varphi\frac{\Delta r}{2\pi}\right)^2 + \frac{\Delta r^2}{4\pi^2}}}{r_0 + \sqrt{r_0^2 + \frac{\Delta r^2}{4\pi^2}}} \right) + \\ & + \left(\frac{r_0\pi}{\Delta r} + \frac{\varphi}{2}\right) \sqrt{\left(r_0 + \varphi\frac{\Delta r}{2\pi}\right)^2 + \frac{\Delta r^2}{4\pi^2}} - \\ & - \frac{r_0\pi}{\Delta r} \sqrt{r_0^2 + \frac{\Delta r^2}{4\pi^2}}. \end{aligned} \tag{3}$$

When $\Delta r << r_0$ then equation (3) can be simplified to

$$L = r_0\varphi + \frac{\Delta r}{4\pi}\varphi^2. \tag{4}$$

2.2 Divider Ratio Modeling

Model of a similar device has been published in [4] but without ratio scalability and statistical modeling. These two important model abilities have been developed and implemented into the new model that is introduced in this paper. The divider ratio is dependent on the drain and source voltage V_D and V_S. The voltage dependency caused by depletion effects in the ptop and nwell layers is modeled by Verilog-A code using nonlinear functions. The increasing of the drain voltage causes the depletion of the ptop and nwell and when the ptop is fully depleted under the spiral polysilicon divider then it causes a change of the ratio voltage dependency slope as is depicted in Fig. 5. The geometrical ratio is based on (4) and can be expressed as

$$\begin{aligned} ratio_{geom} &= \frac{L_1 + L_2}{L_2} = \\ &= \frac{4\pi r_0(\varphi_{tap1} - \varphi_D) + \Delta r(\varphi_{tap1}^2 - \varphi_D^2)}{4\pi r_0(\varphi_{tap1} - \varphi_{tap2}) + \Delta r(\varphi_{tap1}^2 - \varphi_{tap2}^2)} \end{aligned} \tag{5}$$

where φ_D, φ_{tap1} and φ_{tap2} are drain, tap1 and tap2 angles on the spiral, L_1 and L_2 are long and sensing part of the spiral.

The normalized ratio is modeled as

$$ratio_{norm} = \frac{ratio_{el}}{ratio_{geom}} = \frac{V_D/V_{tap2}}{(L_1+L_2)/L_2} = \tag{6}$$

$$= \begin{cases} 1+(\beta_{D1}+\beta_{D2})V_D+\beta_S V_S & \text{for } V_D \le V_P \\ 1+\beta_{D2}V_D+\beta_S V_S & \text{for } V_D > V_P \end{cases}$$

where $ratio_{el}$ is electrical ratio, β_{D1}, β_{D2} and β_S are drain and source voltage dependency model parameters, V_D, V_S and V_{tap2} are voltages on pins drain, source and tap2, and V_P is ptop pinch-off voltage. The voltage dependency coefficient β_{D2} is temperature dependent and can be expressed as

$$\beta_{D2} = \beta_{D2Tnom}\left[1+\alpha_D(T-T_{nom})\right] \tag{7}$$

where T is temperature, model parameter β_{D2Tnom} represents value of β_{D2} at nominal temperature T_{nom} and α_D is temperature coefficient.

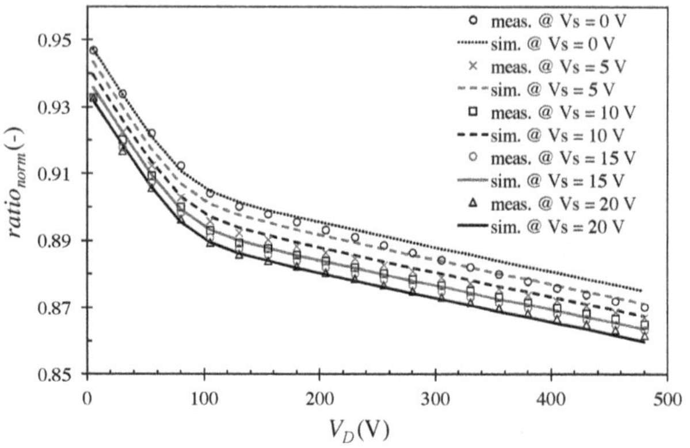

Fig. 5. Drain and source voltage dependency of normalized ratio.

The model is scalable by an editable model argument $ratio_{geom}$ that is refined to

$$ratio'_{geom} = \left(ratio_{mult}\, ratio_{geom} - ratio_{\Delta}\right)\delta_{ratio} \tag{8}$$

where $ratio_{mult}$ and $ratio_{\Delta}$ are model fitting parameters for ratio scalability and δ_{ratio} is relative statistical mismatch model parameter.

2.3 Divider Dynamical Modeling

The AC response is modeled by a distributed RC network. The magnitude and phase of the normalized ratio are depicted in Figs. 6 and 7. The AC measurement setup is described in Appendix. The full macromodel circuit of HV start-up MOSFET is shown in Fig. 12.

The high resistance polysilicon spiral segments are modeled by the Verilog-A code. Each resistor segment is modeled as

$$R_{seg} = \frac{R_{tot}}{N_{seg}}\left(1-\frac{1}{ratio_{norm}\, ratio'_{geom}}\right) \tag{9}$$

where N_{seg} is number of divider segments (excluding sense segment), R_{tot} is total spiral divider resistance and is calculated as

$$R_{tot} = \frac{R_{SH}L_{tot}\left(1+\alpha_{p1}\Delta_T+\alpha_{p2}\Delta_T^2\right)}{W+\delta_W} \tag{10}$$

where R_{SH} is polysilicon spiral sheet resistance, L_{tot} is a total spiral divider length, $\Delta_T = T - T_{nom}$, W is spiral segment width, δ_W is absolute statistical process model parameter, α_{p1} and α_{p2} are temperature coefficients dependent on R_{SH}

$$\begin{aligned} \alpha_{p1} &= \alpha_{rsh1}R_{SH}+\alpha_{rp1} \\ \alpha_{p2} &= \alpha_{rsh2}R_{SH}+\alpha_{rp2} \end{aligned} \tag{11}$$

where α_{rsh1}, α_{rsh2}, α_{rp1}, and α_{rp2} are polysilicon temperature coefficients. The resistance of the sense segment is

$$R_{sense} = \frac{R_{tot}}{ratio_{norm}\, ratio'_{geom}}. \tag{12}$$

The capacitances are modeled by the Verilog-A code and are voltage dependent similarly as resistances. The voltage dependency is caused by depleting effects of very low doped drift drain area due to the high electric field. Each capacitor segment is modeled [4] as

$$C_{seg} = \frac{C_{tot}}{N_{seg}}\left(\frac{1}{\pi}\arctan\left(\frac{V_P-V_D}{2}\right)+c_P\right) \tag{13}$$

where c_P is the pinch-off capacitance coefficient model parameter and C_{tot} is total spiral divider capacitance and is calculated as

$$C_{tot} = L_{tot}(W+\delta_W)C_{pa}+2L_{tot}C_{fr}+C_c \tag{14}$$

where C_{pa} is polysilicon (field oxide) capacitance per unit area, C_{fr} is fringe capacitance per length and C_c is capacitance model fitting parameter.

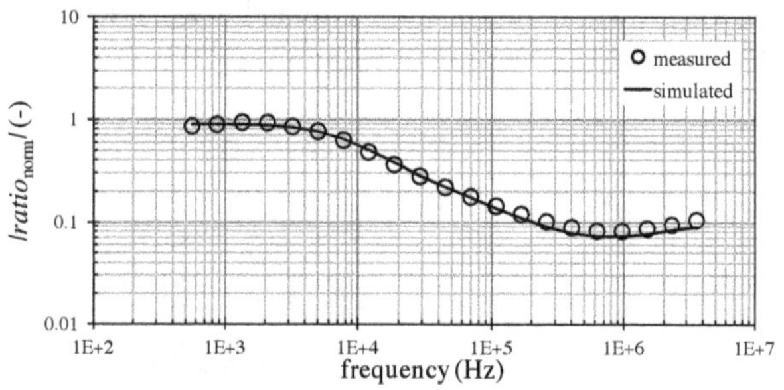

Fig. 6. Magnitude of normalized divider ratio.

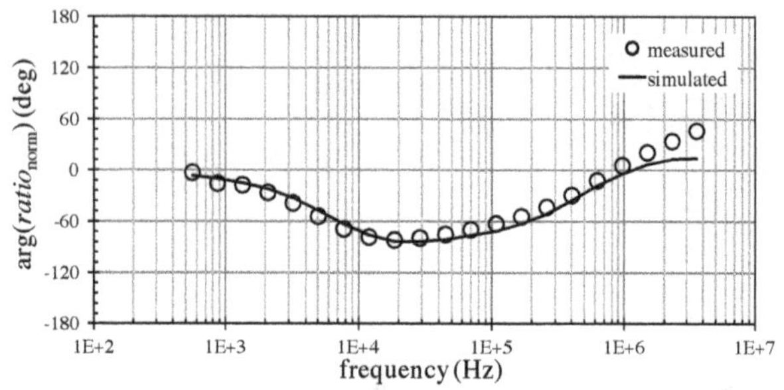

Fig. 7. Phase of normalized divider ratio.

2.4 Divider Statistical Modeling

This HV start-up MOSFET was placed on process control monitoring test chip where this device is measured on each fabricated wafer in a standard production. Data from this test chip are used for statistical process control and also for statistical modeling.

The mismatch modeling [5], [6] is implemented only to the divider ratio. The parameter δ_{ratio} is relative statistical mismatch model parameter and is defined as

$$\delta_{ratio} = 1 + \sigma_{ratio} \cdot VAR_{MATCH\_RATIO} \quad (15)$$

where σ_{ratio} is relative standard deviation of the divider ratio and VAR_{MATCH_RATIO} is random variable of mean 0 and standard deviation 1 that represents the normalized Gaussian distribution for modeling the stochastic variations. The histogram of measured and simulated voltage V_{tap2} and box plot are depicted in Figs. 8 and 9 (one lot typically contains from 20 to 30 wafers and one wafer typically contains 5 test chips). The number of measured devices was 3825. The standard deviation σ_{ratio} is equal to the standard deviation of measured electrical parameter V_{tap2} at $V_D = 100\,\mathrm{V}$, $V_S = V_G = V_{tap1} = 0\,\mathrm{V}$.

The influence of process parameters variation on the device parameters is implemented through master variables by using mapping equations:

$$R_{SH} = R_{SH\_nominal} + \sigma_{RSH} \cdot VAR_{RSH}, \quad (16)$$

$$\delta_W = \delta_{W\_nominal} + \sigma_{DW} \cdot VAR_{DW}, \quad (17)$$

$$C_{pa} = C_{pa\_nominal} + \sigma_{CPA} \cdot VAR_{CPA}, \quad (18)$$

$$C_{fr} = C_{fr\_nominal} + \sigma_{CFR} \cdot VAR_{CFR}, \quad (19)$$

where $R_{SH\_nominal}$, $\delta_{W\_nominal}$, $C_{pa\_nominal}$ and $C_{fr\_nominal}$ are nominal values, σ_{RSH}, σ_{DW}, σ_{CPA} and σ_{CFR} are the standard deviations, and VAR_{RSH}, VAR_{DW}, VAR_{CPA} and VAR_{CFR} are master random variables of mean 0 and standard deviation 1 that represents the normalized Gaussian distribution for modeling the stochastic variations.

As an example, the histogram of measured and simulated electrical process parameter *DW* is depicted in Fig. 10 and the box plot in Fig. 11. The number of measured devices was 29257. The standard deviations of device parameters σ_{RSH}, σ_{DW}, σ_{CPA} and σ_{CFR} can be calculated from the standard deviations of these measured electrical process parameters by using forward and backward propagation of variances [7].

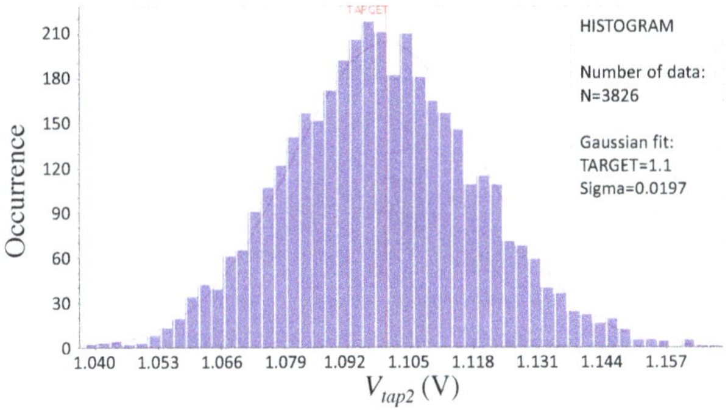

Fig. 8. The histogram of measured and simulated electrical parameter V_{tap2}. The red curve represents modeled Gaussian distribution.

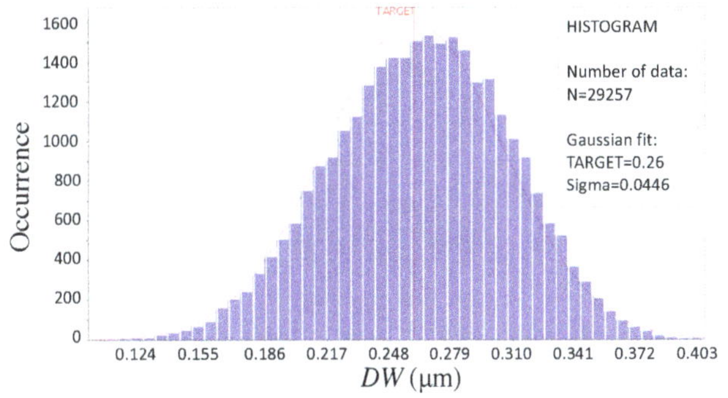

Fig. 10. The histogram of measured and simulated electrical process parameter *DW*. The red curve represents modeled Gaussian distribution.

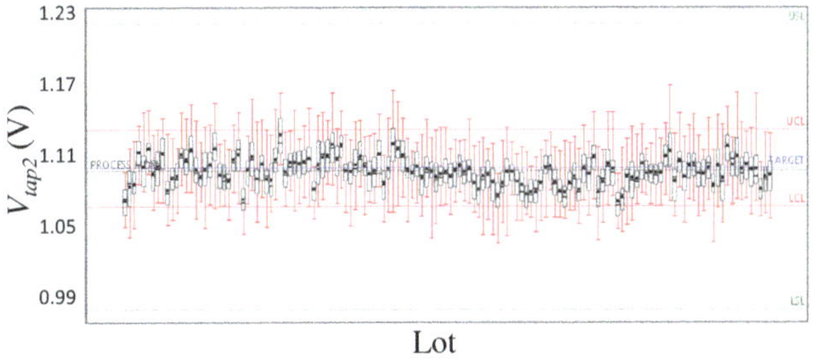

Fig. 9. The boxplot of measured electrical parameter V_{tap2}. The green lines define upper and lower specification limit (USL and LSL) while the red lines define upper and lower control limit (UCL and LCL).

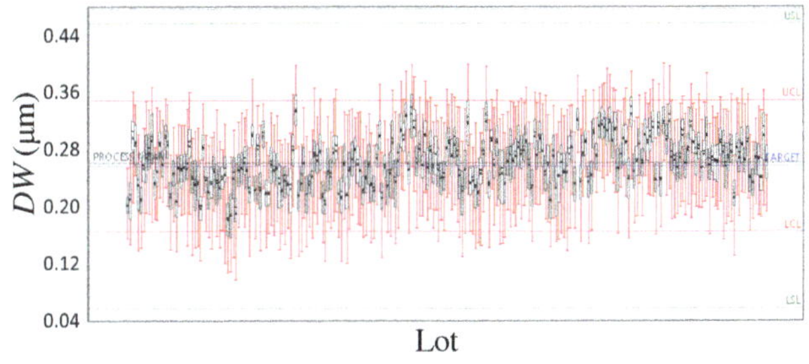

Fig. 11. The boxplot of measured electrical process parameter *DW*. The green lines define upper and lower specification limit (USL and LSL) while the red lines define upper and lower control limit (UCL and LCL).

3. HV Start-Up MOSFET Application

The AC/DC convertor has been selected as an example of typical application of HV start-up MOSFET with polysilicon spiral divider. The simplified AC/DC convertor circuit is depicted in Fig. 13. The HV start-up MOSFET subblock is modeled by the circuit in Fig. 12 and by the equations introduced in this paper.

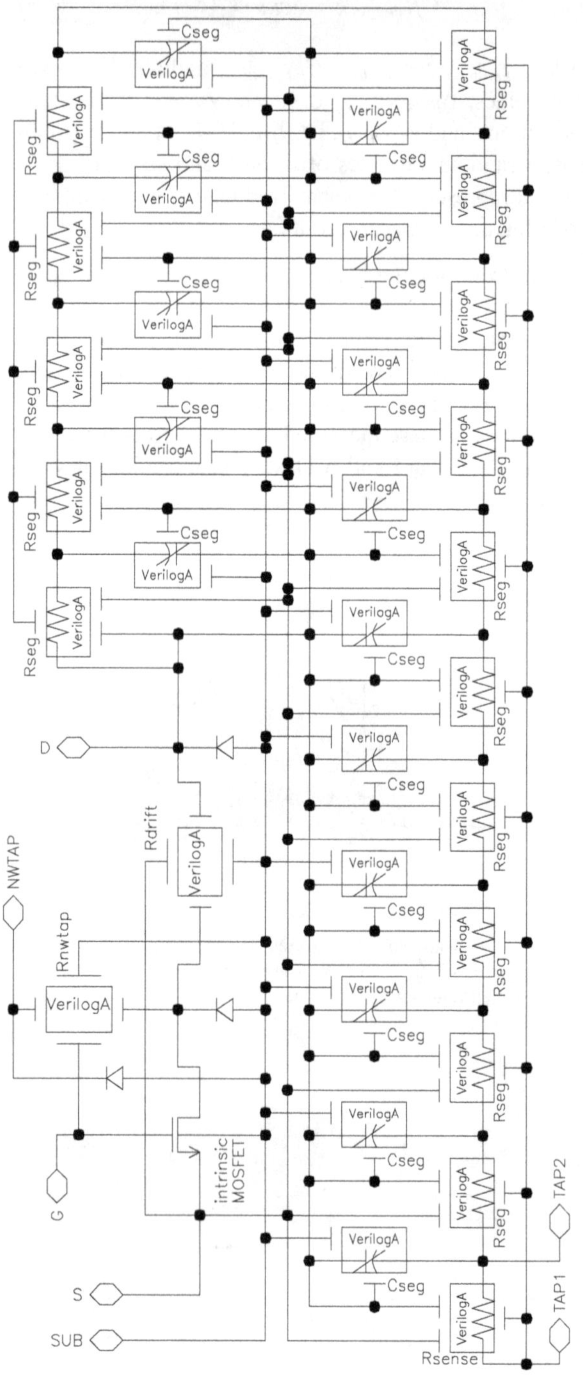

Fig. 12. The full macromodel circuit of HV start-up MOSFET with polysilicon spiral divider.

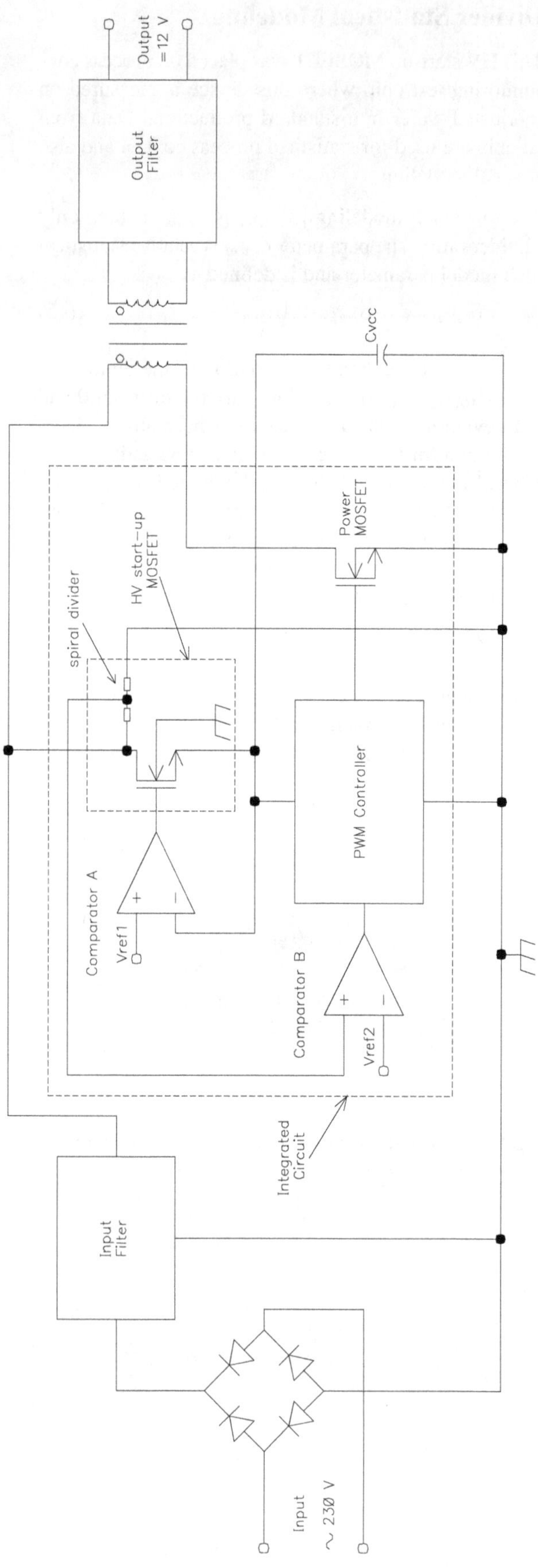

Fig. 13. The simplified circuit of AC/DC convertor. The HV start-up MOSFET subblock is modeled by the circuit in Fig. 12.

The voltage reference V_{ref1} determines the voltage (in Comparator A) to which the external capacitor C_{vcc} is charged up and basically determines the voltage at which the HV start-up MOSFET is switched off. This solution decreases the power consumption in comparison with older solution where the external capacitor C_{vcc} was permanently charged through an external resistor. The solution with HV spiral divider enables to switch off the charging in case the external capacitor C_{vcc} is charged up enough. The voltage reference V_{ref2} determines when the whole AC/DC convertor is switched on based on direct sensing of high voltage by the spiral divider described in this paper. The high input voltage is divided by the spiral divider and compared in Comparator B with the voltage reference V_{ref2}. It means that AC/DC convertor is turned off if the input high voltage is lower than defined value (depends on product specification, e.g. 112 V). See [12] or [13] for more applications.

The circuits that use HV start-up MOSFET with spiral divider allow designing applications with many features such as:

- The dynamic self-supply,
- No need of auxiliary winding [13],
- Low standby-power,
- High voltage sensing,
- Brown-out protection [12],
- Line overvoltage protection [12].

4. Conclusions

The enhanced accurate DC and RF model of nonlinear spiral polysilicon voltage divider in high voltage start-up MOSFET transistor has been created and is presented in this paper. The modeling results are compared with measured data and the maximal relative model error of divider ratio is less than 1.1 %. The intrinsic MOSFET is modeled by the standard BSIM3v3 model described in [8, 9, 10, 11].

A special attention is paid to the ability of the created model to cover the influence of a variation of process parameters on the device characteristics. The statistical process variation model is created based on measurement about 30000 devices and mismatch model is based on measurement about 3000 devices. It has to be pointed out that Monte Carlo simulations represent the most powerful tool to verify the robustness of a designed circuit over natural process and mismatch variations.

The big advantage of this model are smooth derivatives of simulated characteristics. The simulation speed is acceptable and any convergency issue was not observed during the verification realized on several real designs.

Appendix

The AC measurement setup is depicted in Fig. 14. The drain is biased to DC 50, 100, 300, and 400 V and through C_{bias} to AC source. Pin tap1 of the spiral is connected to the ground and pin tap2 is measured output. Gate, source and bulk are connected to the ground.

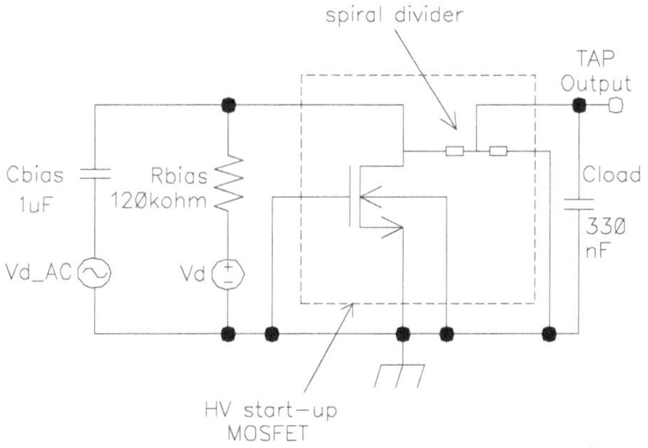

Fig. 14. The AC measurement setup.

Acknowledgements

This paper has been supported by ON Semiconductor company and also by the Grant Agency of the Czech Technical University in Prague, grant No. SGS14/082/OHK3/1T/13 and grant No. SGS13/206/OHK3/3T/13.

References

[1] HALL, J., QUDDUS, M. T., BURTON, R., OIKAWA, K., CHANG, G. *High Voltage Sensor Device and Method Therefor.* US patent 955943. 2011.

[2] HALL, J., QUDDUS, M. T. *Method of Sensing a High Voltage.* US Patent 8349625. 2013.

[3] WEISSTEIN, E. W. Archimedes Spiral. *MathWorld – A Wolfram Web Resource.* [Online] Cited 2014-12-02. Available at: http://mathworld.wolfram.com/ArchimedesSpiral.html

[4] PANKO, V., BANAS, S., PTACEK, K., BURTON, R., DOBES, J. An accurate DC and RF modeling of nonlinear spiral polysilicon voltage divider in high voltage MOSFET transistor. In *Proceedings of the IEEE 11th International Conference on Solid-State and Integrated Circuit Technology (ICSICT).* Xi'an (China), 2012. DOI: 10.1109/ICSICT.2012.6467635

[5] PAPATHANASIOU, K. A designer's approach to device mismatch: Theory, modeling, simulation techniques, applications and examples. *Analog Integrated Circuits and Signal Processing*, 2006, vol. 48, no. 2, p. 95–106. DOI: 10.1007/s10470-006-5367-2

[6] DRENNAN, P. G., McANDREW, C. C. Understanding MOSFET mismatch for analog design. *IEEE Journal of Solid-State Circuits*, 2003, vol. 38, no. 3, p. 450–456. DOI: 10.1109/JSSC.2002.808305

[7] McANDREW, C.C. Statistical modeling using backward propagation of variance (BPV). *Compact Modeling*. Springer Netherlands, 2010, p. 491–520. DOI: 10.1007/978-90-481-8614-3_16

[8] LIU, W. *MOSFET Models for SPICE Simulation: Including BSIM3v3 and BSIM4*. Wiley – IEEE Press, 2001.

[9] YTTERDAL, T., CHENG, Y., FJELDLY, T. A. *Device Modeling for Analog and RF CMOS Circuit Design*. Wiley, 2003.

[10] LIOU, J., ORTIZ-CONDE, A., GARCÍA-SÁNCHEZ, F. *Analysis and Design of MOSFETs: Modeling, Simulation, and Parameter Extraction*. Kluwer Academic Publishers, 1998. DOI: 10.1007/978-1-4615-5415-8

[11] MASSOBRIO, G., ANTOGNETTI, P. *Semiconductor Device Modeling with SPICE*, 2nd edition. McGraw-Hill, 1993.

[12] ON Semiconductor. *Design of a 65 W Adapter Utilizing the NCP1237 PWM Controller*. Application note AND8461/D. October 2014.

[13] ON Semiconductor. *Designing Converters with the NCP101X Family*. Application note AND8134/D. October 2003.

About the Authors ...

Václav PAŇKO was born in Czech Republic in 1979. He received his bachelor's degree in Electronics and Communications in 2006 and master's degree in Radioelectronics in 2008, both at the Czech Technical University in Prague, where he currently works toward the Ph.D. degree in Radioelectronics. In 2000, he joined ON Semiconductor, where he is presently a Senior Modeling and Characterization Engineer. His main interests include device modeling and parameter extraction, behavioral models development, parasitic effects modeling, and modeling and test chip design tools development.

Stanislav BANÁŠ was born in 1970. He received his M.Sc. degree in Electrical Engineering from the Technical University in Brno in 1994. Last year of his study he spent in scholarship in CNRS institute in Grenoble, where he was interested in optoelectronic properties of hydrogenated amorphous silicon. From 1996 he works as a modeling and characterization engineer in Motorola Czech Design Center in Roznov, later transferred to ON Semiconductor SCG Czech Design Center in Roznov. From 2012 he studies for Ph.D. in Technical University in Prague. His research interests include the modeling of high-voltage semiconductor components.

Richard BURTON received the Ph.D. degree in Electrical Engineering at Carnegie Mellon University in Pittsburg in 1994 simulating, designing, and fabricating integrated optoelectronics. From 1994 to 1996 he developed MESFET and HEMT technologies for cell phone power amplifiers. From 1996 to 2006 he developed manufacturable and reliable InGaP HBT technologies for OC-192 fiber optic communications PHY interfaces and highly rugged PAs for multi-band cellular communications. In 2006 he joined ON Semiconductor focusing on high reliable ultra-high voltage IC technology development for AC-DC off-line applications. His research interests include advanced technology development with focus on manufacturability and reliability.

Jan DIVÍN was born in Valašské Meziříčí, Czech Republic, in 1986. He works as characterization engineer of the ON Semiconductor company. He is a post gradual student of Czech Technical University in Prague at the Department of Radio Engineering. He has a M.Sc. in Electronics and Communication from the Brno University of Technology. His Ph.D. study is devoted to characterization of new model types of radio-frequency semiconductor devices.

Karel PTÁČEK received his M.Sc. degree in Electrical Engineering from the Brno University of Technology, Czech Republic, in 2003. Currently, he works as a design engineer at ON Semiconductor, Czech Republic. His research interests include designing of monolithic high voltage devices as well as their ESD protections. He works on his Ph.D. thesis with the topic of communication between galvanically isolated high voltage and low voltage parts of an integrated circuit.

Josef DOBEŠ received the Ph.D. degree in microelectronics at the Czech Technical University in Prague in 1986. From 1986 to 1992, he was a researcher of the TESLA Research Institute, where he performed analyses on algorithms for CMOS Technology Simulators. Currently, he works at the Department of Radio Electronics of the Czech Technical University in Prague. His research interests include the physical modeling of radio electronic circuit elements, especially RF and microwave transistors and transmission lines, creating or improving special algorithms for the circuit analysis and optimization, such as time- and frequency-domain sensitivity, poles-zeros or steady-state analyses, and creating a comprehensive CAD tool for the analysis and optimization of RF and microwave circuits.

A Wilkinson Power Divider with Harmonic Suppression and Size Reduction using High-low Impedance Resonator Cells

Mohsen HAYATI, Ashkan ABDIPOUR, Arash ABDIPOUR

Electrical Engineering Dept., Faculty of Engineering, Razi University, Tagh-E-Bostan, Kermanshah-67149, Iran

mohsen_hayati@yahoo.com, Ashkan_abdipour@yahoo.com, Arash.abdipour@yahoo.com

Abstract. *A miniaturized Wilkinson power divider using high-low impedance resonator cells is designed and fabricated. The proposed power divider occupies 23.7 % of the conventional structure circuit area at the operating frequency of 0.9 GHz and it is also able to suppress harmonics. According to the measured results at 0.9 GHz, the insertion-losses of output ports are 3.087 dB, the return-losses at all ports are more than 30 dB, and the isolation between output ports is better than 35 dB. Also, 2nd to 10th spurious frequencies are suppressed. According to the measured S_{11}, when it is less than -15 dB (from 0.65 GHz to 1.1 GHz) the fractional bandwidth of the proposed structure is 50 %. Good agreement between simulation and measured results is achieved.*

Keywords

Harmonic suppression, high-low impedance resonator, miniaturized Wilkinson power divider

1. Introduction

The power dividers are extremely important devices in microwave and millimeter-wave systems such as mixers, frequency multipliers, and the feeding network for an antenna array. There are several approaches for size reduction and harmonic suppression in the process of designing power dividers. The reported Wilkinson power divider with conventional quarter-wavelength transmission-line (TLIN) in [1], occupies a large area (especially at low operating frequencies) and it is not capable of suppressing spurious frequencies. In order to overcome these disadvantages several methods are used to reduce overall circuit size and suppress unwanted harmonics in the Wilkinson power dividers [2–7]. For instance, for size reduction and harmonic suppression, a kind of Wilkinson power divider based on standard printed-circuit-board (PCB) etching processes is reported in [2]. This structure is designed considering slow-wave loading and reduced the occupied area to 36.5% of the conventional structure at its operating frequency. In [3] and [4], microstrip electromagnetic band-gap (EBG) structure has been applied to the conventional power divider design, which led to the miniaturization and harmonic rejection in the conventional Wilkinson power divider.

Utilizing defected ground structure (DGS) can reject unwanted harmonics and decrease the occupied area in power dividers as it has been reported in [5] and [6]. Since, (DGS) and (EBG) need etching process, so their fabrication process is complex and these methods are not useable on a metal surface. The use of π-equivalent shunt-stub-based artificial transmission lines can effectively decrease the circuit size of conventional power divider [7]. This structure occupies 14.7% of the conventional power divider, but it is not capable for harmonic rejection at its operating frequency of 0.9 GHz.

In this paper, a Wilkinson power divider with harmonic suppression and size reduction is proposed. In order to reduce the circuit size of the conventional Wilkinson power divider, transmission lines with high-low impedance resonators are used instead of conventional quarter-wavelength TLIN sections. This technique not only reduces the occupied area to 23.7% of conventional one at operating frequency of 0.9 GHz, but also suppresses the second up to tenth harmonics.

2. Power Divider Design

2.1 The Procedure of Designing and the Effect of High-low Impedance Resonator on Size Reduction and Harmonic Suppression

Utilizing traditional TLIN in the structure of power dividers results in a large occupied area. Furthermore, this kind of transmission line is not able to suppress spurious frequencies. Using transmission line with loaded capacitance instead of quarter-wavelength transmission-line not only reduces the circuit size, but also can suppress unwanted harmonics. In the first step, a conventional Wilkinson power divider with an operating frequency at 0.9 GHz is designed as it is shown in Fig. 1a. In order to make capacitor loading on each $\lambda/4$ TLIN, four resonators with primary dimensions of W1 = W2 = 0.1 mm, L1 = L2 = 0.1 mm, W3 = 0.1 mm and L3 = 0.1 mm are added inside the free area of the conventional structure. These values are selected to control the effects of changing dimensions on frequency response and determine the operating frequency. The locations of the added resonators are determined with a, b, c and d in Fig. 1b. By increasing the values of W1, W2, L1 and L2 as low impedance TLINs a large loaded capacitance can be obtained. In order to reduce the occupied area of the power divider, the length of the main TLIN can decrease simultaneously, with increasing the dimensions of low impedance TLINs. Note that changing the values of variables does not have to shift the desired operating frequency, i.e. 0.9 GHz. Furthermore, adding these resonators makes a lowpass filter on the each main transmission line of the designed circuit. It appears in the insertion loss (S21), because of high order harmonics suppression in the frequency response.

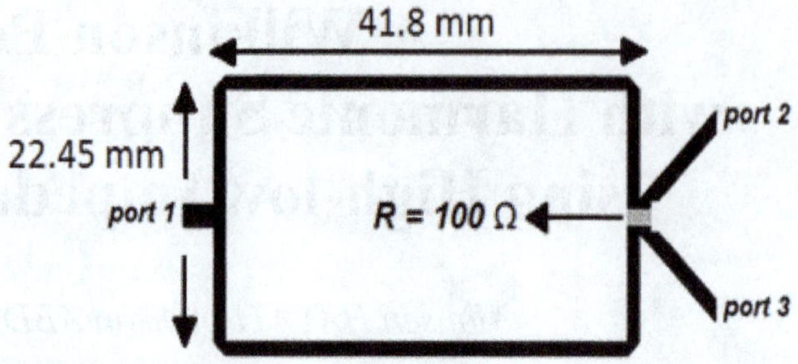

Fig. 1a. Topology of conventional Wilkinson power divider at operating frequency equal to 0.9 GHz.

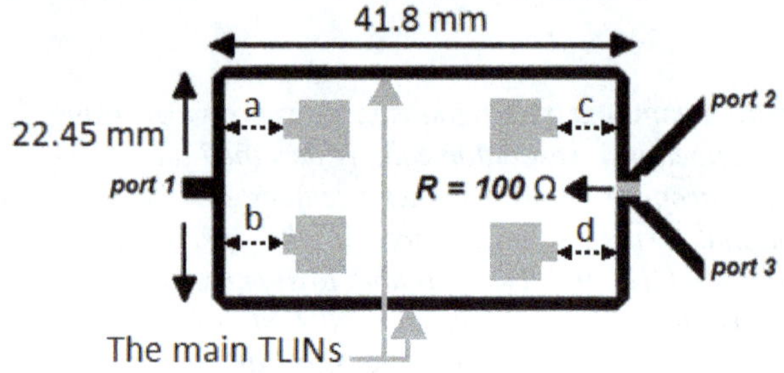

Fig. 1b. Topology of conventional Wilkinson power divider at operating frequency of 0.9 GHz with the locations of added high-low impedance resonators.

The proposed power divider at 0.9 GHz and its equivalent circuit using lumped components are shown in Fig. 2a and Fig. 2b, respectively. In Fig. 2b L_b, L_c and L_d are equivalent inductors caused by the main transmission line. High-low impedance resonators are modeled by L_a, $C1$ and $C2$, where L_a determines high impedance transmission lines of these resonators. $C1$ and $C2$ show low impedance open-circuit transmission lines of resonators 1 and 2, respectively. The gaps g_1, g_2 and g_3 between low impedance open-circuit lines cause coupling effects, which are modeled by C_{g1}, C_{g2} and C_{g3}. Furthermore, C_{p1} and C_{p2} present the capacitance between the microstrip structure and the ground. *Lout* accounts for inductor of output transmission lines. Notice that the coupling capacitances between the main transmission line and open-circuit transmission lines are not included in the LC circuit as they are trivial.

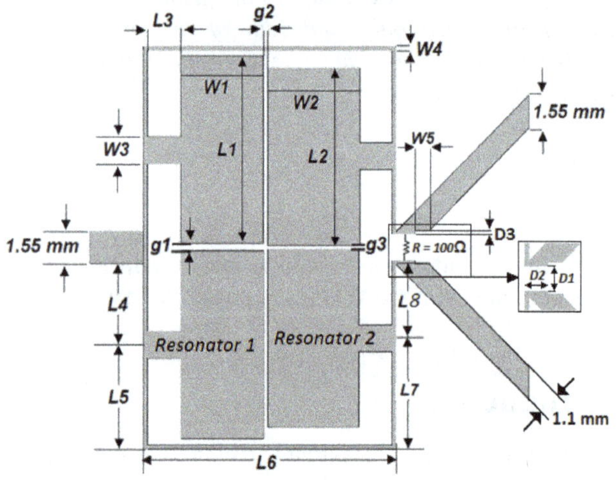

Fig. 2a. Topology of the proposed Wilkinson power divider.

Open-stub loads in the structure of high-low impedance resonators of the proposed power divider, modeled by $C1$ and $C2$ lead to a large shunt capacitance. Therefore, the circuit size of the proposed Wilkinson power divider could reduce because the propagation constant, i.e. β enhances (βproposed/βconventional is about 1.923). The relationship for β is given by:

$$\beta = \omega\sqrt{LC}\ , \tag{1a}$$

$$\beta = \frac{2\pi}{\lambda_g} \tag{1b}$$

where in (1a) L is the total inductance in per length unit of the main transmission line and high impedance line, and C depicts the total capacitance in per length unit of the main transmission line. In (1b) λ_g determines guided wavelength. Since C (the capacitance of proposed power divider) is increased in comparison with the transmission line of a conventional power divider, the propagation constant is enhanced considerably. As a result, the occupied area of circuit will be decreased [1].

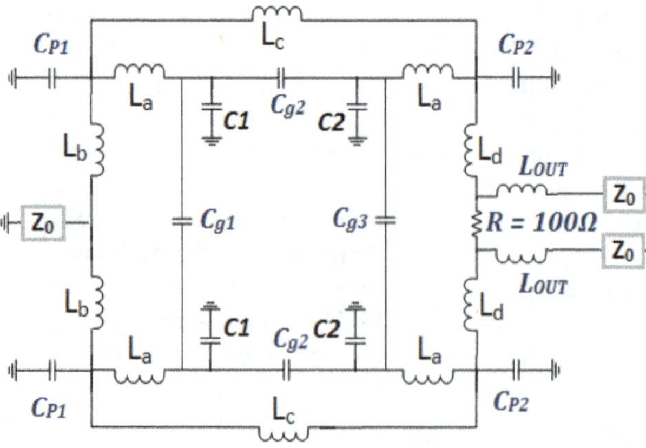

Fig. 2b. Equivalent LC circuit of the proposed Wilkinson power divider.

Moreover, based on insertion loss (S21) of the proposed structure shown in Fig. 6, optimized transmission line with high-low impedance resonators in higher frequencies has features of a lowpass filter. Spurious resonant frequencies of the resonator have been shifted by the high-low impedance resonators from the integer multiples of the basic resonant frequency [1]. So, replacing conventional quarter-wavelength transmission-line with a transmission line loaded by the high-low impedance resonators leads to harmonic suppression.

2.2 The Structure of the Proposed Power Divider

Comparison between the topology of conventional power divider in Fig. 1a and the proposed design, illustrated in Fig. 2a shows that in the proposed power divider, four microstrip high-low impedance resonator cells are used within the free area of the conventional Wilkinson power divider. The low impedance patches with rectangular shapes are microstrip open-stubs, so each of the high-low impedance resonator cells refers to a loaded capacitor. As a result, the capacitor loading not only reduces the circuit size, but also can suppress spurious frequencies.

The designed resonators, i.e. resonators 1 and 2 in both sides (left and right) have the same structure, but their rectangular patches have different dimensions. The circuit size of the proposed power divider and the conventional structure are 18.55 mm × 12 mm and 41.8 mm × 22.45 mm, respectively. It shows that the occupied area is reduced to 23.7% of the conventional power divider at operating frequency of 0.9 GHz. The type of the used 100 Ω isolation resistor is 0603, which is placed between two output ports. The dimensions of the proposed power divider shown in Fig. 2a are: W1 = 3.9, W2 = 4.3, W3 = 1.2, W4 = 0.2, W5 = 0.66, D1 = 1.15, D2 = 1.1, D3 = 0.2, L1 = 8.6, L2 = 8.2, L3 = 1.6, L4 = 3.7, L5 = 4.8, L6 = 12, L7 = 5.1, L8 = 3.4, g1 = 0.3, g2 = 0.2 and g3 = 0.15 (all in millimeter). The calculated values of inductors and capacitors of the shown LC circuit in Fig. 2b are [8]: La = 0.438 nH, Lb = 1.838 nH ,Lc = 12.38 nH, Ld = 1.68 nH, C1 = 1.36 pF, C2 = 1.43 pF, Cg1 = 25 fF, Cg2 = 128 fF, Cg3 = 42 fF,Cp1 = 0.471 pF, Cp2 = 0.466 pF. Note that the values of Cg1, Cg2 and Cg3 are achieved by tuning. Comparison between LC simulation and EM simulation results of the shown circuits in Figs. 2a and b on a substrate with permittivity of 2.2, thickness of 0.508 mm and loss tangent of 0.0009 are depicted in Figs. 3-6.

3. Simulated and Measured Results

The measured and simulated results of the proposed power divider are accomplished using Agilent's ADS Electromagnetic simulator (EM Simulator) software and HP 8720B vector network analyzer, respectively. The operating frequency of the proposed structure is located at 0.9 GHz. The designed microstrip Wilkinson power

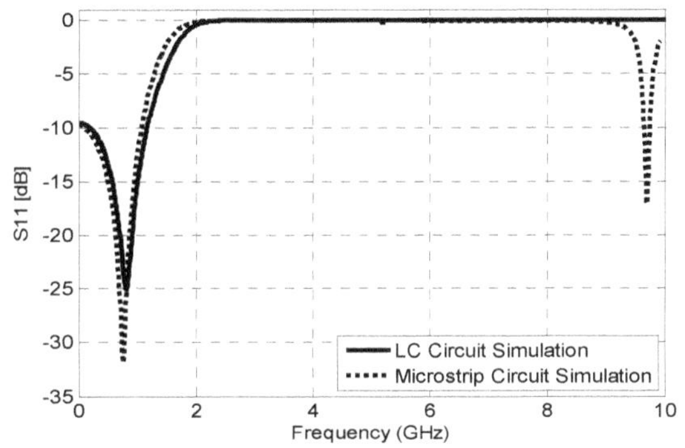

Fig. 3. Comparison between LC simulation and EM simulation results of input return-loss.

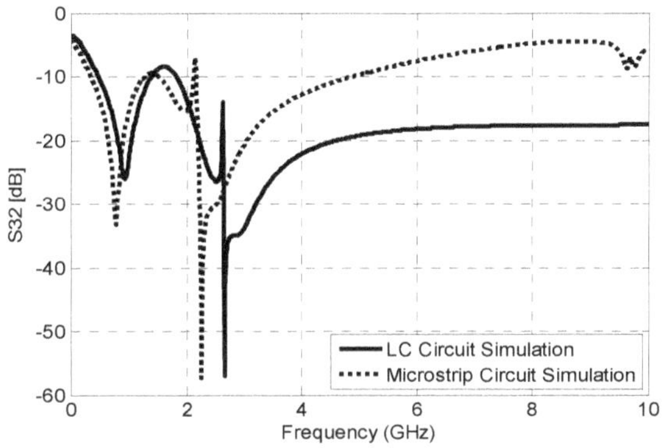

Fig. 4. Comparison between LC simulation and EM simulation results of isolation.

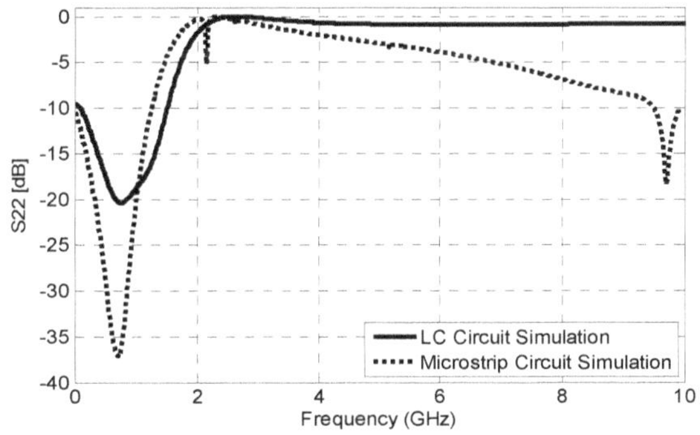

Fig. 5. Comparison between LC simulation and EM simulation results of output return-loss.

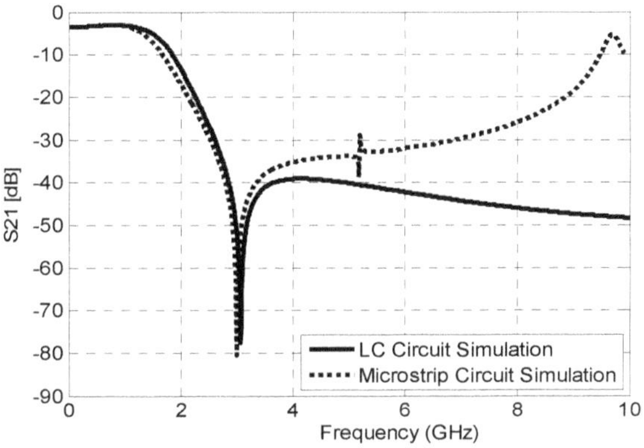

Fig. 6. Comparison between LC simulation and EM simulation results of insertion-loss.

divider is fabricated on RT/Duroid 5880 substrate with the thickness of 0.508 mm, the permittivity of 2.2 and the loss tangent of 0.0009. The results of measurement and simulation of S-parameters are illustrated in Figs. 7–11. As it is shown in Fig. 7, the measured return loss (S11) is at least –15 dB from 0.65 GHz to 1.1 GHz. In Fig. 8, the measurement shows over the frequency range 0.66–1.12 GHz the isolation (S23) is better than –15 dB. According to Fig. 9, the output return loss (S22) less than –15 dB from 0.28 GHz to 1.25 GHz is achieved. It can be observed from the measured insertion loss (S21) in Fig. 10, both even and odd spurious harmonics from 1.8 GHz to 9 GHz, i.e. second to tenth harmonics have been suppressed, where the 3rd to 10th harmonics are suppressed with a level less than –20 dB and the second harmonic is suppressed with a level better than -11 dB. It is to be noted that the suppression of higher order harmonic frequencies is related to S21 and S31. Exactly at operating frequency equal to 0.9 GHz, the measured S11, S32, and S22 are –33 dB, –38.88 dB and –48 dB, respectively. Furthermore, the measured insertion loss shows that S21 at 0.9 GHz is –3.087 dB. The characteristic impedance of all three ports are 50 Ω. Tab. 1 shows the comparison between the proposed power divider and the other published works. Based on the results of measurement shown in Fig. 11, an appropriate phase performance between two output ports around operating frequency is achieved. The measured phase difference of ports 2 and 3 as output ports is ± 0.15°. It shows that the proposed Wilkinson power divider is symmetric, so |S21| = |S31| (thus, harmonic suppression is related to both S21 and S31) and |S22| = |S33|. The photograph of the proposed Wilkinson power divider is shown in Fig. 12.

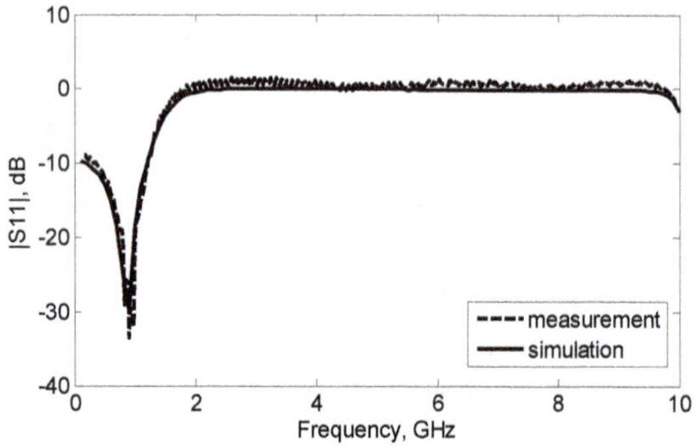

Fig. 7. Simulated and measured input return-loss.

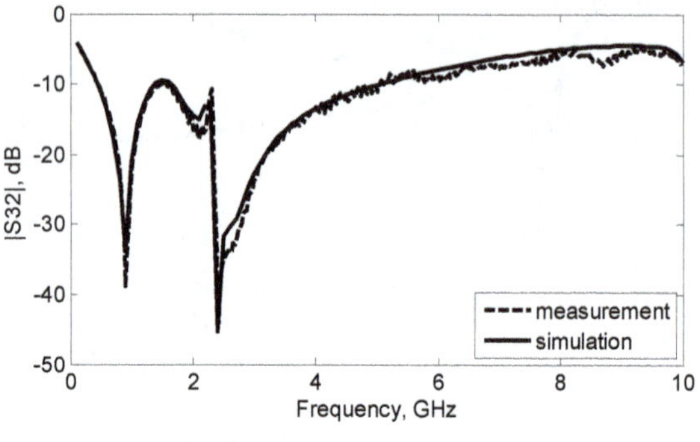

Fig. 8. Simulated and measured isolation.

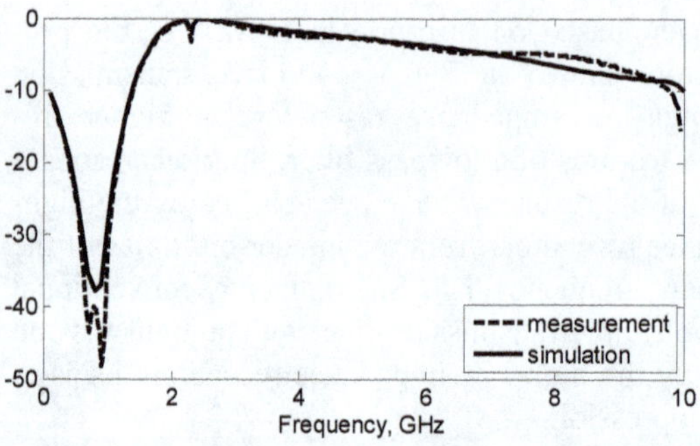

Fig. 9. Simulated and measured output return-loss.

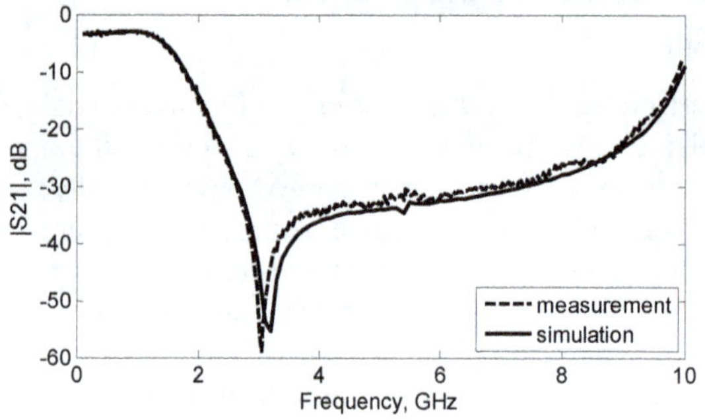

Fig. 10. Simulated and measured insertion-loss.

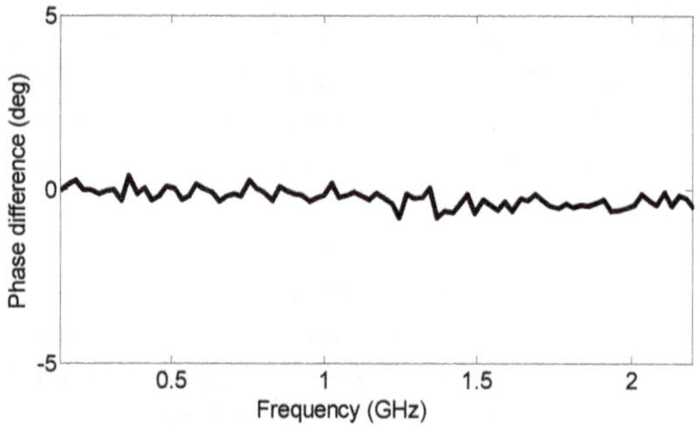

Fig. 11. Measured phase difference between S21 and S31 of the proposed power divider.

Fig. 12. Photograph of the proposed structure.

4. Conclusion

In this paper, a Wilkinson power divider using high-low impedance resonator cells for harmonic suppression and size reduction is proposed. The key features of the proposed structure are:

Ref.	Area reduction	Harmonic suppression (dB)								
		2nd	3rd	4th	5th	6th	7th	8th	9th	10th
[2]	63%	13	29	32	34	-	-	-	-	-
[3]	70%	8	32	10	12	-	-	-	-	-
[4]	39%	26	25	-	-	-		-	-	-
[5]	10%	18	15	-	-	-		-	-	-
[6]	66%	13	35	-	-	-		-	-	-
[7]	85.3%	-	-	-	-	-		-	-	-
This work	76.3%	11.3	31.5	35.5	33.2	32.4	30.1	25.9	25.9	22.4

Tab. 1. Comparison between the performance of the proposed power divider and previous works.

1- Small occupied area, i.e. 18.55 mm × 12 mm at the frequency of 0.9 GHz;

2- At operating frequency low insertion-losses of output ports (3.087 dB) and more than 30 dB return-losses at all ports are obtained. Moreover, better than 35 dB isolation and ±0.15° phase difference between output ports are achieved;

3- In the proposed structure spurious frequencies from 1.8 GHz up to 9 GHz, i.e. second to tenth harmonics are suppressed.

Therefore, the designed circuit with its operating frequency at 0.9 GHz can be used, where a power divider with small size and capable of suppress harmonics is required.

References

[1] POZAR, D. M. *Microwave Engineering*. 3rd ed. New York: Wiley, 2005, ch. 7, p. 333–337.

[2] WANG, J., NI, J., GUO, Y. X., FANG, D. Miniaturized microstrip Wilkinson power divider with harmonic suppression. *IEEE Microwave and Wireless Components Letters*, 2009, vol. 19, no. 7, p. 440–442. DOI: 10.1109/LMWC.2009.2022124

[3] LIN, C. M., SU, H. H., CHIU, J. C., WANG, Y. H. Wilkinson power divider using microstrip EBG cells for the suppression of harmonics. *IEEE Microwave and Wireless Components Letters*, 2007, vol. 17, no. 10, p. 700–702. DOI: 10.1109/LMWC.2007.905595

[4] ZHANG, F., LI, C. F. Power divider with microstrip electromagnetic band gap element for miniaturization and harmonic rejection. *Electronics Letters*, 2008, vol. 44, no. 6, p. 422–423. DOI: 10.1049/el:20083693

[5] WOO, D. J., LEE, T. K. Suppression of harmonics in Wilkinson power divider using dual-band rejection by asymmetric DGS. *IEEE Transactions on Microwave Theory and Techniques*, 2005, vol. 53, no. 6, p. 2139–2144. DOI: 10.1109/TMTT.2005.848772

[6] YANG, J., GU, C. F., WU, W. Design of novel compact coupled microstrip power divider with harmonic suppression. *IEEE Microwave and Wireless Components Letters*, 2008, vol. 18, no. 9, p. 572–574. DOI: 10.1109/LMWC.2008.2002444

[7] TSENG, C.-H., WU, C.-H. Compact planar Wilkinson power divider using pi-equivalent shunt-stub-based artificial transmission lines. *Electronics Letters*, 2010, vol. 46, no. 19, p. 1327–1328. DOI: 10.1049/el.2010.2194

[8] HONG, J.-S., LANCASTER, M. J. *Microstrip Filters for RF/Microwave Applications*. John Wiley & Sons, Inc., 2001.

About the Authors ...

Mohsen HAYATI received the BE in Electronics and Communication Engineering from Nagarjuna University, India, in 1985, and the ME and PhD in Electronics Engineering from Delhi University, Delhi, India, in 1987 and 1992, respectively. He joined the Electrical Engineering Dept., Kermanshah Branch, Islamic Azad University, Kermanshah, as a part time assistant professor in 2004. At present, he is a professor with the Electrical Engineering Dept., Kermanshah Branch, Islamic Azad University. He has published more than 155 papers in international and domestic journals and conferences. His current research interests include microwave and millimeter wave devices and circuits, application of computational intelligence, artificial neural networks, fuzzy systems, neuro-fuzzy systems, electronic circuit synthesis, modeling and simulations.

Ashkan ABDIPOUR received the B.S in Electronics Engineering from Islamic Azad University, Kermanshah Branch, Kermanshah, Iran, in 2009 and M.S degree from the Razi University, Kermanshah, Iran, in 2013. His research interests include microwave and millimeter wave devices and circuits.

Arash ABDIPOUR received the B.S in Electronics Engineering from Islamic Azad University, Kermanshah Branch, Kermanshah, Iran, in 2009 and M.S degree in Electronic Engineering from the University of Science and Research, Kermanshah Branch, Kermanshah, Iran in 2013. His research interests include microwave and millimeter wave devices and circuits.

A 1.2 V and 69 mW 60 GHz Multi-channel Tunable CMOS Receiver Design

Ahmet ONCU

Dept. of Electrical and Electronics Engineering, Bogazici University, Istanbul, 34342, Turkey

ahmet.oncu@boun.edu.tr

Abstract. *A multi-channel receiver operating between 56 GHz and 70 GHz for coverage of different 60 GHz bands worldwide is implemented with a 90 nm Complementary Metal-Oxide Semiconductor (CMOS) process. The receiver containing an LNA, a frequency down-conversion mixer and a variable gain amplifier incorporating a band-pass filter is designed and implemented. This integrated receiver is tested at four channels of center frequencies 58.3 GHz, 60.5 GHz, 62.6 GHz and 64.8 GHz, employing a frequency plan of an 8 GHz-intermediate frequency (IF). The achieved conversion gain by coarse gain control is between 4.8 dB and 54.9 dB. The millimeter-wave receiver circuit is biased with a 1.2 V supply voltage. The measured power consumption is 69 mW.*

Keywords

60 GHz, CMOS, integrated circuit, receiver design, low-power

1. Introduction

Research on the Complementary Metal-Oxide Semiconductor (CMOS) millimeter-wave transceiver operating at 60 GHz has been driven by the availability of the unlicensed bands in many countries, including Europe (57 ÷ 66 GHz), the United States (57 ÷ 64 GHz) and Japan (59 ÷ 66 GHz) as shown in Fig. 1. The available 9 GHz wide millimeter-wave band is also divided into approximately 2 GHz wide four sub-channels. This has resulted in many enabling design blocks and various modeling techniques reported in the literature [1–5]. Potential applications include low-power short-distance consumer applications such as wireless High-Definition Multimedia Interface (HDMI) for high-definition television (HDTV) video streaming and high data-rate wireless personal area networks. The distance limit is due to the 60 GHz electromagnetic waves being attenuated more as a result of oxygen absorption than at other frequencies. However, taking advantage of the wider multi-gigahertz bandwidth, short-distance applications can allow high data rates while allowing more frequency reuse in a limited area with minimal interference. A consequence is that the system offers higher security since signals cannot travel far beyond the intended recipients. The motivation of this work is to develop further such a system using the CMOS process technology. Traditional developments of the 60 GHz system have been largely confined to process technologies such as GaAs or SiGe. With price-sensitive services, lower production cost and possible integrations consumer electronics driving developments, CMOS technology offers the advantages of an accessible foundry with the digital baseband. The present state-of-the art CMOS process nodes at sub-100nm demonstrate device f_t that exceeds 400 GHz [6], thereby providing reasonable gain and other design margins for 60 GHz systems.

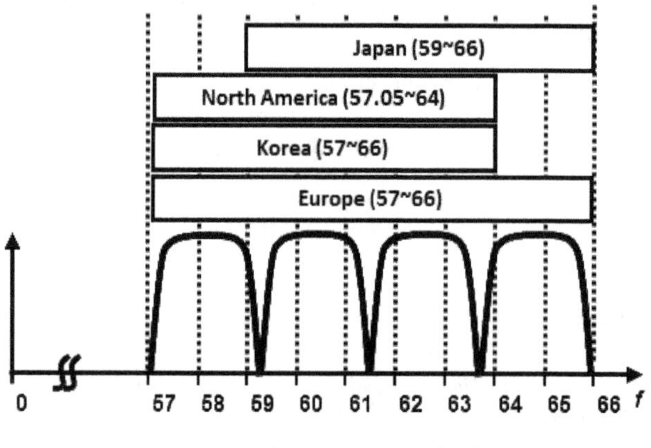

Fig. 1. Worldwide 60 GHz regulations with sub-channels.

A critical building block in the 60 GHz system is the millimeter-wave receiver, which includes the low-noise amplifier (LNA), down conversion mixer, band-pass filter and variable gain amplifier. In this work, a wideband CMOS receiver circuit that covers all 60 GHz bands and selects the sub-channels is proposed. In the following sections the design of the proposed receiver circuit and the performance of the implemented receiver are presented.

2. Design of Wideband Receiver

A block diagram of the proposed multi-channel tunable receiver based on the enhanced wide-band LNA is shown in Fig. 2. The receiver consists of an LNA, down-conversion mixer, an intermediate frequency (IF) variable gain amplifier (VGA) integrating a band-pass filter (BPF)

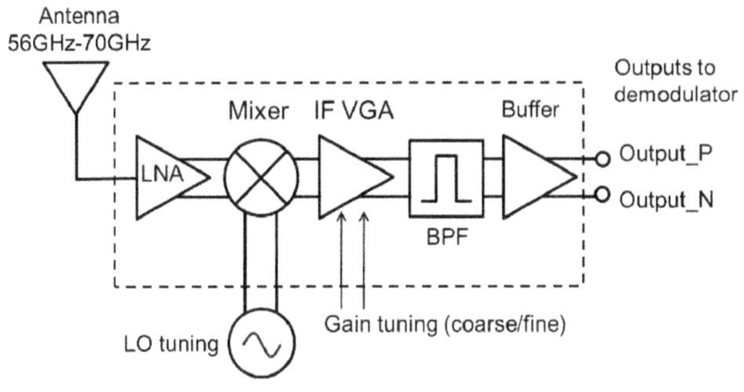

Fig. 2. Proposed receiver architecture.

and an output buffer. To test the performance of the proposed receiver chain, LO signal is externally applied, but in a practical application local oscillator will be implemented on the chip. A wideband LNA with active balun in the first stage to cover the all 60 GHz bands as reported in [9] is designed for this multi-channel receiver. Its schematic is shown in Fig. 3.

The LNA contains an internal balun. It has a single ended input and two outputs with 180° phase difference. The designed wideband LNA has 20 dB gain and 14 GHz bandwidth from 56 GHz to 70 GHz. It covers all 60 GHz bands sub-channels. The noise figure of the LNA is 6.8 dB at 60 GHz.

The main purpose of this receiver is to receive all four sub-channels of the license free 60 GHz bands from 57 GHz to 66 GHz. The LNA can amplify the signals from 56 GHz to 70 GHz frequency range. Therefore it is suitable for our application. The intermediate frequency is chosen as 8 GHz to be able to select the sub-channels and not to have an image rejection filter to have less complex receiver architecture. Since the frequency gap between the desired signal and the image will be 16 GHz, the image signal can be filtered out by the receiver front-end. For lower frequency bands, the image frequency will be far from the receiver's front-end coverage frequency band. When the desired frequency goes higher, image frequency goes closer to the coverage band.

When the receiver receives the 66 GHz signal, the highest frequency of the 60 GHz band sub-channel, the image frequency will be at 49 GHz in the worst case. It is outside of the LNA bandwidth. Therefore, the image frequency will be filtered by LNA. In practical wireless communication, an antenna will be connected to receiver input and it will also contribute to filter out the image signal.

Figure 4 shows the circuit of the implemented mixer. A double-balanced design is realized by the four transistors M10–M13 with each differential set of RF and LO inputs applied 180° out of phase. A modification made to this circuit from the standard Gilbert-cell configuration is the input of the *RF* signal directly into the source of the MOSFETs, instead of first passing through the gates of common source amplifiers. This current mode interface between the LNA and the mixer alleviates the low headroom limitation of the advanced CMOS process to improve the linearity and high-speed operation with a limited supply voltage

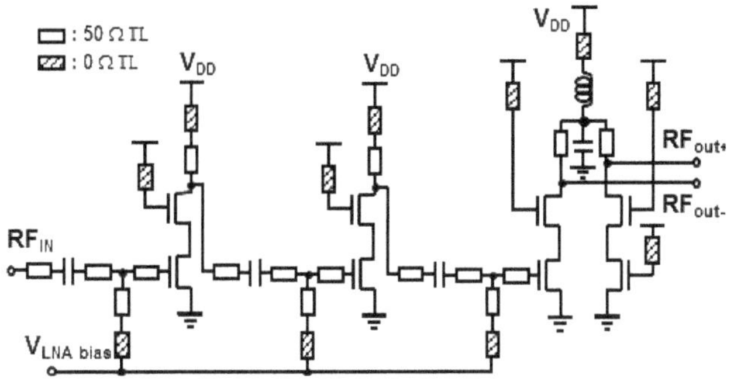

Fig. 3. Schematic of the LNA.

magnitude. A tail current control is then implemented by M9 to provide the appropriate DC current for the differential pairs. To avoid the noise contribution from M9 and its current source, a large bypass capacitance Cbypass is added [16]. Without the capacitor the noise from the M9 and the noise coming from the bias V_{b1} will inject to the mixer tail current in common mode. Due to the non-idealities in pair transistors, M10-M13, the noise will appear at the mixer output that will reduce receiver the sensitivity. A center tapped inductor shown by L5 and L6 are added to resonate out the parasitic capacitance at the source terminals of M10–M13. Thus, a high input impedance at the RF port from 56 GHz to 70 GHz can be achieved. These inductors can also reduce the signal loss caused by the parasitic capacitances. Inductors L1–L4 are chosen instead of resistors for gain and linearity purposes since an increase in resistance results in a decrease in the voltage headroom at the output node. Thus, there is a trade-off between conversion gain and linearity. In practice, L1–L4 and L5-L6 are implemented as single center-tap inductors in order to save chip area. The inductances L1–L4 and the parasitic capacitances of M10–M13 comprises two resonating tanks. The bandwidth of the tanks is equal to the bandwidth of the mixer. The bandwidth of the tank is narrower than the desired one. Therefore, in this design two resistors R1 and R2 are added to ensure 2 GHz bandwidth at 8 GHz IF center frequency.

In comparison with a conventional mixer, the proposed down-conversion mixer aims to achieve lower noise

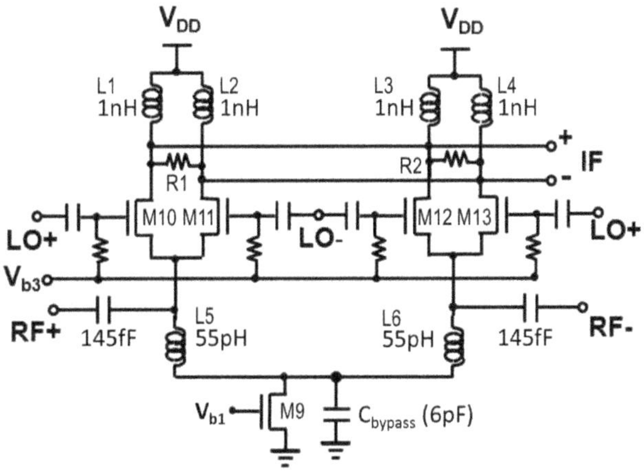

Fig. 4. Schematic of the mixer.

because of the absence of the noise generated from the transconductance stage in a conventional mixer. In addition, a DC current through this transconductance stage is not required. Thus the DC current can be reduced to improve the noise performance further. These considerations were necessary to reduce noise for stringent wideband and high frequency operations.

Figure 5 shows the circuit of the VGA and the BPF implemented. The VGA topology is a two-stage differential cascode. Due to the transconductance characteristic of the MOSFET, coarse tuning of the gain can be made in several discrete steps by selecting values of V_{b3} thereby directly setting M16, M17, M20 and M21 bias points. If desired, fine tuning can be made through V_{b2} at the gate of M14 and M15 to control the DC current through the differential cascodes.

The band-pass filter is implemented by two RLC circuits, each in parallel with the output of an amplifier stage. Fixed inductor sizes are used with capacitances to center the BPF at IF. The capacitances are tunable by varactors designed with a tuning range of 10%. Using this method, the 3-dB bandwidth can be controlled and the flatness of the BPF response can be adjusted with the proper selection of the value of R3 and R4 since the fractional bandwidth F_b of each parallel RLC can be characterized by (1). Note that the quality factor of the resonant tank is the inverse of F_b.

$$F_b = \frac{1}{R}\sqrt{\frac{L}{C}} \,. \qquad (1)$$

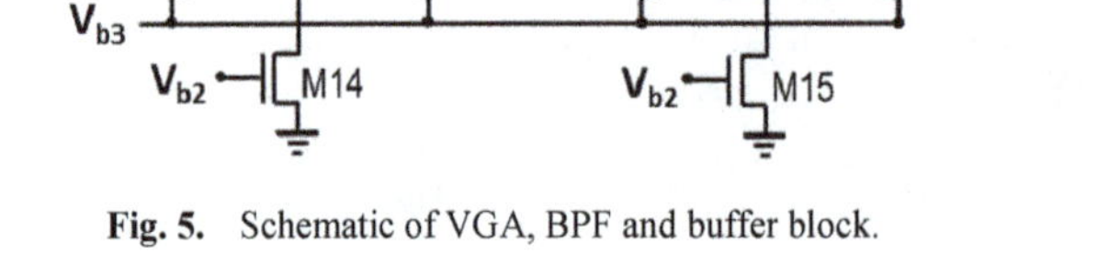

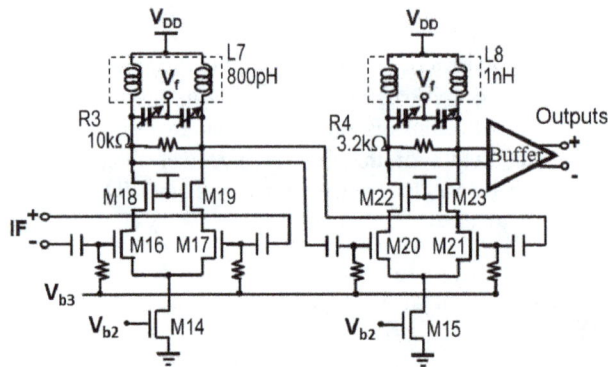

Fig. 5. Schematic of VGA, BPF and buffer block.

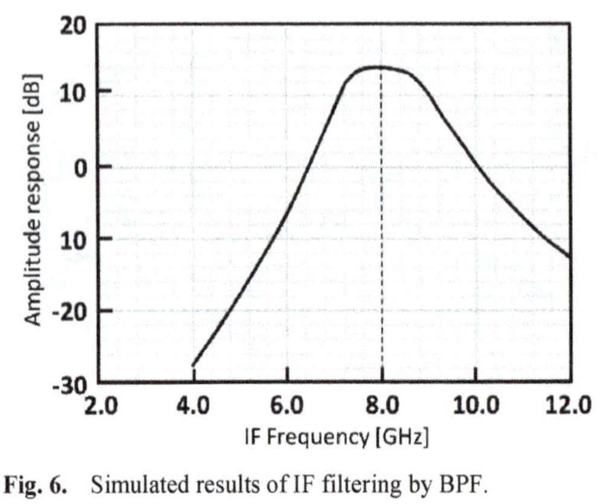

Fig. 6. Simulated results of IF filtering by BPF.

Equation (1) assumes a high-Q inductor. With a realistic CMOS inductor model [17], simulations were performed. Figure 6 shows the simulated results of the IF filtering by the BPF. The BPF is centered at 8 GHz with a bandwidth of 1.8 GHz. The simulated channel rejection at this bandwidth is 14.9 dB. Tuning is available with a range of 1.2 GHz. For measurement considerations, the unity gain open-drain buffers are used to drive the 50 Ω measurement system. It can drive the 50 Ω instruments and obtain the desired output signal. The buffer consumes 12.3 mW.

3. Experimental Result

Figure 7 shows the micrograph of the complete wideband receiver chip design. The LNA, mixer, VGA and band-pass filter are indicated on the micrograph. The receiver was fabricated using the 90 nm 1P8M CMOS process.

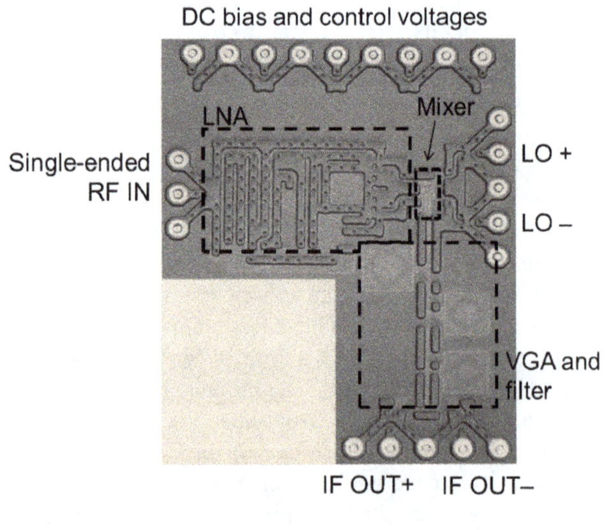

Fig. 7. Micrograph of wideband receiver chip.

The fabricated chip was tested by using a probe station. To test the performance of the fabricated receiver circuit the LO signal was applied from an external single ended millimeter-wave signal source by using an on chip GSG probe. But in practice the differential ended LO signal can be applied on the chip. It would reduce even order nonlinearities. In this measurement, the LO signal was injected at P_{LO} = -1 dBm at four frequencies corresponding to four different channels to evaluate the multi-channel characteristics of the receiver. The IF output was maintained at 8 GHz. Setting the gain control voltages to V_{b2} = 0.65 V and V_{b3} = 0.35 V, the IF output power response as a function of the RF input frequency is shown in Fig. 8. In this figure, a gain of 23 dB is observed with an input-referred 1 dB-compression point at -29 dBm. To measure the nonlinear characteristic of the receiver, two tone millimeter wave signal was applied from two external sources trough the RF input of the receiver by using a GSG probe. The measured fundamental and third order responses are shown in Fig. 9. The input referred third order interception point was measured to be a -20.5 dBm. The measured noise figure of the whole receiver is 8.1 dB. These performances can be acceptable for indoor short-range communications operating at low-power levels.

In such a scenario, adjacent channel interferences at 60 GHz are limited due to attenuation and the absence of large transmitter signals on the same chip, even though the receiver operates wideband. Depending on the linearity requirements of the modulation scheme, the trade-off between current drain for lower power consumption and linearity may be further adjusted. Figure 10(a) shows the results of the coarse tuning of the VGA gain.

As shown in Fig. 10, the coarse gain tuning of the VGA block is set by V_{b3} between 0.2 V and 0.5 V, corresponding to a minimum gain of 4.8 dB and a maximum of 54.9 dB. The coarse tuning demonstrates the limit of the gain obtainable by further fine tunings using V_{b2}. Results shown in Fig. 10(a) was obtained when V_{b2} is set to 0.65 V. It is noted that the gain flatness deteriorates at higher gain settings with a span of several gigahertz, resulting in changing 3-dB bandwidths. This is likely a result of the VGA input impedance changing due to bias changes, causing the mixer-VGA wideband inter-stage matching network. For a system with narrowband channels, it is possible to adjust the band-pass filter bandwidth to achieve better gain flatness over different gain settings. Alternatively, in our envisioned pulse communication system that typically employs On-Off-Keying (OOK) or Amplitude-Shift Keying (ASK), such gain control characteristics may be sufficient.

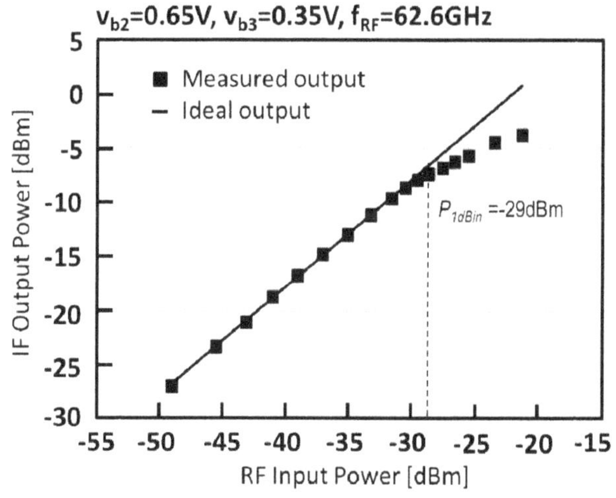

Fig. 8. Conversion gain characteristics of receiver.

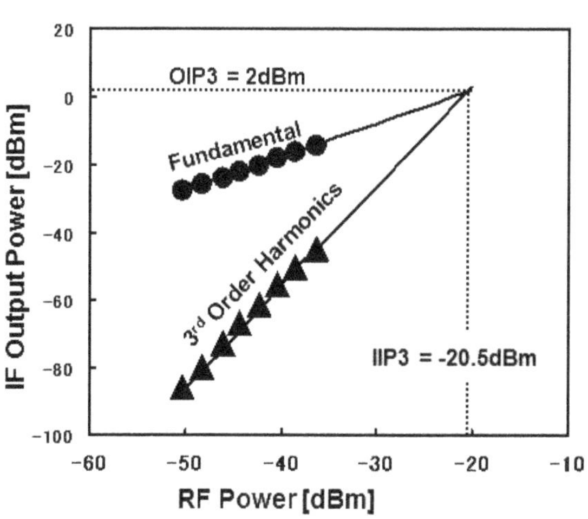

Fig. 9. Measured IIP3 of the receiver.

Measurement of the receiver channel selection by LO tuning is shown in Fig. 10(b). The LO frequency is selected at four frequencies of 50.3 GHz, 52.5 GHz, 54.6 GHz and 56.8 GHz. With the IF at 8 GHz, the corresponding RF-input frequencies for channels 1 to 4 are 58.3 GHz, 60.5 GHz, 62.6 GHz and 64.8 GHz, respectively. The LO power is set to -1 dBm. It is noted that the frequency response at the third channel (CH3) is about 2 dB lower than the other channels. It is believed that this performance is caused by the limited gain flatness of the LNA. Its gain can be compensated by VGA gain tuning. This proposed receiver reports a considerable gain-bandwidth performance. The proposed wide-band single receiver can select all four 60 GHz band sub-channels and its gain can be controlled to improve the communication performance. In Tab. 1, recently reported wireless 60-GHz

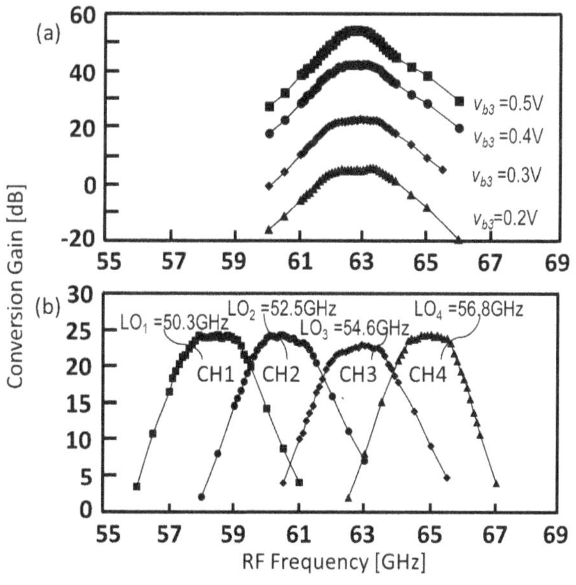

Fig. 10. (a) Receiver gain selection by VGA coarse tuning V_{b3} (V_{b2} set at 0.65 V). (b) Receiver channel selection by LO tuning (PLO = -1 dBm).

	Tech.	Freq. [GHz]	Gain [dB]	Power [mW]
This work	90nm CMOS	56.0-70.0	4.8-54.9	69.0
[18]	65nm CMOS	57.5-66.5	10.0-19.0	74.0
[19]	65nm CMOS	54.0-61.0	4.7-17.3	81.5
[20]	65nm CMOS	55.0-68.0	14.0-35.5	75.0
[21]	40nm CMOS	58.0	20.0	41.0
[22]	90nm CMOS	57-61.0	19.8-22.0	36.0
[23]	65nm CMOS	55.0-65.0	14.7	151.0
[24]	45nm CMOS	56.0-67.0	26.0	11.0-21.0
[25]	130nm SiGe	57.0-65.0	14.0	18.0

Tab. 1. Comparison of recent works.

band millimeter-wave receivers are compared. This proposed receiver using the enhanced LNA has the largest gain-bandwidth among them.

4. Conclusion

A multi-channel receiver operating between 56 GHz and 70 GHz for coverage of all four 60 GHz band sub-channels worldwide has been implemented with a 90 nm CMOS process. The receiver comprises of the millimeter-wave LNA, frequency mixer, VGA and bandpass filter. This enables the receiver to operate at wideband with reference measurements showing a 23 dB gain through LO injections of -1 dBm at each of the four channels tested, corresponding to receiver RF of 58.3 GHz, 60.5 GHz, 62.6 GHz and 64.8 GHz. Tuning of the VGA allows a range of gain to be selected with measured results between 4.8 dB and 54.9 dB. It requires a 1.2 V supply voltage and consumes 69 mW power. The proposed millimeter-wave multi-channel CMOS receiver can be used for low-voltage, low-power consuming and high-speed wireless communication applications.

Acknowledgements

The chip in this work was realized in 90nm CMOS process through Silicon Library Inc. Author thanks Prof. Minoru Fujishima for accessing millimeter-wave measurement facilities, Dr. Lai Chee Hong and Mr. Pong for their valuable discussion on low power-LNA and layout designs.

References

[1] DOAN, C. H., EMAMI, S., NIKNEJAD, A. M., BRODERSEN, R. W. Design of CMOS for 60GHz applications. In *IEEE International Solid-State Circuits Conference ISSCC Digest of Technical Papers*. 2004, p. 440–441. DOI: 10.1109/ISSCC.2004.1332783

[2] REYNOLDS, S. K., FLOYD, B. A., PFEIFFER, U. R., BEUKEMA, T., GRZYB, J., HAYMES, C., GAUCHER, B., SOYUER, M. A silicon 60-GHz receiver and transmitter chipset for broadband communications. *IEEE Journal on Solid-State Circuits*, 2006, vol. 41, no. 12, p. 2820–2831. DOI: 10.1109/JSSC.2006.884820

[3] LAI, I. C. H., KAMBAYASHI, Y., FUJISHIMA, M. 60-GHz CMOS down-conversion mixer with slow-wave matching transmission lines. In *Proceedings of the IEEE Asian Solid-State Circuits Conference ASSCC 2006*. Hangzhou (China), 2006, p. 195–198. DOI: 10.1109/ASSCC.2006.357884

[4] RAZAVI, B. A millimeter-wave CMOS heterodyne receiver with on-chip LO and divider. In *IEEE International Solid-State Circuits Conference ISSCC Digest of Technical Papers*, 2007, p. 188–189. DOI: 10.1109/ISSCC.2007.373357

[5] EMAMI, S., WISER, R. F., ALI, E., FORBES, M. G., GORDON, M. Q., GUAN, X., LO, S., MCELWEE, P. T., PARKER, J., TANI, J. R., GILBERT, J. M., DOAN, C. H. A 60GHz CMOS phased-array transceiver pair for multi-Gb/s wireless communications. In *IEEE ISSCC Digest of Technical Papers*, 2004, p. 164–166.

[6] YASUTAKE, N., OHUCHI, K., FUJIWARA, M., ADACHI, K., HOKAZONO, A., KOJIMA, K., AOKI, N., SUTO, H., WATANABE, T., MOROOKA, T., MIZUNO, H., MAGOSHI, S., SHIMIZU, T., MORI, S., OGUMA, H., SASAKI, T., OHMURA, M., MIYANO, K., YAMADA, H., TOMITA, H., MATSUSHITA, D., MURAOKA, K., INABA, S., TAKAYANAGI, M., ISHIMARU, K., ISHIUCHI, H. A hp22 nm node low operating power (LOP) technology with sub-10 nm gate length planar bulk CMOS devices. In *VLSI Technology Symposium Digest of Technical Papers*, 2004, p. 84–85. DOI: 10.1109/VLSIT.2004.1345407

[7] VARONEN, M., KARKKAINEN, M., KANTANEN, M., HALONEN, K. Millimeter-wave integrated circuits in 65-nm CMOS. *IEEE Journal of Solid-State Circuits*, 2008, vol. 43, no. 9, p. 1991–2002. DOI: 10.1109/JSSC.2008.2001902

[8] COSTANTINI, A., LAWRENCE, B., MAHON, S., HARVEY, J., MCCULLOCH, G., BESSEMOULIN, A. Broadband active and passive balun circuits: Functional blocks for modern millimeter-wave radio architectures. In *Proceedings of the 1st European Microwave Integrated Circuits Conference*. Manchester (UK), 2006, p. 421–424. DOI: 10.1109/EMICC.2006.282672

[9] NATSUKARI, Y., FUJISHIMA, M. 36 mW 63 GHz CMOS differential low-noise amplifier with 14 GHz bandwidth. In *CORD Conference Proceedings*. June 2009, p. 252–253.

[10] RAZAVI, B. A 60GHz direct-conversion CMOS receiver. In *IEEE International Solid-State Circuits Conference ISSCC Digest of Technical Papers*. San Francisco (USA), 2005, p. 400–401. DOI: 10.1109/ISSCC.2005.1494038

[11] MARCU, C., CHOWDHURY, D., THAKKAR, C., KONG, L. K., TABESH, M., PARK, J. D., WANG, Y., AFSHAR, B., GUPTA, A., ARBABIAN, A., GAMBINI, S., ZAMANI, R., NIKNEJAD, A. M., ALON, E. A 90nm CMOS low-power 60GHz transceiver with integrated baseband circuitry. In *IEEE International Solid-State Circuits Conference ISSCC Digest of Technical Papers*. 2009, p. 314–315.

[12] PINEL, S., SARKAR, S., SEN, P., PERUMANA, B., YEH, D., DAWN, D., LASKAR, J. A 90nm CMOS 60GHz radio. In *IEEE International Solid-State Circuits Conference ISSCC Digest of Technical Papers*. San Francisco (USA), 2008, p. 130 –131. DOI: 10.1109/ISSCC.2008.4523091

[13] LEE, J., HUANG, Y., CHEN, Y., LU, H., CHANG, C. A low-power fully integrated 60GHz transceiver system with OOK modulation and on-board antenna assembly. In *IEEE International Solid-State Circuits Conference ISSCC Digest of Technical Papers*. San Francisco (USA), 2009, p. 316 –318. DOI: 10.1109/ISSCC.2009.4977435

[14] WEYERS, C., MAYR, P., KUNZE, J. W., LANGMANN, U. A 22.3dB voltage gain 6.1dB NF 60GHz LNA in 65nm CMOS with differential output. In *International Solid-State Circuits Conference ISSCC Digest of Technical Papers*. San Francisco (USA), 2008, p. 192 –606. DOI: 10.1109/ISSCC.2008.4523122

[15] CHAO-SHIUN WANG, J. W. H. A 0.13μm CMOS fully differential receiver with on-chip baluns for 60GHz broadband wireless communications. In *Proc. of Custom Integrated Circuits Conference CICC 2008*. San Jose (USA), 2008, p. 479–482. DOI: 10.1109/CICC.2008.4672125

[16] CHEN, P.-H., CHEN, M.-C., KO, C.-L., WU, C.-Y. An integrated CMOS front-end receiver with a frequency tripler for V-band applications. *IEICE Transactions on Electronics*, 2010, vol. E93-C, no. 6, p. 877–883. DOI: 10.1587/transele.E93.C.877

[17] BLASCHKE, V., VICTORY, J. A scalable model methodology for octagonal differential and single-ended inductors. In *Proceedings*

of Custom Integrated Circuits Conference CICC 2006. San Jose (USA), 2006, p. 717 –720. DOI: 10.1109/CICC.2006.320897

[18] SILIGARIS, A., RICHARD, O., MARTINEAU, B., MOUNET, C., CHAIX, F., FERRAGUT, R., DEHOS,C., LANTERI, J., DUSSOPT, L., YAMAMOTO, S. D., PILARD, R., BUSSON, P., CATHELIN, A., BELOT, D., VINCENT, P. A 65nm CMOS fully integrated transceiver module for 60GHz wireless HD applications. In *IEEE International Solid-State Circuits Conference ISSCC Digest of Technical Papers 2011.* San Francisco (USA), 2011, p. 162–164. DOI: 10.1109/ISSCC.2011.5746264

[19] OKADA, K., MATSUSHITA, K., BUNSEN, K., MURAKAMI, R., MUSA, A., SATO, T., ASADA, H., TAKAYAMA, N., LI, N., ITO, S., CHAIVIPAS, W., MINAMI, R., MATSUZAWA, A. A 60GHz 16QAM/8PSK/QPSK/BPSK direct-conversion transceiver for IEEE 802.15.3c. In *IEEE International Solid-State Circuits Conference ISSCC 2011 Digest of Technical Papers.* San Francisco (USA), 2011, p. 160–162. DOI: 10.1109/ISSCC.2011.5746263

[20] VECCHI, F., BOZZOLA, S., POZZONI, M., GUERMANDI, D., TEMPORITI, E., REPOSSI, M., DECANIS, U., MAZZANTI, A., SVELTO, F. A wideband mm-wave CMOS receiver for Gb/s communications employing interstage coupled resonators. In *IEEE International Solid-State Circuits Conference ISSCC 2010 Digest of Technical Papers.* San Francisco (USA), 2010, p. 220–221. DOI: 10.1109/ISSCC.2010.5433953

[21] KAWASAKI, K., AKIYAMA, Y., KOMORI, K., UNO, M., TAKEUCHI, H., ITAGAKI, T., HINO, Y., KAWASAKI, Y., ITO, K., HAJIMIRI, A. A millimeter-wave intra-connect solution. In *IEEE International Solid-State Circuits Conference ISSCC 2010 Digest of Technical Papers.* San Francisco (USA), 2010, p. 414 to 415. DOI: 10.1109/ISSCC.2010.5433831

[22] PARSA A., RAZAVI, B. A new transceiver architecture for the 60-GHz band. *IEEE Journal of Solid-State Circuits*, 2009, vol. 44, no. 3, p. 751–762. DOI: 10.1109/JSSC.2008.2012368

[23] TOMKINS, A., AROCA, R. A., YAMAMOTO, T., NICOLSON, S. T., DOI, Y., VOINIGESCU, S. P. A zero-IF 60 GHz 65 nm CMOS transceiver with direct BPSK modulation demonstrating up to 6 Gb/s data rates over a 2 m wireless link. *IEEE Journal of Solid-State Circuits*, 2009, vol. 44, no. 8, p. 2085–2099. DOI: 10.1109/JSSC.2009.2022918

[24] BORREMANS, J., RACZKOWSKI, K., WAMBACQ, P. A digitally controlled compact 57-to-66GHz front-end in 45nm digital CMOS. In *IEEE International Solid-State Circuits Conference ISSCC2009 Digest of Technical Papers.* San Francisco (USA), 2009, p. 492–493. DOI: 10.1109/ISSCC.2009.4977523

[25] NATARAJAN, A., TSAI, M.-D., FLOYD, B. 60GHz RF-path phase-shifting two-element phased-array front-end in silicon. In *Dig. Symposium on VLSI Circuit.* Kyoto (Japan), 2009, p. 250 to 251.

About the Author ...

Ahmet ONCU was born in Istanbul, Turkey in 1979. He received the B.S. degree in Physics and the B.S. degree in Electrical and Electronics Engineering from Middle East Technical University (METU), Ankara, Turkey, in 2001 and 2002, respectively. He received the M.S. degree in Microwave Engineering from the Technical University of Munich, Germany, in 2004. He received the PhD degree in Frontier Sciences from the University of Tokyo, Japan, in 2008. Currently he is an assistant professor at the Department of Electrical and Electronics Engineering, Bogazici University, Istanbul, Turkey. Dr. Oncu received FP7 Marie Curie Reintegration Grant for (project no: 268232) (UWB-IR) Study on Low-power Multi-Gbps Ultra-Wideband Impulse Radio at License-free UWB and 60GHz bands in 2010. His research interests are in designs of high-speed low-power CMOS analog and RF integrated circuits.

Pipelined Two-Operand Modular Adders

Maciej CZYŻAK, Jacek HORISZNY, Robert SMYK

Faculty of Electrical and Control Engineering, Gdansk University of Technology, G Narutowicza 11/12, 80-233

mczyzak@ely.pg.gda.pl, jhor@ely.pg.gda.pl, rsmyk@ely.pg.gda.pl

Abstract. *Pipelined two-operand modular adder (TOMA) is one of basic components used in digital signal processing (DSP) systems that use the residue number system (RNS). Such modular adders are used in binary/residue and residue/binary converters, residue multipliers and scalers as well as within residue processing channels. The structure of pipelined TOMAs is usually obtained by inserting an appropriate number of pipeline register layers within a nonpipelined TOMA structure. Hence the area of pipelined TOMAs is determined by the nonpipelined TOMA structure and by the total number of pipeline registers. In this paper we propose a new pipelined TOMA, that has a considerably smaller area and the attainable pipelining frequency comparable with other known pipelined TOMA structures. We perform comparisons of the area and pipelining frequency with TOMAs based on ripple carry adder (RCA), Hiasat TOMA and parallel-prefix adder (PPA) using the data from the very large scale of integration (VLSI) standard cell library.*

Keywords

Carry-lookahead adder, FPGA, modular adder, parallel-prefix adder, residue number system (RNS), ripple-carry adder, VLSI design

1. Introduction

Modular addition plays an important role in the implementation of digital signal processing systems that use the residue number system [1–4] as well as its derivatives like the quadratic residue number system (QRNS) [5] and modified quadratic residue number system (MQRNS) [6] for processing of complex signals. The RNS is a non-weighted integer number system that is determined by its base $\boldsymbol{B}=\{m_1, m_2, ..., m_n\}$ being the set of positive pairwise prime integers m_i , i = 1, 2,.., n. Each integer $X \in \boldsymbol{Z}_M$, $M = \prod_{i=1}^{n} m_i$ and can be represented as $X \leftrightarrow (x_1, x_2, \ldots, x_n) = (|X|_{m_1}, |X|_{m_2}, \ldots, |X|_{m_n})$ with $x_i \in \boldsymbol{Z}_{m_i}$. This mapping is the bijection and for $X, Y \in \boldsymbol{Z}_M$ and for $x_i, y_i \in \boldsymbol{Z}_{m_i}$, we have $z_i = |x_i \otimes y_i|_{m_i}$,where $\otimes$ denotes addition, subtraction or multiplication.

The reverse conversion from the RNS to a weighted system can be performed using the Chinese remainder theorem (CRT) [1], [2] or the mixed-radix system (MRS) [1], [2]. The main advantage of the RNS comes from the fact that addition, subtraction and multiplication are carry-free and can be performed without carries between individual positions of the number. The principal advantage of the RNS with respect to the high-speed DSP is due to the replacement of large multipliers that limit the pipelining frequency, by small multipliers modulo m_i. If their binary size $l = \lceil(\log_2 m_i)\rceil$, where $\lceil\bullet\rceil$ denotes rounding off to an integer, does not exceed six bits, multiplications by a constant can be performed by look-up with small ROMs or using combinatorial networks. General multiplications are also easier to perform because their standard realizations are small or segmentation of operands can be used for the combinatorial realization. It is worth mentioning that moduli with $l < 7$ may provide for the dynamic ranges over 90 bits [7]. The additional advantage of the RNS is the possibility of reducing power dissipation in CMOS circuits which is due to the lower switching activity and reduction of supply voltages [9]. The RNS has found numerous applications in the DSP, for example, in FIR filters [8–11], FFT processors [12], digital downconversion [13] and image processing [14], [15].

Generally TOMAs can be divided into two main categories determined by the type of the modulus. TOMAs for moduli akin to 2^n represent the first category and those for generic moduli the other. There are several works in the literature that consider the TOMA design.

Banerji [16] presented a look-up approach, Agrawal and Rao [17] proposed a TOMA for moduli of the form $(2^n + 1)$ based on binary adders. Soderstrand [18] introduced a hybrid approach based on look-up table along with the binary adder. Bayoumi and Jullien [19] described TOMAs using the table approach and binary adders approach. Dugdale [20] demonstrated an implementation of TOMAs that used binary adders, Piestrak [21] proposed a TOMA based on the carry-save adder (CSA) and two binary adders. Zimmermann [22] introduced modulo $(2^n \pm 1)$ adders based on parallel prefix-architecture (PPA). Hiasat [23] proposed a TOMA with the reduced area based on the carry-look-ahead (CLA) adder. Also a novel delay-power-area-efficient approach to the TOMA design was given by Patel et al. [24]. Their TOMA structure was based on the cascaded connection of the modified carry-save adder

(CSA) and reduced carry-propagate adder (CPA). The used CPA designs included ELM [25], Kogge-Stone [26] and Ladner Fischer [27] PPA.

In this paper we propose a new TOMA based on a modified CLA adder. This TOMA has the smaller area than other considered TOMAs and allows to derive a new pipelined TOMA that is better than other known pipelined TOMAs in terms of the area and the number of stages of pipeline registers. We shall show the structure of the new pipelined TOMA and, for comparison, TOMAs based on the RCA, PPA in the Brent-Kung form [28] and Hiasat TOMA [23]. Comparisons are made using the data from the VLSI standard cell library. We shall compare structures of individual TOMAs in terms of area, delay and pipelining frequency with the use of the additive method. The method uses summation of areas of individual components expressed in gate equivalents (GE), where 1 GE is the area of the NAND with the fan-out = 1 for the given standard cell library. The propagation delay of an individual element is taken as the worst case delay for all possible inputs. The analysis relies upon the established 130 nm Samsung standard cell library STDH150 [29]. Calculations of areas and delays of individual components are practically technology independent and they can be scaled down for VLSI technologies such as 28 nm or 22 nm. Therefore we may therefore suppose that for comparison of individual digital structures, the assumed technology will give sufficient and dependable information. The paper has the following structure: in Sec. 2 we review the basic TOMA structures, in Sec. 3 we consider the TOMA-RCA, and in Sec. 4 Hiasat TOMA, in Sec. 5 we present the TOMA based on the PPA adder and finally in Sec. 6 a new TOMA. In each section we analyze a nonpipelined and pipelined form.

2. Basic TOMA Structures Based on Binary Adders

In this section we shall shortly describe the basic known TOMA structures that use exclusively binary adders in series and which therefore may be the most suitable for transformation to the pipelined form and not those that use two parallel adders as in [21]. Two-operand modular addition for small m, $\lceil \log m \rceil \leq 6$ can be implemented by using the ROM ($2^{2 \cdot \lceil \log m \rceil} \times \lceil \log m \rceil$), but such approach remarkably reduces the attainable pipelining frequency.

The TOMA computes $r_m = |X+Y|_m$, where r_m is the least nonnegative remainder from the division $X + Y$ by the modulus m. Assuming $Z = 2^{\lceil \log m \rceil} - m$, the computation can be also expressed as

$$r = \begin{cases} |X+Y+Z|_{2^{\lceil \log m \rceil}} & \text{if } X+Y+Z \geq 2^{\lceil \log m \rceil} \\ X+Y & \text{otherwise} \end{cases} \quad (1a)$$

In Fig. 1 to 3 three basic TOMA structures are shown Bayoumi-Jullien, Hiasat, and Piestrak.

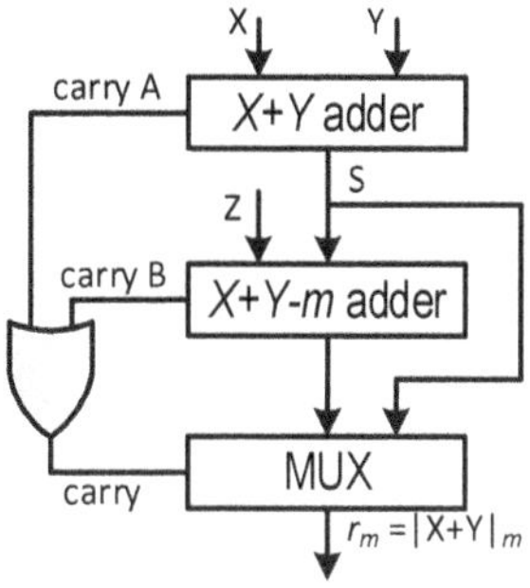

Fig. 1. Bayoumi-Jullien TOMA [19].

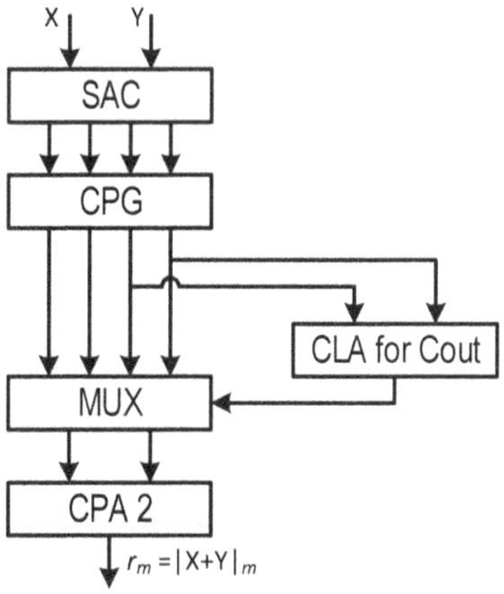

Fig. 2. Hiasat TOMA [23].

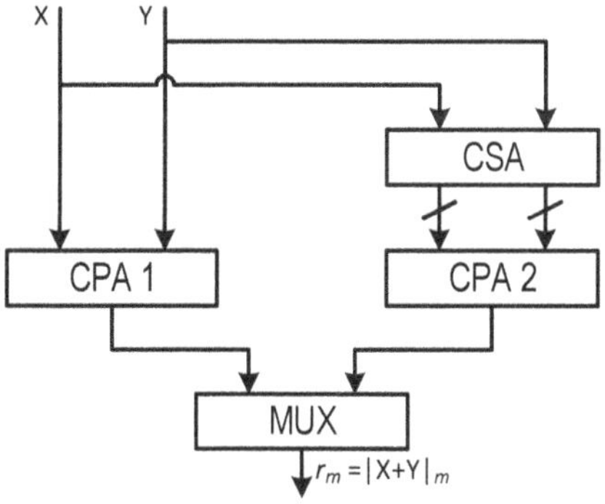

Fig. 3. Piestrak TOMA [21].

We shall shortly analyze the operation of the Bayoumi- Jullien TOMA (Fig. 1) because this structure will be the basis for the design of selected TOMAs. The binary adder in the first stage of this TOMA computes $X + Y$, whereas the second adder $X + Y - m$. The output of the TOMA is selected using $carry = carryA \vee carryB$. For $X + Y < m$, $carry = 0$ and $r_m = X + Y$, whereas for $X + Y \geq m$, $carry = 1$ and $r_m = X + Y - m$.

3. TOMA-RCA

By way of introduction we shall consider the realization of the Bayoumi-Jullien TOMA based on the RCA. In order to obtain a pipelined structure, layers of pipeline registers consisting of flip-flops (FFs) have to be inserted between individual adders as shown in Fig. 5.

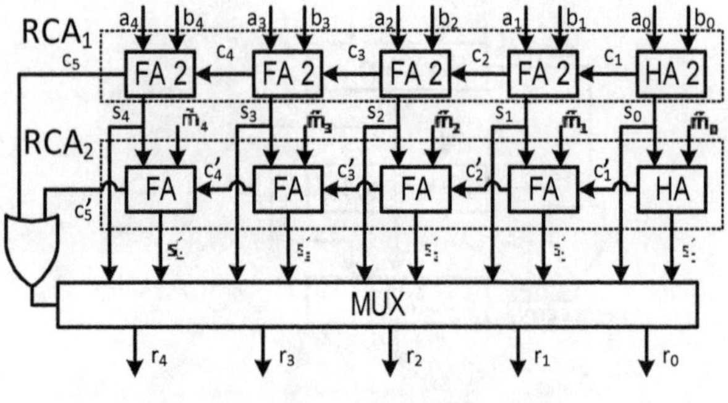

Fig. 4. Bayoumi-Jullien TOMA based on the RCA.

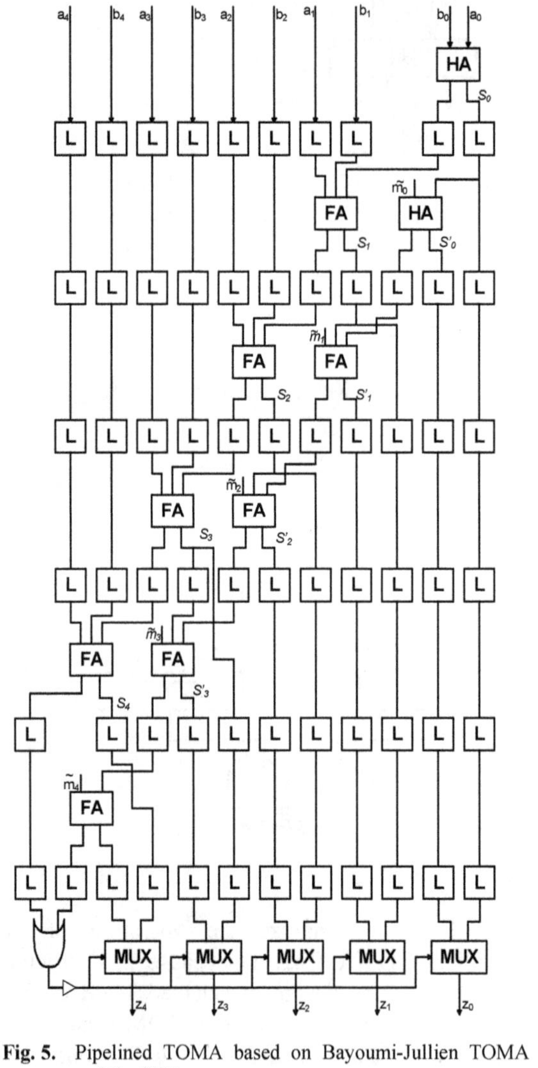

Fig. 5. Pipelined TOMA based on Bayoumi-Jullien TOMA and the RCAs.

In the following we shall analyze the area of the TOMA-RCA expressed in GE, the delay and the maximum attainable pipelining frequency. The area will be estimated using the areas of the individual components from STDH150, the delay for a nonpipelined structure will be evaluated by using the maximum delays for the individual components. In order to estimate the pipelining frequency a structure is divided into balanced layers with respect to the delay and the maximum pipelining frequency is obtained as the inverse of the sum of the delay of the slowest layer and the FF delay.

A. Nonpipelined 5-bit TOMA-RCA area

This area of 5-bit TOMA-RCA can be expressed in the following manner:

$$A_{TOMA\_RCA} = A_{HAd2} + 4 \cdot A_{FAd2} + A_{HAd1} + 4 \cdot A_{FAd1} + A_{OR2d1} + A_{NID6} + 5 \cdot A_{MX2d1} \,. \quad (1)$$

The indices of the individual components come from STDH150. The data of individual components is given in Appendix A. After inserting these data into (1) we obtain $A_{TOMA\_RCA} = 98.68\text{GE}$. The area given by (1) does not depend upon the form the of the two's complement system (TCS) representation of $-m$, $\tilde{m} = (1,\ldots,\tilde{m}_4,\tilde{m}_3,\tilde{m}_2,\tilde{m}_1,\tilde{m}_0)$. The particular form of this representation allows to reduce the area for the given modulus. For example, if $m_i = 0$, the HA reduces to single connection and for $m_i = 1$ to one connection and to one inverter. For the FA and $m_i = 0$, we have one XOR gate and a single AND gate, and for $m_i = 1$ one OR gate and exclusive NOR. For $m = 29$ and $\tilde{m} = (1,\ldots,0,0,0,1,1)$, we obtain $A_{TOMA\text{-}RCA} = 81.68GE$.

B. Nonpipelined 5-bit TOMA-RCA delay

We shall estimate the delay of the structure of Fig. 4 taking into consideration individual delays of signals inside individual HAs and FAs.

The delay of the 5-bit TOMA-RCA can be expressed as

$$t_{TOMA\text{-}RCA} = \max(t_{s_4}, t_{s'_4}, \max(t_{c_5}, t_{c'_5})) + t_{OR2d1} + t_{MX2d1} \,. \quad (2)$$

The delay for s_4 and c_5 bits can be calculated as

$$t_{s_4} = \max(t_{HAd2\_ACO}, t_{HAd2\_BCO}) + 3 \cdot t_{FAd2\_CICO} + t_{FAd2\_CIS},$$

$$t_{c_5} = \max(t_{HAd2\_ACO}, t_{HA\_BCO}) + 4 \cdot t_{FAd2\_CICO} \,.$$

In order to compute c'_5, we shall first calculate t_{c_i} and $t_{c'_i}$, $i = 1, 2, 3, 4$. We have

$$t_{c_1} = \max\left(t_{HAd2\_ACO}, t_{HAd2\_BCO}\right), \quad (3a)$$

$$t_{c_i} = t_{c_{i-1}} + t_{FAd2\_CICO}, \; i = 2,3,4,5 \,. \quad (3b)$$

Consequently

$$t_{c'_1} = \max\left(t_{HAd2\_AS}, t_{HAd2\_BS}\right) + t_{HAd1\_BCO}, \quad (4a)$$

$$t_{c'_i} = \max\left(t_{c_{i-1}} + t_{FAd2\_CIS} + t_{FAd2\_BCO}, t_{c'_{i-1}} + t_{FAd1\_CICO}\right), \; i = 2,3,4,5 \,, \quad (4b)$$

and $t_{s'_4} = t_{c'_4} + t_{FAd1\_CIS}$. (4c)

Example 1. Computation of 5-bit TOMA-RCA delay for components from the STDH150.

We shall first compute t_{s_4} and t_{c_5} as

$$t_{s_4} = \max(0.092\,\text{ns}, 0.074\,\text{ns}) + 3\cdot 0.089\ \text{ns} + 0.150\,\text{ns} =$$
$$= 0.509\,\text{ns},$$
$$t_{c_5} = \max(0.092\,\text{ns}, 0.074\,\text{ns}) + 4\cdot 0.089\,\text{ns} = 0.448\,\text{ns}.$$

Before we can compute $c_5^{'}$, we have to determine c_i, $i = 1, 2, 3, 4$. We have

$$t_{c_1} = \max(0.092\ \text{ns}, 0.074\ \text{ns}) = 0.092\ \text{ns},$$
$$t_{c_2} = 0.181\ \text{ns},$$
$$t_{c_3} = 0.270\ \text{ns},$$
$$t_{c_4} = 0.359\ \text{ns}.$$

Subsequently we obtain

$$t_{c_1^{'}} = \max(0.092\,\text{ns}, 0.074\,\text{ns}) + 0.055\,\text{ns} = 0.147\,\text{ns},$$
$$t_{c_2^{'}} = \max(0.092\,\text{ns} + 0.102\,\text{ns} + 0.152\,\text{ns}, 0.147\,\text{ns} + 0.083\,\text{ns})$$
$$= \max(0.346\,\text{ns}, 0.230\,\text{ns}) = 0.346\,\text{ns},$$
$$t_{c_3^{'}} = \max(0.181\,\text{ns} + 0.102\,\text{ns} + 0.143\,\text{ns}, 0.346\,\text{ns} + 0.083\,\text{ns})$$
$$= \max(0.426\,\text{ns}, 0.429\,\text{ns}) = 0.429\,\text{ns},$$
$$t_{c_4^{'}} = \max(0.270\,\text{ns} + 0.102\,\text{ns} + 0.143\,\text{ns}, 0.429\,\text{ns} + 0.083\,\text{ns})$$
$$= \max(0.515\,\text{ns}, 0.512\,\text{ns}) = 0.429\,\text{ns},$$
$$t_{c_5^{'}} = \max(0.359\,\text{ns} + 0.102\,\text{ns} + 0.143\,\text{ns}, 0.515\,\text{ns} + 0.083\,\text{ns})$$
$$= \max(0.604\,\text{ns}, 0.598\,\text{ns}) = 0.604\,\text{ns},$$
$$t_{s_4^{'}} = 0.515\,\text{ns} + 0.095\,\text{ns} = 0.601\,\text{ns}.$$

Finally, we may determine the TOMA-RCA delay as

$$t_D^{TOMA-RCA} = \max(0.509\ \text{ns}, 0.601\ \text{ns}, 0.604\ \text{ns} + 0.065\ \text{ns})$$
$$+ 0.078\ \text{ns} = 0.747\ \text{ns}.$$

C. The area of pipelined 5-bit TOMA-RCA

In Fig. 5 a pipelined form of the RCA-TOMA is presented. Six flip-flops stages are used with 66 flip-flops. The area is the sum of the nonpipelined 5-bit TOMA-RCA area and the area of pipeline registers. In this case these registers require $n_s = 66$ FFs. Thus the area can be expressed as

$$A_{TOMA\_RCA\_p} = A_{TOMA\_RCA} + n_s \cdot A_{FF}\ . \tag{5}$$

As A_{FF} we shall use the area of the flip-flop FD1Q, A_{FD1Q} from STDH150. For the structure from Fig. 5 we receive $A_{TOMA\text{-}RCAp} = 472.9$ GE.

D. Pipelined 5-bit RCA-TOMA pipelining rate

In order to design a pipelined structure of a TOMA, we have to decompose its nonpipelined structure into a certain number of layers and place pipeline registers between them. The decomposition is, to certain extent, arbitrary. The lower limit of the number of layers is two and the upper limit is determined by a delay of the component that we treat as indivisible. The minimum pipelining rate is approximately the sum of the delay of the layer with the maximum delay and the delay of the pipeline register. In this case we have assumed that after each FA or HA a register layer is placed and the OR gate and the MUXs are in the same layer. Hence we may evaluate the maximum delay of the layer as

$$t_{LD}^{TOMA\_RCA} = \max(t_{FAd1}, t_{OR2d1} + t_{MX2d1}) + t_{FD1Q} \tag{6}$$

where t_{FD1Q} is the maximum delay of the flip-flop.

Using the data from the STDH150, we may evaluate a theoretical maximum pipelining frequency as

$$f_{PF\_\max}^{TOMA\_RCA} = 1/(\max(0.143\,\text{ns}, 0.065\,\text{ns} + 0.078\,\text{ns})$$
$$+ 0.094\,\text{ns}) = 1/0.237\,\text{ns} = 4.22\ \text{GHz}.$$

4. Hiasat TOMA

In the following we shall examine the results of transforming the Hiasat TOMA which requires the smallest hardware amount among known TOMAs. This TOMA consists of the serial connection of five units: the sum-and-carry (SAC), the carry propagate and generate (CPG), CLA for c_{OUT}, multiplexer (MUX), CLA and Summation (CLAS). The SAC is composed of HAs and HALs (the modified HAs in [23]). The SAC performs

$$s_i = x_i \oplus y_i \oplus z_i, \tag{7}$$

$$c_{i+1} = x_i \cdot y_i + x_i \cdot z_i + y_i \cdot z_i, \tag{8}$$

for the individual bits of $X + Y$, and $X + Y - m$, with the assumption that TCS representation of $-m$ without the sign bit is $(z_{n-1}, \ldots, z_0)$ with $n = 5$. Regarding that $z_i = 0$ or $z_i = 1$, the HAL is obtained that implements

$$A_i = x_i \oplus y_i, \tag{9a}$$

$$\hat{A}_i = \overline{x_i \oplus y_i}, \tag{9b}$$

$$B_{i+1} = x_i \cdot y_i, \tag{9c}$$

$$\hat{B}_{i+1} = x_i + y_i\,. \tag{9d}$$

As $(z_{n-1}, \ldots, z_0)$ may have w bits for which $z_i = 0$ and $n - w$ bits for which $z_i = 1$. Hence the SAC has w HAs and $n - w$ HAL cells. The CFG computes the carry generate and carry propagate vectors as in the standard CLA $P_i = A_i \oplus B_i$, $G_i = A_i\ \ B_i$ and $p_i = \hat{A}_i \oplus \hat{B}_i$, $g_i = \hat{A}_i \cdot \hat{B}_i$. This unit has at most $2k - 2$ HAs. In the CLAS p_i and g_i are used to compute c_{OUT}, that controls the selection of $X + Y$ or $X + Y - m$. Regarding that $c_0 = 0$, $g_0 = 0$, c_{OUT} can be computed for the five-bit Hiasat adder as

$$c_{OUT} = B_5 + g_4 + g_3 \cdot p_4 + g_2 \cdot p_3 \cdot p_4 + g_1 \cdot p_2 \cdot P_3 \cdot P_4 . \quad (10)$$

The following stage, MUX selects using c_{OUT} the carry's and generate's $p_i' = p_i$ or $p_i' = P_i$ and $g_i' = g_i$ or $g_i' = G_i$, $i = 0, 1, ..., 4$.

The final stage, the five-bit CLA adder computes the carries

$$c_1 = g_0', \quad (11)$$

$$c_2 = g_1' + g_0' \cdot p_1', \quad (12)$$

$$c_3 = g_2' + g_1' \cdot p_2' + g_0' \cdot p_1' \cdot p_2', \quad (13)$$

$$c_4 = g_3' + g_2' \cdot p_3' + g_1' \cdot p_2' \cdot p_3' + g_0' \cdot p_1' \cdot p_2' \cdot p_3' . \quad (14)$$

In the next step the sum bits are calculated as

$$s_i' = c_i \oplus p_i', \; i = 0, 1, 2, 3, 4. \quad (15)$$

First we shall determine the area for components of the Hiasat five-bit TOMA and then the area for $m = 29$.

A. 5-bit Hiasat TOMA area

The area of the five-bit Hiasat TOMA can be computed as follows

$$A_{TOMA\_Hiasat} = A_{SAC\_5} + A_{CFG\_5} + A_{CLA\_Cout\_5} + A_{MUX\_5} + A_{CLAS\_5}. \quad (16)$$

The areas of the individual blocks from (16) can be expressed as:

$$A_{SAC\_5} = 2 \cdot A_{HAd1} + A_{HAd2} + A_{HAL}, \quad (17)$$

with

$$A_{HAL} = A_{OR2d1} + A_{AND2d1} + A_{XOR2d1} + A_{IVd1}. \quad (18)$$

In general, the area of the CFG_5 can be expressed as

$$A_{CFG\_5} = 5 \cdot A_{HAd2} + A_{HAd1}, \quad (19)$$

$$A_{CLA\_out\_5} = A_{AND2d1} + A_{AND3d1} + A_{AND4d1} + A_{OR5d1} + A_{NID6}, \quad (20)$$

$$A_{MUX\_5} = A_{MX2d1} + A_{MX2d2} + 3 \cdot A_{MX4d1}. \quad (21)$$

The CLAS block consists of the five-bit Propagate-Generate Unit (PGU_5), Carry-Generate Unit (CGU_5) and Summation Unit (SU_5). Its hardware amount can be estimated as

$$A_{CLAS\_5} = A_{CGU\_5} + A_{SU\_5}, \quad (22)$$

with the fan-outs 1, 3, 3, 4, 2. We get

$$A_{CGU\_5} = A_{AND2d1} + A_{OR2d1} + A_{AND2d2} + A_{AND3d1} + A_{OR3d1}, \quad (23)$$

and

$$A_{SU\_5} = 5 A_{XOR2d1} = 15.0 \text{ GE}. \quad (24)$$

Example 2. Area of the five-bit Hiasat TOMA for $m = 29$.

The TCS representation of $(-m)$ is equal to 100011, hence $w = 3$, and $k - w = 2$ (the sign bit is excluded). Thus we obtain

$$A_{SAC\_5} = 2A_{HAd1} + A_{HAd2} + 2A_{HAL}$$
$$= 2 \; 4.67 \text{ GE} + 5.67 \text{ GE} + 7.34 \text{ GE}$$
$$= 22.350 \text{ GE},$$

$$A_{CFG\_5} = 5 \cdot A_{HAd2} + A_{HAd2} = 5 \cdot 5.67\text{GE} + 4.67\text{GE} = 33.02\text{GE},$$

$$A_{CLA\_Cout\_5} = 1.67 \text{ GE} + 2 \text{ GE} + 2.33 \text{ GE} + 3.33 \text{ GE} + 3.67 \text{ GE} = 13 \text{ GE},$$

$$A_{MUX\_5} = 3 \text{ GE} + 3.33 \text{ GE} + 3 \cdot 6.33 \text{ GE} = 25.32 \text{ GE},$$

$$A_{CGU\_5} = 1.67\text{GE} + 1.67\text{GE} + 2\text{GE} + 2\text{GE} + 2\text{GE} + 1.67\text{GE} + 2\text{GE} + 2.33\text{GE} + 3\text{GE} = 18.34\text{GE},$$

$$A_{SU\_5} = 5 \cdot 3\text{GE} = 15\text{GE},$$

$$A_{CLAS\_5} = 18.34\text{GE} + 15\text{GE} = 33.34\text{GE}.$$

In effect we obtain the area of the five-bit Hiasat TOMA as

$$A_{TOMA\_Hiasat\_5} = 22.35\text{GE} + 33.02\text{GE} + 13\text{GE} + 25.32\text{GE} + 33.34\text{GE} = 127.03\text{GE}.$$

B. 5-bit Hiasat TOMA delay

The Hiasat five-bit TOMA delay, t_H can be expressed as

$$t_D^{Hiasat-TOMA} = t_{HAL} + t_{HAd2} + t_{AND4d1} + t_{OR5d1} + t_{NID6} + t_{MX2d4} + t_{AND4d1} + t_{OR2d1} + t_{OR4d1} + t_{XORd1} =$$
$$0.119\text{ns} + 0.092\text{ns} + 0.082\text{ns} + 0.094\text{ns} + 0.054\text{ns} + 0.092\text{ns} + 0.082\text{ns} + 0.090\text{ns} + 0.076\text{ns} + 0.090\text{ns} = 0.871\text{ns},$$

with

$$t_{HAL} = t_{XOR2d1} + t_{IVd1} = 0.090 \text{ ns} + 0.029 \text{ ns} = 0.119 \text{ ns}.$$

C. Pipelined 5-bit Hiasat TOMA area

The area of the Hiasat pipelined 5-bit TOMA can be expressed as

$$A_{TOMA\_Hiasat\_p} = A_{TOMA\_Hiasat} + n_h A_{FF} \quad (25)$$

where n_h is a number of flip-flops used in pipeline registers. For example, for the structure from Fig. 6 we obtain

$$A_{TOMA\_Hiasat\_p} = 127.03\text{GE} + 64 \cdot 5.67\text{GE} = 489.91\text{GE}.$$

D. Pipelining frequency of pipelined 5-bit Hiasat TOMA

In Fig. 6, a pipelined form of the Hiasat TOMA is presented. Five pipeline register stages are used with 58 flip-flops.

In this case we have adopted a decomposition into six layers that leads to a balanced structure. In order to evaluate the maximum pipelining frequency we shall calculate delays of the adopted individual layers. The maximum pipelining frequency will depend on the delay of the layer with the maximum delay and the delay of the assumed pipeline register. These layers have the following delays:

layer 1 $t_D^{L1,H}$: $t_{HAL} = 0.119\,\text{ns}$,

layer 2 $t_D^{L2,H}$: $t_{HAd1} = 0.088\,\text{ns}$,

layer 3 $t_D^{L3,H}$: $t_{AND4d1} + t_{OR5d1} = 0.176\,\text{ns}$,

layer 4 $t_D^{4,H}$: $t_{MX2d1} + t_{NID6} = 0.132\,\text{ns}$,

layer 5 $t_D^{L5,H}$: $t_{AND4d1} + t_{XOR2d1} = 0.172\,\text{ns}$,

layer 6 $t_D^{L6,H}$: $t_{XOR2d1} + t_{OR4d1} = 0.166\,\text{ns}$.

Using $t_D^{L3,H}$ as the maximum layer delay, we may evaluate the maximum pipelining frequency as

$$f_{PF\_max}^{TOMA\_Hiasat} = 1/(0.176\,\text{ns} + 0.094\,\text{ns}) = 1/0.27\,\text{ns} = 3.7\,\text{GHz}.$$

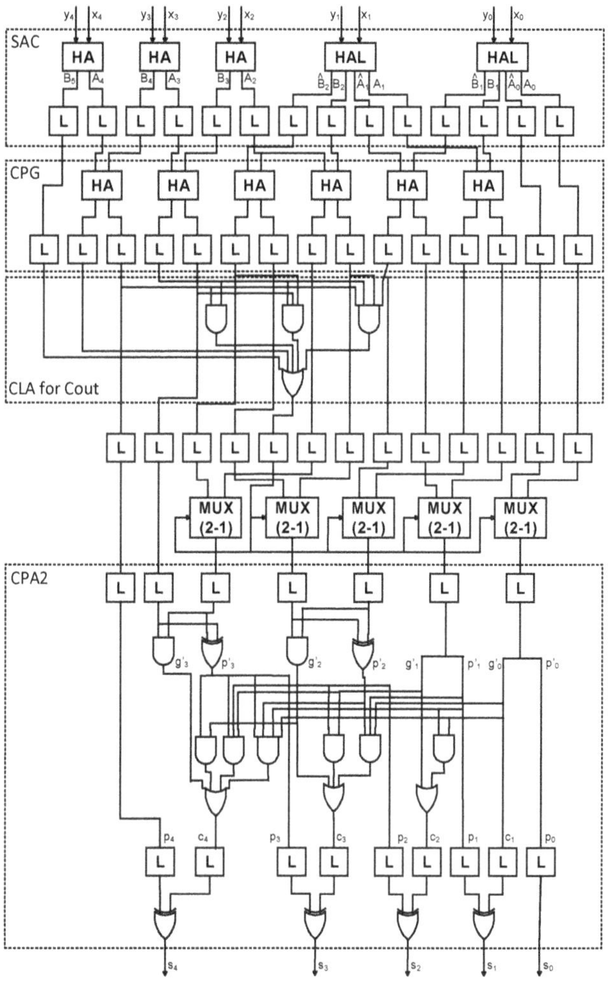

Fig. 6. Pipelined TOMA based on Hiasat TOMA.

5. PPA-based TOMA

As the next structure we shall consider the TOMA based on a PPA. As the PPA the Brent-Kung (BK) [28] adder has been selected. The Brent-Kung TOMA can be relatively easy transformed to the pipelined form, moreover the use of the Brent-Kung PPA allows one to simplify the adder used in the second stage when one of addends is a constant. The prefix operator ϕ is defined as

$$(g,p) = (g',p')\,\phi\,(g'',p''), \tag{26}$$

where

$$g = g'' + g' \cdot p'', \tag{27a}$$

$$p = p' \cdot p''. \tag{27b}$$

The block that implements (27a-b) will be denoted as BK_i. Subsequently we shall analyze the area and delay of the TOMA based on two BK adders.

The area of the TOMA BK A_{TOMA_BK} can be expressed as

$$A_{TOMA\_BK} = A_{BK} + A_{BK-m} \tag{28}$$

where A_{BK}, A_{BK-m} represent the area of the *BK* adder and the modified *BK-m* adder that subtracts *m*, respectively.

A. The area of BK adder

A_{BK} can be calculated as

$$A_{BK} = 4 \cdot A_{HAd2} + A_{HAd1} + A_{BK0} + A_{BK1} + A_{BK2} + A_{BK3} + \;\; + A_{BK4} + 4 \cdot A_{XOR2d1}. \tag{29}$$

The area of the first two terms is

$$4 \cdot A_{HAd2} + A_{HAd1} = 25.99\ \text{GE}. \tag{30}$$

After transforming the logic functions used for the realization of individual adders in (29), we receive the following areas

$$A_{BK_0} = A_{IVd1} + A_{NAND2d1} + A_{NAND2d2} + A_{AND2d2} = 6\ \text{GE}, \tag{31a}$$

$$A_{BK_1} = A_{IVd1} + A_{NAND2d1} + A_{NAND2d1} + A_{AND2d1} = = 4.67\,\text{GE}, \tag{31b}$$

$$A_{BK_2} = A_{IVd1} + A_{NAND2d1} + A_{NAND2d1} = 3\ \text{GE}, \tag{31c}$$

$$A_{BK_4} = A_{IVd1} + A_{NAND2d1} + A_{NAND2d2} + A_{AND2d2} = = 6\ \text{GE}, \tag{31d}$$

$$A_{BK_4} = A_{BK_2} = 3\ \text{GE}. \tag{31e}$$

Using (29), (30) and (31a-e) we obtain

$$A_{BK} = 62.02\ \text{GE}. \tag{31f}$$

B. The delay of BK adder

The BK adder delay can be expressed as

$$t_{BK} = t_{HA2} + \max(t_{BK_0}, t_{BK_1}) + \max(t_{BK_2}, t_{BK_3}) + t_{BK_4} + t_{XOR2d1}, \quad (32)$$

where

$t_{BK_0} = 2 \cdot t_{NAND2d1}$, $t_{BK_1} = t_{NAND2d1} + t_{NAND2d2}$, $t_{BK_2} = t_{BK_1}$, $t_{BK_3} = t_{BK_1}$, $t_{BK_4} = t_{BK_0}$.

Using the data from the STDH150, we have $t_{BK_0} = 0.074$ ns and $t_{BK_1} = 0.068$ ns.

Finally we obtain

$$t_{BK} = 0.092 \text{ ns} + 0.074 \text{ ns} + 0.068 \text{ ns} + 0.074 \text{ ns} + 0.09 \text{ ns} = 0.398 \text{ ns}.$$

C. The area of BK-m adder

The form of the first layer of the BK-*m* adder depends on the TCS representation of $-m$, $\widetilde{m}$. We shall analyze the prefix operator computation for a pair of bits ($\widetilde{m}_i$, $\widetilde{m}_{i+1}$).

(27a-b) can be expressed as

$$g_{\widetilde{m}} = s_{i+1} \cdot \widetilde{m}_{i+1} + s_i \cdot \widetilde{m}_i \cdot (s_{i+1} \oplus \widetilde{m}_{i+1}), \quad (33a)$$

$$p_{\widetilde{m}} = (s_{i+1} \oplus \widetilde{m}_{i+1}) \cdot (s_i \oplus \widetilde{m}_{i1}). \quad (33b)$$

For individual combinations of $(\widetilde{m}_i, \widetilde{m}_{i+1})$ we get

$(\widetilde{m}_i, \widetilde{m}_{i+1}) = (0,0)$ $g_{\widetilde{m}} = 0$ and $p_{\widetilde{m}} = s_i \cdot s_{i+1}$,

$(\widetilde{m}_i, \widetilde{m}_{i+1}) = (0,1)$ $g_{\widetilde{m}} = \overline{s}_i \cdot s_{i+1}$ and $p_{\widetilde{m}} = s_i \cdot s_{i+1}$,

$(\widetilde{m}_i, \widetilde{m}_{i+1}) = (1,0)$ $g_{\widetilde{m}} = s_i$ and $p_{\widetilde{m}} = s_i \cdot \overline{s}_{i+1}$,

$(\widetilde{m}_i, \widetilde{m}_{i+1}) = (1,1)$ $g_{\widetilde{m}} = s_{i+1} + s_i \cdot \overline{s}_{i+1}$ and $p_{\widetilde{m}} = \overline{s}_0 \cdot \overline{s}_1$.

The HA's become reduced, for we have $g_i = 0$, and the XOR gate that computes p_i, is reduced to the direct connection, i.e. $p_i = s_i$. For $\widetilde{m}_i = 1$, $g_i = s_i$, the XOR gate that computes p_i .becomes an inverter, i.e. $p_i = \overline{s}_i$. The form of $g_{\widetilde{m}}$ and $p_{\widetilde{m}}$ influences the form of BK_0 and BK_1.

Next we shall analyze the BK-*m* adder for $m = 29$ in order to have a comparison with the adder presented by Hiasat [23]. The TCS representation of $m = 29$ has the form 100011, then for HA_0, g_0 - connection, p_0 - inversion, for HA_1 g_1 - connection, p_1 - inversion, for HA_2 $g_2 = 0$, p_2- connection, for HA_3 $g_3 = 0$, p_3 - connection, for HA_4 $g_4 = 0$, p_4 - connection.

Moreover, regarding that $\widetilde{m}_0 = 1$ and $\widetilde{m}_1 = 1$, we may transform BK_0, to obtain $BK_{0\text{-}m}$ as

$$g_{\widetilde{m},BK_0} = s_{i+1} + s_i \cdot \overline{s}_{i+1} \quad (34a)$$

and

$$p_{\widetilde{m},BK_0} = \overline{s}_i \cdot \overline{s}_{i+1} = \overline{s_i + s_{i+1}}, \quad (34b)$$

and the $A_{BK_{0-m}}$ can be calculated as

$$A_{BK_{0-m}} = A_{IVd1} + A_{AND2d1} + A_{OR2d1} + A_{NOR2d2} = 6.34\,\text{GE} \quad (35)$$

and the delay

$$t_{BK_{0-m}} = t_{IVd1} + t_{ANDd1} + t_{ORd1} = 0.05 \text{ ns} + 0.105 \text{ ns} + 0.111 \text{ ns} = 0.266 \text{ ns}. \quad (36)$$

For BK_1 $\widetilde{m}_2 = 0$, $\widetilde{m}_3 = 0$, hence $g_{\widetilde{m}} = 0$ and $p_{\widetilde{m}} = s_i \cdot s_{i+1}$.

Assuming the direct realization we receive

$$A_{BK_{1-m}} = A_{AND2d1}, \quad (37)$$

$$t_{BK_{1-m}} = t_{AND2d1}. \quad (38)$$

For other blocks we have

$$A_{BK_{2-m}} = A_{BK_2}, \; A_{BK_{3-m}} = A_{BK_3}, \; A_{BK_{4-m}} = A_{BK_4}. \quad (39)$$

We finally receive for the BK-*m* adder

$$A_{BK-m} = 2A_{IV1d2} + A_{BK0m} + A_{BK1m} + A_{BK2} + A_{BK3} + A_{BK4} + 4A_{XOR2d1} = 32.34\,\text{GE}, \quad (40)$$

and for TOMA for $m = 29$ based on BK adders

$$A_{TOMA\ BK} = A_{BK} + A_{BK\ m} = 59.99\,\text{GE} + 31.68\,\text{GE} = 91.67\,\text{GE}. \quad (41)$$

The *BK-m* delay can be calculated as

$$t_{BK-m} = t_{IV1} + t_{BK0\text{-}m} + \max(t_{BK2\text{-}m}, t_{BK3\text{-}m}) + t_{BK4m} + t_{XOR2d1} + t_{MX2d1}. \quad (42)$$

Hence

$$t_{BK-m} = 0.03 \text{ ns} + 0.15 \text{ ns} + 0.07 \text{ ns} + 0.07 \text{ ns} + 0.09 \text{ ns} + 0.08 \text{ ns} = 0.49 \text{ ns}.$$

Finally we obtain

$$t_{TOMA\_BK} = t_{BK} + t_{BK\text{-}m} = 0.398 \text{ ns} + 0.490 \text{ ns} = 0.888 \text{ ns}.$$

D. The area of the pipelined TOMA BK

This area can be evaluated as

$$A_{TOMA\_BK\_p} = A_{BK} + A_{BK\ m} + n_{BK} \cdot A_{FF} \quad (43)$$

where n_{BK} is the number of flip-flops in pipeline registers. For the structure from Fig. 7 with $n_{BK} = 51$ and $A_{FF} = A_{FD1Q} = 5.67$ GE, we get $A_{TOMA\_BK\_p} = 380.84\,\text{GE}$.

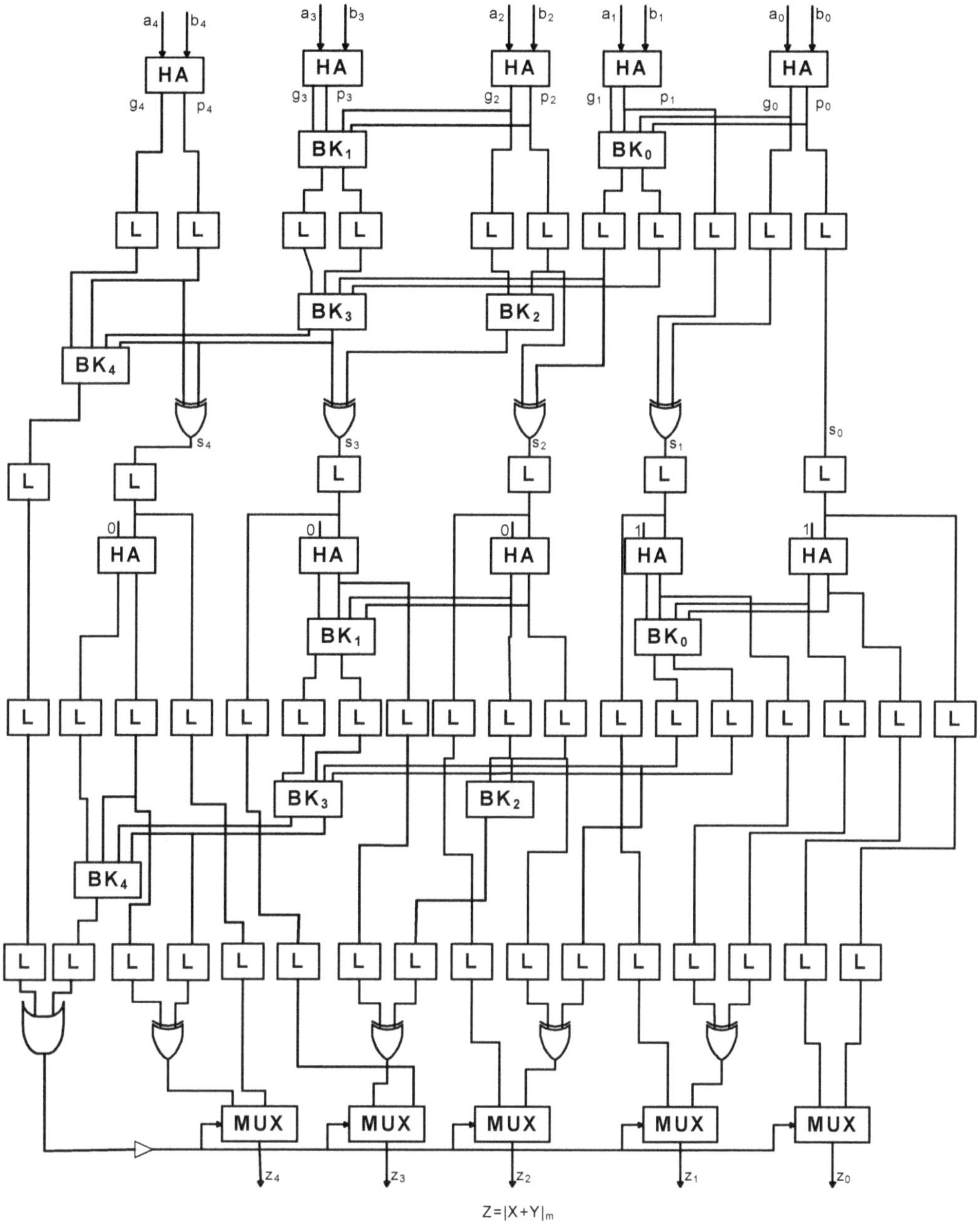

Fig. 7. Pipelined TOMA based on Brent-Kung adder.

E. Pipelining frequency of pipelined TOMA BK:

layer 1

$t_D^{L1,BK}$: $t_{HAd2} + t_{BK1} = 0.092 \text{ ns} + 0.068 \text{ ns} = 0.160 \text{ ns}$,

layer 2

$t_D^{L2,BK}$:

$t_{BK3} + t_{BK4} + t_{XOR2d1} =$

$= 0.068 \text{ ns} + 0.074 \text{ ns} + 0.090 \text{ ns} = 0.232 \text{ ns}.$

layer 3

$t_D^{L3,BK}$: $t_{HAd1} + t_{BK1} = 0.088 \text{ ns} + 0.074 \text{ ns} = 0.162 \text{ ns}$,

layer 4

$t_D^{4,BK}$: $t_{BK3} + t_{BK4} = 0.152 \text{ ns}$,

layer5

$t_D^{L5,BK} = t_{XOR2d1} + t_{NID6} + t_{MX2d1} = 0.222 \text{ ns}.$

Using the $t_D^{L2,BK}$ as the maximum layer delay, we receive the maximum pipelining frequency

$$f_{PF\_max}^{TOMA\_BK} = 1/(t_D^{L2,BK} + t_{FD1Q}) = 1/(0.232 \text{ ns} + 0.094 \text{ ns})$$
$$= 1/0.326 \text{ ns} = 3.06 \text{ GHz}.$$

6. New Five-bit TOMA

In this section we shall show a new TOMA structure and its pipelined form that requires smaller area than other TOMA structures. The TOMA is configured as a serial connection $X+Y$ adder and $X+Y-m$ adder that are designed in such a manner that leads to a substantial simplification and thus to a smaller delay or a smaller number of pipeline levels. Both adders are modifications of the standard CLA adder. In the first stage of the proposed structure the propagate's and generate's and transfer functions [30] $t_i = a_i + b_i$ are used. The first three carries c_1, c_2 and c_3 are computed simultaneously, and c_3 is used to generate c_4 and c_5.

Generally, the computation of the carry c_i can be expressed, assuming $c_0 = 0$, as

$$c_1 = g_0, \tag{44a}$$

$$c_2 = g_1 + c_1 \cdot t_1, \tag{44b}$$

$$c_3 = g_2 + c_2 \cdot t_2, \tag{44c}$$

$$c_4 = g_3 + c_3 \cdot t_3, \tag{44d}$$

$$c_5 = g_4 + c_4 \cdot t_4. \tag{44e}$$

In the above formulas instead of p_i, the transfer function $t_i = a_i + b_i$ is used, which is justified as follows

$$c_{i+1} = g_i + c_i \cdot p_i, \tag{45}$$

$$c_{i+1} = g_i + c_i \cdot p_i + c_i \cdot g_i = g_i + c_i \cdot (g_i + p_i) = g_i + c_i \cdot t_i, \tag{46}$$

with $t_i = a_i + b_i$, $g_i = a_i b_i$ and $p_i = a_i \oplus b_i$.

We may express c_2 and c_3 as the functions of g_i and t_i as

$$c_2 = g_1 + c_1 \cdot t_1, \tag{47}$$

$$c_3 = g_2 + g_1 \cdot t_2 + g_0 \cdot t_1 \cdot t_2. \tag{48}$$

Consequently, we receive

$$c_4 = g_1 + c_3 \cdot t_3, \tag{49}$$

and

$$c_5 = g_4 + g_3 \cdot t_4 + c_3 \cdot t_3 \cdot t_4. \tag{50}$$

In the adder realization the above equations are transformed to the NAND form. The sum bits are generated using $s_i = p_{i-1} \oplus c_{i\cdot}$, 1, 2, 3, 4 with $s_0 = p_0$. The second stage of the TOMA implements the subtraction of $-m$ making use of the TCS representation of $-m$, $\widetilde{m} = (1, \widetilde{m}_4, \widetilde{m}_3, \widetilde{m}_2, \widetilde{m}_1, \widetilde{m}_0)$.

Regarding that the second operand of the $X+Y-m$ adder is $\widetilde{m}$, we can write

$$c_1 = s_0 \cdot \widetilde{m}_0, \tag{51a}$$

$$c_2 = s_1 \cdot \widetilde{m}_1 + s_0 \cdot s_1 \cdot \widetilde{m}_0 + s_0 \cdot \widetilde{m}_0 \cdot \widetilde{m}_1, \tag{51b}$$

$$\begin{aligned} c_3 = {} & s_2 \cdot \widetilde{m}_2 + s_1 \cdot s_2 \cdot \widetilde{m}_1 + s_0 \cdot s_1 \cdot s_2 \cdot \widetilde{m}_0 + \\ & + s_0 \cdot s_2 \cdot \widetilde{m}_0 \cdot \widetilde{m}_1 + + s_1 \cdot s_2 \cdot \widetilde{m}_1 \cdot \widetilde{m}_2 \\ & + s_0 \cdot s_1 \cdot \widetilde{m}_0 \cdot \widetilde{m}_2 + s_0 \cdot \widetilde{m}_0 \cdot \widetilde{m}_1 \cdot \widetilde{m}_2 \end{aligned} \tag{51c}$$

$$c_4 = s_3 \cdot \widetilde{m}_3 + c_3 \cdot s_3 + c_3 \cdot \widetilde{m}_3, \tag{51d}$$

$$\begin{aligned} c_5 = {} & s_4 \cdot \widetilde{m}_4 + s_3 \cdot s_4 \cdot \widetilde{m}_3 + c_3 \cdot s_3 \cdot s_4 + c_3 \cdot s_4 \cdot \widetilde{m}_3 + \\ & + s_3 \cdot \widetilde{m}_3 \cdot \widetilde{m}_4 + c_3 \cdot s_3 \cdot \widetilde{m}_4 + c_3 \cdot \widetilde{m}_3 \cdot \widetilde{m}_4 \end{aligned} \tag{51e}$$

We may simplify the above equations by substituting $\widetilde{m}$ values of the individual five-bit moduli. The results of this simplification are given in Tab. 1.

m	c_1	c_2	c_5	c_4	c_5
17	s_0	$s_1 + s_0$	$s_2 + s_1 + s_0$	$s_3 + c_3$	$s_4 \cdot (s_3 + c_3)$
19	0	s_1	$s_2 + s_1$	$s_3 + c_3$	$s_4 \cdot (s_3 + c_3)$
21	s_0	$s_1 + s_0$	$s_2 \cdot (s_1 + s_0)$	$s_3 + c_3$	$s_4 \cdot (s_3 + c_3)$
23	s_0	$s_1 \cdot s_0$	$s_2 \cdot s_1 \cdot s_0$	$s_3 + c_3$	$s_4 \cdot (s_3 + c_3)$
25	s_0	$s_1 + s_0$	$s_2 + s_1 + s_0$	$s_3 \cdot c_3$	$c_3 \cdot s_4 \cdot s_3 \cdot$
27	s_0	$s_1 \cdot s_0$	$s_2 + s_1 s_0$	$s_3 \cdot c_3$	$c_3 \cdot s_4 \cdot s_3 \cdot$
29	s_0	$s_1 + s_0$	$s_2 \cdot (s_1 + s_0)$	$s_3 \cdot c_3$	$c_3 \cdot s_4 \cdot s_3 \cdot$
31	s_0	$s_1 \cdot s_0$	$s_2 \cdot s_1 \cdot s_0$	$s_3 \cdot c_3$	$c_3 \cdot s_4 \cdot s_3 \cdot$

Tab. 1. Logical functions for realizations of the carries of $X+Y-m$ adder.

In Fig. 8, the TOMA based on the new principle for $m = 29$ is depicted.

A. 5-bit new TOMA area

We shall analyze the area and delay of the new TOMA for $m = 29$. The area of the new TOMA can be computed as

$$A_{TOMA\_New} = A_{X+Y} + A_{X+Y-m}. \tag{52}$$

The hardware amount of the $X+Y$ adder can be expressed as

$$A_{X+Y} = A_{HA-stage} + A_{t_i} + A_{c_2} + A_{c_3} + A_{c_4} + A_{c_5} + A_{SU} \tag{53}$$

where $A_{HA-stage}$ is the area of the input summation stage (HAs and ORs), A_{c_i} are the areas of circuits generating the individual carries c_i.

Subsequently we have

$$A_{HA\_stage} = 2 \cdot A_{HAd1} + 3 \cdot A_{HAd2} + 4 A_{OR2d1} = 27.35 \text{ GE},$$

$$A_{c_2} = A_{IVd1} + 2 \cdot A_{NAND2d1} = 3 \text{ GE},$$

$$A_{c_3} = A_{IVd1} + 2 \cdot A_{NAND2d2} + A_{NAND3d1} + A_{NAND3d2} = 9.67 \text{ GE}$$

$$A_{c_4} = A_{IVd1} + 2 \cdot A_{NAND2d1} = 3 \text{ GE},$$

$$A_{c_5} = A_{IVd1} + A_{NAND2d1} + 2 A_{NAND3d1} = 4.67 \text{ GE},$$

$$A_{SU} = 5 \cdot A_{XOR2d1} = 15 \text{ GE}.$$

In effect, we receive

$$A_{X+Y} = 55.69 \text{ GE}.$$

For the $X + Y - m$ adder we have

$$A_{X+Y-m} = A_{c_2-m} + A_{c_3-m} + A_{c_4-m} + A_{c_5-m} + A_{SU-m},$$

where

$A_{c_1-m} = 0$ (direct connection),

$A_{c_2-m} = A_{OR2d2} = 1.67 \text{ GE},$

$A_{c_3-m} = A_{AND2d2} = 2 \text{ GE},$

$A_{c_4-m} = A_{AND2d1} = 1.67\text{GE},$

$A_{c_5-m} = A_{AND3d1} = 2 \text{ GE},$

$A_{SU-m} = A_{OR2d1} + A_{NID6} + 5 \cdot A_{MX2d1} = 20.34 \text{ GE}.$

We receive $A_{X+Y-m} = 27.68 \text{ GE}$.

The total hardware amount is $A_{TOMA\_New} = 110.06\text{GE}$.

B. 5-bit New TOMA delay

The delay of the new TOMA can be written as

$$t_{TOMA\_New} = t_{X+Y} + t_{X+Y-m}, \quad (54)$$

where

$t_{X+Y} = t_{HA\_Stage} + t_{c_1} + \max(t_{c_2}, t_{c_3}) + \max(t_{c_4}, t_{c_5}) + t_{SU}$ and t_{c_i}, $i = 1, 2, \ldots, 5$, denote the individual carry generator delays

$t_{c_1} = \max(t_{HAd2\_ACO}, t_{HAd2\_BCO}),$

$t_{c_2} = \max(t_{NAND2d1}, t_{IVd1}) + t_{NAND2d1},$

$t_{c_3} = \max(t_{NAND2d1}, t_{NAND3d1}, t_{IVd1}) + t_{NAND3d1},$

$t_{c_4} = \max(t_{NAND2d1}, t_{IVd1}) + t_{NAND2d1},$

$t_{c_5} = \max(t_{NAND2d1}, t_{NAND3d1}, t_{IVd1}) + t_{NAND3d1},$

$t_{X+Y} = t_{HAd2} + t_{NAND3d1} + t_{NAND3d2} + 2 \cdot t_{NAND2d1} + t_{XOR2d1},$

$t_{X+Y} = 0.092\text{ns} + 0.052\text{ns} + 0.044\text{ns} + 2 \cdot 0.037\text{ns} + 0.09\text{ns}$
$= 0.352\text{ns}$

and for $X + Y - m$ adder we have

$t_{X+Y-m} = t_{c_5'} + \max(t_{OR2d1} + t_{NID6}, t_{XOR2d1})$, where

$t_{c_5'} = t_{OR2d1} + t_{AND2d1} + t_{AND3d1},$

$t_{c_4\_c_5} = \max(t_{c_4}, t_{c_5}),$

$t_{X+Y-m} = t_{NAND2d1} + t_{NAND3d2} + t_{AND3d1} + t_{XOR2d1} + t_{MX2d1},$

$t_{X+Y-m} = 0.037 \text{ ns} + 0.044 \text{ ns} + 0.066 \text{ ns} +$
$+ 0.09 \text{ ns} + 0.078 \text{ ns} = 0.315 \text{ ns},$

$t_{TOMA\_New} = 0.352 \text{ ns} + 0.315 \text{ ns} = 0.667 \text{ ns}.$

D. The area of the pipelined new TOMA

This area is expressed as

$$A_{TOMA\_New\_p} = A_{X+Y} + A_{X+Y-m} + n_N \cdot A_{FF}$$

where n_N is the number of flip-flops in pipeline registers. For the structure from Fig. 8 with $n_N = 30$ and $A_{FF}=A_{FD1Q}=5.67\text{GE}$, we get . $A_{TOMA\_New} = 280.82\text{GE}$

E. Pipelining frequency of the pipelined new TOMA

For the individual layers in the pipelined structure of the new TOMA, shown in Fig. 8, we have the following delays:

layer 1:
$t_D^{L1,N} = t_{HAd1} + 2 \cdot t_{NAND3d1} = 0.192 \text{ ns},$

layer 2:
$t_D^{L2,N} = 2 \cdot t_{NAND3d1} + t_{XOR2d1} = 0.194 \text{ ns},$

layer 3:
$t_D^{L3,N} = t_{OR2d1} + t_{AND2d1} + t_{AND3d1} = 0.185 \text{ ns},$

layer 4:
$t_D^{L4,N} = t_{XOR2d1} + t_{NID6} + t_{MX2d1} = 0.222 \text{ ns}.$

The design of the pipelined structure aimed at the minimization of the number of pipeline stages while preserving possibly high pipelining frequency. The structure allows one to employ only three pipeline register stages with 30 flip-flops with the maximum pipelining frequency equal to

$$f_{PF\_max}^{new\_TOMA} = 1/(0.222 \text{ ns} + .094 \text{ ns}) = 1/0.316 \text{ ns} = 3.16 \text{ GHz}.$$

In Tab. 2 the summary of the obtained TOMA parameters is given.

	TOMA-RCA	TOMA-BK	TOMA-Hiasat	New TOMA I
Area [GE] (nonpipelined)	81.68	99.01	127.03	110.72
Delay[ns]	0.747	0.888	0.886	0.667
Area x delay	61.01	87.64	112.55	73.41
Number of pipeline layers	6	4	5	3
Number of FFs	66	58	64	30
Area [GE] (pipelined)	472.90	380.84	489.91	280.82
Pipelining frequency max [GHz]	4.22	3.06	3.7	3.16

Tab. 2. TOMA parameters for $m = 29$.

It is seen that the area-delay product has the best values for the TOMA-RCA and the new TOMA, moreover the new TOMA requires the smallest area for the pipelined structure but at the cost of the reduced maximum pipelining frequency. In general the new pipelined TOMA calls for about 35% less area than the TOMA-BK, the best of three other considered structures.

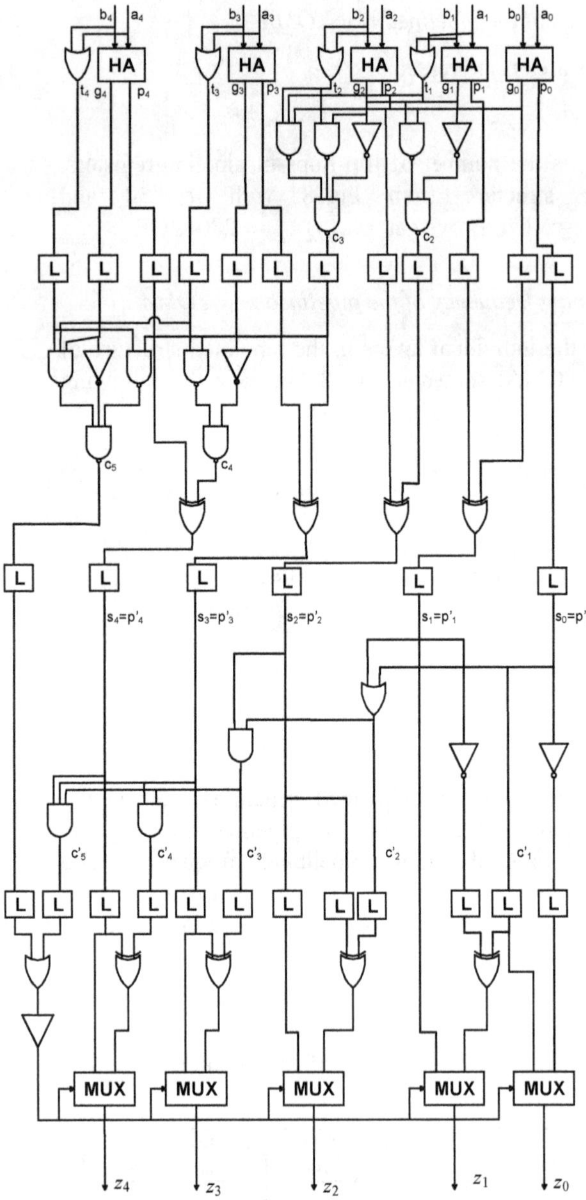

Fig. 8. New five-bit TOMA for $m = 29$.

7. Conclusions

The structures of pipelined two-operand modular adders for five-bit moduli based on ripple carry-adder, Brent-Kung adder and Hiasat adder have been presented and analyzed with respect to the area, number of layers and attainable pipelining frequency. Also a new structure of the two-operand modular adder based on the modified carry-look ahead adder has been proposed. It has been shown that the new pipelined adder has the smallest number of pipeline layers as well as the area smaller by about 35% than the best of other considered structures.

References

[1] SZABO, N. S., TANAKA, R. I. *Residue Arithmetic and its Applications to Computer Technology*. McGraw-Hill Inc., 1967.

[2] SODERSTRAND, M. A., JENKINS, M., JULLIEN, G. A., TAYLOR, F. J. *Residue Number System Arithmetic: Modern Applications in Digital Signal Processing*. IEEE Press, 1986.

[3] OMONDI, A., PREMKUMAR, B. *Residue Number Systems, Theory and Implementation*. Imperial College Press, 2007.

[4] ANANDA MOHAN, P.V. *Residue Number Systems, Algorithms and Architectures*. Kluwer Academic Publishers, 2002.

[5] JENKINS, W. K., KROGMEIER, J. V. The design of dual-mode complex signal processors based on quadratic modular number codes. *IEEE Transactions on Circuits and Systems*, 1987, vol. 34, no. 4, p. 354–364. DOI: 10.1109/TCS.1987.1086154

[6] KRISHAN, R., JULLIEN, G. A., MILLER, W. C. The modified quadratic residue number system (MQRNS) for complex high-speed signal processing. *IEEE Transactions on Circuits and Systems*, 1986, vol. 33, no. 3, p. 325–327. DOI: 10.1109/TCS.1986.1085897

[7] ULMAN, Z., CZYZAK, M. Highly parallel, fast scaling of numbers in nonredundant residue arithmetic. *IEEE Transactions on Signal Processing*, 1998, vol. 46, no. 2, p. 487–496. DOI: 10.1109/78.655432

[8] FRERKING, W. I., PARHI, K. K. Low-power FIR digital filters using residue arithmetic. In *Proceedings* of *the 31st Asilomar Conference on Signals, Systems and Computers ACSSC1997*. Pacific Grove (CA, USA), November 2 - 7, 1997, p. 739–743. DOI: 10.1109/ACSSC.1997.680542

[9] CARDARILLI, G. C., DEL RE, A., LOJACONO, R., NANNARELLI, A., RE, M. RNS implementation of high performance filter for satellite demultiplexing. In *Proceedings of the 2003 IEEE Aerospace Conference*. Big Sky (Montana, USA), March 8 - 15, 2003, vol. 3, p. 1365–1379. DOI: 10.1109/AERO.2003.1235253

[10] CARDARILLI, G. C., DEL RE, A. NANNARELLI, A., RE, M. Low-power implementation of polyphase filters in quadratic residue number system. In *Proceedings of the IEEE International Symposium on Circuits and Systems (ISCAS 2004)*. Vancouver (Canada), 2004, p. 725–728. DOI: 10.1109/ISCAS.2004.1329374

[11] CZYZAK, M., SMYK, R. FPGA implementation of the two-stage high-speed FIR filter in residue arithmetic. *Elektronika*, 2011, no. 12, p. 90–92.

[12] CZYZAK, M., SMYK, R. Radix-4 DFT butterfly realization with the use of the modified quadratic residue number system. *Poznan University of Technology Academic Journals. Electrical Engineering*, 2010, no. 63, p. 39–51.

[13] GARCIA, A., MEYER-BAESE, U., TAYLOR, F. J. Pipelined Hogenauer CIC filters using field-programmable logic and residue number system. In *Proceedings of the 1998 IEEE International Conference on Acoustics, Speech and Signal Processing.* Seattle (Washington, USA), May 12 - 15, 1998, p. 3085–3088. DOI: 10.1109/ICASSP.1998.678178

[14] BARRACLOUGH, S. R., SOTHERAN, M., BURGIN, K., WISE, A. P., et al. The design and implementation of the IMS A110 image and signal processor. In *Proceedings of the 1989 IEEE Custom Integrated Circuits Conference* (*ICICC 1989*). San Diego (CA, USA), May 15 - 18, 1989, p. 24.5.1–24.5.4. DOI: 10.1109/CICC.1989.56826

[15] WEI, W., SWAMY, M.N.S., AHMAD, M.O. RNS application for digital image processing. In *Proceedings of the 4th IEEE International Workshop System-on-Chip for Real-Time Applications*. Banff (Alberta, Canada), July 19 - 21, 2004, p. 77 to 80. DOI: 10.1109/IWSOC.2004.1319854

[16] BANERJI, D. K. A novel implementation method for addition and subtraction in residue number systems. *IEEE Transactions on Computers*, 1974, vol. 23, no. 1, p. 106–109. DOI: 10.1109/T-C.1974.223790

[17] AGRAWAL, D. P., RAO, T. R. N. Modulo 2^n +1 arithmetic logic. *IEEE Journal on Electronic Circuits and Systems*, 1978, vol. 2, no. 6, p. 186–188. DOI: 10.1049/ij-ecs.1978.0037

[18] SODERSTRAND, M. A. A new hardware implementation of modulo adders for residue number systems. In *Proceedings of the 26th IEEE Midwest Symposium on Circuits and Systems.* Puebla (Mexico), August 15 - 16, 1983, p. 412–415.

[19] BAYOUMI, M., JULLIEN. G. A. VLSI implementation of residue adders. *IEEE Transactions Circuits and Systems*, 1987, vol. 34, no. 3, p. 284–288. DOI: 10.1109/TCS.1987.1086130

[20] DUGDALE, M. VLSI implementation of residue adders based on binary adders. *IEEE Transactions on Circuits and Systems II: Analog and Digital Signal Processing*, 1992, vol. 39, no. 5, p. 325–329. DOI: 10.1109/82.142036

[21] PIESTRAK, S. J. Design of high-speed residue-to-binary number system converter based on the chinese remainder theorem. In *Proceedings of the International Conference on Computer Design ICCD'94, VLSI in Computers and Processors.* Cambridge (MA, USA), 1994, p. 508–511. DOI: 10.1109/ICCD.1994.331962

[22] ZIMMERMANN, R. Efficient VLSI implementation of modulo ($2^n \pm 1$) addition and multiplication. In *Proceedings of the 14th IEEE Symposium on Computer Arithmetic.* Adelaide (Australia), April, 1999, p. 158–167. DOI: 10.1109/ARITH.1999.762841

[23] HIASAT, A. A. High-speed and reduced-area modular adder structures for RNS. *IEEE Transactions on Computers*, 2002. vol. 51, no. 1, p. 84–89. DOI: DOI: 10.1109/12.980018

[24] PATEL, R. A., BENAISSA, M., POWELL, N., BOUSSAKTA, S. Novel power-delay-area-efficient approach to generic modular addition. *IEEE Transactions on Circuits and Systems–I Regular Papers*, June 2007, vol. 54, no. 6, p. 1279–1292. DOI: 10.1109/TCSI.2007.895369

[25] KELLIHER, T. P., OWENS, R.M., IRWIN, M. J., HWANG, T.-T. ELM - a fast addition algorithm discovered by a program. *IEEE Transactions on Computers*, 1992, vol. 41, no. 9, p. 1181–1184. DOI: 10.1109/12.165399

[26] KOGGE, P. M., STONE, H. S. A parallel algorithm for the efficient solution of a general class of recurrence equations. *IEEE Transactions on Computers*, 1973, vol. 22, no. 8, p. 786–793. DOI: 10.1109/TC.1973.5009159

[27] LADNER, R. E., FISCHER, M. J. Parallel-prefix computation. *Journal of Association of Computing Machinery*, 1980, vol. 27, no. 4, p. 831–838. DOI: 10.1145/322217.322232

[28] BRENT, R. P., KUNG, H. T. A regular layout for parallel adders. *IEEE Transactions on Computers*, 1982, vol. 31, no. 3, p. 260 to 264.

[29] *0.13 micron Standard Cell Logic Library STDH 150.* Samsung Inc., 2004.

[30] PARHAMI, B. *Computer Arithmetic: Algorithms and Hardware Designs.* New York: Oxford University Press, 2000.

Appendix A

Area [GE]	Delay [ns]
$A_{AND2d1} = 1.67$	t_{AND2d1}=0.054
$A_{AND2d2} = 2.00$	t_{AND2d2}=0.055
A_{AND3d1} =2.00	t_{AND3d1} = 0.066
A_{AND3d2} =2.67	t_{AND3d2} = 0.068
A_{AND4d1} =2.33	t_{AND4d1}=0.082
A_{AND4d2} =2.67	t_{AND4d2}=0.085
A_{FAd1}=8.00	t_{FAd1}=0.143
A_{FAd2}=9.00	t_{FAd2}=0.150
A_{HAd1}=4.67	t_{HAd1}=0.088
A_{HAd2}=5.67	t_{HAd2}=0.092
$A_{NAND2d1}$=1.00	$t_{NAND2d1}$=0.037
$A_{NAND2d2}$=2.00	$t_{NAND2d2}$=0.031
$A_{NAND3d1}$=1.67	$t_{NAND3d1}$=0.052
$A_{NAND3d2}$=3.00	$t_{NAND3d2}$=0.044
$A_{NAND4d1}$=2.00	$t_{NAND4d1}$=0.067
$A_{NAND4d2}$=3.67	$t_{NAND4d2}$=0.059
A_{NOR2d1} =1.33	t_{NOR2d1}= 0.050
A_{NOR2d2} =2.00	t_{NOR2d2}= 0.040
A_{OR2d1} =1.67	t_{OR2d1}=0.065
A_{OR2d2} = 2.00	t_{OR2d2}=0.069
A_{OR3d1} =2.00	t_{OR3d1}=0.090
A_{OR3d2} =2.67	t_{OR3d2}=0.090
A_{OR4d1} =3.00	t_{OR4d1}= 0.076
A_{OR4d2} =3.33	t_{OR4d2}= 0.082
A_{OR5d1} =3.33	t_{OR5d1}=0.094
A_{OR5d2} =3.67	t_{OR5d2}= 0.105
A_{XOR2d1}=3.00	t_{XOR2d1}=0.090
A_{IVd1} =1	t_{IVd1}=0.029
A_{NID6} =3.67	t_{NID6} =0.054
A_{MX2d1} =3.00	t_{MX2d1}=0.078
A_{MX2d2} =3.33	t_{MX2d2}=0.076
A_{MX2d4} = 4.33	t_{MX2d4} = 0.092
A_{MX4d1} =6.33	t_{MX4d1}=0.105
A_{FD1Q} = 5.67	$t_{FD1Q\ SU}$ = 0.094

Tab. 3. Hardware amount and time delays for STDH150 basic elements.

Half-adder (HA) delays [ns]	Full-adder (FA) delays [ns]
$t_{HAd1\ ACO}$=0.054	$t_{FAd1\ CICO}$=0.083
$t_{HAd1\ BCO}$=0.055	$t_{FAd1\ ACO}$=0.122
$t_{HAd1\ AS}$= 0.088	$t_{FAd1\ BCO}$=0.143
$t_{HAd1\ BS}$= 0.073	$t_{FAd1\ AS}$=0.121
$t_{HAd2\ ACO}$=0.057	$t_{FAd1\ BS}$=0.139
$t_{HAd2\ BCO}$=0.058	$t_{FAd2\ CICO}$=0.089
$t_{HAd2\ AS}$= 0.092	$t_{FAd2\ ACO}$=0.130
$t_{HAd2\ BS}$= 0.074	$t_{FAd2\ BCO}$=0.150
	$t_{FAd2\ AS}$=0.129
	$t_{FAd2\ BS}$=0.150

Tab. 4. Individual delays between input and output nodes for FAs and HAs (STDH150).

About the Authors ...

Maciej CZYŻAK was born in 1950, Chorzow, Poland. He received his M. Sc. in Automatic Control and Computer Engineering from the Faculty of Electronics, Gdansk University of Technology (GUT) in 1973, and his Ph.D. in Automatic Control from the Faculty of Electrical Engineering, GUT in 1985. He authored and co-authored over 60 technical papers. His interests encompass digital signal processing with the use of residue arithmetic, VLSI and FPGA design.

Jacek HORISZNY was born in 1962, Gdansk Poland. He received his M. Sc. in Electrical Engineering from the Faculty of Electrical Engineering, Gdansk University of Technology (GUT) in 1986, and his Ph.D. in Electrical Engineering from the Faculty of Electrical and Control Engineering, GUT in 1996. He is the author and co-author of over 40 technical papers. His current interests include the transformer design and applications of digital signal processing for transformer protection.

Robert SMYK was born in 1978, Malbork, Poland. He received his M. Sc. in Automatic Control from the Faculty of Electrical and Control Engineering, Gdansk University of Technology (GUT) in 2008, and his Ph.D. in Automatic Control from the Faculty of Electrical and Control Engineering, GUT in 2008. He co-authored and authored over 20 technical papers. His interests encompass digital signal processing, residue arithmetic, and FPGA design.

Design of Passive Analog Electronic Circuits using Hybrid Modified UMDA Algorithm

Josef SLEZAK, Tomas GOTTHANS

Dept. of Radio Electronics, Brno University of Technology, Technická 12, 616 00 Brno, Czech Republic

xsleza08@stud.feec.vutbr.cz, gotthans@feec.vutbr.cz

Abstract. *Hybrid evolutionary passive analog circuits synthesis method based on modified Univariate Marginal Distribution Algorithm (UMDA) and a local search algorithm is proposed in the paper. The modification of the UMDA algorithm which allows to specify the maximum number of the nodes and the maximum number of the components of the synthesized circuit is proposed. The proposed hybrid approach efficiently reduces the number of the objective function evaluations. The modified UMDA algorithm is used for synthesis of the topology and the local search algorithm is used for determination of the parameters of the components of the designed circuit. As an example the proposed method is applied to a problem of synthesis of the fractional capacitor circuit.*

Keywords

Evolutionary algorithm, optimization, estimation of distribution algorithm, EDA, fractional capacitor

1. Introduction

In many areas of the engineering applications Estimation of Distribution Algorithms (EDA) have proved excellent capabilities of dealing with various kinds of problems of different nature. However, in the area of the evolutionary electronics utilization of the EDA algorithms has not been sufficient. Several papers focused on the evolutionary design of the whole analog circuit (the topology and the parameters of the components) have been published. In [2] Zinchenko published method of synthesis of the passive analog circuits using Univariate Marginal Distribution Algorithm (UMDA) algorithm however the method has two drawbacks. Parallel connection of the components in the branches of the circuits is not possible and with increasing number of the nodes of the designed circuit the number of the components is increasing more than necessary. Another paper focused on the evolutionary synthesis of the passive analog circuits using EDA algorithms was published by Torres [3] however the method employs UMDA algorithm only for determination of the number of the resistors, capacitors and inductors in the designed circuit. Minimal Switching Graph Problem solved using hybrid EDA algorithm method was presented in [4]. The method employs the UMDA to sample start search points and a hill-climbing algorithm to find local optimum of the search space.

The proposed EDA algorithm solves the drawbacks of the method [2] mentioned above. The maximum number of the nodes and the maximum number of the components of the synthesized circuit can be set independently. Also encoding of the parallel components is possible. Furthermore the proposed method employs efficient hybrid approach which significantly reduces the number of the objective function evaluations. For the same optimization problem the number of the evaluations of the objective function of the proposed method is almost four times lower than for simulated annealing algorithm [12].

2. Hybrid Modified UMDA Algorithm

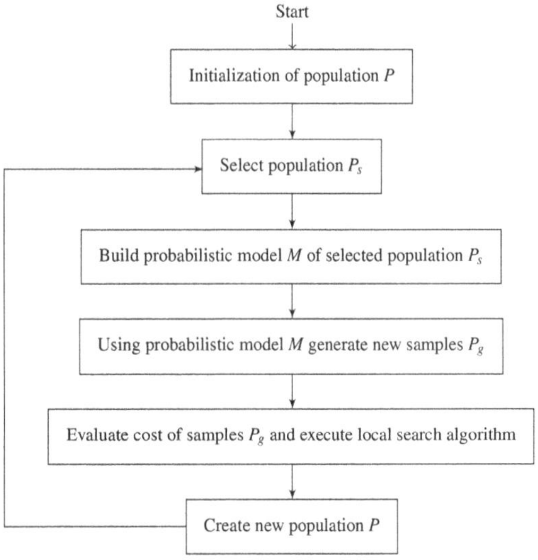

Fig. 1. Principal flowchart of the proposed method [1].

The flowchart of the proposed method is presented in Fig. 1. Population P is formed of binary vectors of length 135 bits which are initialized randomly with seeding of 10 bits ($n_c = 10$). Parameters storage e_{ps} is formed of a vector of length 135 consisting of real numbers in the range $\langle 0,1 \rangle$. Parameters storage e_{ps} is dynamically optimized during the whole synthesis process and it is adapted to the selected topologies in the selection phase of the algorithm. The vector includes a component value for every single admittance of the used expanded fully connected admittance network (see Sec. 6). During initialization phase e_{ps} is set randomly with uniform distribution.

In the next step selected population P_s is formed of the good individuals of the previous population P.

Probabilistic model M of selected population P_s is built. For more information on building of the probabilistic model in UMDA algorithm please refer to [5].

Probabilistic model M built in the previous step is used for generation of new samples of solutions P_g. The new samples are generated using Stochastic Universal Sampling method (SUS) and are repaired using the repairing method described in Sec. 3.

The generated samples are evaluated using the topological information stored in the individuals of P_g and the parameters of the components stored in e_{ps}. If the condition of execution of the local search algorithm is fulfilled, the local search algorithm tries to optimize the parameters of the current solution. If the accuracy of the current solution is improved, then storage of the parameters e_{ps} is updated according to the results of the local search algorithm. Detailed description of the cost evaluation phase and the local search algorithm is presented in Sec. 4.

In the next step new population P is formed of the best individuals of P_g and selected population P_s. The described process is repeated until one of the termination criteria of the algorithm is met.

3. The Unitation Constraints

Generally desired specifications of an analog circuit are easier to reach using an analog circuit of higher complexity. Due to the fact the evolutionary analog circuits synthesis methods tend to evolve analog circuits with complexity as large as possible. Without restriction of the number of the components of the evolved circuit its complexity becomes higher than necessary.

Therefore the number of the components of the evolved circuit should be restricted to a user define value. Since in the proposed encoding method (section 6) the number of the components of the encoded circuit is determined by the number of the "ones" of the binary characteristic vector c the restriction of the number of the components leads to a problem with unitation constraints [6].

Definition 1. Let's define vector $x = (x_1, x_2, \ldots, x_n) \in \Omega$. Then the unitation value of x is defined as

$$u(x) := \sum_{i=1}^{n} x_i. \tag{1}$$

Value of unitation function $u(x)$ depends only on the number of the "ones" in an input binary vector x. The unitation values of two vectors with the same numbers of "ones" are equal.

A problem with unitation constraints is defined as solution e in which unitation value $u(e)$ (the number of the "ones" in solution e) is restricted to a defined number [6].

As was described above the analog circuit synthesis problem has to be viewed as a problem with unitation constraints [6]. Modification of Factorized Distribution Algorithm (FDA) [7] which enables solving problems with unitation constraints was described in [6]. Modification of the UMDA algorithm which is able to handle the problems with unitation constraints is proposed in the text bellow. Pseudocode of the original UMDA algorithm [5] is presented in Fig. 2.

step0: Set $k = 1$. Generate $n_i \gg 0$ points randomly.
step1: Select $n_s \leq n_i$ points. Compute the marginal frequencies $r_{k;i}(x_i)$ of the selected set.
step2: Generate n_i new points according to the distribution $q_{k+1}(x) = \prod_{i=1}^{n} r_{k;i}(x_i)$. Set $k = k+1$.
step3: If not terminated, go to **step1**.

Fig. 2. Pseudo-code of the original UMDA algorithm [1].

In [6] to handle unitation constraints problems only generation phase (sampling) of FDA algorithm was modified. The presented approach was adopted also in the proposed modification of the UMDA algorithm. The UMDA algorithm was implemented using toolbox MATEDA 2.0 [8]. Pseudo-code of the modified UMDA algorithm is presented in Fig. 3.

step0: Set $k = 1$. Generate $n_i \gg 0$ points randomly.
step1: Select $n_s \leq n_i$ points. Compute the marginal frequencies $r_{k;i}(x_i)$ of the selected set.
step2: Generate n_i new points according to the distribution $q_{k+1}(x) = \prod_{i=1}^{n} r_{k;i}(x_i)$. Set $k = k+1$.
step3: With regard to unitation constraints repair generated points.
step4: If not terminated, go to **step1**.

Fig. 3. Pseudo-code of the modified UMDA algorithm [1].

One additional step (**step3**) was added to the original UMDA algorithm. In **step3** the generated samples are repaired to satisfy the desired unitation constraints.

The repairing function is applied to every single individual of the population of the generated samples in **step2**. Only n_c "ones" with the highest marginal frequencies $r_{k;i}$ of every generated sample are accepted. The rest of the "ones" of the samples are set to zero. If the number of the "ones" of the sample is equal or lower than n_c then the sample is accepted without any modification and no repairing is performed. This way the number of the "ones" (which corresponds to the number of the components of the encoded analog circuit) of every generated sample never exceeds n_c.

Note that the repairing function is applied only for the part of the encoding vector e which encodes the topology of the solution (characteristic vector c in Fig. 13).

4. The Local Search Algorithm

Evaluation of the cost values and using of the local search algorithm will be discussed in the section. Principal flowchart of this phase is presented in Fig. 4.

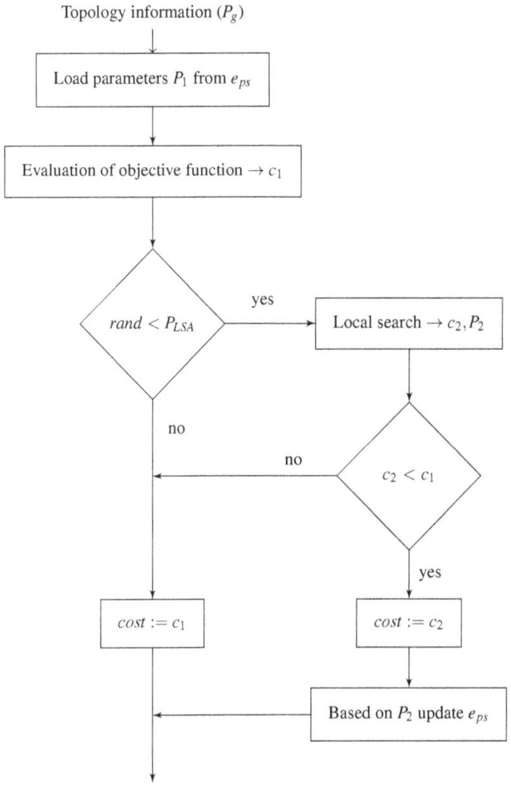

Fig. 4. Evaluation of the cost value and employing of the local search algorithm [1].

The following procedure which is described bellow is performed for every single individual $P_g(i)$ of the population of the generated samples P_g.

Based on the topology information stored in the binary vector of individual $P_g(i)$ appropriate set of parameters P_1 is loaded from parameters storage e_{ps} and cost value c_1 of individual $P_g(i)$ is evaluated.

If the condition of execution of the local search algorithm (LSA) is fulfilled ($rand < P_{LSA}$) the LSA tries to improve accuracy of individual $P_g(i)$. The probability of execution of the LSA is set to $P_{LSA} = 0.02$. The initial point of the search of the LSA is set to parameters set P_1. The number of the objective function evaluations of the LSA is set to $MaxFunEvals = 100$. The results of the LSA are the cost value of the optimized solution c_2 and set of the optimized parameters P_2.

If the LSA is successful in improving of the accuracy of individual $P_g(i)$ and its cost value was improved ($c_2 < c_1$) then value $cost$ of individual $P_g(i)$ is set to c_2 ($cost := c_2$) and the appropriate parameters of parameters storage e_{ps} are updated according to parameters set P_2.

If the condition of execution of the LSA was not fulfilled ($rand > P_{LSA}$) or the LSA was not able to improve the cost value of individual $G(i)$ ($c_2 \geq c_1$) the resulting value $cost$ of individual $P_g(i)$ is set to c_1 ($cost := c_1$).

After performing of the described process cost value $cost$ of individual $P_g(i)$ and updated parameters storage e_{ps} are obtained. During the run of the algorithm the parameters stored in e_{ps} are adapted to the topological information of the good individuals selected in the selection phase of the algorithm (Fig. 1). This way the information about the topology stored in P_g and the information about the parameters stored in e_{ps} are mutually optimized and the whole synthesis process is directed towards the promising areas of the solution space.

5. Application of the Method

A problem of synthesis of a fractional capacitor circuit was adopted from [9] and will be used for demonstration of the synthesis capabilities of the proposed method. The goal is to synthesize a circuit with input impedance (2)

$$Z_{in} = s^{-0.6}. \tag{2}$$

In Fig. 5 there is a circuit realization of function (2) as presented in [9]. For the rest of the section, the circuit will be called original approximation circuit.

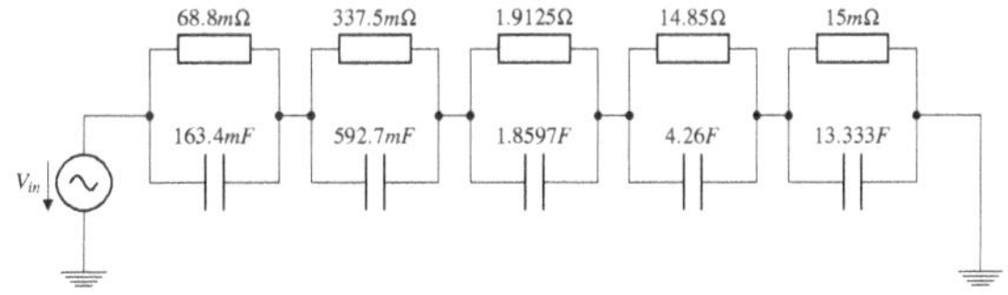

Fig. 5. Schematic of the original approximation circuit [1].

Comparison of the magnitude and the phase of Z_{in} of the original approximation circuit and (2) is presented in Fig. 6 and Fig. 7 respectively. Deviation of the magnitude and the phase of Z_{in} of the original approximation circuit and (2) is presented in Fig. 8 and Fig. 9 respectively.

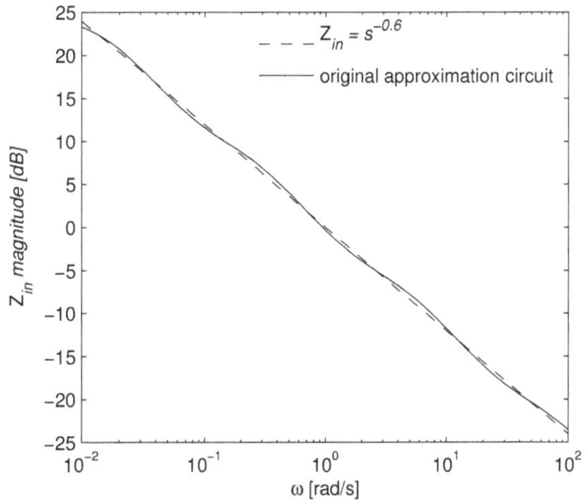

Fig. 6. Comparison of the magnitude characteristics of Z_{in} regarding (2) and the original approximation circuit [1].

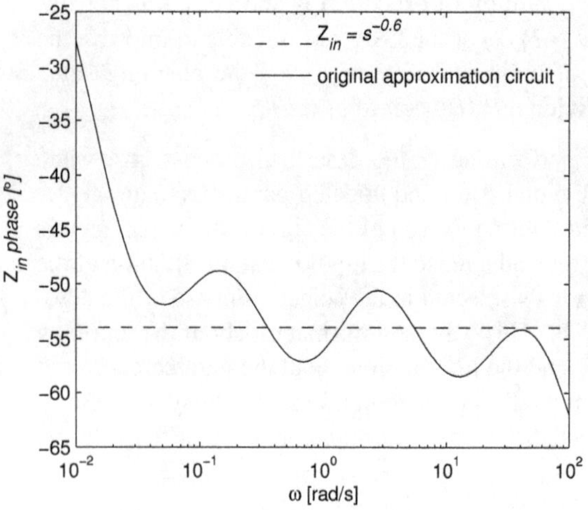

Fig. 7. Comparison of the phase characteristics of Z_{in} closer (2) and the original approximation circuit [1].

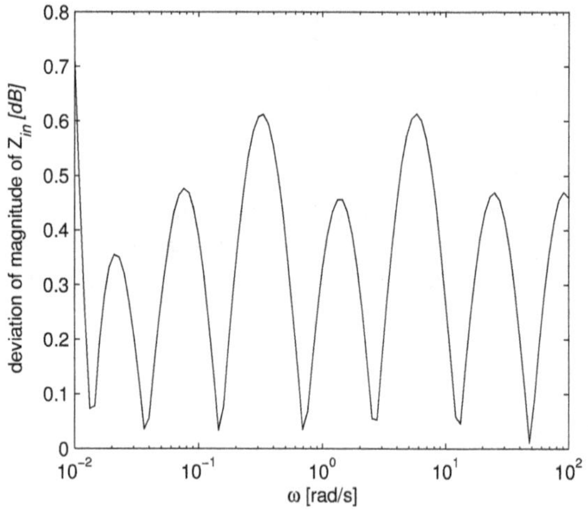

Fig. 8. Deviation of the magnitude of Z_{in} of the original approximation circuit [1].

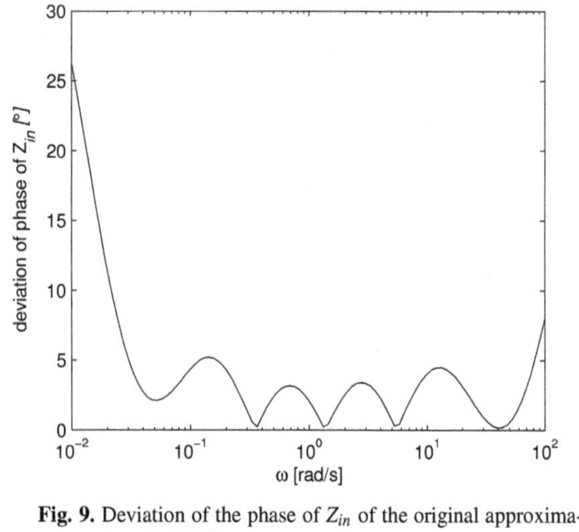

Fig. 9. Deviation of the phase of Z_{in} of the original approximation circuit [1].

The highest deviations of the magnitude and the phase of Z_{in} of the original approximation circuit are presented in Tab. 1.

Magnitude			
ω[rad/s]	0.01	100	0.33
Δ_m[dB]	0.71	0.46	0.61

Phase			
ω[rad/s]	0.01	100	0.14
Δ_p[∘]	26.3	7.98	5.2

Tab. 1. The highest deviations of the magnitude and the phase of Z_{in} of the original approximation circuit.

The following sections will be focused on synthesis of the fractional capacitor circuit problem using the proposed hybrid modified UMDA method.

6. The Encoding Method

The used encoding method is based on the idea of fully connected admittance network. For chosen number of the nodes the fully connected admittance network is formed by connecting the admittances between all combinations of the nodes of the network. The number of the admittances of the fully connected admittance network with n_n nodes can be calculated according to (3)

$$n_{adm} = \binom{n_n}{2} = \frac{n_n!}{2!(n_n-2)!}. \qquad (3)$$

Every single admittance of the fully connected admittance network can be replaced by resistor, capacitor, inductor or their parallel combination. Therefore the largest circuit which can be for chosen number of the nodes n_n obtained is the circuit where every single admittance of the fully connected admittance network is replaced by parallel combination of resistor, capacitor and inductor. In this paper such circuit is denoted as expanded fully connected admittance network and includes $3n_{adm}$ components. Example of the expanded fully connected admittance network is presented in Fig. 10.

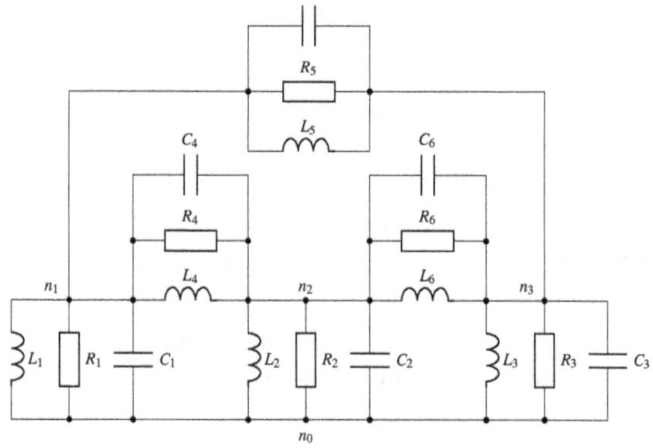

Fig. 10. Expanded fully connected admittance network N_c ($n_c = 4$) [1].

The expanded fully connected admittance network can be represented using complete multigraph with three multiple edges at the most [10]. Complete multigraph G_c corresponding to expanded fully connected admittance network

N_c is presented in Fig. 11. Nodes n_0 to n_3 of network N_c correspond to vertices v_0 to v_3 of complete multigraph G_c. Branches of network N_c correspond to edges of complete multigraph G_c. For example edges $e_5(1), e_5(2), e_5(3)$ (on complete multigraph G_c) correspond to components L_5, R_5, C_5 (in network N_c) respectively. Then the problem of searching of the topology of the analog RLC circuits can be defined as searching of subgraph G_s on complete multigraph G_c [10].

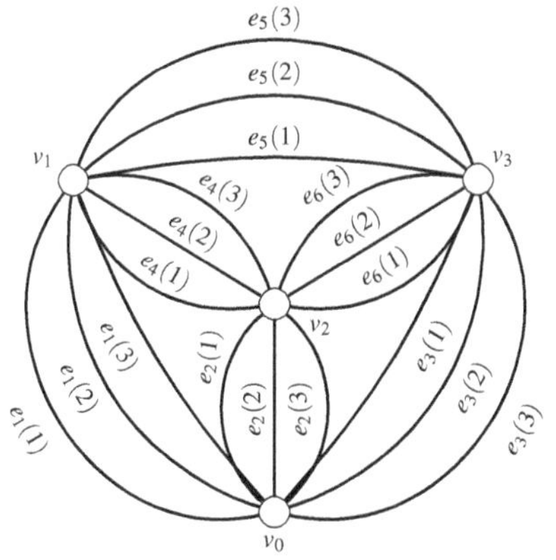

Fig. 11. Complete multigraph G_c representing expanded fully connected admittance network N_c [1].

Subgraph G_s can be encoded using binary characteristic vector c of length $3n_{adm}$. Every single bit of characteristic vector c represents including or not including of the corresponding edge of the complete multigraph G_c in subgraph G_s. For example complete multigraph G_c is encoded using characteristic vector c of length 18 bits where $c(i) = 1$ for $i \in \{1, 2, \ldots, 18\}$. Example of subgraph G_s, corresponding analog circuit and its characteristic vector c are presented in Fig. 12.

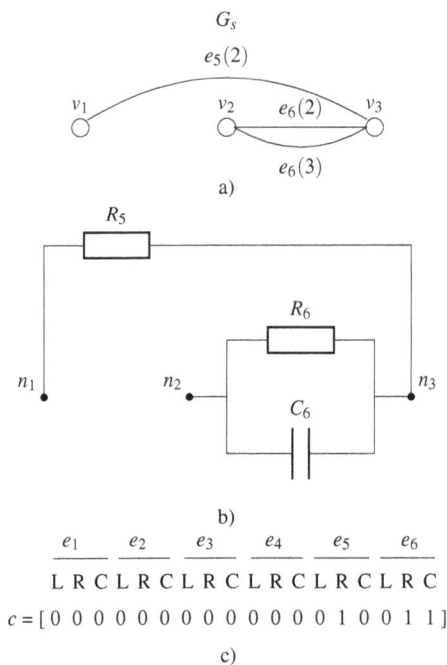

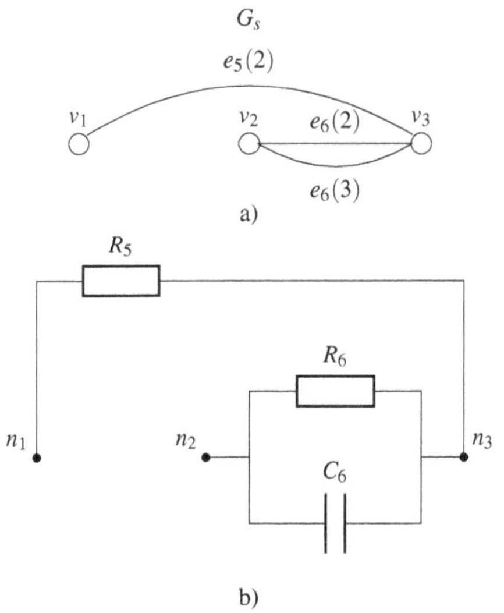

$$
\begin{array}{cccccc}
e_1 & e_2 & e_3 & e_4 & e_5 & e_6 \\
\text{L R C} & \text{L R C} & \text{L R C} & \text{L R C} & \text{L R C} & \text{L R C}
\end{array}
$$

c = [0 0 0 0 0 0 0 0 0 0 0 0 0 1 0 0 1 1]

c)

Fig. 12. a) Example of subgraph G_s b) analog circuit corresponding to G_s c) characteristic vector c of G_s [1].

Encoding vector e of every single solution is formed of two parts. The first one represents the topology and is encoded using characteristic graph c approach as described above. The number of the nodes of the expanded fully connected admittance network was experimentally chosen $n_n = 10$. To enable direct comparison of the accuracy of the circuits synthesized using the proposed method and the original approximation circuit presented in [9] the maximum number of the components of the synthesized circuit was set to the number of the components of the original approximation circuit ($n_c = 10$). According to (3) the topology is encoded using binary characteristic vector c of length 135 bits ($3n_{adm}$).

The second part of encoding vector e represents information about the parameters of the components and is represented by vector of real numbers p of length n_c.

Schematic of the used encoding vector e is presented in Fig. 13. The topological information is encoded using binary characteristic vector c = {b1, b2, b3, ..., b135} and the parameters (the values of the components) are encoded using vector of real numbers p = {dbl1, dbl2, db3, ..., dbl10}.

$$e = \overbrace{[\text{ b1 b2 b3} \ldots \text{b135 }]}^{c} \; \overbrace{[\text{ dbl1 dbl2 dbl3} \ldots \text{dbl10 }]}^{p}$$

Fig. 13. Schematic diagram of encoding vector e [1].

Based on the components selected in the topological part of the information (characteristic vector c) corresponding values of the parameters are loaded from storage of the parameters e_{ps} and copied to vector p. For example let's assume that the topology is encoded using characteristic vector $c(j) = 1$ for $j \in \{1, 3, 15, 18, 28, 52, 78, 92, 107, 115\}$. Based on the information in characteristic vector c corresponding parameters of storage of the parameters e_{ps} will be loaded to vector of the parameters p as follows: $p(k) = e_{ps}(j)$ for $k \in \{1, 2, \ldots, N_c\}$ and $j \in \{1, 3, 15, 18, 28, 52, 78, 92, 107, 115\}$.

Based on the parameters in vector p the values of the components are calculated using formula (4)

$$v = \frac{2 \times 10^6}{1 + e^{(-1.4(10r - 14))}} \tag{4}$$

where r are values of the parameters loaded from vector of the parameters p. Formula (4) was formed to map the values of the parameters stored in vector p to suitable range of the values of the components. Since the parameters in vector p are set in the range <0,1> corresponding values of the components for the lowest $r = 0$ and for the highest $r = 1$ are $v_{min} = 0.0061$ and $v_{max} = 7.3685 \times 10^3$ respectively. Note that formula (4) is used for all three types of components RLC. Since the used angular frequency range is from 0.01 rad/s to 100 rad/s, the values of the components are set to non-realistic values.

7. The Objective Function

Cost value *cost* is according to (7) computed as weighted summation of the magnitude and the phase differences. Difference of magnitude Δ_m is according to (5) calculated as weighted absolute value of differences of desired magnitude function f_{md} and magnitude of current solution f_{mc} over $m = 101$ frequency points in the range 0.01 rad/s to 100 rad/s. Similarly difference of phase Δ_p is according to (6) calculated as weighted absolute value of differences of desired phase function f_{pd} and phase of current solution f_{pc}.

$$\Delta_m = \frac{1}{m}\sum_{i=1}^{m} w_{dm}(i)|f_{md}(i) - f_{mc}(i)| \tag{5}$$

$$\Delta_p = \frac{1}{m}\sum_{i=1}^{m} w_{dp}(i)|f_{pd}(i) - f_{pc}(i)| \tag{6}$$

$$cost = \Delta_m w_{cm} + \Delta_p w_{cp}. \tag{7}$$

Weights w_{cm} and w_{cp} were set to 1 and 2 respectively. Setting of weights w_{dm}, w_{dp} is presented in Tab. 2. All weight coefficients were set experimentally.

angular frequency range:	w_{dm}	w_{dp}
0.01 rad/s to 0.0398 rad/s	1.3	1.3
0.0437 rad/s to 20.8930 rad/s	1	1
22.9087 rad/s to 100 rad/s	1.3	1.3

Tab. 2. Setting of weights w_{dm} and w_{dp}.

Frequency responses of the current solution f_{mc} and f_{pc} are obtained using nodal analysis method implemented in Matlab.

8. Settings of the Proposed Algorithm

The goal of the synthesis is to design a circuit which approximates function (2). The only information supplied to the system is desired magnitude and phase characteristics (2), maximum number of the used components ($n_c = 10$), maximal number of the nodes ($n_n = 10$) and types of the used components (resistors, capacitors, inductors).

The number of the objective function evaluations (evals) required by the proposed algorithm consists of the number of the evaluations required for calculation of the cost values of all individuals of the population (*PopEvals*) and the number of the evaluations required by the used local search algorithm (*LSevals*). Population size was set $PopSize = 200$ and number of generations was set $MaxGen = 200$. Therefore $PopEvals = 40e3$. The local search method requires $MaxFunEvals = 100$ evaluations in each its run and it is executed with probability $P_{LSA} = 0.02$ (2% Lamarckian approach [11]). The condition of execution of LSA is tested during every single objective function evaluation. Therefore $LSevals = 80e3$ and the total number of the objective function evaluations required by the proposed algorithm is 120e3 ($PopEvals + LSevals$). Local search algorithm was realized using Matlab function *fmincon*.

All parameters of the synthesized problem and settings of the proposed algorithm are summarized in Tab. 3.

maximum number of the nodes (n_n)	10 (0 to 9)
maximum number of the components (n_c)	10
used types of the components	R,L,C
negative resistors	not allowed
angular frequency range	0.01 rad/s to 100 rad/s
number of the points (m)	101 (25 points/decade)
population size (*PopSize*)	200
number of generations (*MaxGen*)	200
probability of execution of LSA (P_{LSA})	0.02
w_{cm}	1
w_{cp}	2

Tab. 3. Settings of the proposed algorithm.

The algorithm UMDA is realized using Matlab toolbox MATEDA 2.0 [8]. MATEDA initialization file is presented in Fig. 14

```
PopSize = 200; n = 135; cache  = [0,0,0,0,0];
Card = 2*ones(1,n); MaxGen = 200; MaxVal = -1e-2;
stop_cond_params = {MaxGen,MaxVal};
Cliques = CreateMarkovModel(n, 0);
edaparams{1} = {'seeding_pop_method','seedingFract',25};
edaparams{2} = {'learning_method','LearnFDA',{Cliques}};
edaparams{3} = {'sampling_method','SampleFDAmodif',{PopSize}};
edaparams{4} = {'stop_cond_method','maxgen_maxval',stop_cond_params};
[AllStat,Cache]=RunEDAfractal(PopSize, n, F, Card, cache, edaparams);
```

Fig. 14. Configuring of MATEDA 2.0 toolbox for realization of the proposed algorithm [1].

The whole program flow of the UMDA algorithm is realized using the functions of MATEDA 2.0 toolbox. The only exception is the sampling phase of the algorithm which is modified to enable dealing with the problems with unitation constraints. All the parameters in Tab. 2 and Tab. 3 were set experimentally after high number of experiments. The experiments were performed on a computer with processor AMD Athlon II X2 245, 6GB RAM and operational system Centos 6.5.

9. Experiments and The Solutions

There were executed 20 instances of the proposed algorithm. Average running time of a single execution was 11 min. The cost values of the solutions are presented in Tab. 4.

id of run	1	2	3	4	5
cost value	3.805	2.442	3.805	2.431	2.428
id of run	6	7	8	9	10
cost value	2.437	2.443	3.653	2.433	2.436
id of run	11	12	13	14	15
cost value	2.432	3.662	5.197	5.663	5.353
id of run	16	17	18	19	20
cost value	6.071	3.805	5.691	6.072	2.429

Tab. 4. Results of 20 runs of the proposed algorithm.

As can be seen in Tab. 4 the best solution was achieved in run 5 with cost value 2.428. The schematic diagram is presented in Fig. 15. Schematic diagrams of another three good solutions (run 4, run 11, run 20) are presented in Fig. 16 to Fig. 18.

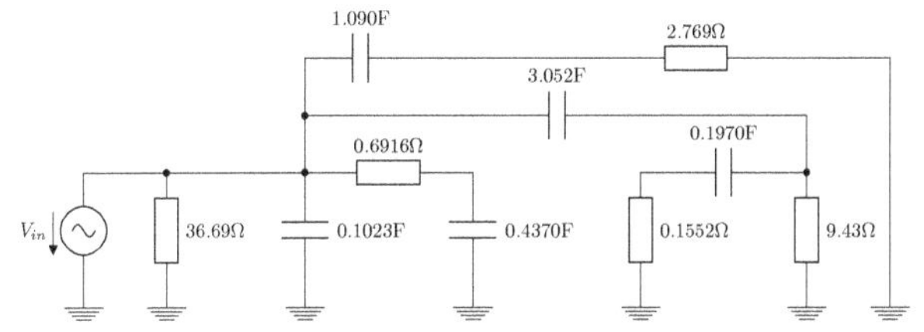

Fig. 15. Schematic of the circuit synthesized in run 5 [1].

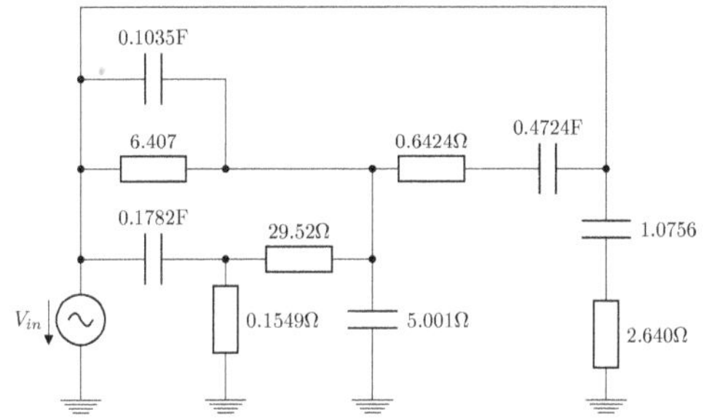

Fig. 16. Schematic of the circuit synthesized in run 4 [1].

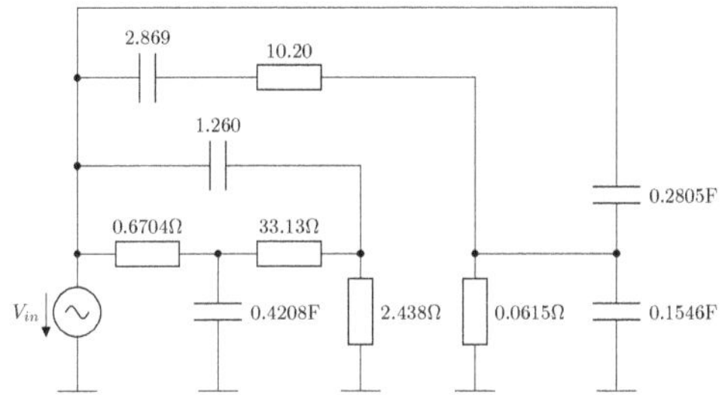

Fig. 17. Schematic of the circuit synthesized in run 11 [1].

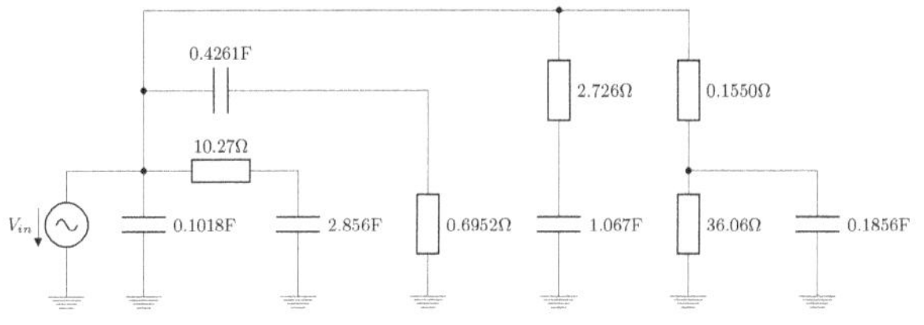

Fig. 18. Schematic of the circuit synthesized in run 20 [1].

Comparison of the magnitude and the phase characteristics of Z_{in} of the best found approximation circuit and desired function (2) are presented in Fig. 19 and Fig. 20 respectively.

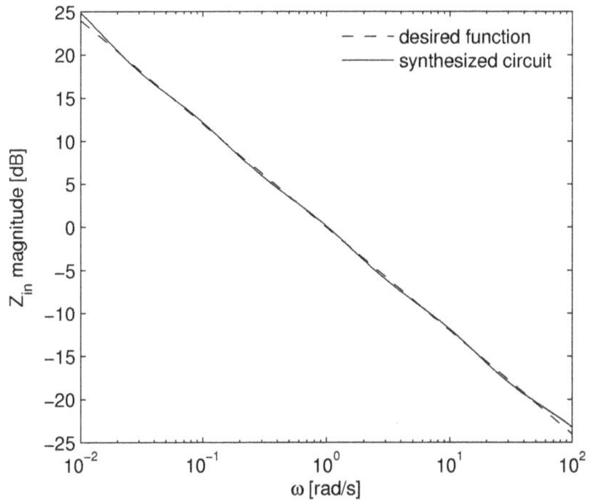

Fig. 19. Comparison of the magnitude of Z_{in} of the best found approximation circuit and (2) [1].

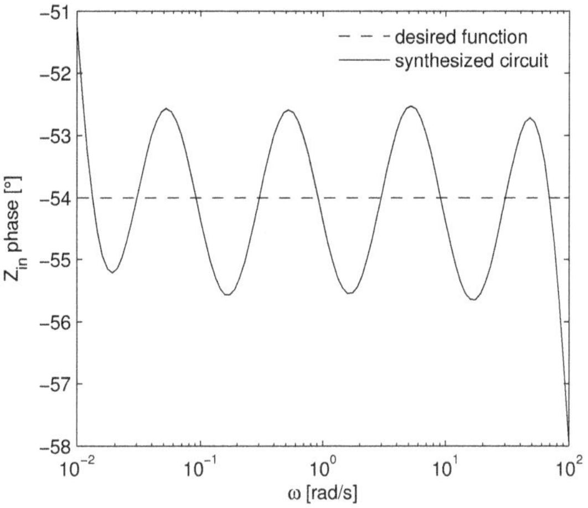

Fig. 20. Comparison of the phase of Z_{in} of the best found approximation circuit and (2) [1].

Absolute values of the deviations of the magnitude and the phase of Z_{in} of the best synthesized circuit are presented in Fig. 21. and Fig. 22 respectively.

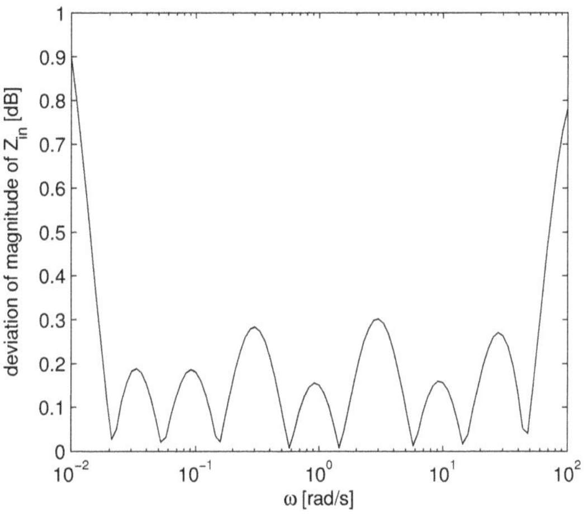

Fig. 21. Deviation of the magnitude of Z_{in} of the best found approximation circuit [1].

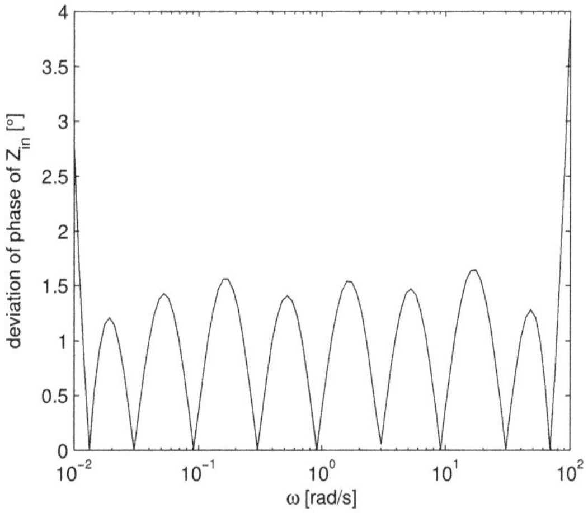

Fig. 22. Deviation of the phase of Z_{in} of the best found approximation circuit [1].

Three highest deviations of the magnitude and the phase characteristics of Z_{in} of the best synthesized circuit are summarized in Tab. 5.

Magnitude			
$\omega[rad/s]$	0.01	100	3.02
$\Delta_m[dB]$	0.9	0.78	0.30

Phase			
$\omega[rad/s]$	0.01	100	17.4
$\Delta_p[\circ]$	2.78	3.97	1.64

Tab. 5. The highest deviations of the magnitude and the phase of Z_{in} of the best synthesized approximation circuit.

As can be seen in Fig. 19 to Fig. 22 the maximum deviations of the magnitude and phase responses are located at the boundaries of the used frequency range. At these frequencies the unoptimized areas of the frequency response (0 to 10^{-2} rad/s and 10^2 to ∞ rad/s) affect behavior of the circuit in the area where the optimization was performed. The zeros and poles diagram of the synthesized circuit is presented in Fig. 23. All coefficients of the denominator of the approximation function of Z_{in} are positive therefore stability of the synthesized circuits is guaranteed.

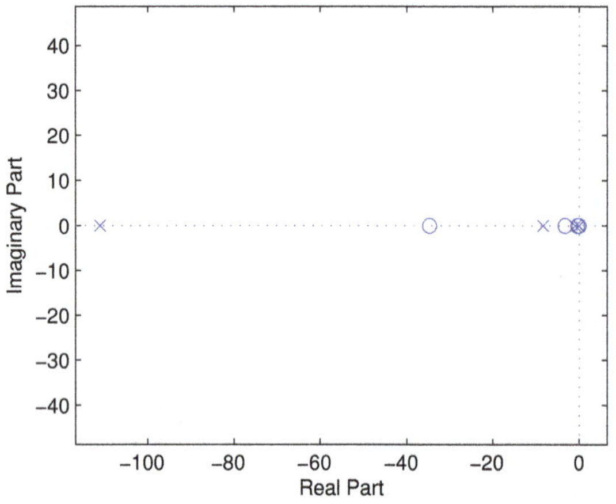

Fig. 23. The zeros and poles diagram of the best synthesized circuit [1].

Although the probability of using of all three component types (resistors, capacitors, inductors) was equal during the synthesis process, none of the circuits presented in Fig. 15 to Fig. 18 include any inductors. As the proposed algorithm was constrained to use only $n_c = 10$ components, it seems that using only capacitors and resistors allows the method to reach lower cost values than in solutions where inductors are included.

10. Comparison of The Results

In the section the best synthesized approximation circuit obtained using the proposed algorithm will be compared to the original approximation circuit designed in [9] by a classical method of the analog circuits design.

Since the proposed evolutionary synthesis method was configured to use the same circuit complexity (10 components at the most) as the original approximation circuit, accuracy of both circuits can be directly compared.

Except the deviations at the boundaries of the used frequency range (as was commented above) for the original approximation circuit the highest deviation of the magnitude is $\Delta_m = 0.61$ dB at angular frequency 0.33 rad/s. For the best solution of the proposed method the highest deviation of the magnitude is $\Delta_m = 0.27$ dB at angular frequency 0.30 rad/s. Thus in terms of deviation of the magnitude the accuracy of the synthesized circuit is more than twice better than the original approximation circuit. Comparison of the deviations of the magnitude of Z_{in} for both circuits is presented in Tab. 6.

original circuit	ω[rad/s]	0.01	100	0.33
	Δ_m[dB]	0.71	0.46	0.61
synthesized circuit	ω[rad/s]	0.01	100	0.30
	Δ_m[dB]	0.93	0.92	0.27

Tab. 6. Comparison of the deviations of the magnitude of Z_{in} of the original approximation circuit and the best synthesized circuit.

The highest phase deviation inside the used frequency range is for original circuit $\Delta_p = 5.2°$ at angular frequency 0.14 rad/s and for the synthesized circuit it is $\Delta_p = 1.5°$ at angular frequency 0.58 rad/s. Thus phase accuracy of the best synthesized circuit is more than three times better than the original approximation circuit. Comparison of the maximum deviations of the phase of Z_{in} is presented in Tab. 7.

original circuit	ω[rad/s]	0.01	100	0.14
	$\Delta_p[\circ]$	26.3	7.98	5.2
synthesized circuit	ω[rad/s]	0.01	100	0.58
	$\Delta_p[\circ]$	3.0	4.2	1.5

Tab. 7. Comparison of the deviations of the phase of Z_{in} of the original approximation circuit and the best synthesized circuit.

In [12] the same problem of synthesis of the fractional capacitor circuit was solved using simulated annealing method. The method was able to reach solutions of the same accuracy however the number of required evaluations of the objective function was almost four times higher. Comparison of accuracy of the best solutions and numbers of evaluations of the objective function of the proposed EDA method and simulated annealing method [12] is presented in Tab. 8.

simulated annealing method [12]	best cost	evals
	2.45	440e3
the proposed EDA method	best cost	evals
	2.43	120e3

Tab. 8. Comparison of the proposed EDA method and simulated annealing method presented in [12].

11. Conclusion

Automated analog circuit synthesis approach based on hybrid evolutionary method employing modified UMDA algorithm and a local search algorithm was presented in the paper. Used hybrid approach enables to employ specialized methods for both sub problems of different nature (synthesis of the topology and determination of the parameters).

Synthesis of the topology which is combinational optimization problem was solved using modified UMDA algorithm. Determination of the parameters which is continuous optimization problem was solved using a local search algorithm. The principle of the method is based on mutual interaction of synthesis of the topology phase (modified UMDA algorithm) and determination of the parameters phase (the local search algorithm) of the desired solution. Modification of the UMDA algorithm which allows solving problems with unitation constrains was proposed in the paper.

The proposed method was verified on the problem of synthesis of the fractional capacitor circuit introduced in [9]. Presented experiments have shown that the proposed algorithm is capable to synthesize solutions with accuracy overperforming solutions obtained using a classical method of the analog circuit design given in [9]. Accuracy of the magnitude of Z_{in} of the best obtained solution was more than twice better than the original approximation circuit. Accuracy of the phase of Z_{in} of the best obtained solution was more than three times better than the original approximation circuit.

In [12] the problem of fractional capacitor circuit realization was solved using simulated annealing method (SA). The accuracy of the circuits synthesized using SA was the same as the solutions produced using the proposed EDA method.

However SA required much higher number of the objective function evaluations. While the proposed EDA method required 120e3 objective function evaluations for synthesis of the same problem SA required 440e3 objective function evaluations.

For demonstration purposes the presented algorithm was verified using nodal analysis circuit simulator implemented in Matlab. Another improvement of the efficiency of the algorithm can achieved using Model Order Reduction Techniques [13].

Acknowledgements

The presented research was financed by the Czech Ministry of Education in frame of the National Sustainability Program, the grant LO1401 INWITE. For the research, infrastructure of the SIX Center was used.

References

[1] SLEZÁK, J. *Evolutionary Synthesis of Analog Electronic Circuits Using EDA Algorithms.* Ph.D. Thesis, Brno: Brno University of Technology, The Faculty of Electrical Engineering and Communication, 2014. 123 pages.

[2] ZINCHENKO, L., MÜHLENBEIN, H., KUREICHIK, V., MAHNING, T. Application of the univariate marginal distribution algorithm to analog circuit design. In *Proceedings of NASA/DoD Conference on Evolvable Hardware*, 2002, p. 93–101. DOI: 10.1109/EH.2002.1029871

[3] TORRES, A., PONCE, E. E., TORRES, M. D., DIAZ, E., PADILLA, F. Comparison of two evolvable systems in the automated analog circuit synthesis. In *Proceedings of Eighth Mexican International Conference on Artificial Intelligence (MICAI 2009)* Mexico, 2009, p. 3–8, ISBN 978-0-7695-3933-1. DOI: 10.1109/MICAI.2009.25

[4] TANG, M., LAU, R. Y. K. A hybrid estimation of distribution algorithm for the minimal switching graph problem. In *Proceedings of International Conference on Computational Intelligence for Modelling, Control and Automation and International Conference on Intelligent Agents, Web Technologies and Internet Commerce*, Austria 2005, p. 708–713. DOI: 10.1109/CIMCA.2005.1631347

[5] MÜHLENBEIN, H., PAAß, G. From recombination of genes to the estimation of distributions I. Binary parameters. *Lecture Notes in Computer Science 1411: Parallel Problem Solving from Nature - PPSN IV*, p. 178–187. DOI: 10.1007/3-540-61723-X_982

[6] SANTANA, R., OCHOA, A., SOTO, M. R. Factorized distribution algorithms for functions with unitation constraints. In *Proceedings of the Third Symposium on Adaptive Systems (ISAS-2001)*, 2001, p. 158–165.

[7] MÜHLENBEIN, H., MAHNIG, T., OCHOA, A. Schemata, distributions and graphical models in evolutionary optimization. *Journal of Heuristics*, 1999, p. 215–247.

[8] SANTANA, R., ECHEGOYEN, C., MENDIBURU, A., BIELZA, C., LOZANO, J. A., LARRAÑAGA, P., ARMAÑANZAS, R., SHAKYA, S. MATEDA: A suite of EDA programs in Matlab. *Technical Report EHU-KZAA-IK-2/09*, University of the Basque Country, 2009.

[9] CARLSON, G. E., HALIJAK, C. A. Approximation of fractional capacitors $(1/s)^{(1/n)}$ by a regular Newton process. *IEEE Transactions on Circuit Theory*, 1964, vol. 11, no. 2, p. 210–213, ISSN 0018-9324. DOI: 10.1109/TCT.1964.1082270

[10] BALAKRISHNAN, V. K. *Graph Theory*, McGraw-Hill, 1997, ISBN 0-07-005489-4.

[11] EL-MIHOUB, T. A., HOPGOOD, A. A., NOLLE, L., BATTERSBY, A. Hybrid genetic algorithms: A review. *Engineering Letters*, 2006, vol. 13, no. 2, p. 124-137, ISSN: 1816-093X.

[12] SLEZAK, J., GOTTHANS, T., DRINOVSKY, J. Evolutionary Synthesis of Fractional Capacitor Using Simulated Annealing Method. *Radioengineering*, 2012, vol. 21, no. 4, p. 1252–1259, ISSN 1210-2512.

[13] FELDMANN, P. Model order reduction techniques for linear systems with large numbers of terminals. In *Proceedings of Design, Automation and Test in Europe Conference and Exhibition*, 2004, vol. 2.1, p. 944-947. DOI: 10.1109/DATE.2004.1269013

About the Authors...

Josef SLEZAK was born in Zlin, Czech Republic, in 1982. He received the MSc. degree at the Brno University of Technology in 2007. In 2014 he received Ph.D. degree from the Brno University of Technology. His research interests include evolutionary synthesis of analog electronic circuits, design automation and optimization.

Tomas GOTTHANS was born in Brno, Czech Republic, in 1985. He received the MSc. degree at the Brno University of Technology (BUT) in 2010. In 2014 he received Ph.D. degree from the Université Paris-Est (UPE) and from the Brno University of Technology. He was working in the École Supérieure d'Ingénieurs en Électronique et Électrotechnique de Paris (ESIEE) in the ESYCOM Laboratory. He is presently a researcher in the Sensor, Information and Communication Systems (SIX) research laboratory, BUT. His research interests include programming, wireless communications and non-linear phenomenons.

Level Crossing Rate of Macrodiversity System in the Presence of Multipath Fading and Shadowing

Branimir JAKSIC [1], *Dusan STEFANOVIC* [2], *Mihajlo STEFANOVIC* [1], *Petar SPALEVIC* [3], *Vladeta MILENKOVIC* [1]

[1] Faculty of Electrical Engineering, University of Nis, Aleksandra Medvedeva 14, 18000 Nis, Serbia
[2] College of Applied Technical Sciences, Aleksandra Medvedeva 20, 18000 Nis, Serbia
[3] Faculty of Technical Sciences, University of Pristina, Knjaza Milosa 7, 38220 Kosovska Mitrovica, Serbia

branimir.jaksic@pr.ac.rs, dusan.stefanovic@itcentar.rs, misa.profesor@gmail.com, petarspalevic@yahoo.com, vladeta.milenkovic@elfak.ni.ac.rs

Abstract. *Macrodiversity system including macrodiversity SC receiver and two microdiversity SC receivers is considered in this paper. Received signal experiences, simultaneously, both, long term fading and short term fading. Microdiversity SC receivers reduce Rayleigh fading effects on system performance and macrodiversity SC receivers mitigate Gamma shadowing effects on system performance. Closed form expressions for level crossing rate of microdiversity SC receivers output signals envelopes are calculated. This expression is used for evaluation of level crossing rate of macrodiversity SC receiver output signal envelope. Numerical expressions are illustrated to show the influence of Gamma shadowing severity on level crossing rate.*

Keywords

Macrodiversity selection combining (SC) receiver, Rayleigh multipath fading, Gamma shadowing, level crossing rate, correlation

1. Introduction

Small scale fading and large scale fading degrade system performance and limit quality of service (QoS). Received signal experiences multipath fading resulting in signal envelope variation and shadowing resulting in signal envelope power variation. Reflection and refraction of radio wave cause short term fading and large obstacles between transmitter and receiver cause long term fading. It is important to determine how small scale fading and large scale fading affect performance of wireless communication system as outage probability and bit error probability. There are more distributions which can be applied to describe short term signal envelope variation depending on propagation environment. The most used statistical models are Rayleigh, Rician, Nakagami-*m*, Weibull and α-μ distributions [1]. Rayleigh and Nakagami-*m* distributions can be applied to describe signal envelope variation in linear non line-of-sight (LOS) multipath fading channel. In line-of-sight multipath fading environment, small scale signal envelope variation can be described by using Rician distribution. Weibull and α-μ distributions can be used to describe small scale signal envelope variation in nonlinear multipath fading channel dependent on the number of clusters in propagation environment [2].

There are two distributions which can be applied to describe large scale signal envelope power variation in shadowing fading channels. These statistical models are log-normal distribution and Gamma distribution. Log-normal distribution well describes large scale signal envelope power variation but it does not lead to closed form expression for probability density function of output signal envelope. The expression for outage probability of wireless communication system subjected to long term fading has a closed form, when Gamma distribution describes signal envelope power variation.

Macrodiversity system has one macrodiversity receiver and two or more microdiversity receivers. Macrodiversity receiver reduces large scale fading effects on outage probability and microdiversity receivers reduce small scale fading effects on outage probability. Signal envelope at output of macrodiversity receiver is equal to signal envelope at output of microdiversity receiver with greater signal envelope power at inputs [3].

In this paper macrodiversity selection combining (SC) receiver with two microdiversity SC receivers is considered. There are several combining techniques that can be used to mitigate the influence long term fading and short term fading on system performance. The most frequently combining techniques are maximal ratio combining (MRC), equal gain combining (EGC) and selection combining (SC). MRC receiver enables the best performance and it has the highest implementation complexity. The MRC receiver requires channel state information on each diversity branch and need provide receiver train on each diversity branch. The EGC provides performance comparable to MRC and it has lower implementation complexity than the MRC receiver. The SC receiver has the least implementation complexity due to the processing being performed only on one diversity branch. The SC receiver selects and outputs branch with the strongest signal envelope [4–5].

The first order performances of wireless communication systems are outage probability, bit error probability and channel capacity. The second order performances of wireless systems are level crossing rate and average fade duration. Level crossing rate can be calculated as the average value of the first derivative of output signal envelope. Average fade duration can be calculated as a ratio of outage probability and level crossing rate. Outage probability is defined as probability that signal envelope falls below the threshold.

There are more works considering macrodiversity system with correlated branches [6] and level crossing rate of macrodiversity systems [7–9]. In [10], macrodiversity system with macrodiversity SC receiver and two microdiversity MRC receivers are analyzed. Received signal is subjected simultaneously to Nakagami-*m* multipath fading and Gamma shadowing. Macrodiversity SC receiver is applied to reduce Gamma shadowing and microdiversity MRC receivers are used to reduce Nakagami-*m* multipath fading. Closed form expressions for average level crossing rate and average fade duration are calculated. In [11], level crossing rate and average fade duration of macrodiversity system with macrodiversity SC receiver and two microdiversity MRC receivers operating over Gamma shadowed Rician multipath fading environment are evaluated.

Macrodiversity system with macrodiversity SC receiver and two microdiversity SC receivers is considered. Received signal experiences short term fading and long term fading resulting in system performance degradation. Microdiversity SC receivers are used to combat Rayleigh short term fading and macrodiversity SC receiver is used to combat long Gamma term fading. Closed form expression for level crossing rate considering macrodiversity system is calculated. To the best author's knowledge, wireless communication system with macro and micro structures operating over Gamma shadowed Rayleigh multipath fading channels is not reported in open technical literature. Obtained results in this paper can be used in performance analysis and designing macrodiversity systems with macrodiversity SC receiver and two or more microdiversity SC receivers in the presence of Gamma shadowing and Rayleigh multipath fading.

2. Rayleigh Random Variable Level Crossing Rate

Squared Rayleigh random variable can be expressed as a sum of two independent, zero mean Gaussian random variables x_1 and x_2 with the same variance:

$$x^2 = x_1^2 + x_2^2 \tag{1}$$

where x is Rayleigh random variable. The first derivative of x is

$$\dot{x} = \frac{1}{x}\left(x_1\dot{x}_1 + x_2\dot{x}_2\right) \tag{2}$$

where $\dot{x}_1$ and $\dot{x}_2$ are independent, zero mean Gaussian random variables. Therefore, $\dot{x}$ is Gaussian random variable as linear transformation of Gaussian random variables. The average value of $\dot{x}$ is:

$$\overline{\dot{x}} = \frac{1}{x}\left(x_1\overline{\dot{x}_1} + x_2\overline{\dot{x}_2}\right). \tag{3}$$

The variance of the first derivative of Rayleigh random variable is

$$\sigma_{\dot{x}}^2 = \frac{1}{x^2}\left(x_1^2\sigma_{\dot{x}_1}^2 + x_2^2\sigma_{\dot{x}_2}^2\right) \tag{4}$$

where

$$\sigma_{\dot{x}_1}^2 = \sigma_{\dot{x}_2}^2 = \pi^2 f_m^2 \sigma^2 = f_1^2 \tag{5}$$

where f_m is maximal Doppler frequency and f_1 is normalized Doppler frequency. After substituting (5) in (4), the expression for variance becomes:

$$\sigma_{\dot{x}}^2 = \frac{f_1^2}{x^2}\left(x_1^2 + x_2^2\right) = f_1^2. \tag{6}$$

The probability density function of $\dot{x}$ is

$$p_{\dot{x}}(\dot{x}) = \frac{1}{\sqrt{2\pi} f_1} e^{-\frac{\dot{x}^2}{2f_1^2}}. \tag{7}$$

Rayleigh random variable and the first derivative of Rayleigh random variable are independent. Therefore, joint probability density function of Rayleigh random variable and the first derivative of Rayleigh random variable is

$$p_{x\dot{x}}(x\dot{x}) = p_x(x)p_{\dot{x}}(\dot{x}) = \frac{2x}{\Omega} e^{-\frac{x^2}{\Omega}} \frac{1}{\sqrt{2\pi} f_1} e^{-\frac{\dot{x}^2}{2f_1^2}}. \tag{8}$$

The level crossing rate of Rayleigh random variable can be calculated as average value of the first derivative of Rayleigh random variable:

$$\begin{aligned} N_x &= \int_0^\infty p_{x\dot{x}}(x\dot{x}) \cdot \dot{x} \cdot d\dot{x} \\ &= \frac{2x}{\Omega} e^{-\frac{x^2}{\Omega}} \int_0^\infty \dot{x} \frac{1}{\sqrt{2\pi} f_1} e^{-\frac{\dot{x}^2}{2f_1^2}} d\dot{x} = \frac{2x}{\Omega} e^{-\frac{x^2}{\Omega}} f_1 \end{aligned} . \tag{9}$$

The expression for level crossing rate of Rayleigh random variable can be used in performance analysis of wireless communication system operating over Rayleigh multipath fading environment.

3. Level Crossing Rate of SC Receiver Output Signal Envelope

The average level crossing rate (LCR) is a measure [12] that clearly reflects the performances of fading affected system and is used for modeling of wireless communication systems. LCR is related to criterion used to assess error probability of packets of distinct length [4], and to determinate parameters of equivalent channel, modeled by a Markov chain with defined number of states. LCR is used for determining of the rate at which the envelope of the received signal crosses a specified defined level.

Double SC receiver operating over Rayleigh multipath fading channel is considered. Signal envelopes at inputs of SC receiver are denoted with x_1 and x_2 and signal envelope at output of SC receiver is denoted with x. At inputs of SC receiver identical and independent Rayleigh multipath fading is present. Probability density function of SC receiver output signal is

$$p_x(x) = p_{x_1}(x)F_{x_2}(x) + p_{x_2}(x)F_{x_1}(x) = 2p_{x_1}(x)F_{x_2}(x) \quad (10)$$

where $F_{x2}(x)$ is cumulative distribution function of Rayleigh random variable:

$$F_{x_2}(x) = 1 - e^{-\frac{x^2}{\Omega}}. \quad (11)$$

After substituting (11) in (10), the expression for $p_x(x)$ becomes

$$p_x(x) = \frac{2x}{\Omega}e^{-\frac{x^2}{\Omega}}\left(1 - e^{-\frac{x^2}{\Omega}}\right) \quad (12)$$

where Ω is signal envelope power.

The joint probability density function of x and $\dot{x}$ is

$$\begin{aligned} p_{x\dot{x}}(x\dot{x}) &= p_{x_1\dot{x}_1}(x\dot{x})F_{x_2}(x) + p_{x_2\dot{x}_2}(x\dot{x})F_{x_1}(x) \\ &= 2p_{x_1\dot{x}_1}(x\dot{x})F_{x_2}(x) \\ &= \frac{2x}{\Omega}e^{-\frac{x^2}{\Omega}}\left(1 - e^{-\frac{x^2}{\Omega}}\right)\frac{1}{\sqrt{2\pi}f_1}e^{-\frac{\dot{x}^2}{2f_1^2}}. \end{aligned} \quad (13)$$

Level crossing rate of SC receiver output signal envelope is

$$\begin{aligned} N_x &= \int_0^\infty p_{x\dot{x}}(x\dot{x}) \cdot \dot{x} \cdot d\dot{x} \\ &= \frac{2x}{\Omega}e^{-\frac{x^2}{\Omega}}\left(1 - e^{-\frac{x^2}{\Omega}}\right)\int_0^\infty \dot{x}\frac{1}{\sqrt{2\pi}f_1}e^{-\frac{\dot{x}^2}{2f_1^2}}d\dot{x} \\ &= \frac{2xf_1}{\Omega}e^{-\frac{x^2}{\Omega}}\left(1 - e^{-\frac{x^2}{\Omega}}\right). \end{aligned} \quad (14)$$

The expression for level crossing rate can be applied for calculation of average fade duration of wireless communication system with dual SC receiver operating over Gamma shadowed Rayleigh multipath fading channels.

4. Level Crossing Rate of Macrodiversity SC Receiver Output Signal Envelope

Macrodiversity system with macrodiversity SC receiver and two microdiversity SC receivers operating over shadowed multipath fading environment is considered. Short term Rayleigh fading and Gamma correlated long term fading are presented at inputs of microdiversity SC receivers. Macrodiversity SC receiver reduces signal envelope power variation and microdiversity SC receivers reduce signal envelope variation on system performance. Signal envelopes at output of microdiversity receivers are denoted with x_1 and x_2 and macrodiversity SC receiver output signal envelope is denoted with x.

Signal envelopes powers at inputs in microdiversity receivers are correlated. Signal envelope powers Ω_1 and Ω_2 follow Gamma distribution [11]:

$$\begin{aligned} p_{\Omega_1\Omega_2}(\Omega_1\Omega_2) &= \frac{1}{\Gamma(c)(1-\rho^2)\rho^{\frac{c-1}{2}}\Omega_0^{c+1}} \times \\ &(\Omega_1\Omega_2)^{\frac{c-1}{2}}e^{-\frac{\Omega_1+\Omega_2}{\Omega_0(1-\rho^2)}}I_{c-1}\left(\frac{2\rho}{\Omega_0(1-\rho^2)}(\Omega_1\Omega_2)^{\frac{1}{2}}\right) \end{aligned} \quad (15)$$

where c is fading severity, ρ is correlation coefficient and Ω_o is average power of Ω_1 and Ω_2 ($\Omega_1 \geq 0$, $\Omega_2 \geq 0$). Level crossing rate of signal envelopes x_1 and x_2 can be calculated by using expression (14).

The level crossing rate of macrodiversity SC receiver output signal envelope is

$$\begin{aligned} N_x &= \int_0^\infty d\Omega_1 \int_0^{\Omega_1} d\Omega_2 N_{x_1/\Omega_1} p_{\Omega_1\Omega_2}(\Omega_1\Omega_2) \\ &+ \int_0^\infty d\Omega_2 \int_0^{\Omega_2} d\Omega_1 N_{x_2/\Omega_1} p_{\Omega_1\Omega_2}(\Omega_1\Omega_2) \\ &= 2\int_0^\infty d\Omega_1 \int_0^{\Omega_1} d\Omega_2 N_{x_1/\Omega_1} p_{\Omega_1\Omega_2}(\Omega_1\Omega_2) \\ &= \frac{4f_1x}{\Gamma(c)(1-\rho^2)\rho^{\frac{c-1}{2}}\Omega_0^{c+1}}\sum_{i=0}^{\infty}\left(\frac{\rho}{\Omega_0(1-\rho^2)}\right)^{c+2i-1} \\ &\times \frac{(\Omega_0(1-\rho^2))^{c+i}}{\Gamma(c+i)i!}\int_0^\infty d\Omega_1\Omega_1^{c+i-2}e^{-\frac{x^2}{\Omega_1}}e^{-\frac{\Omega_1}{\Omega_0(1-\rho^2)}} \\ &\times\left(1 - e^{-\frac{x^2}{\Omega_1}}\right)\gamma\left(c+i, \frac{\Omega_1}{\Omega_0(1-\rho^2)}\right). \end{aligned} \quad (16)$$

The incomplete Gamma function $\gamma(n,x)$ is [13]:

$$\gamma(n,x)=\Gamma(n)-\frac{1}{n}x^n e^{-x}{}_1F_1(1,n+1,x)=\Gamma(n)-\frac{1}{n}x^n e^{-x}\sum_{i=0}^{\infty}\frac{x^i}{(n+1)_i} \tag{17}$$

where $(a)_n$ denotes the Pocchammer symbol.

After substituting (17) in (16), the expression for level crossing rate of macrodiversity SC receiver output signal envelope becomes:

$$N_x=\frac{4f_1 x}{\Gamma(c)\left(1-\rho^2\right)\rho^{\frac{c-1}{2}}\Omega_0^{c+1}}\sum_{i=0}^{\infty}\left(\frac{\rho}{\Omega_0\left(1-\rho^2\right)}\right)^{c+2i-1}\times\frac{\left(\Omega_0\left(1-\rho^2\right)\right)^{c+i}}{\Gamma(c+i)i!}\left(I_1-I_2-I_3+I_4\right) \tag{18}$$

where

$$I_1=\int_0^{\infty}d\Omega_1\Omega_1^{c+i-2}e^{-\frac{x^2}{\Omega_1}-\frac{\Omega_1}{\Omega_0\left(1-\rho^2\right)}}\Gamma(c+i), \tag{19}$$

$$I_2=\int_0^{\infty}d\Omega_1\Omega_1^{c+i-2}e^{-\frac{x^2}{\Omega_1}-\frac{2\Omega_1}{\Omega_0\left(1-\rho^2\right)}}\frac{1}{c+i}\left(\frac{\Omega_1}{\Omega_0\left(1-\rho^2\right)}\right)^{c+i}\times\sum_{j=0}^{\infty}\left(\frac{\Omega_1}{\Omega_0\left(1-\rho^2\right)}\right)^{j}\frac{1}{(c+i+1)_j}, \tag{20}$$

$$I_3=\int_0^{\infty}d\Omega_1\Omega_1^{c+i-2}e^{-\frac{2x^2}{\Omega_1}-\frac{\Omega_1}{\Omega_0\left(1-\rho^2\right)}}\Gamma(c+i) \tag{21}$$

and

$$I_4=\int_0^{\infty}d\Omega_1\Omega_1^{c+i-2}e^{-\frac{2x^2}{\Omega_1}-\frac{2\Omega_1}{\Omega_0\left(1-\rho^2\right)}}\frac{1}{c+i}\left(\frac{\Omega_1}{\Omega_0\left(1-\rho^2\right)}\right)^{c+i}\times\sum_{j=0}^{\infty}\left(\frac{\Omega_1}{\Omega_0\left(1-\rho^2\right)}\right)^{j}\frac{1}{(c+i+1)_j}. \tag{22}$$

After processing and solving integrals (19), (20), (21) and (22), expression (18) becomes:

$$\begin{aligned}N_x=&\frac{4fx_1}{\Gamma(c)\left(1-\rho^2\right)\rho^{\frac{c-1}{2}}\Omega_0^{c+1}}\times\sum_{i=0}^{\infty}\left(\frac{\rho}{\Omega_0\left(1-\rho^2\right)}\right)^{c+2i-1}\frac{\left(\Omega_0\left(1-\rho^2\right)\right)^{c+i}}{\Gamma(c+i)i!}\\&\times\Bigg[\Gamma(c+i)\left(x^2\Omega_0\left(1-\rho^2\right)\right)^{\frac{c+i-1}{2}}\times K_{c+i-1}\left(2\sqrt{\frac{x^2}{\Omega_0\left(1-\rho^2\right)}}\right)-\\&-\frac{1}{c+i}\sum_{j=0}^{\infty}\frac{1}{(c+i+1)_j}\frac{1}{\left(\Omega_0\left(1-\rho^2\right)\right)^{c+i+j}}\times\left(\frac{1}{2}x^2\Omega_0\left(1-\rho^2\right)\right)^{c+i+\frac{j}{2}-\frac{1}{2}}\\&\times K_{2c+2i+j-1}\left(2\sqrt{\frac{2x^2}{\Omega_0\left(1-\rho^2\right)}}\right)-\\&-\Gamma(c+i)\left(2x^2\Omega_0\left(1-\rho^2\right)\right)^{\frac{c+i-1}{2}}\times K_{c+i-1}\left(2\sqrt{\frac{2x^2}{\Omega_0\left(1-\rho^2\right)}}\right)+\\&+\frac{1}{c+i}\sum_{j=0}^{\infty}\frac{1}{(c+i+1)_j}\frac{1}{\left(\Omega_0\left(1-\rho^2\right)\right)^{c+i+j}}\times\left(x^2\Omega_0\left(1-\rho^2\right)\right)^{c+i+\frac{j}{2}-\frac{1}{2}}\\&\times K_{2c+2i+j-1}\left(2\sqrt{\frac{4x^2}{\Omega_0\left(1-\rho^2\right)}}\right)\Bigg].\end{aligned} \tag{23}$$

where $K_n(x)$ is modified Bessel function of the second kind, order n and argument x.

5. Numerical Results

In Fig. 1 and Fig. 2, normalized average level crossing rate of macrodiversity SC receiver output signal envelope versus macrodiversity SC receiver output signal envelope is plotted. For lower values of SC receiver output signal envelope, average level crossing rate increases as signal envelope increases and for higher values of signal envelope, level crossing rate decreases as signal envelope increases. The influence of signal envelope on level crossing rate is greater for higher values of signal envelope. Level crossing rate decreases as Gamma fading severity increases. For higher values of Gamma shadowing severity, long term fading is of less severity. Graphical results are shown; average level crossing rate increases as correlation coefficient increases. Diversity gain decreases as correlation coefficient increases.

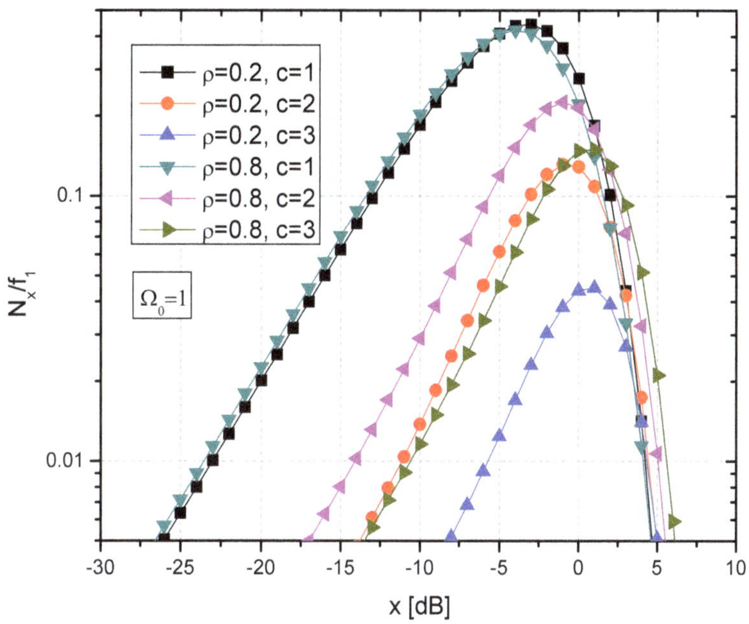

Fig. 1. Level crossing rate of macrodiversity SC receiver output signal envelope for different values of Gamma shadowing severity parameter c.

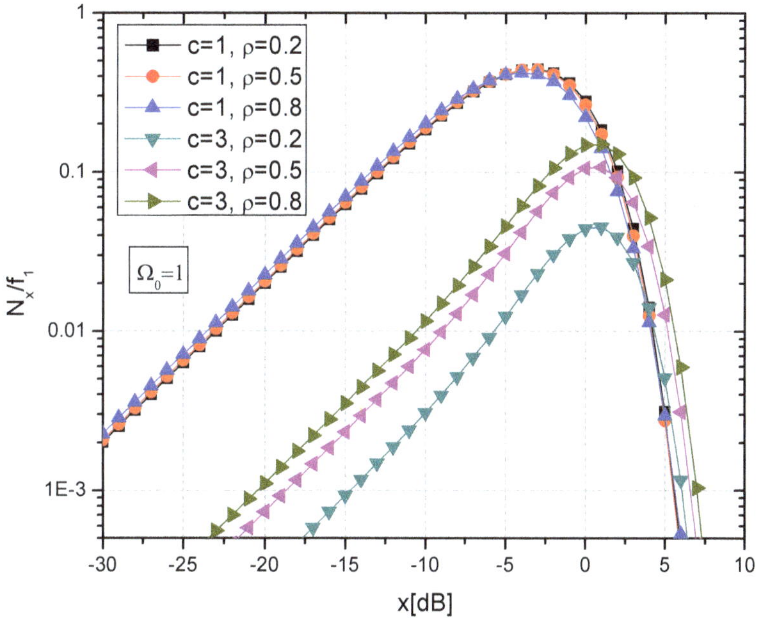

Fig. 2. Level crossing rate of macrodiversity SC receiver output signal envelope for different values of correlation coefficient ρ.

In Fig. 3, normalized average level crossing rate of macrodiversity SC receiver output signal envelope versus Gamma shadowing severity parameter *c* for several values of correlation coefficient is plotted. Average level crossing rate decreases as Gama fading severity increases. When Gamma fading severity parameter *c* goes to infinity composite Gamma shadowed, Rayleigh multipath channel goes to Rayleigh multipath fading channel.

Normalized average level crossing rate of macrodiversity SC receiver output signal envelope versus correlation coefficient ρ for several values of Gamma shadowing severity is illustrated in Fig. 4. Average level crossing rate increases as coefficient of correlation increases. Outage probability is better for lower values of average level crossing rate. When correlation coefficient goes to 1, level crossing rate has maximum and system performance is the worst. For the case, the least value of signal envelope occurs at both antennas.

6. Conclusion

Macrodiversity system with macrodiversity SC receiver and two microdiversity SC receivers subjected simultaneously to Rayleigh multipath fading and Gamma shadowing has been analyzed in this paper. Closed form

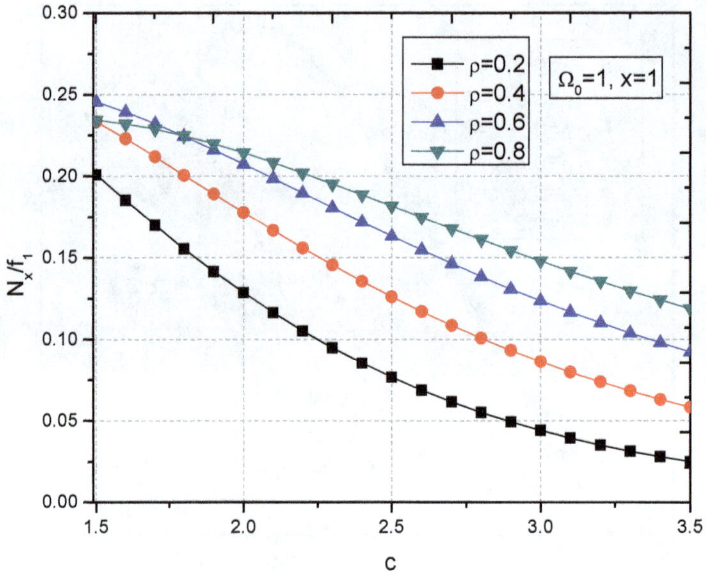

Fig. 3. Level crossing rate of macrodiversity SC receiver output signal envelope versus Gamma shadowing severity parameter c.

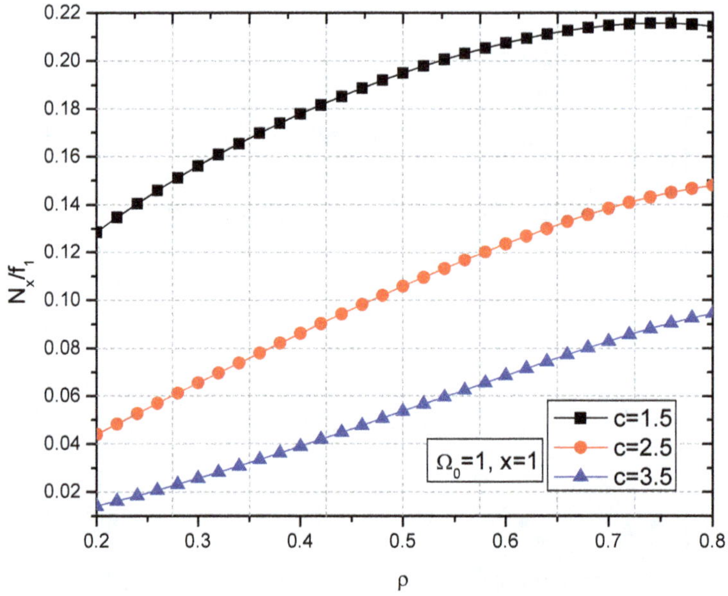

Fig. 4. Level crossing rate of macrodiversity SC receiver output signal envelope versus correlation coefficient ρ.

expressions for LCR of macrodiversity SC receiver output signal envelope are derived. Capitalizing on obtained expressions numerical results are presented graphically to show Gamma shadowing severity effects and correlation coefficient effects on LCR of macrodiversity system output signal envelope. LCR reduction has been analyzed as a function of Gamma fading severity decrease and correlation coefficient decrease.

Presented performance evaluation for proposed communication scenarios provides insight into the performance dependence on key system parameters. Presented LCR analysis allows system designers to perform trade-off studies among the various communication type/drawback combinations in order to determine the optimal choice in the presence of given constraints.

References

[1] SUBER, G. L. *Mobile Communication.* 2nd ed. Dordrecht: Kluwer Academic Publisher, 2003.

[2] PROAKIS, J. *Digital Communications.* 4nd ed. New York: McGraw-Hill, 2001.

[3] MUKHERJEE, S., AVIDOR, D. Effect of microdiversity and correlated macrodiversity on outages in a cellular system. *IEEE Transactions on Wireless Technology*, 2003, vol. 2, no. 1, p. 50 to 58. DOI: 10.1109/TWC.2002.806363

[4] PANIC, S., STEFANOVIC, M., ANASTASOV, J., SPALEVIC, P. *Fading and Interference Mitigation in Wireless Communications.* USA: CRC Press, 2013.

[5] SIMON, M. K., ALOUINI, M. S. *Digital Communication over Fading Channels.* USA: John Wiley & Sons, 2000.

[6] SHANKAR, P. M. Analysis of microdiversity and dual channel macrodiversity in shadowed fading channels using a compound fading model. *International Journal of Electronics and Communications (AEUE)*, 2008, vol. 62, no. 6, p. 445–449. DOI: 10.1016/j.aeue.2007.06.008

[7] KLINGENBRUNN, T., MOGENSEN, P. Modelling cross-correlated shadowing in network simulations. In *Proceedings of the 50th IEEE Vehicular Technology Conf. (VTC 1999 Fall)*. Amsterdam (The Netherlands), 1999, vol. 3, p. 1407–1411. DOI: 10.1109/VETECF.1999.801494

[8] ZHANG, J., AALO, V. Effect of macrodiversity on average-error probabilities in a Rician fading channel with correlated lognormal shadowing. *IEEE Transactions on Communications*, 2001, vol. 49, no. 1, p. 14–18. DOI: 10.1109/26.898244

[9] SAFAK, A., PRASAD, R. Effects of correlated shadowing signals on channel reuse in mobile radio systems. *IEEE Transactions on Vehicular Technology*, 1991, vol. 40, no. 4, p. 708–713. DOI: 10.1109/25.108381

[10] STEFANOVIC, D., PANIC, S., SPALEVIC, P. Second-order statistics of SC macrodiversity system operating over Gamma shadowed Nakagami-m fading channels. *AEU - International Journal of Electronics and Communications,* 2011, vol. 65, no. 5, p. 413–418. DOI: 10.1016/j.aeue.2010.05.001

[11] SEKULOVIC, N., STEFANOVIC, M. Performance analysis of system with micro- and macrodiversity reception in correlated Gamma shadowed Rician fading channels. *Wireless Personal Communications,* 2012, vol. 65, no. 1, p. 143–156. DOI: 10.1007/s11277-011-0232-8

[12] ISKANDER, C. D., MATHIOPOULOS, P. T. Analytical level crossing rate and average fade duration in Nakagami fading channels. *IEEE Transactions on Communications*, 2002, vol. 50, no. 8, p 1301–1309. DOI: 10.1109/TCOMM.2002.801465

[13] GRADSHTEYN, I., RYZHIK, I. *Tables of Integrals, Series, and Products*. New York: Academic Press, 1994.

About the Authors ...

Branimir JAKSIC was born in Kosovska Mitrovica, Serbia, in 1984. He received B.Sc. and M.Sc. degrees in Electrical Engineering from the Faculty of Technical Sciences in Kosovska Mitrovica, University of Pristina, Serbia. He is Ph.D. candidate in the Faculty of Electronic Engineering, University of Nis, Serbia. Areas of research include statistical communication theory and optical telecommunications. He has authored several scientific papers on the above subject.

Dusan STEFANOVIC was born in Nis, Serbia, in 1979. He received the M.Sc. in Electrical Engineering from the Faculty of Electronic Engineering (Dept. of Telecommunications), University of Nis, Serbia, in 2005 and Ph.D. in the same department in 2012. He is interested in optical and wireless communication. Now he is working in College of Applied Technical Sciences as professor on the following subjects: Database administration, computer networks, administration of computer networks and opto-laser technologies.

Mihajlo STEFANOVIC was born in Nis, Serbia in 1947. He received B.Sc., M.Sc. and Ph.D. degrees in Electrical Engineering from the Faculty of Electronic Engineering (Dept. of Telecommunications), University of Nis, Serbia, in 1971, 1976 and 1979, respectively. His primary research interests are statistical communication theory, optical and satellite communications. He has written or co-authored a great number of journal publications. Dr. Stefanovic is a full-time professor with the Dept. of Telecommunications, Faculty of Electronic Engineering, University of Nis, Serbia.

Petar SPALEVIC was born in Kraljevo, Serbia, in 1973. He received the B.S. degree from the Faculty of Electronic Engineering, University of Pristina, in 1997, and M.S. and Ph.D. degrees from the Faculty of Electronic Engineering, University of Nis in 1999 and 2003, respectively. He is a Professor with the Faculty of Technical Sciences in Kosovska Mitrovica. His primary research interests are statistical communications theory, wireless communications, applied probability theory and optimal receiver design.

Vladeta MILENKOVIC was born in Nis, Serbia. He received B. Sc. and M. Sc. degrees in Electrical Engineering from the Faculty of Electronic Engineering, University of Nis, Serbia. He is Ph.D. candidate in the same faculty. He has authored several scientific papers on wireless communications, metrology and measurement techniques.

ARQ Protocols in Cognitive Decode-and-Forward Relay Networks: Opportunities Gain

Zongsheng ZHANG[1,2], *Qihui WU*[1] , *Jinlong WANG*[1]

[1]College of Communications Engineering, PLA University of Science and Technology, Street YuDao, Nanjing, China
[2]College of National Information Science, JieFangGongyuan Road, Wuhan, China

zhangzongsheng1984@163.com, 532692271@qq.com, 15951803143@163.com

Abstract. *In this paper, two novel automatic-repeat-request (ARQ) based protocols are proposed, which exploit cooperation opportunity inherent in secondary retransmission to create access opportunities. If the signal was not decoded correctly by destination, another user can be acted as a relay to reduce retransmission rounds by relaying the signal. For comparison, we also propose a Direct ARQ Protocol. Specifically, we derive the exact closed-form outage probability of three protocols, which provides an effective means to evaluate the effects of several parameters. Moreover, we propose a new metric to evaluate the performance improvement for cognitive networks. Finally, Monte Carlo simulations were presented to validate the theory analysis, and a comparison is made among the three protocols.*

Keywords

Cognitive relay networks, outage probability, automatic-repeat-request

1. Introduction

Since the 1920s, each wireless communications system has been required to have an exclusive license from government in order to mitigate the interference from each other. Over the past decades, we have witnessed an increasing development and popularity of wireless communications, which has turned the limited spectrum resources into a scarce resource, which consequently imposes increasing stress on the fixed and limited radio spectrum. In practice, recent researches by the Federal Communications Commission (FCC) indicated that the utilization of spectrum is very low varying form 6% to 85%, whereas only 2% of spectrum would be used in the US at any given moment [1]. The inefficient use of the radio spectrum resources under the current fixed policy, as well as the demand for more and better services has contributed to the reconsideration of the way the spectrum is used today and has led to decision of the FCC to allow access of unlicensed users to the broadcast television spectrum at locations where that spectrum is not being used by licensed users [2].

In order to increase the efficiency of spectrum utilization, diverse types of technologies have been deployed. The key technology towards efficient spectrum usage is Cognitive Radio, which was first introduced by J. Mitola III [3]. Cognitive radio is typically built on the software-defined radio technology, in which the transmitter's operating parameters, such as the frequency range, modulation type, and maximum transmission power can be dynamically adjusted by software [4]– [7]. Cognitive radio allows unlicensed users to access licensed users free from harmful interference. Specifically, the unlicensed (Secondary) users can access the spectrum resources originally licensed to primary users through spectrum overlay, interweave, and underlay ways. In overlay spectrum way, the unlicensed users need allocate a fraction of resources to maintain or improve the transmission of the primary users using sophisticated signal processing and coding. In interweave spectrum way, the unlicensed users share the licensed spectrum when the spectrum is idle, known as spectrum hole. Contrary to overlay and interweave ways, the unlicensed users control their transmit power to satisfy interference power constraint which licensed user can tolerate in underlay way. Due to its simplicity, we focus on underlay spectrum sharing in this paper.

Direct transmission, which demands large transmit power, ends up with small opportunity of access and hence low spectrum sharing efficiency. At the same time, cooperative diversity [8] has emerged as a promising technique to combat fading in wireless communications. Therefore, cognitive radio combined with cooperative diversity technique, referred to as cognitive relay networks (CRNs) [9]–[10], appears as an attractive solution to boost the spectrum sharing efficiency. In [9], a linear cooperative sensing frame work has been proposed based on the combination of local statistics from individual cognitive users. A distributed algorithm for channel access and power control was proposed for cognitive multi-hop relays in [10]. In [11], the authors have considered the application of cooperative diversity to spectrum sensing, and shown that the sensing performance is improved by exploiting the user cooperation. The authors analyzed the delay of a cognitive relay assisted multi-access networks, however, they did not consider the impact of primary user activities and dynamic spectrum-sharing in [12]. Most recently, some studies [13]– [17] focused on the outage performance in cognitive relay networks.

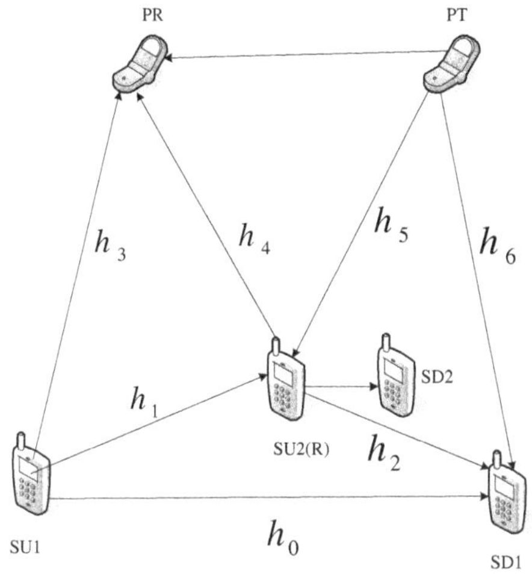

Fig. 1. System model.

The above work in cognitive relay networks has greatly improved our understanding of cognitive relay networks, such as protocol design, outage performance. However, they all have not considered the automatic-repeat-request (ARQ) [18], [19] technique in cognitive relay networks. To the best of our knowledge, the performance analysis combined cognitive relay networks and ARQ technique is almost unexplored from the analytical point view. As such, we will fill this important gap in this paper. As such, we propose two ARQ protocols in this paper, and a direct ARQ protocol for comparison.

The main contributions of this paper can be summarized as follows. First, we propose two novel ARQ protocols in cognitive relay networks, which can exploit additive transmission opportunities by cooperation with each other. Second, we derived the closed-form outage probability expressions which can provide efficient means to evaluate the effects of several parameters. Finally, we propose a new metric to evaluate the improvement of the two proposed ARQ protocols. The remained of this paper is organized as follows. In Sec. 2, the system model is presented. Followed by Sec. 3, we propose three ARQ protocols and outage probability analysis is presented. Sec. 4 provides some simulation results to validate the theory analysis. Finally, in Sec. 5, some concluding remarks are presented.

2. System Model

Consider a cognitive radio system with the coexistence of one primary user pair and two secondary user pairs, as depicted in Fig. 1. In the secondary network, we assume that only one pair secondary user[1] can concurrently transmit with primary user. When the SD1 can not decode the packet transmitted by SU1 and SU2 can decode the packet correctly, the SU2 will participate in transmitting in next transmission round if the maximum allowable transmission round is not reached. The decode-and-forward (DF) protocol is considered throughout this paper. In this paper, we considered three ARQ protocols: Cooperative ARQ I Protocol, Cooperative ARQ II Protocol, and Direct ARQ Protocol, respectively.

The Direct ARQ protocol works as follows. First, the SU1 transmits its signal to its destination SD1 in the first time round. SD1 indicates success or failure of receiving the signal by feeding back a single bit of acknowledgement (ACK) or negative-acknowledgement (NACK). If an ACK is received or the transmission rounds reaches the maximum allowed number M, the system stops transmitting the current message and starts transmitting a new signal. Otherwise, if a NACK is received and the transmission rounds has not reached the maximum number M, SU1 retransmits it. When the maximum transmission rounds are reached, SD1 still can not decode the signal, an outage is declared.

The main difference between Direct ARQ protocol and Cooperative ARQ I Protocol is that when a NACK is fed back, we should take the SU2 into consideration. If SU2 decodes the signal correctly, then at the next transmission round, SU1 and SU2 should both transmit the signal to the destination SD1. Similarly, the main difference between Direct ARQ Protocol and Cooperative ARQ II Protocol is that when a NACK is fed back, we should also take the SU2 into consideration, If SU2 decodes the signal correctly, then at the next transmission round, the better node between SU1 and SU2 will be selected to retransmit. Basically speaking, if the channel between SU1 and SD1 is in deep fading, the Cooperative ARQ I Protocol and Cooperative ARQ II Protocol will reduce the rounds of retransmission greatly.

The channel between any node is assumed Rayleigh fading throughout this paper. h_0, h_1, h_2, h_3, h_4, h_5, h_6 denote the channels between SU1$\rightarrow$ SD1, SU1$\rightarrow$SU2, SU2$\rightarrow$SD1, SU1$\rightarrow$PD, SU2$\rightarrow$PD, PT$\rightarrow$SU2, PT$\rightarrow$SD1, respectively. As such, the effective channel gains $|h_i|^2, (i = 0, 1, \cdots, 6)$ follow the exponential distribution with parameters $a_i, (i = 0, 1, \cdots, 6)$. Therefore, the probability density function (PDF) and cumulative distribution function (CDF) of X , for $X \in \{|h_0|^2, |h_1|^2, |h_2|^2, |h_3|^2, |h_4|^2, |h_5|^2, |h_6|^2\}$, can be formulated as

$$f_X(x) = \frac{1}{a} e^{-\frac{x}{a}}, \tag{1}$$

and

$$F_X(x) = 1 - e^{-\frac{x}{a}}. \tag{2}$$

3. Performance Analysis

3.1 Direct ARQ Protocol

In this protocol, in order to satisfy the constraint of primary user, SU1 should control its transmit power. As such,

[1]For simplicity of presentation, the working pair of secondary users is denoted as SU1 and SD1, another pair of secondary users is represented as SU2 and SD2.

the transmit power at SU1 is constrained as

$$P_{S1} = \frac{Q}{|h_3|^2}, \tag{3}$$

where Q denotes the maximum allowable interference power of primary user. As such, the received signal-to-interference ratio (SIR)[2] can be written as

$$SIR_{SD,1} = \frac{Q|h_0|^2}{P_P|h_3|^2|h_6|^2}, \tag{4}$$

where P_P denotes the transmit power of primary user. According to the definition of outage in ARQ protocol, the outage probability of Direct ARQ Protocol can be mathematically derived as

$$\begin{aligned} P_{out}^{Dir}(L) &= [\Pr\{SIR_{SD,1} < \gamma\}]^M \\ &= [\underbrace{\Pr\{\frac{Q|h_0|^2}{P_P|h_3|^2|h_6|^2} < \gamma\}}_{I_1}]^M. \end{aligned} \tag{5}$$

Therefore, the main task is to derive I_1. I_1 can be derived as

$$\begin{aligned} I_1 &= \Pr\{\frac{Q|h_0|^2}{P_P|h_3|^2|h_6|^2} < \gamma\} \\ &= \int_0^\infty f_{|h_3|^2}(x)\int_0^\infty f_{|h_6|^2}(y)\int_0^{\frac{P_P\gamma xy}{Q}} f_{|h_0|^2}(z)dzdydx \\ &= \int_0^\infty f_{|h_3|^2}(x)\int_0^\infty f_{|h_6|^2}(y)F_{|h_0|^2}(\frac{P_P\gamma xy}{Q})dydx \\ &= \frac{1}{a_3a_6}\int_0^\infty\int_0^\infty e^{-\frac{x}{a_3}}e^{-\frac{y}{a_6}}(1-e^{-\frac{P_P\gamma xy}{a_0Q}})dydx \\ &= 1-\frac{1}{a_3a_6}\int_0^\infty\int_0^\infty e^{-\frac{x}{a_3}}e^{-\frac{y}{a_6}}e^{-\frac{P_P\gamma xy}{a_0Q}}dydx \\ &= 1+\frac{a_0Q}{a_3a_6P_P\gamma}e^{\frac{a_0Q}{a_3a_6P_P\gamma}}Ei(-\frac{a_0Q}{a_3a_6P_P\gamma}). \end{aligned} \tag{6}$$

In this paper, we defined a new metric – average transmission rounds to represent the goodness of the protocol. Table 1[3] presents the probability that after m-th[4] transmission round the SD1 decodes the signal successfully for Direct ARQ Protocol.

Transmission Round	Probability
1	$1-P_{out}^{Dir}(1)$
2	$P_{out}^{Dir}(1)-P_{out}^{Dir}(2)$
$\cdots$	$\cdots$
$M-1$	$P_{out}^{Dir}(M-2)-P_{out}^{Dir}(M-1)$
M	$P_{out}^{Dir}(M-1)$

Tab. 1. Probability of the considered system stop transmitting after m-th transmission round based on Direct ARQ Protocol.

As such, the average transmission rounds of Direct ARQ Protocol can be given as

$$\begin{aligned} T_{out}^{Dir} &= 1\times(1-P_{out}^{Dir}(1))+2\times(P_{out}^{Dir}(1)-P_{out}^{Dir}(2)) \\ &\quad +\cdots+L\times P_{out}^{Dir}(L-1) \\ &= 1+P_{out}^{Dir}(1)+P_{out}^{Dir}(2)+\cdots+P_{out}^{Dir}(M-1) \\ &= 1+\sum_{i=1}^{M-1}P_{out}^{Dir}(i). \end{aligned} \tag{7}$$

3.2 Cooperative ARQ I Protocol

Due to coexistence of SU1 and SU2 in the retransmission stage, we divided the maximum interference power Q equally to SU1 and SU2 for simplicity of analysis. As such, the transmit power of SU1 and SU2 can be expressed as

$$P_{S1} = \frac{Q}{2|h_3|^2}, \tag{8}$$

and

$$P_{S2} = \frac{Q}{2|h_4|^2}. \tag{9}$$

As such, the received SIR at SD1 from SU1 and SU2 in the next transmission can be expressed as

$$SIR_{SD,2} = \frac{Q|h_0|^2}{2P_P|h_3|^2|h_6|^2}, \tag{10}$$

and

$$SIR_{RD,2} = \frac{Q|h_2|^2}{2P_P|h_4|^2|h_6|^2}. \tag{11}$$

We define $\{Tr = m\}$ denoting the event that SU2 successfully decodes the signal during the m-th transmission round for any $m = 1,2,\cdots,M-1$. Specifically, $\{Tr = M\}$ denotes the event that SU2 can not decode the signal in the first $M-1$ rounds, which means that no matter SU2 decodes successfully or not at the M-th transmission round, it has no chance to help in forwarding the signal. Specifically, we define $Pr(Outage|Tr = m)$ to represent the conditional probability that SD1 can not decode the signal after m rounds given the event $\{Tr = m\}$ occurred. As such, the outage probability of the considered system after the maximum transmission round can be given as

$$P_{out}^{Coop,I}(L) = \sum_{m=1}^{M}\Pr\{Tr = m\}\Pr\{Outage|Tr = m\}. \tag{12}$$

Consequently, according to the definition of $\{Tr = m\}(m < M)$, the probability of event $\{Tr = m\}$ can be derived as (13) at the top of the following page.

[2] In this paper, we focus on the interference-limited scenario where the interference power from the primary user is dominant relative to the additive white Gaussian noise. As such, the additive white Gaussian noise is not considered [13].

[3] The probability means that m-th transmission round, the SD1 decodes the packet successfully. As such, the system will begin to transmit new signal.

[4] In the case $m = M$, no matter the system transmits successfully or not in the last transmission round, it should stop transmitting.

$$\begin{aligned}\Pr\{Tr=m\} &= [\Pr\{\frac{Q|h_1|^2}{P_P|h_3|^2|h_5|^2}<\gamma\}]^{m-1}\Pr\{\frac{Q|h_1|^2}{P_P|h_3|^2|h_5|^2}>\gamma\}\\ &= [\Pr\{\frac{Q|h_1|^2}{P_P|h_3|^2|h_5|^2}<\gamma\}]^{m-1}(1-\Pr\{\frac{Q|h_1|^2}{P_P|h_3|^2|h_5|^2}<\gamma\})\\ &= -[1+\frac{a_1Q}{a_3a_5P_P\gamma}e^{\frac{a_1Q}{a_3a_5P_P\gamma}}Ei(-\frac{a_1Q}{a_3a_5P_P\gamma})]^{m-1}\frac{a_1Q}{a_3a_5P_P\gamma}e^{\frac{a_1Q}{a_3a_5P_P\gamma}}Ei(-\frac{a_1Q}{a_3a_5P_P\gamma}).\end{aligned} \quad (13)$$

The conditional outage probability $Pr(Outage|Tr=m)$ can be derived as

$$P\{Outage|Tr=m\} = \underbrace{[\Pr\{SIR_{SD,1}<\gamma\}]^m}_{I_2} \underbrace{[\Pr\{(SIR_{SD,2}+SIR_{RD,2})<\gamma\}]^{M-m}}_{I_3}. \quad (14)$$

The exact I_2 can be easily deduced from I_1 by substituting the parameters of I_1 with their respective counterparts. Therefore, the main focus is to derive I_3. I_3 can be written as

$$\begin{aligned}I_3 &= \Pr\{(\frac{Q|h_0|^2}{2P_P|h_3|^2|h_6|^2}+\frac{Q|h_2|^2}{2P_P|h_4|^2|h_6|^2})<\gamma\}\\ &= \Pr\{(\frac{|h_0|^2}{|h_3|^2}+\frac{|h_2|^2}{|h_4|^2})<\frac{2P_P\gamma|h_6|^2}{Q}\}.\end{aligned} \quad (15)$$

For simplicity of representation, we set $X=\frac{|h_0|^2}{|h_3|^2}$, $Y=\frac{|h_2|^2}{|h_4|^2}$ and $Z=\frac{2P_P\gamma|h_6|^2}{Q}$. As such, PDF and CDF of them can be expressed as

$$\begin{aligned}F_X(x) &= \Pr\{X<x\} = \Pr\{\frac{|h_0|^2}{|h_3|^2}<x\}\\ &= \int_0^\infty f_{|h_3|^2}(u)\int_0^{xu} f_{|h_0|^2}(v)dvdu\\ &= \int_0^\infty f_{|h_3|^2}(u)F_{|h_0|^2}(xu)du\\ &= \int_0^\infty \frac{1}{a_3}e^{-\frac{u}{a_3}}(1-e^{-\frac{xu}{a_0}})du\\ &= 1-\frac{1}{a_3}\int_0^\infty e^{-(\frac{1}{a_3}+\frac{x}{a_0})u}du\\ &= \frac{a_3x}{a_0+a_3x},\end{aligned} \quad (16)$$

$$f_X(x) = \frac{a_0a_3}{(a_0+a_3x)^2}, \quad (17)$$

$$F_Y(y) = \frac{a_4y}{a_2+a_4y}, \quad (18)$$

$$f_Y(y) = \frac{a_2a_4}{(a_2+a_4y)^2}, \quad (19)$$

$$F_Z(z) = 1-e^{-\frac{2P_P\gamma z}{Q}}, \quad (20)$$

and

$$f_Z(z) = \frac{Q}{2P_P\gamma a_6}e^{-\frac{Qz}{2P_P\gamma a_6}}, \quad (21)$$

respectively. As such, I_3 can be re-written as

$$\begin{aligned}I_3 &= \Pr\{(X+Y)<Z\}\\ &= \int_0^\infty f_X(z)\int_0^\infty f_Y(y)\int_{x+y}^\infty f_Z(x)dzdydx\\ &= \int_0^\infty \frac{a_0a_3}{(a_0+a_3x)^2}\int_0^\infty \frac{a_2a_4}{(a_2+a_4y)^2}(1-F_Z(x+y))dydx\\ &= \int_0^\infty \frac{a_0a_3}{(a_0+a_3x)^2}e^{-\frac{2P_P\gamma x}{Q}}dx\int_0^\infty \frac{a_2a_4}{(a_2+a_4y)^2}e^{-\frac{2P_P\gamma y}{Q}}dy\\ &= \frac{a_0a_2}{a_3a_4}\underbrace{\int_0^\infty \frac{1}{(x+\frac{a_0}{a_3})^2}e^{-\frac{2P_P\gamma x}{Q}}dx}_{I_4}\underbrace{\int_0^\infty \frac{1}{(y+\frac{a_2}{a_4})^2}e^{-\frac{2P_P\gamma y}{Q}}dy}_{I_5}.\end{aligned} \quad (22)$$

According to [20 (3.353.2)], we have

$$\begin{aligned}I_4 &= \int_0^\infty \frac{1}{(x+\frac{a_0}{a_3})^2}e^{-\frac{2P_P\gamma x}{Q}}dx\\ &= \frac{a_3}{a_0}+\frac{2P_P\gamma}{Q}e^{\frac{2P_P\gamma a_0}{Qa_3}}Ei(-\frac{2P_P\gamma a_0}{Qa_3}),\end{aligned} \quad (23)$$

$$\begin{aligned}I_5 &= \int_0^\infty \frac{1}{(y+\frac{a_2}{a_4})^2}e^{-\frac{2P_P\gamma y}{Q}}dy\\ &= \frac{a_4}{a_2}+\frac{2P_P\gamma}{Q}e^{\frac{2P_P\gamma a_2}{Qa_4}}Ei(-\frac{2P_P\gamma a_2}{Qa_4}).\end{aligned} \quad (24)$$

Similarly, the probability of the event that after m-th transmission round the SD1 decodes the signal successfully for Cooperative ARQ I Protocol is presented in Tab. 2.

Transmission Round	Probability
1	$1-P_{out}^{Coop,I}(1)$
2	$P_{out}^{Coop,I}(1)-P_{out}^{Coop,I}(2)$
...	...
$M-1$	$P_{out}^{Coop,I}(M-2)-P_{out}^{Coop,I}(M-1)$
M	$P_{out}^{Coop,I}(M-1)$

Tab. 2. Probability of the considered system stop transmitting after m-th transmission round based on Cooperative ARQ I Protocol.

As such, the average transmission rounds of Cooperative ARQ I Protocol can be given as

$$\begin{aligned} T_{ave}^{Coop,I} &= 1\times(1-P_{ave}^{Coop,I}(1))+2\times(P_{ave}^{Coop,I}(1)-P_{ave}^{Coop,I}(2)) \\ &+\cdots+M\times P_{ave}^{Coop,I}(M-1) \\ &= 1+P_{ave}^{Coop,I}(1)+P_{ave}^{Coop,I}(2)+\cdots+P_{ave}^{Coop,I}(M-1) \\ &= 1+\sum_{i=1}^{M-1} P_{ave}^{Coop,I}(i). \end{aligned} \tag{25}$$

Through cooperation from SU2, the additive access opportunities can be expressed as

$$\begin{aligned} G_{Dir}^{Coop,I} &= T_{out}^{Dir} - T_{ave}^{Coop,I} \\ &= \sum_{i=1}^{M-1}(P_{out}^{Dir}(i)-P_{ave}^{Coop,I}(i)) > 0. \end{aligned} \tag{26}$$

3.3 Cooperative ARQ II Protocol

Based on the Cooperative ARQ II Protocol, we find that the main difference between Cooperative ARQ II Protocol and Cooperative ARQ I Protocol is that when the SU2 can decode the signal correctly, the system selects the better node between SU1 and SU2 to retransmit. As such, if the following condition is satisfied, we will select SU2 to retransmit. Otherwise, the SU1 is selected.

$$\frac{Q|h_2|^2}{P_P|h_4|^2|h_6|^2} > \frac{Q|h_0|^2}{P_P|h_3|^2|h_6|^2} \Rightarrow \frac{|h_2|^2}{|h_4|^2} > \frac{|h_0|^2}{|h_3|^2}. \tag{27}$$

Consequently, the main difference is I_3 when we derive the outage probability of the Cooperative ARQ II Protocol. I_3 will be replaced by I_6 which is defined as follows.

$$I_6 = \Pr\{\max(SNR_{SD,2}, SNR_{RD,2}) < \gamma\}. \tag{28}$$

Therefore, I_6 can be mathematically derived as

$$\begin{aligned} I_6 &= \Pr\{\max(\frac{Q|h_0|^2}{P_P|h_3|^2|h_6|^2}, \frac{Q|h_2|^2}{P_P|h_4|^2|h_6|^2}) < \gamma\} \\ &= \Pr\{\frac{Q}{P_P|h_6|^2}\max(\frac{|h_0|^2}{|h_3|^2}, \frac{|h_2|^2}{|h_4|^2}) < \gamma\} \\ &= \underbrace{\Pr\{\frac{Q}{P_P|h_6|^2}\frac{|h_0|^2}{|h_3|^2} < \gamma\}\Pr\{\frac{|h_0|^2}{|h_3|^2} > \frac{|h_2|^2}{|h_4|^2}\}}_{I_7} \\ &+ \underbrace{\Pr\{\frac{Q}{P_P|h_6|^2}\frac{|h_2|^2}{|h_4|^2} < \gamma\}\Pr\{\frac{|h_0|^2}{|h_3|^2} < \frac{|h_2|^2}{|h_4|^2}\}}_{I_8}. \end{aligned} \tag{29}$$

The first part of I_7 and I_8 can be easily deduced from I_1. Therefore, another part of I_7 can be derived as

$$\begin{aligned} &\Pr\{\frac{|h_0|^2}{|h_3|^2} > \frac{|h_2|^2}{|h_4|^2}\} = \Pr\{|h_2|^2 < \frac{|h_0|^2|h_4|^2}{|h_3|^2}\} \\ &= 1 - \frac{1}{a_0 a_3 a_4}\int_0^\infty\int_0^\infty\int_0^\infty e^{-\frac{u}{a_0}} e^{-\frac{v}{a_3}} e^{-\frac{w}{a_4}} e^{-\frac{uw}{a_2 v}} dw dv du \\ &= 1 - \frac{1}{a_4} - \underbrace{\frac{1}{a_0 a_2 a_3 a_4}\int_0^\infty u e^{(\frac{1}{a_2 a_3 a_4}-\frac{1}{a_0})u} Ei(-\frac{u}{a_2 a_3 a_4}) du}_{I_9}. \end{aligned} \tag{30}$$

According to [20 (6.227.1)] and after some simple mathematical manipulations, we have

$$\begin{aligned} I_9 &= \frac{1}{a_0 a_2 a_3 a_4}\int_0^\infty u e^{(\frac{1}{a_2 a_3 a_4}-\frac{1}{a_0})u} Ei(-\frac{u}{a_2 a_3 a_4}) du \\ &= \theta(\frac{1}{\theta(\theta-1)} - \frac{1}{(\theta-1)^2}\ln\theta) \end{aligned} \tag{31}$$

where $\theta = \frac{a_2 a_3 a_4}{a_0}$. Similarly, the probability of event that after m-th transmission round the SD1 decodes the signal successfully for Cooperative ARQ II Protocol is presented in Tab. 3.

Transmission Round	Probability
1	$1 - P_{out}^{Coop,II}(1)$
2	$P_{out}^{Coop,II}(1) - P_{out}^{Coop,II}(2)$
···	···
$M-1$	$P_{out}^{Coop,II}(M-2) - P_{out}^{Coop,II}(M-1)$
M	$P_{out}^{Coop,II}(M-1)$

Tab. 3. Probability of the considered system stop transmitting after m-th transmission round based on Cooperative ARQ II Protocol.

As such, the average transmission rounds of Cooperative ARQ II Protocol can be given as

$$T_{ave}^{Coop,II} = 1 + \sum_{i=1}^{M-1} P_{ave}^{Coop,II}(i). \tag{32}$$

Through Cooperative ARQ II Protocol, the additive access opportunities can be expressed as

$$\begin{aligned} G_{Dir}^{Coop,II} &= T_{out}^{Dir} - T_{ave}^{Coop,II} \\ &= \sum_{i=1}^{M-1}(P_{out}^{Dir}(i)-P_{ave}^{Coop,II}(i)) > 0. \end{aligned} \tag{33}$$

Remarks: According the exact expression of the cooperation gain, we conclude that the Cooperative ARQ I Protocol and Cooperative ARQ II Protocol can achieve additive opportunities. Specifically, Cooperative ARQ II Protocol will use least energy consumption to achieve more spectrum access. However, Cooperative ARQ II Protocol needs information exchange between SU1 and SU2.

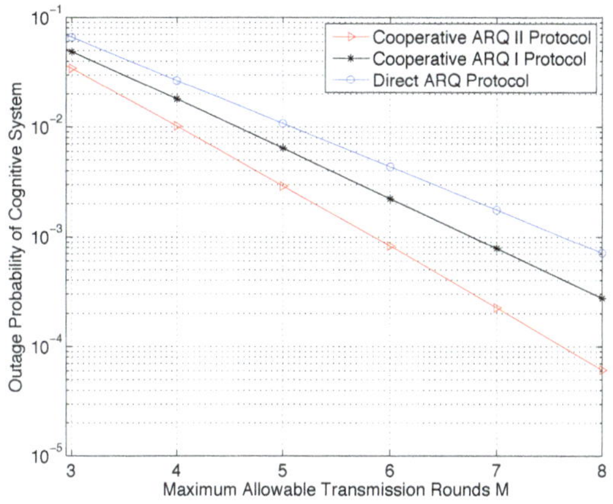

Fig. 2. Outage probability verse the maximum allowable transmission rounds when $Q = 10$ dB, $P_P = 10$ dB.

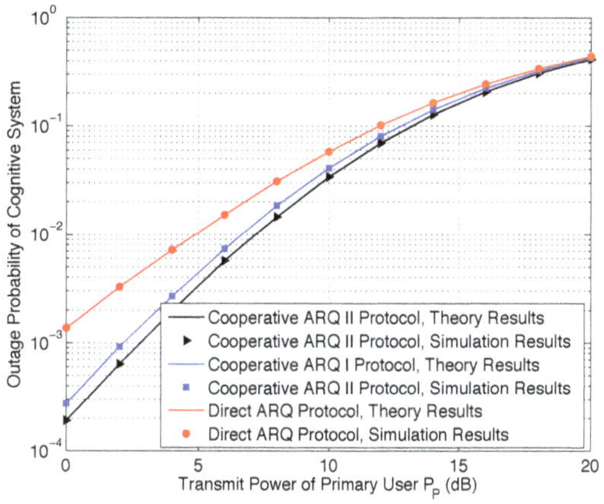

Fig. 3. Outage probability verse the transmit power of primary user when $Q = 10$ dB, $M = 10$.

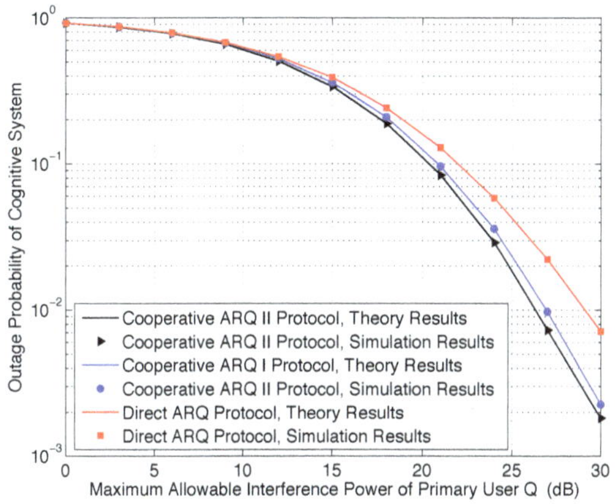

Fig. 4. Outage probability verse the maximum allowable interference power of primary user when $P_P = 10$ dB, $M = 10$.

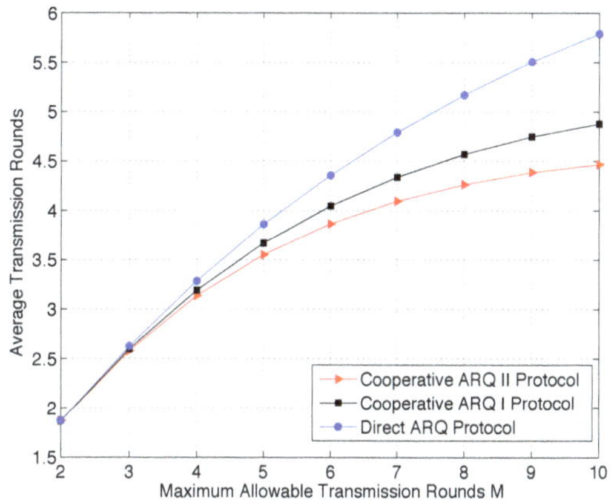

Fig. 5. Average transmission rounds verse the maximum allowable transmission rounds when $P_P = 10$ dB, $Q = 10$ dB.

4. Numerical Results

In this section, we provide some insight into the benefits of the proposed two ARQ-based protocols. Hereafter, we assume $a_0 = 1$, $a_1 = 2$, $a_2 = 2$, $a_3 = 1$, $a_4 = 1$, $a_5 = 1$, $a_6 = 1$.

Fig. 2, Fig. 3, and Fig. 4 evaluate the impact of maximum allowable transmission rounds, transmit power of primary user and maximum allowable interference power of primary user on the outage performance of considered system, respectively. From Fig. 2, we can observe that the outage probability will decrease when the maximum allowable transmission rounds increases. Obviously, the outage performance will deteriorate when transmit power of primary user increase. Similarly, the outage performance will improve when the maximum allowable interference power of primary user increases. From Fig. 2, Fig. 3, and Fig. 4, we conclude that the best performance protocol among three protocols is Cooperative ARQ II Protocol. However, in Cooperative ARQ II Protocol, information exchange between SU1 and SU2 is needed. Specifically, the theory results and the Monte Carlo simulation results match perfectly, which validates the correctness of the analytical results.

Fig. 5 and Fig. 6 evaluate the average transmission rounds and additive transmission rounds gain verse the maximum allowable transmission rounds. From Fig. 5 and Fig. 6, we clearly deduce that the average transmission rounds will increase when the maximum allowable transmission rounds increases. Moreover, Fig. 6 indicates that the proposed two novel protocols can obtain some additive access opportunities by reducing the average transmission rounds of SU1 and SD1. Specifically, due to selecting the better node to transmit in retransmission stage, the energy consumption will also reduce.

Fig. 7 and Fig. 8 show the relation between average transmission rounds and transmit power of primary user, additive transmission rounds gain and transmit power of primary user, respectively. We can conclude from Fig. 7 that the average transmission rounds will increase when the transmit power of primary user increases. Interestingly, from Fig. 8, we find that the additive transmission gain is high in medium region, and low in low and high transmit power region. This can be explained as follows. When the transmit power of

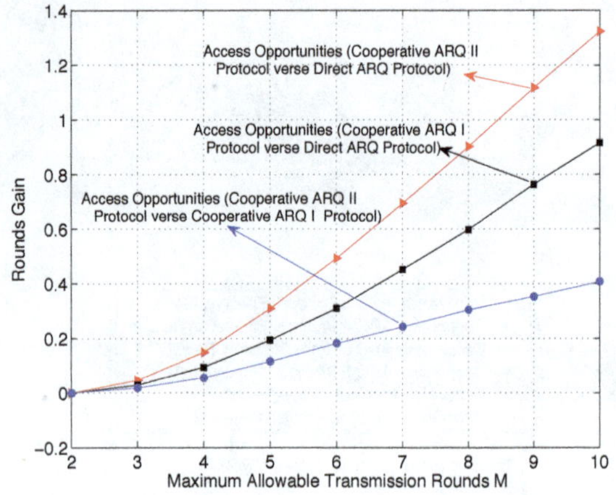

Fig. 6. Additive access opportunities verse the maximum allowable transmission rounds when $P_P = 10$ dB, $Q = 10$ dB.

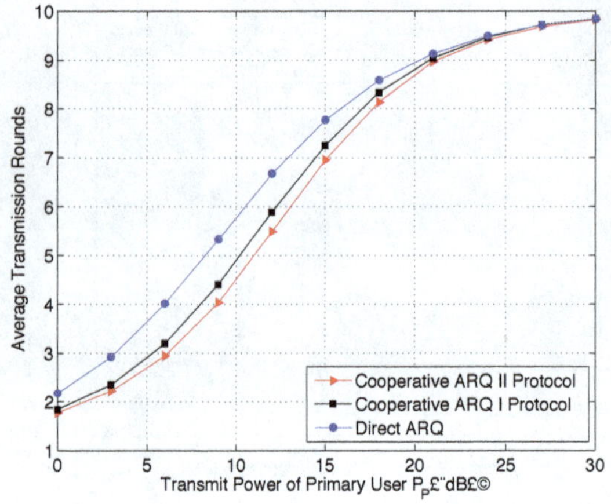

Fig. 7. Average transmission rounds verse the transmit power of primary user when $Q = 10$ dB, $M = 10$.

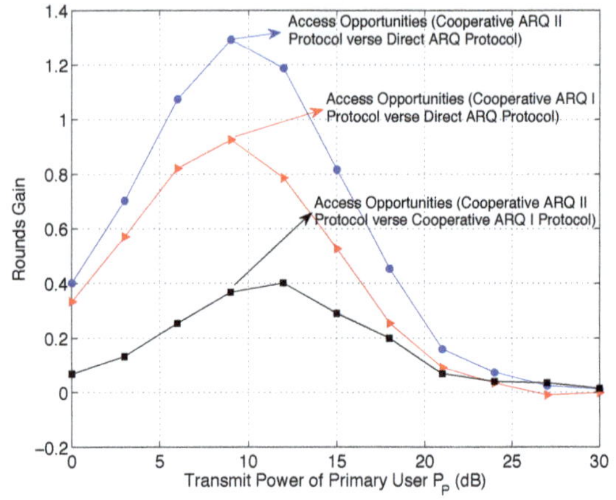

Fig. 8. Additive access opportunities verse the transmit power of primary user when $Q = 10$ dB, $M = 10$.

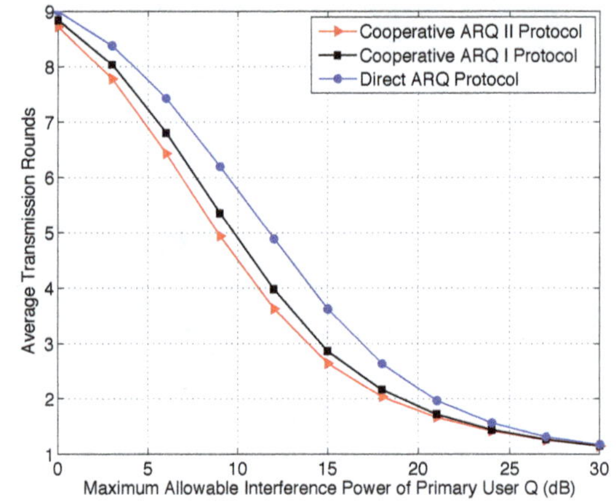

Fig. 9. Average transmission rounds verse the maximum allowable interference power of primary user when $P_P = 10$ dB, $M = 10$.

the primary user is in low region, SD1 will decode the signal correctly in high probability. As such, even SU2 can decode the signal successfully, it may have little cooperation opportunity in this area. In the high transmit power region, SD1 and SU2 can not decode the signal correctly with high probability. Therefore, SU2 may have no ability to forward the signal in high probability.

Fig. 9 and Fig. 10 depict the impacts of maximum allowable interference power of primary user on the average transmission rounds and additive transmission rounds gain, respectively. Contrary to the Fig. 7 and Fig. 8, the average transmission rounds will decrease when the maximum allowable interference power of primary user increases. Similarly, the additive transmission rounds gain is high in medium region of Q, and low in low and high of Q. This can be explained as follows. When the maximum allowable interference power of primary user is low, the SD1 and SU2 can not decode the signal in high probability. As a result, the SU2 has no ability to forward the signal in the retransmission stage. In high region, SD1 can decode the signal successfully in high probability. As such, it needs less help from SU1 in each transmission.

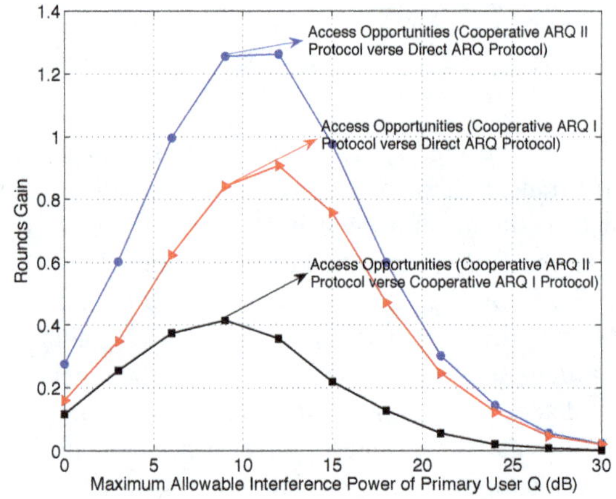

Fig. 10. Additive access opportunities verse the maximum allowable interference power of primary user when $P_P = 10$ dB, $M = 10$.

5. Conclusion

In this paper, we have proposed two novel ARQ-based protocols to achieve additive access opportunities in cognitive decode-and-forward relay networks. First, we have derived the closed-form outage probability expressions of the proposed protocols which provide an efficient means to evaluate the impacts of several parameters. Moreover, we define a new metric to evaluate the improvement of cooperative ARQ protocols. We can observe that, via cooperating with SU2, the outage performance and average transmission rounds can be improved greatly. As such, SU2 can obtain some access opportunities by reducing the transmission rounds of SU1. We also find that Cooperative ARQ II Protocol performs better than Cooperative ARQ I Protocol. However, there exists information exchange in Cooperative ARQ II Protocol. Finally, numerical simulation results were presented to validate the correctness of the proposed protocols.

Acknowledgements

This work was supported by the National Natural Science Foundation of China under grant no. 60932002, the National Natural Science Fund of China under grant no. 61172062, and the Natural Science Fund of Jiangsu, China under grant no. BK2011116.

References

[1] Federal Communications Commission. *Spectrum Policy Task Force Report, FCC 02-155*. 2002.

[2] Federal Communications Commission. *Second Report and Order, FCC 08-260*. 2008.

[3] MITOLA, J., MAGUIRE, G. Q. Jr. Cognitive radios: making software radio more personal. *IEEE Personal Communications*, 1999, vol. 6, no. 4, p. 13–18. DOI: 10.1109/98.788210

[4] MITOLA, J. *Cognitive Radio: An Integrated Agent Architecture for Software Defined Radio*, Ph. D. dissertation. Stockholm (Sweden) KTH Royal Institute of Technology, 2000.

[5] XU, Y., WANG, J., WU, Q., ZHANG, Z., ANPALAGAN, A., SHEN, L. Optimal energy-efficient channel exploration for opportunistic spectrum usage. *IEEE Wireless Commmunication Letters*, 2012, vol. 1, no. 2, p. 77–80. DOI: 10.1109/WCL.2012.012012.110257

[6] ZHANG, Z., HUI, Q., WANG, J. Energy-efficient power allocation strategy in cognitive relay networks. *Radioengineering*, 2012, vol. 21, no. 3, p. 809–814.

[7] WU, Q., HAN, H., WANG, J., ZHAO, Z., ZHANG, Z. Sensing task allocation for heterogeneous channels in cooperative spectrum sensing. *Radioengineering*, 2010, vol. 19, no. 4, p. 544–551.

[8] BLETSAS, A., SHIN, H., WIN, M. Z., LIPPMAN, A. A simple cooperative diversity method based on network path selection. *IEEE Journal on Selected Areas in Communications*, 2006, vol. 24, no. 3, p. 659–672. DOI: 10.1109/JSAC.2005.862417

[9] ZHI, Q., CUI, S., SAYED, A. H. Optimal linear cooperation for spectrum sensing in cognitive radio networks. *IEEE Journal of Selected Topics in Signal Processing*, 2008, vol. 2, no. 1, p. 28–40. DOI: 10.1109/JSTSP.2007.914882

[10] HOU, Y. T., SHI, Y., SHERALI, H. D. Spectrum sharing for multi-hop networking with cognitive radios. *IEEE Journal of Selected Areas in Communications*, 2008, vol. 26, no. 1, p. 146–155. DOI: 10.1109/JSAC.2008.080113

[11] GANNESAN, G., LI, Y. G. Cooperative spectrum sensing in cognitive radio – part I: Two user networks. *IEEE Transactions on Wireless Communications*, 2007, vol. 6, no. 6, p. 2204–2213. DOI: 10.1109/TWC.2007.05775

[12] SADEK, A. K., LIU, K. J. R., EPHREMIDES, A. Cognitive multiple access via cooperation: Protocol design and performance analysis. *IEEE Transactions on Information Theory*, 2007, vol. 53, no. 10, p. 3677–3696. DOI: 10.1109/TIT.2007.904784

[13] YANG, P., LUO, L., QIN, J. Outage performance of cognitive relay networks with interference from primary user. *IEEE Communications Letters*, 2012, vol. 16, no. 10, p. 1695–1698. DOI: 10.1109/LCOMM.2012.081612.121086

[14] DUONG, T. Q., ET AL. Cognitive relay networks with multiple primary transceivers under spectrum-sharing. *IEEE Signal Processing Letters*, 2012, vol. 19, no. 11, p. 741–744. DOI: 10.1109/LSP.2012.2217327

[15] ZHONG, C., RATNARAJAH, T., WONG, K. K. Outage analysis of decode-and-forward cognitive dual-hop systems with the interference constraint in Nakagami-*m* fading channels. *IEEE Transactions on Vehicular Technology*, 2011, vol. 60, no. 6, p. 2875–2879. DOI: 10.1109/TVT.2011.2159256

[16] DUONG, T. Q., ET AL. Outage and diversity of cognitive relaying systems under spectrum sharing environments in Nakagami-*m* fading. *IEEE Communications Letters*, 2012, vol. 16, no. 12, p. 2075–2078. DOI: 10.1109/LCOMM.2012.100812.121859

[17] DUONG, T. Q., DA COSTA, D. B., ELKASHLAN, M., BAO, V. N. Q. Cognitive amplify-and-forward relay networks over Nakagami-m fading. *IEEE Transactions on Vehicular Technology*, 2012, vol. 61, no. 5, p. 2368–2374. DOI: 10.1109/TVT.2012.2192509

[18] TANNIOUS, R., NOSRATINIA, A. Coexistence through ARQ restransmission in fading cognitive radio channels. In *IEEE International Symposium on Information Theory Proceedings*. Austin (TX, USA), 2010, p. 2078–2082. DOI: 10.1109/ISIT.2010.5513394

[19] GUAN, X., CAI, Y., SHENG, Y., YANG, W. Exploiting primary retransmission to improve secondary throughput by cognitive relaying with best-relay selection. *IET Communications*, 2012, vol. 6, no. 2, p. 1769–1780. DOI: 10.1049/iet-com.2011.0940

[20] GRADSHTEYN, I. S., RYZHIK, I. M. *Table of Integrals, Series, and Products*, 5th ed. Orlando (FL, USA): Academic Press, 1994.

About the Authors ...

Zongsheng ZHANG was born in 1986. He received his B.S. degree in Communications Engineering from Institute of Communications, Nanjing, China, in 2009. He is currently pursuing the Ph.D. degree in Communications and Information Systems from Institute of Communications, PLA University of Science and Technology. His research interests focus on wireless communications, cognitive radio and channel coding techniques.

Jinlong WANG was born in 1963. He received his B.S. degree in Wireless Communications, M.S. degree and Ph.D. degree in Communications and Electronic Systems from Institute of Communications Engineering, Nanjing, China, in 1983, 1986 and 1992, respectively. He is currently professor at the PLA University of Science and Technology, China. He is also the co-chairman of IEEE Nanjing Section. He has published widely in the areas of signal processing for communications, information theory, and wireless networks. His current research interests include wireless communication, cognitive radio, soft-defined radio and ultra-wide bandwidth (UWB) systems.

Qihui WU was born in 1970. He received his B.S. degree in Communications Engineering, M.S. degree and Ph.D. degree in Communications and Information Systems from Institute of Communications Engineering, Nanjing, China, in 1994, 1997 and 2000, respectively. He is currently professor at the PLA University of Science and Technology, China. His current research interests are algorithms and optimization for cognitive wireless networks, soft-defined radio and wireless communication.

26

Bandwidth Efficient Root Nyquist Pulses for Optical Intensity Channels

Sabeena FATIMA, S. Sheikh MUHAMMAD, A. D. RAZA

Dept. of Electrical Engineering, National University of Computer and Emerging Sciences, Lahore, Pakistan

sabeena.fatima@nu.edu.pk, sm.sajid@nu.edu.pk, ad.raza@nu.edu.pk

Abstract. *Indoor diffuse optical intensity channels are bandwidth constrained due to the multiple reflected paths between the transmitter and the receiver which cause considerable inter-symbol interference (ISI). The transmitted signal amplitude is inherently non-negative, being a light intensity signal. All optical intensity root Nyquist pulses are time-limited to a single symbol interval which eliminates the possibility of finding bandlimited root Nyquist pulses. However, potential exists to design bandwidth efficient pulses. This paper investigates the modified hermite polynomial functions and prolate spheroidal wave functions as candidate waveforms for designing spectrally efficient optical pulses. These functions yield orthogonal pulses which have constant pulse duration irrespective of the order of the function, making them ideal for designing an ISI free pulse. Simulation results comparing the two pulses and challenges pertaining to their design and implementation are discussed.*

Keywords

Nyquist Pulse, optical intensity signaling, PSWF (Prolate Spheroidal Wave Functions), MHPF (Modified Hermite Polynomial Function), ISI (Inter Symbol Interference)

1. Introduction

The capacity of Radio Frequency (RF) systems is limited due to scarcity and cost of licensing. Therefore, to meet the need of growing data rate requirements in wireless communications optical bands are being excessively exploited. Optical signals have numerous advantages over RF signals, the most attractive ones being that they are unlicensed and have many THz of bandwidth naturally associated to them.

Wireless optical intensity channel transmit information by modulating the instantaneous optical intensity. A Light Emitting Diode (LED) or a Laser Diode (LD) converts the input electrical signal to an optical signal. A photodiode detector at the receiver end converts the incident optical intensity signal back to an electrical signal. The optical intensity channel commonly employs Intensity Modulation and Direct Detection (IM/DD) [2]. Multipath distortion arising due to the multiple reflected paths between the transmitter and the receiver in indoor environment cause severe Inter Symbol Interference (ISI), necessitating the use of pulse shaping. Diffuse links [3] are attractive as they allow user mobility and eliminate the need of alignment but at the expense of bandwidth. The diffuse indoor optical links are severely bandwidth limited.

For indoor optical applications, signaling design varies significantly from the conventional electrical channels as optical channels impose the additional requirement of signal being non-negative. Secondly optical links require that for the case of indoor optical channels the transmitted signal be Class I eye safe under all conditions. As a result the peak transmitted power of the light signal remains constrained. Received signal is corrupted by free space losses, shot noise and ambient light noise. For the case of indoor optical channels distortion due to shot noise is most significant. Shot noise is modeled as uncorrelated additive white Gaussian noise [3].

There exists a wealth of literature on design of pulses that maximize in-band energy [4, 5, 6] to mitigate the effect of ISI; however these pulses have been designed for the electrical channel. Halpern [7] proposed a design for finite duration Nyquist pulses for mean power constrained channels but not subject to amplitude constraints. Considerable information on optical intensity channels and their numerous advantages have been investigated [3, 8, 9]. However, the design of optical intensity pulse to minimize ISI still requires a much deeper study [9]. Steve Hranilovic [1] proved that all optical intensity root Nyquist pulses must be time limited to a single symbol interval. The time-limitedness of the root Nyquist pulse eliminates the possibility of finding a bandlimited pulse. Steve [1] proposed Prolate Spheroidal Wave pulse to be used since it is time limited and also has better spectral concentration among all time limited signals. For signals limited to a single symbol interval PSWF is the optimum signal choice [7, 11, 12]. From studies on pulse shaping techniques for time limited systems [13, 14, 15] Prolate Spheroidal Wave Functions (PSWF) and Modified Hermite Polynomial Function (MHPF) have emerged as potential pulse shaping functions as they provide the optimal spectral concentration. Individual discussion on these two contender pulses is widely available but no comparison between the two has been drawn. Design of a root Nyquist pulse which maximizes the in-band fractional energy has been proposed in [4, 7], in context of electrical domain. A linear combination of PSWF pulses proposed in [4] seeks to maximize the in-band fractional energy and to satisfy the Nyquist criteria. The problem was reduced to solving the linear objective function. This design can be extended to the optical domain by introducing an additional constraint of optical signal remaining non-negative.

In this work we propose root Nyquist pulse designs. The non-negativity constraint limits the root Nyquist pulses to be time-limited [1]. Although this eliminates the possibility of finding bandwidth limited root Nyquist pulses, potential still exists to find bandwidth efficient pulses for these channels. Pulses can be shaped such that most of the energy is concentrated in the band of interest [16]. Section 2 presents the discussion on root Nyquist pulse and shows that they are strictly time limited to a single symbol interval. Sections 3 and 4 focus on the discussion of PSWF and MHPF pulses. In Section 5 a comparison is drawn between the two pulses. Conclusions are presented in Section 6.

2. Optical Intensity Root Nyquist Pulses

For the case where receiver filter is matched to the transmitted pulse, to ensure zero ISI the pulse shape at the output must satisfy the Nyquist criteria [15] given by:

$$\int x(\tau)x(\tau - kT)d\tau = \delta_{k0} \tag{1}$$

where $\delta_{kl} = \{ \begin{matrix} 1 & k = l \\ 0 & otherwise. \end{matrix}$

An optical intensity root Nyquist pulse satisfies the non-negativity constraint in addition to the Nyquist criteria. All practical optical intensity root Nyquist pulses are time-limited to single symbol duration [1]. The root Nyquist pulse is required to be of small duration so that the number of interfering neighbors is small. This eliminates the possibility of obtaining band-limited root Nyquist pulses.

The search therefore reduces to bandwidth efficient pulses. Most of the energy needs to be concentrated in a finite band. The main objective is to maximize the in-band fractional energy of the time-limited root Nyquist pulse. In-band fractional energy is defined as the ratio of the energy of the pulse in the given bandwidth to the total energy of the time limited pulse. Its expression is given as [9]:

$$\frac{\int_{-W}^{W} |X(f|^2 df)}{\int_{-\infty}^{\infty} |X(f|^2 df)} = K \tag{2}$$

where $K \in (0,1)$.

The (1-ε) fractional energy bandwidth of a transmitted symbol $x(t)$ with Fourier Transform $X(f)$ is defined as [9]:

$$W\varepsilon(x) = W\varepsilon[0,\infty) : \int_{-W}^{W} |X(f|^2 df) \geq (1-\varepsilon) \int_{-\infty}^{\infty} |X(f|^2 df)$$

If ε is chosen such that out of band energy is small, then $x(t)$ can be thought of as efficiently band-limited to $W\varepsilon(x)$.

The prolate spheroidal wave functions (PSWF) and modified hermite polynomial functions (MHPF) give the highest spectral concentration of all time limited signals [14, 15] and are discussed in detail in the next sections.

3. Prolate Spheroidal Wave Functions

Prolate spheroidal wave functions (PSWF) have the highest spectral concentration of all time-limited signals. They are strictly time-limited to [0, T] and have maximum energy in the band of interest, i.e. [-Ω, Ω] of all unit energy functions. These functions are solution to the integral equation [15]:

$$\int_{-T/2}^{T/2} \frac{\sin\Omega(t-s)}{\pi(t-s)} \psi_i(s)ds = \lambda_i \psi_i(t). \tag{3}$$

Here ψ(t) represents the PSWF, λ is the amount of energy contained in $[-T/2, T/2]$, Ω is the bandwidth and T is the pulse duration.

PSWF are an improvement on other pulses as they satisfy the double orthogonality property. This double orthogonality property guarantees unique demodulation at the receiver.

$$\int_{-\infty}^{\infty} \psi_m(t).\psi_n(t) = \delta_{mn}, \tag{4}$$

$$\int_{-T/2}^{T/2} \psi_m(t).\psi_n(t) = \lambda_m \delta_{mn} \tag{5}$$

where $\psi_n(t)$ is the PSWF of order n.

ψ_n(t) is written in terms of prolate angular function of first kind [15] as:

$$\psi_n(t) = \psi_n(\Omega, T, t) = \frac{(2\lambda_n(c)/T)^{1/2} S_{0n}^1(c, 2t/T)}{\{\int_{-1}^{1} [S_{0n}^1(c,x)]^2 dx\}^{1/2}} \tag{6}$$

where $\{\int_{-1}^{1}[S_{0n}^1(c,x)]^2 dx\}^{1/2} = \frac{2}{2n+1}$ where S_{0n}^1 is the prolate angular function of first kind and λ_n is the fraction of the energy of ψ_n(t) that lies in the interval [-1, 1].

The prolate angular function of the first kind is given by [15]:

$$S_{0n}^1(c,t) = \begin{matrix} \sum_{k=0,2,\dots}^{\infty} d_k(c) P_k(c,t) & n & \text{even} \\ \sum_{k=1,3,\dots}^{\infty} d_k(c) P_k(c,t) & n & \text{odd} \end{matrix} \tag{7}$$

where c is the time bandwidth product given by $c = \Omega T/2$, $P_k(c,t)$ is the Legendre polynomials [13][18] and d_k(c) satisfy the recursion relation[14]:

$$\alpha_k d_{k+2}^n(c) + (\beta_k - \chi_n(c)) d_k^n(c) + \gamma_k d_{k-2}^n(c) = 0 \tag{8}$$

where the coefficients α, β, γ are given by:

$$\alpha_k = \frac{(k+1)(k+2)c^2}{(2k+3)(2k+5)}, \tag{9}$$

$$\beta_k = \frac{(2k^2+2k-1)c^2}{(2k-1)(2k+3)} + k(k+1), \tag{10}$$

$$\gamma_k = \frac{(k)(k-1)c^2}{(2k-1)(2k-3)}. \tag{11}$$

In order to compute $d_k^n(c)$ the following equation [15] is solved:

$$Od^n = \chi_n d^n. \tag{12}$$

Here d^n are the eigenvector and χ_n are the eigenvalues.

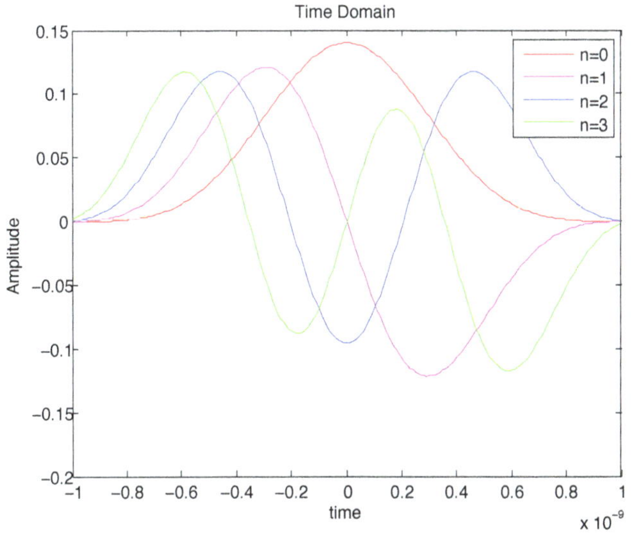

Fig. 1. Time Domain Representation of Prolate Spheroidal Wave Functions of order 0, 1, 2, 3. Note that the pulse duration is same for all orders and the number of zero crossings is equal to the pulse order.

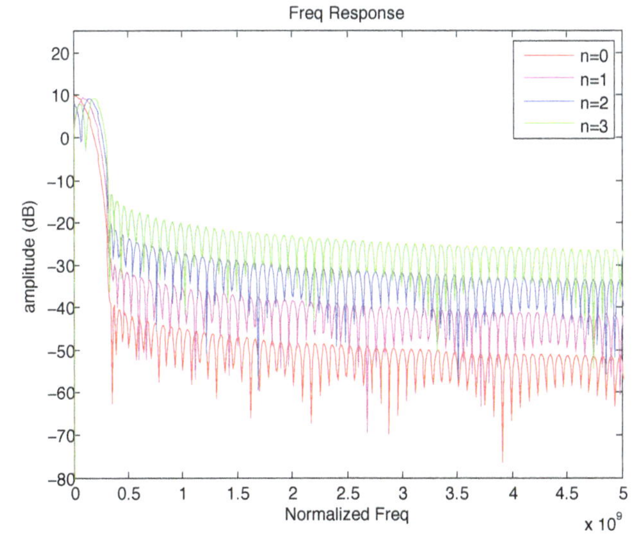

Fig. 2. Frequency Domain Representation of Prolate Spheroidal Wave Functions of order 0, 1, 2, 3.

Figures 1 and 2 represent the time and frequency characteristics for the PSWFs respectively. From the figures it can be deduced that the pulse duration and bandwidth is exactly the same for all orders n. Another interesting property being that number of zero crossings equals n. These pulses are double-orthogonal. Pulse duration and bandwidth can be bartered by utilizing the time-bandwidth product.

For wireless optical intensity channel only PSWF order 0 pulse can be utilized as it fulfills the amplitude non-negativity constraint. However, making use of the interesting property of PSWF pulses that the number of zero crossings on time axis equals the pulse order n and the pulse duration is same for all pulse orders, one approach could be to make a linear combination pulse of these orders and analyze its properties.

3.1 Time Bandwidth Product

To get a better understanding of time-bandwidth product c, PSWF order 0 pulse was plotted for different values of c. As time was kept confined to duration of 2 nanoseconds, the variation of c had a direct impact on bandwidth. The c value $\Omega T/2$ effect the width of main lobe of PSWF pulse, as the c value increases; the main lobe becomes compressed causing an expansion in frequency domain (as seen in the Figs. 3 and 4).

3.2 Linear Combination Pulse

Since we have limitations on channel bandwidth we need to concentrate most of the energy in our band of interest. We wish to maximize the in-band fractional energy which is given by:

$$\frac{\int_{-W}^{W} |X(f|^2 df)}{\int_{-\infty}^{\infty} |X(f|^2 df)} = K. \tag{13}$$

Nigam [4] proposed a linear combination of PSWF pulses which maximize the in-band fractional energy. The linear combination pulse is given as:

$$g(t) = \sum_{0}^{2N+2} a_j \psi_j(t). \tag{14}$$

Here $\psi_j(t)$ is the PSWF of j-th order; N is the number of PSWFs used and a_j are the coefficients which need to be determined.

Maximizing the in-band fractional energy reduces the problem [4] of determining the coefficients a_j to

$$\text{Maximize: } \sum_{0}^{2N-2} a_j \psi_j(0)\lambda_j \tag{15}$$

Subj to Nyquist Criteria:

$$\sum_{0}^{2N-2} a_j \psi_j(kT) = \delta_k. \tag{16}$$

This ensures that the overall pulse is zero at multiples of T except $k = 0$.

However, this pulse was designed for transmission in the electrical domain. Extending these results to the optical domain introduces an additional constraint that the transmitted pulse remains non-negative. Taking into account all

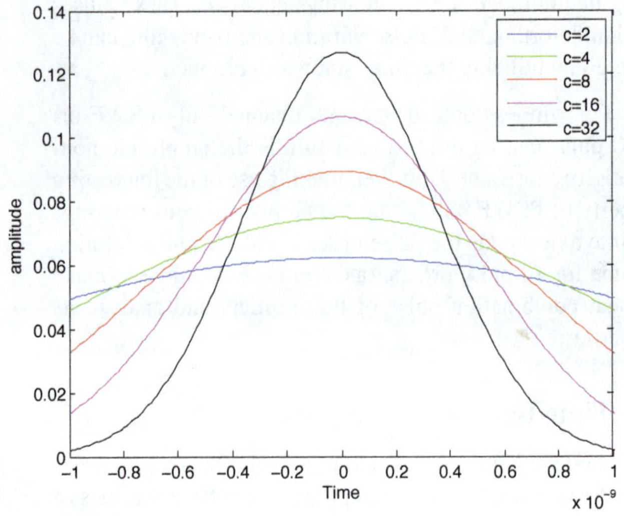

Fig. 3. Effect of varying time-bandwidth product c (time representation).

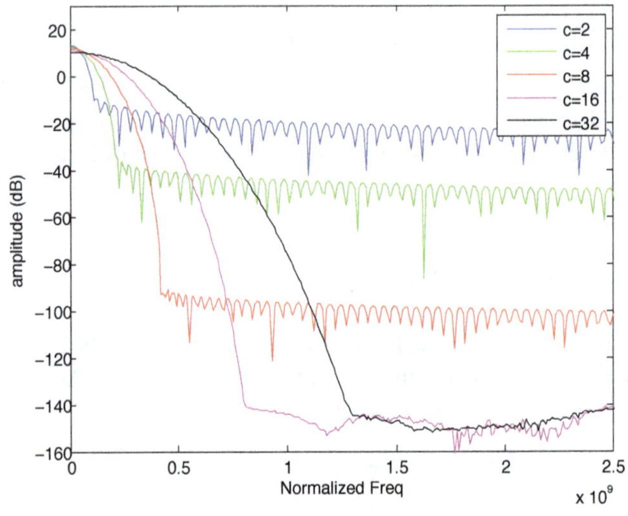

Fig. 4. Effect of varying time-bandwidth product (frequency representation). As the time-bandwidth product value increases the bandwidth occupied by the pulse increases.

these constraints the problem reduces to evaluating the linear objective function subject to linear constraints which can readily be solved by applying the simplex method. The results show that the coefficients for higher order PSWF pulse approach zero.

Firstly, we constrain the PSWF pulse to have unit value at time zero, then solving (13) and (14) for the coefficients a_j we get the values shown in Tab. 1. As can be observed the PSWF order 0 pulse makes the major contribution to the linear combination pulse, in fact up to $c = 8$ PSWF 0 pulse is the only contributor, beyond $c = 8$ other even-ordered PSWF pulse shapes come into play. This proves that PSWF order 0 pulse is the optimum pulse which fulfills the non-negativity criteria.

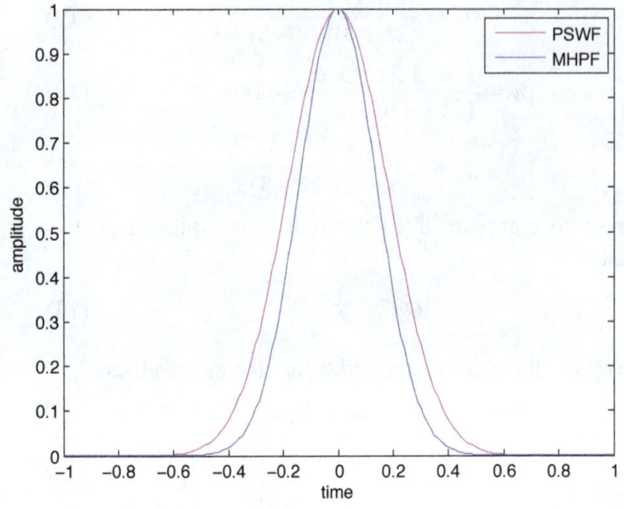

Fig. 5. Comparison between the optimal linear combination pulse and PSWF order 0 pulse proves that the optimal pulse follows the shape of the PSWF order 0.

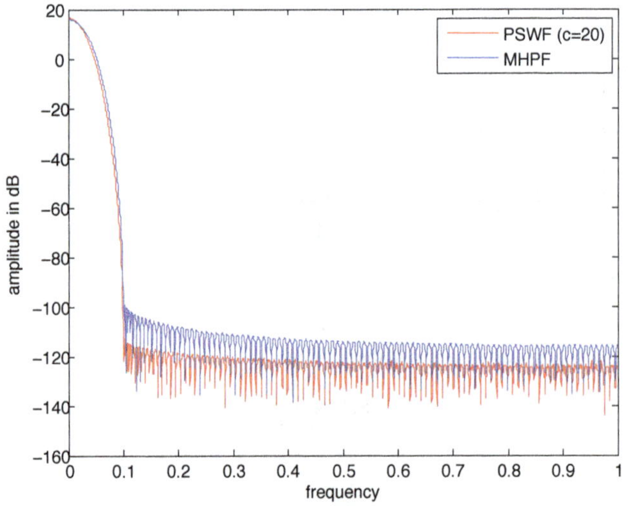

Fig. 6. No appreciable change in bandwidth between the two is observed proving that the major input is from the order 0 pulse in the optimal pulse design.

c	a_0	a_1	a_2	a_3	a_4	a_5
2	1	0	0	0	0	0
4	1	0	0	0	0	-0
8	1	0	0	0	0	-0
10	0.9396	0	0.0860	0	0	0
16	0.9442	0	0.0795	0	0	0
20	0.9479	0	0.0668	0	0.0085	-0

Tab. 1. Coefficient values computed for the linear combination pulse for different time-bandwidth product using Simplex Methods. Note in order to ensure uniformity the PSWF pulses were scaled to have unit value at time 0.

Figure 5 compares the optimal pulse with the PSWF order zero pulse, the linear combination pulse has a quicker decay rate and from the plot in Fig. 6 it can be observed that there is no significant change in band occupied by the two.

4. Modified Hermite Polynomial Function

The Hermite polynomials are defined by [15][19]:

$$h_{en}(t) = (-1)^n e^{\frac{t^2}{2}} \frac{d^n}{dt^n}(e^{\frac{-t^2}{2}}) \tag{17}$$

where $n = 0, 1, 2, .. -\infty < t < \infty$. Hermite polynomials are not orthogonal [15]. However, they can be modified to become orthogonal as follows:

$$h_n(t) = e^{\frac{-t^2}{4}} h_{en}(t) = (-1)^n e^{\frac{-t^2}{4}} \frac{d^n}{dt^n}(e^{\frac{-t^2}{2}}). \tag{18}$$

The general formula for defining Modified Hermite Polynomial Functions (MHPF) is given by [15]:

$$h_n(t) = k_n e^{\frac{-t^2}{4}} n! \sum_{i=0}^{n/2} (-\frac{1}{2})^i \frac{t^{n-2i}}{(n-2i)!i!}. \tag{19}$$

The constant k_n where $n = 0, 1, 2, \ldots$ is indicative of the energy of the pulse given by [15]:

$$k_n = \sqrt{\frac{E_n}{n!\sqrt{2\pi}}}. \tag{20}$$

Figure 7 and Figure 8 represent the time and frequency response for the MHPFs. It can be observed that the pulse duration and bandwidth is exactly the same for different orders n with the number of zero crossings in time domain being equal to n. These pulses are orthogonal to one another.

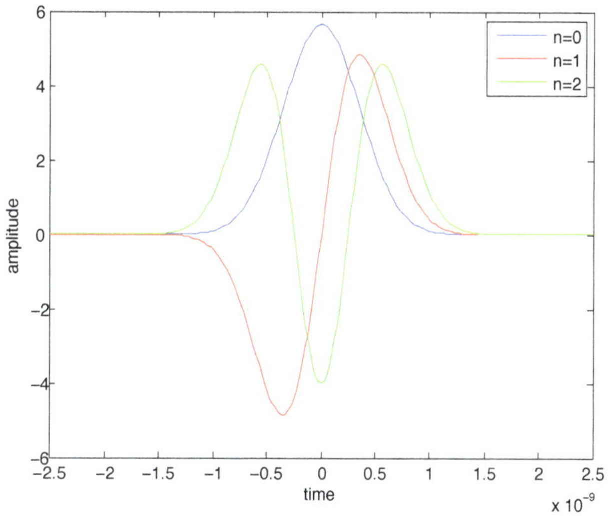

Fig. 7. Time domain representation of MHPF pulse for order 0, 1, 2. Different order pulses have identical pulse duration.

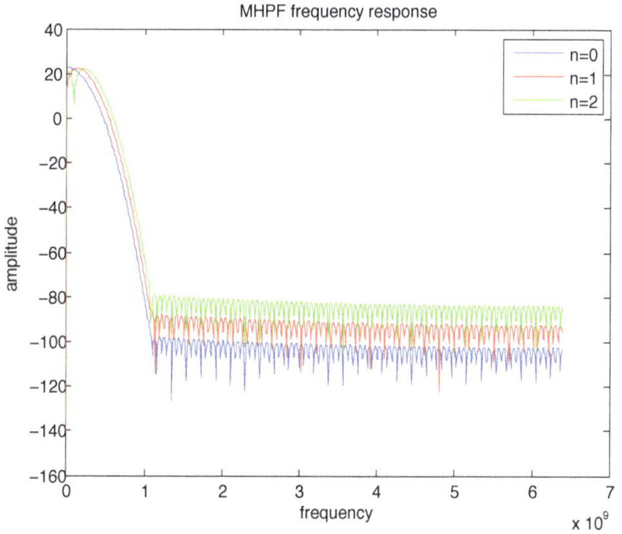

Fig. 8. Frequency representation for MHPF order 0, 1, 2 pulse. Note the constant identical bandwidth for different orders of the pulse.

5. Comparison

The MHPF and PSWF are very different functions so drawing a comparison between the two is fairly complex. We confined the PSWF and MHPF pulse to similar time durations and observed the pulse shapes obtained.

PSWF pulse's duration and bandwidth can be bartered using a time-bandwidth product c where $c = \Omega T/2$. The MHPF pulses have no such constant c so only one pulse shape is obtained for a given pulse duration. However, the MHPF pulse time duration and bandwidth are inversely related as $\Omega T/\pi$ [20].

The curves obtained when the MHPF and PSWF generated pulse shapes have identical time domain characteristics were plotted side by side to draw a better comparison. For instance, in the current scenario when the time bandwidth product $c = 16$ with pulse duration $T = [-1, 1]$, the MHPF and PSWF pulse have identical time domain characteristics (also shown in Fig. 9). While comparing the frequency domain plots obtained it was observed that the band occupied by the MHPF pulse is 1.3 times the band occupied by PSWF pulse which is roughly equal to $\pi/2$ (Fig. 10).

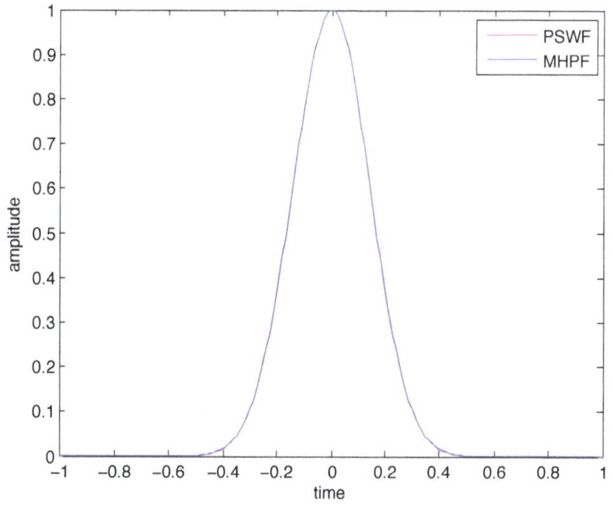

Fig. 9. PSWF (c=16) and MHPF pulse. The two pulse shapes appear to be exactly similar.

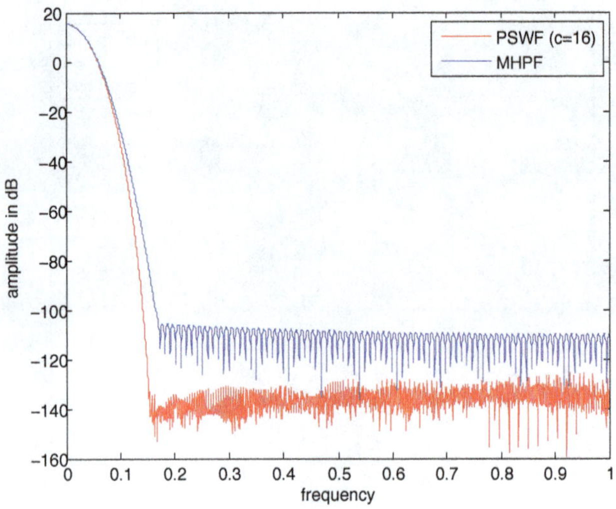

Fig. 10. Comparison between MHPF pulse and PSWF pulse ($c = 16$). Both occupy similar bandwidth as is expected from the similar time response.

Table 2 compares the band occupied by the PSWF and MHPF generated pulse while having similar parameters. c-PSWF denotes the value of time-bandwidth product to obtain a PSWF pulse shape similar to MHPF for the same time-duration. It can be observed that the band occupied by MHPF pulse is $1.2-1.3$ times the one of PSWF pulse, therefore it can be safely said that the PSWF is most efficient in terms of bandwidth occupancy.

Time	c-PSWF	BW-PSWF	BW-MHPF	Factor diff.
$[-1,1]$	16	0.154	0.196	$\simeq 1.3$
$[-\frac{3}{4},\frac{3}{4}]$	10	0.1295	0.138	$\simeq 1.1$
$[-\frac{1}{2},\frac{1}{2}]$	4	0.081	0.104	$\simeq 1.3$

Tab. 2. Comparison of band occupancy of PSWF and MHPF generated pulse shapes under similar time duraction.

To validate the benefits of the pulse design technique utilized, optical data transmission in indoor environments has been simulated. The unequalized OOK system has been adapted for validation. The input bits, assumed to be independent identically distributed (i.i.d) and uniform on [0,1] are passed through a transmitter filter whose impulse response allows pulse shaping. The block diagram of the simulative setup is shown in Fig. 11 which uses the ceiling bounce model to cater for indoor light propagation [21]. The input signal $x(t)$ is passed through a multipath channel impulse response $h(t)$ and noise $n(t)$ is added. On the receiver end matched filter detection is employed as depicted in Fig. 11. To characterize the system performance, the eye diagram analysis (Fig. 12 and 13) has been utilized. The effect of multipath propagation and intersymbol interference is catered for better with our self designed pulse as the opening of the eye (Fig. 13) on an arbitrary amplitude scale is wider compared to the scenario where a rectangular pulse is used (Fig. 12).

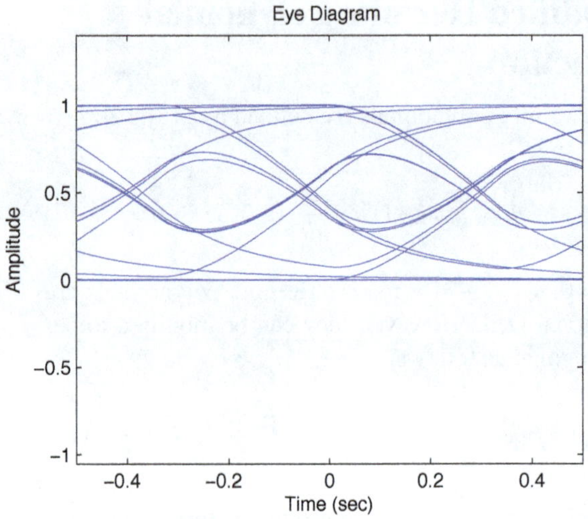

Fig. 12. Eye diagram using rectangular pulse for the Ceiling Bounce model.

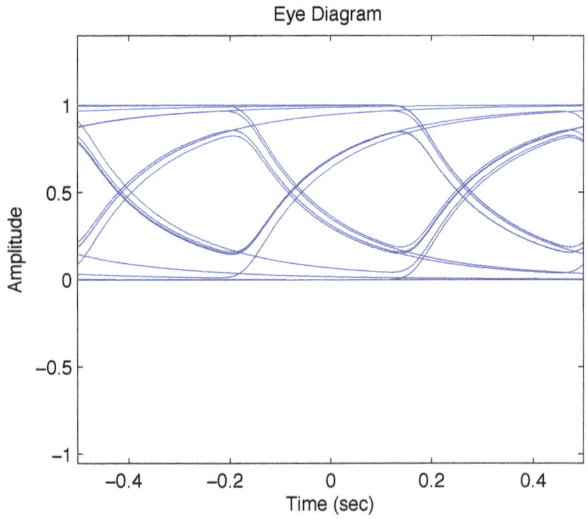

Fig. 13. Eye diagram using self designed pulse for the Ceiling Bounce model.

6. Conclusion

This paper investigates ISI free pulses satisfying the non-negative amplitude constraint of optical intensity channels for high speed optical wireless communications [22]. The optical intensity channels constrain the pulse to be non-negative; also there is a limitation on channel bandwidth. This non-negativity constraint limits the root Nyquist pulse to be time limited [1]. Modified Hermite polynomial and Prolate Spheroidal wave functions have been proposed as potential pulse shaping functions. Both are time-limited and have most of the energy concentrated in a finite band. PSWF pulses' time and bandwidth can be bartered using a constant c (the time-bandwidth product i.e. $\Omega T/2$). Varying values for c it is possible to arrive at a point where the MHPF and PSWF pulse have identical curves for any given time-duration. Comparing the bandwidth occupancy

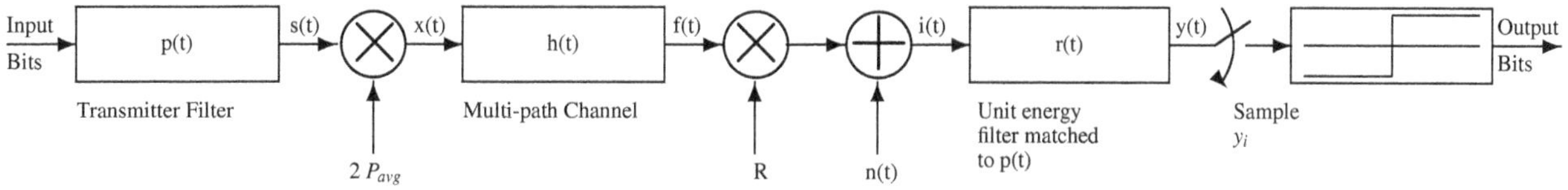

Fig. 11. Block diagram for the simulation setup.

of MHPF and PSWF under this scenario it was observed that the MHPF pulse occupies a band roughly 1.2 times the band occupied by the PSWF pulse. For a major segment (range of *c* values), the MHPF pulse occupies more than twice the band when compared to the PSWF pulse. Also from optimizing the linear objective function PSWF order 0 pulse emerges as most efficient in terms of bandwidth concentration. MHPF pulses hold the advantage of being easier to generate as they have a closed form expression, contrary to the PSWF pulse which lack a closed form solution. Since both PSWF and MHPF pulses satisfy the optical channel constraints, they are well suited for ease of implementation and fulfill system bandwidth requirements.

References

[1] HRANILOVIC, S. Minimum bandwidth optical intensity nyquist pulses. *IEEE Transactions on Communications*, 2007, vol. 55, no. 3, p. 574–583. DOI: 10.1109/TCOMM.2006.888878

[2] HRANILOVIC, S., KSCHISCHANG, F. R. Optical intensity-modulated direct detection channels: signal space and lattice codes. *IEEE Transactions on Information Theory*, 2003, vol. 49, no. 6, p. 1385–1399. DOI: 10.1109/TIT.2003.811928

[3] KAHN, J. M., BARRY, J. R. Wireless infrared communications. *Proceedings of the IEEE*, 1997, vol. 85, no. 2, p. 263–298. DOI: 10.1109/5.554222

[4] NIGAM, G., SINGH, R., CHATURVEDI, A. K. Finite duration root nyquist pulses with maximum in-band fractional energy. *IEEE Communications Letters*, 2010, vol. 14, no. 9, p. 797–799. DOI: 10.1109/LCOMM.2010.09.100314

[5] BEAULIEU, N.C., TAN, C.C., DAMEN, M.O. A better than Nyquist pulse. *IEEE Communications Letters*, 2001, vol. 5, no. 9, p. 367–368. DOI: 10.1109/4234.951379

[6] SOOD, R., XIAO, H. Root Nyquist pulses with an energy criterion. In *IEEE International Conference on Communications*. Glasgow (UK), 2007, p. 2711–2716. DOI: 10.1109/ICC.2007.450

[7] HALPERN, P. Optimum finitie duration nyquist signals. *IEEE Transactions on Communications*, 1979, vol. 27, no. 6, p. 884–888. DOI: 10.1109/TCOM.1979.1094486

[8] RAMIREZ-INIGUIEZ, R., GREEN, R. J. Indoor optical wireless communications. In *IEEE Colloquium on Optical Wireless Communications*. London (UK), 1999, p. 14/1–14/7. DOI: 10.1049/ic:19990705

[9] TAVAN, M., AGRELL, E., KAROUT, J. Bandlimited intensity modulation. *IEEE Transactions on Communications*, 2012, vol. 60, no. 11, p. 3429–3439. DOI: 10.1109/TCOMM.2012.091712.110496

[10] HRANILOVIC, S. *Wireless Optical Communication Systems*. Boston (USA): Springer Science + Business Media, 2005. DOI: 10.1007/b99592

[11] SELPIAN, D., POLLAK, H. O. Prolate spheroidal wave functions, fourier analysis and uncertainty — I. *Bell Systems Technical Journal*, 1961, p. 43–63.

[12] SELPIAN, D., POLLAK, H. O. Prolate spheroidal wave functions, fourier analysis and uncertainty — II. *Bell Systems Technical Journal*, 1962, p. 65–84.

[13] ALLEN, B., GHORASHI, S. A., GHAVAMI, M. A review of pulse design for impulse radio. In *Proceedings of IEEE Ultra Wideband Communications Technologies and System Design*. 2004, p. 93–97.

[14] GHAVAMI, M., MICHAEL, L., KOHNO, R. Hermite function based orthogonal pulses for ultra wideband communication. *Wireless Personal Multimedia Communications Symposium*, 2001.

[15] DILMAGHANI, R. S., GHAVAMI, M., ALLEN, B., AGHVAMI, H. Novel UWB pulse shaping using prolate spheroidal wave functions. In *Proceedings of IEEE Personal, Indoor and Mobile Radio Communications*. Beijing (China), 2003, vol. 1, p. 602–606. DOI: 10.1109/PIMRC.2003.1264343

[16] TAVAN, M., AGRELL, E. , KAROUT, J. Strictly bandlimited ISI-free transmission over intensity-modulated channels. In *IEEE Globecom*. Houston (TX, USA), 2011, p. 1–6. DOI: 10.1109/GLOCOM.2011.6133734

[17] PROAKIS, J. G. *Digital Communications*. 4th ed. New York: McGraw-Hill, 2001.

[18] TSAI, C. Y., JENG, S. K. Design of Legendre polynomial based orthogonal pulse generator for ultra wideband communications. In *IEEE Antennas and Propagation Society International Symposium*. 2005, vol. 2B, p. 680–683. DOI: 10.1109/APS.2005.1552105

[19] GHAVAMI, M., MICHAEL, L.B., KOHNO, R. *Ultra Wideband Signals and Systems in Communication Engineering*. 2nd ed. Wiley, 2007.

[20] OUERTANI, M., BESBES, H., BOUALLEGUE, A. Modified Hermite functions for designing new optimal UWB pulse-shapers. In *European Signal Processing Conference (EUSIPCO)*. Antalya (Turkey), 2005.

[21] GHASEEMLOOY, Z., POPOOLA, W., RAJBHANDARI, S. *Optical Wireless Communications: System and Channel Modelling with Matlab*. London: CRC Press, 2013.

[22] TAVAN, M., AGRELL, E., KAROUT, J. Indoor high-speed optical wireless communications: Recent developments. In *Proceedings of IEEE International Conference on Transparent Optical Networks (ICTON)*. Graz (Austria), 2014.

About the Authors ...

Sabeena FATIMA received her BS in Telecommunications Engineering and MS in Electrical Engineering from the National University of Computer and Emerging Science, Lahore, Pakistan, in 2009 and 2013 respectively. She joined National University of Computer and Emerging Science in the year 2009 where she currently works as lecturer in Dept of Electrical Engineering. Her research interests include optical communications, signal estimation and detection techniques, MIMO systems and adaptive signal processing.

S. Sheikh MUHAMMAD completed his Bachelors in Electrical Engineering with Honors in 2001 and received Masters in Electrical Engineering in 2003 the from University of Engineering and Technology, Lahore, Pakistan whereby he also remained in the faculty for around 3 years. He completed his PhD in Electrical Engineering in 2007 with Excellence at the Graz University of Technology conducting research on coded modulation techniques for free space optical systems. He has Guest Edited 2 special issues in Optical Wireless Communications in 2009 and 2012 and has organized regular IEEE Colloquiums on Optical Wireless in 2008 (Austria), 2010 (UK) and 2012 (Poland). He has published more than 50 peer reviewed papers in journals and conferences of repute and has chaired number of international conference sessions. He is currently working as an Associate Professor of Electrical Engineering at the National University of Computer and Emerging Sciences (FAST-NU) in Lahore and his current research revolves around application of network information theory to optical wireless and optical wave propagation through random media.

A. D. RAZA received his Bachelors in Electrical Engineering with Honors in 1973 and Masters in Electrical Engineering in 1984 from the University of Engineering and Technology, Lahore, Pakistan. He remained with Pakistan's state owned Telecom operator from 1975 to 2008 and rose to the level of Executive Vice President. He joined the National University of Computer and Emerging Sciences in 2008 and is pursuing his doctoral studies besides his teaching responsibilities as an Assistant Professor in the Department of Electrical Engineering. His current research interest revolves around application of network information theory to optical wireless networks.

Permissions

All chapters in this book were first published in Radioengineering, by Spolecnost pro radioelektronicke inzenyrstvi; hereby published with permission under the Creative Commons Attribution License or equivalent. Every chapter published in this book has been scrutinized by our experts. Their significance has been extensively debated. The topics covered herein carry significant findings which will fuel the growth of the discipline. They may even be implemented as practical applications or may be referred to as a beginning point for another development.

The contributors of this book come from diverse backgrounds, making this book a truly international effort. This book will bring forth new frontiers with its revolutionizing research information and detailed analysis of the nascent developments around the world.

We would like to thank all the contributing authors for lending their expertise to make the book truly unique. They have played a crucial role in the development of this book. Without their invaluable contributions this book wouldn't have been possible. They have made vital efforts to compile up to date information on the varied aspects of this subject to make this book a valuable addition to the collection of many professionals and students.

This book was conceptualized with the vision of imparting up-to-date information and advanced data in this field. To ensure the same, a matchless editorial board was set up. Every individual on the board went through rigorous rounds of assessment to prove their worth. After which they invested a large part of their time researching and compiling the most relevant data for our readers.

The editorial board has been involved in producing this book since its inception. They have spent rigorous hours researching and exploring the diverse topics which have resulted in the successful publishing of this book. They have passed on their knowledge of decades through this book. To expedite this challenging task, the publisher supported the team at every step. A small team of assistant editors was also appointed to further simplify the editing procedure and attain best results for the readers.

Apart from the editorial board, the designing team has also invested a significant amount of their time in understanding the subject and creating the most relevant covers. They scrutinized every image to scout for the most suitable representation of the subject and create an appropriate cover for the book.

The publishing team has been an ardent support to the editorial, designing and production team. Their endless efforts to recruit the best for this project, has resulted in the accomplishment of this book. They are a veteran in the field of academics and their pool of knowledge is as vast as their experience in printing. Their expertise and guidance has proved useful at every step. Their uncompromising quality standards have made this book an exceptional effort. Their encouragement from time to time has been an inspiration for everyone.

The publisher and the editorial board hope that this book will prove to be a valuable piece of knowledge for researchers, students, practitioners and scholars across the globe.

List of Contributors

A. D. RAZA
Department of Electrical Engineering, National University of Computer and Emerging Sciences (FAST-NU), Lahore, Pakistan

S. Sheikh MUHAMMAD
Department of Electrical Engineering, National University of Computer and Emerging Sciences (FAST-NU), Lahore, Pakistan

Václav PRAJZLER
Dept. of Microelectronics, Czech Technical University, Technická 2, 168 27 Prague, Czech Republic

Pavla NEKVINDOVÁ
Institute of Chemical Technology, Technická 5, 166 27 Prague, Czech Republic

Petr HYPŠ
Dept. of Microelectronics, Czech Technical University, Technická 2, 168 27 Prague, Czech Republic

Oleksiy LYUTAKOV
Institute of Chemical Technology, Technická 5, 166 27 Prague, Czech Republic

Vítězslav JEŘÁBEK
Dept. of Microelectronics, Czech Technical University, Technická 2, 168 27 Prague, Czech Republic

Andrzej A. KUCHARSKI
Telecommunications and Teleinformatics Department, Wrocław University of Technology,
Wybrzeze Wyspianskiego 27, 50-370 Wrocław, Poland

Xiangbin YU
Key Laboratory of Radar Imaging and Microwave Photonics, Ministry of Education,
College of Electronic and Information Engineering, Nanjing University of Aeronautics and Astronautics, Nanjing, China
National Mobile Communications Research Laboratory, Southeast University, Nanjing, China

Yan LIU
Key Laboratory of Radar Imaging and Microwave Photonics, Ministry of Education,
College of Electronic and Information Engineering, Nanjing University of Aeronautics and Astronautics, Nanjing, China

Yang LI
Key Laboratory of Radar Imaging and Microwave Photonics, Ministry of Education,
College of Electronic and Information Engineering, Nanjing University of Aeronautics and Astronautics, Nanjing, China

Qiuming ZHU
Key Laboratory of Radar Imaging and Microwave Photonics, Ministry of Education,
College of Electronic and Information Engineering, Nanjing University of Aeronautics and Astronautics, Nanjing, China

Xin YIN
Key Laboratory of Radar Imaging and Microwave Photonics, Ministry of Education,
College of Electronic and Information Engineering, Nanjing University of Aeronautics and Astronautics, Nanjing, China

Kecang QIAN
Key Laboratory of Radar Imaging and Microwave Photonics, Ministry of Education,
College of Electronic and Information Engineering, Nanjing University of Aeronautics and Astronautics, Nanjing, China

Vandana NIRANJAN
Dept. of Electronics & Communication Engineering, Indira Gandhi Delhi Technical University for Women, Kashmere Gate, Delhi-110006, India

Ashwani KUMAR
Dept. of Electronics & Communication Engineering, Indira Gandhi Delhi Technical University for Women, Kashmere Gate, Delhi-110006, India

Shail Bala JAIN
Dept. of Electronics & Communication Engineering, Indira Gandhi Delhi Technical University for Women, Kashmere Gate, Delhi-110006, India

Xiaomu MU
Dept. of Automation, TNList, Tsinghua University, Qinghuayuan 1, 100084 Beijing, People's Republic of China

Juntang YU
Dept. of Automation, TNList, Tsinghua University, Qinghuayuan 1, 100084 Beijing, People's Republic of China

Shuning WANG
Dept. of Automation, TNList, Tsinghua University, Qinghuayuan 1, 100084 Beijing, People's Republic of China

Mohamed E. FOUDA
Dept. of Engineering Mathematics, Faculty of Engineering, Cairo University, Giza, Egypt,12613

Ahmed G. RADWAN
Dept. of Engineering Mathematics, Faculty of Engineering, Cairo University, Giza, Egypt,12613
Nano-electronic Integrated Systems Center (NISC), Nile University, Giza, Egypt

Fulya BAGCI
Dept. of Engineering Physics, Ankara University, 06100 Besevler, Ankara, Turkey

Sultan CAN
Dept. of Electrical-Electronics Engineering, Ankara University, 06830 Golbasi, Ankara, Turkey

Baris AKAOGLU
Dept. of Engineering Physics, Ankara University, 06100 Besevler, Ankara, Turkey

A. Egemen YILMAZ
Dept. of Electrical-Electronics Engineering, Ankara University, 06830 Golbasi, Ankara, Turkey

Muhammad Taha JILANI
Dept. of Electrical and Electronics Engineering, Universiti Teknologi Petronas, Perak, Malaysia

Wong Peng WEN
Lee Yen CHEONG
Dept. of Fundamental and Applied Sciences, Universiti Teknologi Petronas, Perak, Malaysia

Mohd Azman ZAKARIYA
Dept. of Physics, COMSATS Institute of Information Technology, Islamabad, Pakistan

Muhammad Zaka Ur REHMAN
Dept. of Physics, COMSATS Institute of Information Technology, Islamabad, Pakistan

George SOULIOTIS
Dept. of Physics, University of Patras, 26504 Patras, Greece

Costas LAOUDIAS
Dept. of Physics, University of Patras, 26504 Patras, Greece

Nikolaos TERZOPOULOS
Oxford Brookes University, Wheatley Campus, Oxford, OX33 1HX, UK

Şuayb Çağrı YENER
Dept. of Electrical and Electronics Engineering, Sakarya University, Sakarya, Turkey

H. Hakan KUNTMAN
Dept. of Electronics and Communication Engineering, Istanbul Technical University, Istanbul, Turkey

Mohamed A. BARBARY
Dept. of Electronic and Information Engineering, Nanjing University of Aeronautics and Astronautics, Nanjing, China

Peng ZONG
Dept. of Astronautics, Nanjing University of Aeronautics and Astronautics, Nanjing, China

Miroslav GALABOV
Dept. of Computer Systems and Technologies, University of Veliko Turnovo, street T.Tarnovski 2, 5000 Veliko Turnovo, Bulgaria

Stanislav ZVANOVEC
Dept. of Electromagnetic Field, Czech Technical University in Prague, Technicka 2, 166 27 Prague, Czech Republic

Petr CHVOJKA
Dept. of Electromagnetic Field, Czech Technical University in Prague, Technicka 2, 166 27 Prague, Czech Republic

Paul Anthony HAIGH
Faculty of Engineering, University of Bristol, Bristol, BS8 1TR, UK

Zabih GHASSEMLOOY
Optical Communications Research Group, Faculty of Engineering and Environment, Northumbria University, Newcastle-upon-Tyne NE1 8ST, UK

Nafiseh KHAJAVI
Dept. of Electrical Engineering, Dezful Branch, Islamic Azad University, Dezful, Iran

Seyed Vahab AL-Din MAKKI
Dept. of Electrical Engineering, Razi University, Kermanshah, Iran

Sohrab MAJIDIFAR
Dept. of Electrical Engineering, Kermanshah University of Technology, Kermanshah, Iran

Hassan ELKAMCHOUCHI
Dept. of Electrical Engineering, University of Alexandria, Alhuria Street, Alexandria, Egypt

Mohamed HASSAN
Dept. of Electrical Engineering, University of Alexandria, Alhuria Street, Alexandria, Egypt

Cong-hui QI
School of Electronic Engineering, University of Electronic Science and Technology of China,
Xiyuan Ave 2006, Chengdu, Sichuan, China

Zhi-qin ZHAO
School of Electronic Engineering, University of Electronic Science and Technology of China,
Xiyuan Ave 2006, Chengdu, Sichuan, China

Siming PENG
Dept. of Wireless Communications, PLA University of Science and Technology, 210014 Nanjing, China

Zhigang YUAN
Dept. of Wireless Communications, PLA University of Science and Technology, 210014 Nanjing, China

Jun YOU
Dept. of Command Information System, PLA University of Science and Technology, 210014 Nanjing, China

Yuehong SHEN
Dept. of Wireless Communications, PLA University of Science and Technology, 210014 Nanjing, China

Wei JIAN
Dept. of Wireless Communications, PLA University of Science and Technology, 210014 Nanjing, China

Václav PA NKO
Dept. of Radio Engineering, Czech Technical University in Prague, Technická 2, 166 27 Praha 6, Czech Republic
ON Semiconductor, SCG Czech Design Center, 1. maje 2594, 75661 Roznov p. R., Czech Republic

Stanislav BANÁŠ
Dept. of Radio Engineering, Czech Technical University in Prague, Technická 2, 166 27 Praha 6, Czech Republic
ON Semiconductor, SCG Czech Design Center, 1. maje 2594, 75661 Roznov p. R., Czech Republic

Richard BURTON
ON Semiconductor, 5005 East McDowell Road, Phoenix, AZ 85008, USA

Karel PTÁČEK
Dept. of Microelectronics, Brno University of Technology, Technicka 3058/10, 61600 Brno, Czech Republic
ON Semiconductor, SCG Czech Design Center, 1. maje 2594, 75661 Roznov p. R., Czech Republic

Jan DIVÍN
Dept. of Radio Engineering, Czech Technical University in Prague, Technická 2, 166 27 Praha 6, Czech Republic
ON Semiconductor, SCG Czech Design Center, 1. maje 2594, 75661 Roznov p. R., Czech Republic

Josef DOBEŠ
Dept. of Radio Engineering, Czech Technical University in Prague, Technická 2, 166 27 Praha 6, Czech Republic

Mohsen HAYATI
Electrical Engineering Dept., Faculty of Engineering, Razi University, Tagh-E-Bostan, Kermanshah-67149, Iran

Ashkan ABDIPOUR
Electrical Engineering Dept., Faculty of Engineering, Razi University, Tagh-E-Bostan, Kermanshah-67149, Iran

Arash ABDIPOUR
Electrical Engineering Dept., Faculty of Engineering, Razi University, Tagh-E-Bostan, Kermanshah-67149, Iran

Ahmet ONCU
Dept. of Electrical and Electronics Engineering, Bogazici University, Istanbul, 34342, Turkey

Maciej CZYŻAK
Faculty of Electrical and Control Engineering, Gdansk University of Technology, G Narutowicza 11/12, 80-233

Jacek HORISZNY
Faculty of Electrical and Control Engineering, Gdansk University of Technology, G Narutowicza 11/12, 80-233

Robert SMYK
Faculty of Electrical and Control Engineering, Gdansk University of Technology, G Narutowicza 11/12, 80-233

Josef SLEZAK
Dept. of Radio Electronics, Brno University of Technology, Technická 12, 616 00 Brno, Czech Republic

Tomas GOTTHANS
Dept. of Radio Electronics, Brno University of Technology, Technická 12, 616 00 Brno, Czech Republic

Branimir JAKSIC
Faculty of Electrical Engineering, University of Nis, Aleksandra Medvedeva 14, 18000 Nis, Serbia

Dusan STEFANOVIC
College of Applied Technical Sciences, Aleksandra Medvedeva 20, 18000 Nis, Serbia

Mihajlo STEFANOVIC
Faculty of Electrical Engineering, University of Nis, Aleksandra Medvedeva 14, 18000 Nis, Serbia

Petar SPALEVIC
Faculty of Technical Sciences, University of Pristina, Knjaza Milosa 7, 38220 Kosovska Mitrovica, Serbia

Vladeta MILENKOVIC
Faculty of Electrical Engineering, University of Nis, Aleksandra Medvedeva 14, 18000 Nis, Serbia

Zongsheng ZHANG
College of Communications Engineering, PLA University of Science and Technology, Street YuDao, Nanjing, China
College of National Information Science, JieFangGongyuan Road, Wuhan, China

Qihui WU
College of Communications Engineering, PLA University of Science and Technology, Street YuDao, Nanjing, China

Jinlong WANG
College of Communications Engineering, PLA University of Science and Technology, Street YuDao, Nanjing, China

Sabeena FATIMA
Dept. of Electrical Engineering, National University of Computer and Emerging Sciences, Lahore, Pakistan

S. Sheikh MUHAMMAD
Dept. of Electrical Engineering, National University of Computer and Emerging Sciences, Lahore, Pakistan

A. D. RAZA
Dept. of Electrical Engineering, National University of Computer and Emerging Sciences, Lahore, Pakistan

www.ingramcontent.com/pod-product-compliance
Lightning Source LLC
Chambersburg PA
CBHW050442200326
41458CB00014B/5040